THE COMPLETE
ENCYCLOPEDIA OF THE
ANIMAL
WORLD

THE COMPLETE
ENCYCLOPEDIA OF THE
ANIMAL
WORLD

EDITED BY
DAVID M. BURN

OCTOPUS

Contents

First published 1980 by
Octopus Books Limited
59 Grosvenor Street
London W1

© 1980 Octopus Books Limited

ISBN 0 7064 0760 1

Produced by
Mandarin Publishers Limited
22a Westlands Road
Quarry Bay, Hong Kong

Printed in Hong Kong

About the Authors

Professor R. McNeill Alexander holds the Chair of Zoology in the Department of Pure and Applied Zoology at the University of Leeds, England. He has made a special study of animal mechanics particularly in fish and mammals.

Dr R. Robin Baker is a Lecturer in the Department of Zoology, University of Manchester, England. He is an expert on migration and has recently written 'The Evolutionary Ecology of Animal Migration', a definitive work on the subject.

Michael Boorer is the Education Officer for the Zoological Society of London. In addition to his work of organizing the zoo's educational activities he has written many articles and several books. Wild cats are one of his particular interests.

Dr Alan Brafield is Senior Lecturer in Zoology in the Department of Biology, Queen Elizabeth College, University of London. He works especially on the physiology of marine animals; his most recent book is 'Life in Sandy Shores'.

Dr Percy M. Butler was formerly Professor of Zoology at Royal Holloway College, University of London, and is now an Honorary Research Fellow. He has worked extensively on mammals and has a special interest in their teeth.

Dr Andrew Campbell is a Lecturer in Zoology at Queen Mary College, University of London. He specializes in echinoderms and is the author of 'The Hamlyn Guide to the Sea Shore and Shallow Seas of Britain and Europe'.

Professor J.L. Cloudsley-Thompson holds a research chair at the Department of Zoology, Birkbeck College, University of London. An authority on deserts, he edits the 'Journal of Arid Environments' and has written numerous books.

Dr P. Cornelius heads the Coelenterate Section in the Department of Zoology, British Museum (Natural History). He has participated in an expedition to Oman and has contributed articles on marine invertebrates to several encyclopedias.

P.M. David is a Senior Principal Scientific Officer at the Institute of Oceanographic Sciences, Wormley, England, where he is head of the Biological Division. He has made a special study of arrow worms.

E.G. Easton is a member of the Annelid Section in the Department of Zoology, British Museum (Natural History). He works extensively on earthworms and has a particular interest in those of the Oriental region.

Dr Malcolm Edmunds is Head of the Division of Biology, Preston Polytechnic, England. One of his previous posts was at the University of Ghana. He has published numerous papers and is the author of 'Defence in Animals'.

Dr Peter L. Forey works in the Fossil Fish Section of the Department of Palaeontology, British Museum (Natural History). He has worked extensively on the evolution and anatomy of primitive fish, both in Britain and in Canada.

William G. Fry is Senior Lecturer in Biology in the Department of Science, Luton College, England. Formerly a member of the British Museum (Natural History), his special interests are sponges and sea spiders.

Dr P.E. Gibbs is on the staff of the Marine Biological Association's laboratory at Plymouth, England. He works on marine worms, notably sipunculans, and has participated in expeditions to the western Pacific and the Caribbean.

Dr David I. Gibson heads the Parasitic Worms Section of the Department of Zoology, British Museum (Natural History). He has written extensively on the systematics of roundworms and digenean flukes of marine fish.

Dr Ray Gibson is Reader in Marine Biology in the Department of Biology, Liverpool Polytechnic, England. He is an authority on ribbon worms and his many publications include a definitive monograph on the group.

John Gooders writes extensively on birds and other natural history subjects. He has also led natural history tours to various parts of the world and has been involved in the making of some twenty natural history films.

Dr T.R. Halliday is a Lecturer in the Department of Biology at the Open University, England. He is a specialist in animal behaviour and has a particular interest in amphibians. He is the author and illustrator of 'Vanishing Birds'.

Dr P.J. Hayward lectures in marine biology at the Department of Zoology, University College of Swansea, Wales. He has published numerous scientific papers and has co-authored two books on the British bryozoans.

Dr R.P.S. Jefferies heads the Fossil Echinoderm Section in the Department of Palaeontology, British Museum (Natural History). He has a particular interest in very early echinoderm fossils and their relationship to the chordates.

Dr T.S. Kemp is Curator of the Zoological Collections in the University Museum at Oxford. He has published numerous scientific papers, especially on mammal-like reptiles, and is the editor of 'Studies in Vertebrate Evolution'.

Dr W.J. Kennedy is a Lecturer in Palaeontology and also Curator of the Geological Collections in the University Museum at Oxford. He works chiefly on invertebrate fossils and evolution, and has participated in expeditions to Aldabra.

Dr Michael Land is Reader in the School of Biological Sciences at the University of Sussex, England. He specializes in animal behaviour and physiology, on which he has written numerous papers, and has contributed to several books.

Professor D.L. Lee has the Chair of Agricultural Zoology and is Head of the Department of Pure and Applied Zoology at the University of Leeds, England. An expert on roundworms, he is President of the British Society for Parasitology.

Dr R.H. Millar has recently retired as Deputy Director of the Dunstaffnage Marine Research Laboratory, Oban, Scotland. He has worked on various aspects of marine biology and is a leading authority on the tunicates.

Dr Clive I. Morgan is a Lecturer in Biology at Glasgow College of Technology, Scotland. He specializes in freshwater biology and is an authority on water bears. His publications include a monograph of the British species.

Dr Christopher Perrins is Director of the Edward Grey Institute of Field Ornithology at Oxford University, where he has worked intensively on the population dynamics of the Great Tit and several seabird species.

Dr John Rostron is Senior Lecturer in Biology, North East London Polytechnic. He specializes in evolutionary genetics and his other interests include the use of computers in biology and the population biology of sea anemones.

Bryan Sage is an ecologist and until recently worked for British Petroleum on the environmental and conservation aspects of company operations. Lecturing, writing and photography are some of his many accomplishments.

Dr Bernard Stonehouse is Chairman of the Postgraduate School of Studies in Environmental Science at Bradford University, England. He is a research biologist and writer with special interests in marine and polar ecology.

Dr Stephen Sutton lectures in zoology at the University of Leeds, England. He works chiefly on arthropod ecology and is a regular contributor to the BBC's natural history programmes. He was a member of the 1974/5 Zaire River Expedition.

Dr R.C. Tinsley lectures in parasitology at the Zoology Department, Westfield College, University of London. He specializes in monogenean flukes and their host relations. Much of his field work has been carried out in Africa.

Professor Keith Vickerman is a member of the Department of Zoology, University of Glasgow, Scotland. He is an expert on parasitic diseases, particularly those caused by protozoans, and is a World Health Organisation consultant.

Dr Lynda Warren is a Research Fellow in the Department of Zoology, Bedford College, University of London. She works on marine worms and their respiration and is also engaged on a project to map the distribution of bristle worms.

Dr Phil Whitfield lectures in zoology at King's College, University of London. He has carried out parasitic public health consultancy work in Egypt and Venezuela and is the author of 'The Biology of Parasitism'.

Professor V.C. Wynne-Edwards is Emeritus Professor of Natural History at the University of Aberdeen, Scotland. He is one of the world's leading ecologists and a former Chairman of the Natural Environment Research Council.

The Editor

David M. Burn studied zoology at Queen Mary College, University of London, and at the University of Leeds where his postgraduate research was concerned with the population ecology of bird parasites. After a period of teaching he became involved in scientific publishing and for a time was attached to the staff of 'Nature'. He is currently the natural history editor for a leading London publishing house.

Editorial Consultants

Australia
Dr Peter Crowcroft, Metro Toronto Zoo; formerly Director of Taronga Zoo, Sydney, New South Wales.
Europe
Professeur Jean Dorst, Member of the Academy of Sciences and Director-General of the National Museum of Natural History, Paris.
North America
Dr Porter M. Kier, Director of the National Museum of Natural History, Smithsonian Institution, Washington, DC.

Foreword by Roger Caras

Men and women have always found the other members of the animal kingdom a source of wonder. The earliest known cave paintings and rock carvings depicted animals, and ancient pictograph writing used animal symbols. From very earliest times literature has recorded what was currently known about animal life.

Things have not changed. Now, as then, we wonder about other animals and seek information about them. Indeed, so intense is man's interest in the animal kingdom that natural history books are outdated more quickly than books in many other fields of interest, for facts are seen in new ways and new information is being gathered all the time.

What are examples of change? When I was a child I was taught that rabbits are rodents; they are not—we now know they are lagomorphs. As a teenage boy I was told there were 64 different kinds of grizzly and brown bears in North America alone; not so—there may be no more than one or two species. That ferocious man-eater of my youth, the great Orca or Killer Whale, has never killed a man, so far as is known, and the deadly dangerous Portuguese Man-o-War, while a nuisance to swimmers, has never killed a human being with its venom, according to the best information now available. And the best today is better than it used to be.

There is now a need for a new encyclopedia that explores the world of animals in light of current knowledge. The only way to provide such a book is to gather together the best of the probing minds of our time and see what they have found and what they have synthesized from the discoveries of their peers.

Anyone with even a passing interest in the animal kingdom, and almost everyone has at least that and many of us much more, can turn the pages of this book and learn, perhaps to their surprise, that much has indeed changed. That fact is due to continuing interest, constant research and enormous curiosity, not just the passage of time.

A book that can best serve a naturalist, amateur or professional, is a book that can be read, enjoyed and referred to again and again. A book that professes to be an encyclopedia about animals should be a retreat in the odd leisure hour and a place to go when information is needed in a hurry. That is what this book is. It serves both tasks and serves them well.

The fact that this book, too, will be outdated some decades hence will be in part a result of the minds it helps build now, the interest it kindles and keeps alive and the way it serves a need that goes back to our earliest consciousness – the need to know about and to understand the other creatures with which we share our planet.

Roger Caras

Introduction

by Professor V. C. Wynne-Edwards CBE, FRS

The life sciences have been moving so fast in the last 30 years that biologists sometimes speak of the rate of change as a revolution. It is most important therefore to have books in which experts collaborate to review a substantial part of the scene, and give non-experts a chance to bring themselves up to date. No one book can cover all the aspects but this encyclopedia embraces what is probably the most popular of all the sectors of biology—the one where the emphasis falls on 'whole animal' zoology.

The number of living species of animals is extremely large. It is believed that more than a million species have already been described and named. Many more are added annually, and the total that remains undiscovered could easily be another million, or even more. For plants the figure of half a million described already is probably getting nearer the final goal. For the micro-organisms it is still too early to make any forecast, beyond saying that they are likely to comprise fewer species, and perhaps much fewer, than the plants.

The Thomson's Gazelle lives on the plains of East Africa. It is an abundant member of a grazing association that includes zebra and wildebeest. They are preyed on by Cheetahs, Lions, Hyaenas and wild dogs. How does this three-stage food web, of carnivores eating herbivores eating grasses, remain stable?

Thus there does not seem much doubt that animals are by far the most diverse of the three great kingdoms.

The reason for this is not obvious. There are biochemical arguments for thinking that life originated only once, about 3,500 million years ago. If that is so, then the ancestral lines of all the organisms alive today must go back to the same original event. Evolution has led to their divergence—to the repeated branching of the tree of life. As we shall see, plants had to come before animals, and we are left to conclude that since animals first appeared they have diversified several times as fast as plants, and faster still than micro-organisms.

In the survey of the animal kingdom that forms the largest part of this book there is plenty of evidence that similar disparities in diversity exist between the different phyla. The most striking example is the class of insects, part of the great arthropod phylum. The known insects by themselves begin to approach a million species and any one of the dozen largest orders already contains 100 to 1,000 times as many known species as there are in some phyla.

The fossil record shows that there were groups of plants and animals in past ages that have since declined because more efficient types evolved and usurped their ecological roles or 'niches'. One might ask whether the present greater diversity of animals compared with that of plants is a recent development, or perhaps even a temporary one. The fossil evidence suggests that this is not so. The rates at which species numbers have grown have no doubt fluctuated somewhat with time, but the difference in speciation rate between animals and plants seems to have ecological causes and is likely to have existed from early times.

Most plants contain chlorophylls, the pigment molecules that give them their green colour. Chlorophylls have the life-sustaining property of absorbing sunlight and converting it into chemical energy by the process known as photosynthesis. They give plants the power to combine simple inorganic substances, especially carbon dioxide and water, into all manner of organic compounds, some of which, particularly oils and carbohydrates, are high-energy materials that can be retained as fuels for future use. No organic substances other than the chlorophylls have this property and, since all forms of life depend on expending energy, this amounts to saying that all other organisms depend on green plants as their energy source. Plants are the primary *producers* of chemical energy, and they also produce almost all the amino acids needed by animals for building proteins. Animals are essentially *consumers*, obtaining energy and amino acids either by feeding on plants and their products or, at second hand, by eating other animals that have acquired them. All organisms produce detritus and dead materials still retaining some remnants of chemical energy; and there is a third great 'guild' of organisms, the *decomposers*, including many small animals, fungi and especially bacteria, that live on detritus and in the end break everything down once more into the simple substances that plants use as raw materials.

Thus the living world contrives to recycle carbon, oxygen, nitrogen, water and most of the other 40 or so elements that get incorporated into living matter. It is yet another of the extraordinary properties of life, or of living organisms, to conserve their own environment on land, in the air and water. Evolution has led to a universal blind partnership, a division of labour, uniting the fortunes of three different guilds of organisms, none of which can survive without one or both the others.

It is possible to conceive a viable ecosystem, consisting of green plants and micro-organisms, without any animals. The primary producers and the ultimate decomposers could in theory operate a recycling system on their own. On that basis, the whole animal kingdom can be regarded as a burden that has to be carried on the backs of the plants and micro-organisms, and gives little (and none of it indispensable to the ecosystem) in return. The plants, in order to secure their own survival, have been forced to produce enough surplus fuels and primary compounds to satisfy the needs of the entire animal world.

To resolve the problem about differences in their speciation rates we must look at the different conditions for life that plants and animals require. All green plants use the same basic method of making a living, by photosynthesis, and their primary need is for suitable pieces of ground on which to carry out this process. Seeds and spores that fall on unsuitable sites are expendable but competition for space has led to the evolution of plants that can make do with sites that are far from optimal and consequently they have adapted to fill a large range and variety of ecological niches. But wherever they grow they are all doing the same thing —photosynthesising—and space that fulfils their physical and chemical requirements is the only resource they need compete for.

Animals, on the other hand, because they are consumers, are not confined like plants to a single industry—they can secure their nourishment from innumerable sources by many methods. There is for them a whole extra dimension therefore, when it comes to specializing. Herbivores can use different kinds of plants for food, to say nothing of different parts of the same plant—roots, stems, leaves, buds, flowers, seeds and sap. It must be commonplace to find five, ten or more animal species that feed on the parts and products of the same species of plant.

The problem of why there are several times as many species of animals as there are of plants has already begun to diminish; and when one reflects that almost every herbivore provides prey for insectivores or carnivores, and that there are also decomposer animals and countless parasites, the answer seems perfectly clear: there are many times more ecological niches available for animals than for plants.

On the question of the unique diversity of insects a few lines must suffice. Insects are all constructed as variants on a most excellent design, with a hard but flexible skin that doubles as armour and skeleton. It is made of a plastic-like material that can be shaped into almost any form, including wings and internal breathing tubes (which carry air not blood to the tissues). The design is not capable of being adapted to large animals, and this has decreed that insects must be small. Consequently, the majority can survive in micro-habitats, with the result that an area that would provide a territory for a single pair of birds, or the space occupied by a single tree, can provide distinct, separate habitats for quite a

Green chlorophyll in the leaves of plants is the key substance in the living world. Using the energy of sunlight and non-living materials, chlorophyll can make living molecules. These contain energy stored as chemical fuels. All animals ultimately rely on the energy and organic materials that green plants provide.

A larva of a hoverfly (*Syrphus*) seizes a greenfly nymph that has been sucking sap from a leaf. Greenfly (Aphidae) can multiply fast but they are programmed to control their own numbers and do not destroy their supply of food. Neither do hoverflies.

host of insects. The insects have attained a climax of ecological diversification, unapproached elsewhere in the living world, but it is notable just the same that there are very few marine species.

We can turn again now to that remarkable concert of the living orchestra which is responsible for the balanced, time-less, biogeochemical cycles on which the continuation of life depends. We have seen that all animals derive their energy supplies from their food and that its ultimate source is green plants. That is to say, all animals are parasitic, in the literal sense, on plants. An animal is by definition a feeder dependent on other organisms as food, whereas the great majority of plants are self-sufficient. This implies that plants or other green organisms must have existed before any animals could evolve. Once animals did appear they presumably caused the plants to work harder, in order still to fulfil the plants' own needs and provide the levy the animals took as well.

Realising this, it comes as no great surprise to find that whereas there are perhaps four times as many animal as there are plant species, when it comes to total bulk or biomass the very opposite is true. The world's green plants outweigh its animals many times over. It is possible to estimate roughly how much energy the herbivores do levy from the green plants, and it turns out to be fairly small, say 5 to 15 per cent. It is a relatively light burden, therefore, and this bears out the observable fact that, in the wild,

herbivores eating the living tissues or (like aphids) sucking the sap of plants do not usually kill the plant. Only now and again do we see, for example, a forest tree completely defoliated by caterpillars. Nevertheless the fact that cater-pillars *can* defoliate a tree in exceptional circumstances leads one to ask what factors normally forestall it. Why do the herbivores not multiply till the vegetation is destroyed?

We are confronting here one of the great unresolved questions of ecology, of how the 'balance of nature' is preserved in an ecosystem, between the three harmonious guilds that share the cycling process. Let us consider each of them in turn, and ask what it is that sets the upper limit to their numbers. The first answer is the easiest to give, for we observe that, normally, dead organic material does not pile up over the years, which means that the decomposer populations must be sufficiently large and voracious to keep pace with it. This at once suggests that their numbers are food-limited, since it must imply that where the supply of dead organic material is greater there are more decomposers, and *vice versa*. When the supply is all gone the survivors must migrate, hibernate or starve.

Take next the producers. Here it is only a little more complicated. Where the site is physically adequate for them, the number of plants present is usually limited by how many can find the standing room they need to receive sufficient light, or the rooting space for reaching adequate supplies of water or essential minerals in the soil. This suggests that plants tend to be resource-limited.

It is the consumers—the animals—that pose the difficult problem. Thinking first of herbivores, it has been argued already that those which feed on the non-expendable tissues of plants do not normally do serious permanent damage although they are usually capable of it. This suggests that some other limiting factor must exist that keeps their numbers down, well below the level that would damage the well-being and renewability of the food-plant resource. The most obvious and general candidates as population limiters are the carnivores, including the numerous insect 'parasit-oids'. It is known that some herbivore populations are regularly kept in check in this way, and the commonest reason why herbivorous insects sometimes get out of hand and do damage is that the carnivores and parasitoids have temporarily failed to materialise. Carnivores, in their turn, generally take only a moderate annual toll of their prey, which the prey can likewise sustain without danger.

There are no good examples from nature of carnivores that have endangered or exterminated their prey, and this prompts the question, what controls the 'top' carnivores? There must be some mechanism limiting their numbers so that there continues to be a sufficiency of prey, otherwise their own survival would be in jeopardy. The killing out of prey is not impossible; it has happened on rather rare occasions with the biological control of agricultural pests and it has happened over and over again when avaricious Man has adopted the role of predator on wild animal stocks, in whaling, commercial fishing, and the fur, ivory and reptile-leather trades. If it were a common natural occur-rence, however, it would certainly have been detected but it seems not to be so.

The most general controlling factor keeping both carni-vores and herbivores within the carrying capacity of their

ABOVE: the biomass of herbivores on the African plains is unrivalled but not inexhaustible. The key to survival for carnivores such as this Cheetah (*Acinonyx jubatus*) is keeping their own populations sufficiently low for the prey to carry without depletion.

BELOW: Red Grouse are mature at five months and each autumn they take territories that form a mosaic over the moor. There is usually room for less than half of them; the ones which fail to secure a place are squeezed out and die during the winter.

food resources, will probably turn out to be self-regulation, operating within the consumer populations themselves. There is evidence that such automatic controls, with feedback from food resource to consumer, exist in all the vertebrate classes and most if not all the arthropod classes too. For instance, in some bird species that have food-territory systems, the individual is programmed to vary its territory size in relation to the food resources its territory holds, being content with a smaller area when food is abundant and demanding more space when it is scarce. This has been conclusively demonstrated for the Red Grouse (*Lagopus lagopus scoticus*) in Scotland in recent years. The grouse not only adjusts its territory size when its food (which is heather) has been experimentally improved by applying a mineral fertiliser, but its territory system fills the heather moor like a mosaic, and each autumn the system crowds out the excess birds for which there is no room, and dooms them to die. It is a remarkable adaptation, and seems likely to reveal the real key to the restraint on their numbers that animals have had to evolve, as the condition for their permanent place in the balance of nature.

The million or more extant species have survived every problem of existence and coexistence that has faced them down the long evolutionary road—otherwise they would not be here. On balance, the gain of new species by diversification must have greatly exceeded the loss of old ones by extinction, and we are probably living now in a world that has never been richer in forms of animal life. My own experience is that the more one learns about any kind of animal, the more fascinating it becomes. If we will just open our eyes to see, even the least of them will reward us.

The Distribution of Animals

Nobody could say that nature has been ungenerous in stocking the world with plants and animals. There are far too many organisms to count or even estimate. In addition, the number fluctuates too rapidly for counts to be meaningful, and so we can only estimate biomass—a poor substitute for living creatures. Even the number of living species is astronomical. There may well be nearly two million different kinds of plants in the world, and at least as many different kinds of animals. Again, we could never hope to count them accurately, and their numbers, too, may change too fast for us to keep up with them. And for every species alive today, how many have evolved, flourished, and fallen away to extinction in the past? This is yet another figure we shall never know accurately, but there were probably some hundreds of millions during the three thousand million years or more since life began. Scattered all over the busy, bustling surface of our planet, these are the plants and animals, living and long-dead, that we try to account for in a section on distribution in time and space. Fortunately for our sanity, we have detected several kinds of order in this mass of living material. Taxonomists and systematists have reduced the immense chaos of individuals and species to much smaller and more manageable numbers of taxonomic groups. With fewer, larger units to think about, we can begin to look for significant patterns of distribution among them—for example, how genera and families, rather than species, are scattered about the earth.

Studying the distribution of organisms in time, by reference to the fossil record, is mainly the business of palaeontologists. From the patchy, contorted sequence they find in the rocks, palaeontologists have built up a remarkably complete and self-consistent picture of the history of plants and animals, on a time scale covering nearly six hundred million years. The distribution in space of organisms alive today is the business of biogeographers, who record where plants and animals are to be found and try to explain their presence, both in terms of recent history—where they came from—and in terms of ecology—why they are there, and what is their role in relation to other organisms in the same environment.

Not surprisingly, the fields of study overlap in many respects. Fossil organisms show interesting and often puzzling patterns of distribution in space as well as time, and their ecology can be as intriguing as that of modern organisms. Conversely, the geographical distribution of many modern species, genera and families has changed radically since they first appeared on earth, both preceded and followed by profound ecological changes. Making sense of the distribution of organisms in time and space is an engrossing four-dimensional jig-saw puzzle—a problem that scientists have pursued systematically for two centuries and more. Like all the best problems in science, it has no simple solution.

To orthodox minds of the eighteenth and early nineteenth centuries the problem did not exist. Animals lived, as they had always lived since the time of the Creation, in their appointed places within the framework of an unchanging world. Fossils were remnants of recently-living animals destroyed by floods and other catastrophes; they might even have been inventions of the devil, designed to confuse thinking and upset the stability of the orthodox position.

Then, from the great voyages of discovery of the eighteenth and early nineteenth centuries, came descriptions and specimens of thousands of new plants and animals from all over the world. This massive evidence of genetic bounty stimulated the growth of museums and zoos, and contributed to the new sciences of taxonomy and classification. But above all it challenged the old orthodoxy, posing a number of problems. Could this huge diversity of species really be accounted for by thousands of special acts of creation? On the contrary, the striking family resemblances between animal groups suggested kinship from common ancestry, and more subtle, evolutionary forces at work. In particular there was that most puzzling and disconcerting resemblance of all—man's similarity to the apes. What was the cause of the curiously disjointed distribution of animals about the world? For example, rhinoceroses are found in the East Indies but not in nearby Australia; camel-like creatures in Asia and South America but not in Europe, and marsupials in two of the great southern continents but not in the third.

And, as if these problems were not sufficient, new questions were being posed by the geologists, whose studies were wringing startling information from the distribution of fossils in the earth's crust. Stimulated by early nineteenth century road building, railway engineering and quarrying, the systematic study of fossil-bearing rocks was starting to tell a complex story of a changing, unsettled world—one far removed from the orthodox concept of permanence and immutability. That many of today's mountains were laid down as part of ancient sea beds, that coal measures were formerly live swamp-forests, that glaciers once covered most of Britain and northern Europe—these were some of the simpler propositions put forward by the geologists. More detailed examination of the fossil record uncovered shells, bones and teeth representing thousands of animals no longer in existence, packing the strata in orderly time-sequences. The rise, geographical spread and decline of whole groups of animals could be plotted with certainty.

Clearly, neither the earth nor the organisms on it—and certainly not their patterns of distribution—were comfortably static, as the old orthodoxy would have them. Change and evolution were the governing rules, which even the rocks obeyed. Now that we know the continents, too, have changed, many of the earlier problems of distribution are simpler than they seemed, and a new orthodoxy is rapidly appearing to replace the old ones.

Just as the distribution of Polar Bears (*Thalarctos maritimus*) is restricted to the Arctic, so the range of any species, past or present, is constrained by a complex interaction of geographical and ecological factors

Distribution in Time

Stony artefacts, victims of the Flood—these were among the ways in which fossils were once explained. We now believe that the fossil record shows the variety of life and patterns of evolution that have occurred during the last 600 million years.

There are perhaps, 4.5 million species of plants and animals alive today, the last page of a history of life that extends back more than 3,000 million years. Estimates suggest that perhaps 9,820 million species lived in this interval, most of which did not survive into the fossil record—less than 300,000 fossil species have been described to date. But the length and diversity of the record of extinct organisms gives us a clear idea of the ancestry of living animals. Palaeontology—the study of fossils—is thus of great help in understanding the origin of present day faunas. Furthermore, only from the fossil record can we fully comprehend the major features of evolution, and whilst it tells nothing of evolutionary mechanisms, it records the patterns of evolution.

Fossils have been collected as curiosities and charms since the days of ancient man, but their true nature was not generally appreciated. Some of the Greeks, such as the traveller and historian Herodotus (c. 484–425 BC), recognized the significance of certain apparent anomalies, such as the occurrence of fossil marine shells far inland. But their ideas were lost in the myths and religious dogma that surrounded men in the Middle Ages. Even in the 16th century fossils were regarded as stony concretions. Fossils that looked like animals were interpreted as deceiving products of devils or 'practice creations' of God.

It was not until the 17th century that fossils were accepted as being the remains of once-living animals and plants. Even then they were regarded as remains of organisms destroyed by the Biblical Flood. Subsequently, scientists began to think that one such catastrophe was not enough. The great French palaeontologist Georges Cuvier (1769–1832) proposed a series of four catastrophes. His followers later increased the number to twenty-seven.

However, the 'Principle of Catastrophism', as it was called, was insufficient to explain all the data that were rapidly accumulating, and in the 18th and 19th centuries emerged the modern ideas of fossils and the rocks in which they are found.

The Scottish geologist James Hutton (1726–97) proposed the 'Principle of Uniformitarianism'. This argued that the

BELOW: the trilobite *Pseudosphaerexocus octolobatus* from the Ordovician of Girvan, Ayrshire, Scotland. One of the first groups of animals with skeletons to appear in the fossil record, these arthropods died out in the Permian.

RIGHT: sedimentary rocks exposed at Zion National Park, Utah, demonstrate William Smith's principle of superposition of strata. That lower rocks are older than those above forms the basis for studying earth history.

processes that are shaping the Earth today are the same as those that have always been acting, and that these processes act very slowly. When the Earth was formed many millions of years ago the molten material hardened into igneous rock. The forces of erosion, such as wind, water and ice, have since been wearing down these rocks. Small particles were carried by rivers into lakes and seas, where they formed sediments. Thick layers of sediment subsequently became hardened into sedimentary rocks, trapping the remains of any organisms within them.

Era	Period	Epoch	Millions of years ago	
CAENOZOIC	QUATERNARY	Recent (Holocene)		Modern man emerged. Dawn of civilization. World's fauna much as it is today.
			0.01	
		Pleistocene		Ice ages in northern hemisphere. Woolly animals survived while others retreated south. Giant marsupials in Australia. *Homo erectus* and *Homo sapiens*.
			2	
	TERTIARY	Pliocene		True elephants evolved. Marsupials in Australia. Hominids appeared. Cold weather caused many extinctions.
			7	
		Miocene		Herds of ungulates. Marsupials diversified in South America. Fissipeds and pinnipeds appeared.
			26	
		Oligocene		Old and New World primates. The largest ever land mammal, *Baluchitherium*.
			38	
		Eocene		Perissodactyls flourished. Rodents, small elephants, artiodactyls and whales appeared.
			54	
		Palaeocene		Invertebrate groups similar to those of today. Mammals became major terrestrial group. Carnivores, primates, perissodactyls appeared.
			65	
MESOZOIC	CRETACEOUS			Dinosaurs and ammonites became extinct at the close of this period.
			135	
	JURASSIC			Giant dinosaurs and flying reptiles. Ammonites abundant. Molluscs and echinoderms diversified. First birds appeared.
			190	
	TRIASSIC			Modern corals, ceratitic ammonoids, mammal-like reptiles. First dinosaurs and mammals appeared.
			225	
PALAEOZOIC	PERMIAN			Many reptiles, including mammal-like reptiles and first ichthyosaurs. Many invertebrates, including goniatites and trilobites became extinct.
			280	
	CARBONIFEROUS*			Many amphibians. First flying insects and reptiles. Trilobites, brachiopods and cephalopods declined. Graptolites became extinct.
			345	
	DEVONIAN			First ammonoids (goniatites) appeared. Freshwaters colonized by fish and the first amphibians. First crabs and insects.
			395	
	SILURIAN			Brachiopods, corals and crinoids abundant. Eurypterids, armoured jawless fish, acanthodians and cirripedes appeared.
			440	
	ORDOVICIAN			Abundant corals and crinoids. Trilobites, brachiopods, bryozoans, gastropods, bivalves, graptolites and cephalopods were all common.
			530	
	CAMBRIAN			Inarticulate brachiopods, gastropods, monoplacophorans, echninoderms, bivalves, trilobites and ostracods.
			570	
PROTEROZOIC	PRECAMBRIAN			First multicellular animals: jellyfish, sea pens, worms, proto-arthropods. Ancestral molluscs and echinoderms may have appeared.
			1850	
ARCHAEOZOIC				First algal plants.
			3500	
AZOIC				Formation and cooling of the earth. First life?
			4600	

*In America, geologists recognize two periods in place of the Carboniferous; the Mississippian and Pennsylvanian (345–325 and 325–280 million years ago respectively).

The history of life is recorded in the succession of faunas in sedimentary rocks from the late Pecambrian onwards. Division into eras is based on the main phases in the evolution of plants and animals. The periods, mostly named after areas where the rocks were first typified, are dated radiometrically.

William Smith (1769–1839), an English geologist, took Hutton's ideas a stage further. He earned the nickname 'Strata' Smith because of his 'Law of Superposition of Strata', which states that under normal circumstances the younger sedimentary rocks will rest on the older. Smith was a catastrophist, but he realised that certain fossils only occurred in particular rocks. From this it followed that rocks containing the same fossils were the same age. Armed with Smith's law and a knowledge of these index fossils his successors recognized the whole sequence of rocks and their faunas that make up the great divisions of the history of the Earth. At first this geological time scale showed only the relative ages of the rocks. But since the discovery of radio-activity it has become possible to measure the absolute age of rocks, and to place the history of life against actual dates in millions of years.

In the early 1800s fossils were still regarded as the products of a series of catastrophes. Charles Bonnet (1720–93), a Swiss naturalist, had used the term 'evolution', but he believed that animals evolved by stepping up one rung of the evolutionary ladder after each catastrophe. The English naturalist Charles Darwin (1809–82), on the other hand, took a different view. By the time he set off on his famous expedition to the Galapagos Islands he had read *The Principles of Geology* by the Scottish geologist Sir Charles Lyell (1797–1875). This work popularized the work of Hutton, and it convinced Darwin that the principle of uniformitarianism led to the idea of the continuity of life and its great antiquity. This, combined with his zoological observations, led him to publish in 1859 his great work *On the Origin of Species by Natural Selection*. In this Darwin coordinated a series of elementary observations into an easily understandable theory, often summed up by the phrase 'the survival of the fittest'. He observed that individuals within any species vary to some extent, and concluded that in the intense competition for survival, those with characteristics best suited to their environment would be the most likely to survive and reproduce. Natural selection was the mechanism that eliminated those individuals with characteristics that departed from the optimum.

For evolution to take place, however, new characteristics must arise. The exact way in which new characteristics appear was not solved until many years later. At that time the French naturalist Jean Lamarck (1744–1829) believed that the characteristics a creature acquired during its life were passed on to the next generation. But Darwin's theory was that animals were born with new characteristics. If such characteristics were beneficial then natural selection would ensure that the animals would survive and the characteristics would be passed on.

Any feature of an animal which helps it to escape competition with others is favoured by natural selection. The elongation of a Giraffe's neck has clearly been of advantage in enabling it to reach food inaccessible to others. In this way evolution has produced species adapted to fill almost every available ecological niche.

A group of Pleistocene molluscs from the Red Crag on the Suffolk coast of England. Altered only by iron-staining, their original mineralogy and structure are perfectly preserved.

Of the millions of animals and plants that have ever existed relatively few have survived as fossils. Particular conditions are necessary for the formation of a fossil, and once formed fossils have often been destroyed by changes that have later taken place in the rock matrix.

The majority of fossils are formed from the hard parts of animals. The soft parts generally decay too quickly after the animal dies, although there are cases where the soft parts have been preserved. Thus most fossils consist of the shells, bones and teeth of animals.

The first requirement for fossilization is rapid burial of the animal in a sediment, such as mud or sand. The next stage depends on the chemical conditions present in the sediment and the nature of the animal's skeleton. Bones and teeth, for instance, contain many pores, and the process known as permineralization occurs when dissolved chemicals in the sediment are deposited in these pores. The fossil becomes stony in appearance, but it is basically unchanged and merely seems heavier than the original tooth or bone. Many parts of animals that lived during the Caenozoic era were fossilized in this way.

A shell that has been buried for a longer time becomes included in rock, as the sediment hardens. Subsequently, it may be slowly dissolved and in a hardened sediment this leaves a cavity inside the rock. In many cases this cavity has remained empty, leaving an external mould of the shell. If the inside of a shell was filled with sediment before the shell dissolved away, an internal mould may also be preserved, like a kernel inside a nut. Alternatively, minerals from the rock slowly replace the material of the shell, producing a cast, with none of the original material remaining. Some replacement fossils formed very slowly indeed, and this has resulted in the preservation of very fine detail.

Because hard parts are all that generally survive to become fossilized, the record is biased towards animals like molluscs, echinoderms and corals or vertebrates with bones and teeth. Many wholly soft-bodied groups such as sea anemones or flatworms have a poor or no fossil record as a result. There are, however, some remarkable examples of fossilization of soft tissues (usually reduced to carbonaceous films, or replaced by minerals) or impressions of soft parts. Mammoths and also woolly rhinoceroses have been preserved in frozen ground in Siberia and Alaska for hundreds of thousands of years. We find butterflies and spiders in amber 35 million years old, and dinosaur eggs and young mummified in ancient desert sands over 100 million years old.

In addition to these kinds of fossil there are those that give indications of past life. Such trace fossils include footprints, animal droppings and burrows. Thus, fragmentary and incomplete as it is, the fossil record gives remarkable insight into, and evidence for, the history of life.

The oldest dated rocks on earth are 3,800 million years old and at this time the atmosphere was devoid of free oxygen, consisting of some or all of the following: hydrogen, methane, ammonia and possibly carbon dioxide, carbon monoxide and water. Modern experiments suggest that with this unpromising starting material it is in fact possible to produce many of the basic organic materials that are the building

Specimens of the rhynchonellid brachiopod *Calcirhynchia* from the Jurassic rocks of Dorset, England. They are preserved in the limestone matrix as moulds and calcitic casts of the shell.

A line of dinosaur footprints shows how the animal moved. Sediments in which such trace fossils are found also provide clues about the environment in which the animal lived.

Originally sticky resin oozing from the bark of a Tertiary coniferous tree some 50 million years ago, this piece of amber has trapped within it many insects of the time. Sealed from the air, they are quite perfectly preserved and still show all the details of legs, wings and eyes that were present in life.

blocks of life—either as a result of electrical discharges, volcanic heating, or other events. As a result, we believe that the first, albeit very simple, organisms appeared perhaps as long as 3,500 million years ago. However, we can only guess at the complexity of events needed to turn simple molecules into life.

What were these early forms like? Most were probably tiny spherical bodies, initially aquatic (water protects from the strong radiation the early earth must have suffered in the absence of an ozone layer), without an organized nuclear membrane, and feeding on pre-existing organic molecules.

From these early forms, arose organisms which manufactured their own food-stuffs by processes such as photosynthesis. At this time, oxygen became available all over the world. This is indicated by the appearance of sediments rich in iron oxide 3,500–1,800 million years ago, as the free oxygen reacted with iron in solution around the colonies of primitive plants. Indeed, many unicellular and filamentous algae occur as fossils in such rocks. As oxygen built up atmospheric conditions changed, and land surfaces oxidized, giving red sediments, the first of which appeared 2,600 million years ago. At this time, an ozone layer first developed, shielding early life forms from radiation, and enabling them to move into more exposed habitats. Organisms with a nuclear membrane and with nucleic acids and chromatin packaged into chromosomes developed, and some time between 1,000 and 700 million years ago, these evolved into sexually reproducing multicellular animals. Offspring produced by sexual reproduction are more variable than those produced by asexual reproduction. Thus the appearance of sexual reproduction is probably the one event which triggered the great radiation of life from this time onwards. Through this mechanism, evolutionary change can proceed at much faster rates than in asexually reproducing creatures.

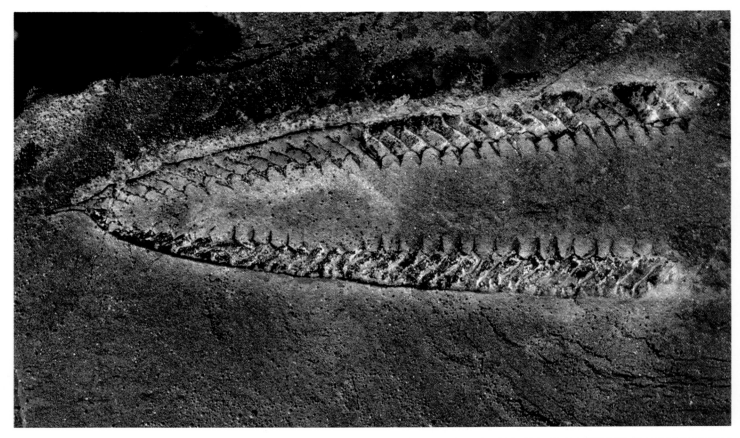

ABOVE: *Didymograptus bifidus*, Ordovician graptolite, from Wales, clearly showing the cup-shaped thecae. The fossil is preserved in pyrites, an iron sulphide mineral.

RIGHT: a close-up view of the surface of the rugose coral *Palaeosmilia reginum* from the Carboniferous, showing several corallites and details of their walls and septa.

BELOW: *Woodocrinus*, a Palaeozoic crinoid (sea lily) in the phylum Echinodermata. The specimen is from Carboniferous limestone, some beds of which are largely built up of crinoid ossicles.

The earliest record of multicellular animals takes two forms. First, there are the trace-fossils: tracks, trails, and burrows produced on the sea bed. These record the slow progressive diversification of soft-bodied worms, arthropods and molluscs. Second, there are true body fossils, known from Australia, South Africa and Europe in rocks 600 million years old, and including worm-like animals, proto-arthropods, jellyfish, sea pens and possibly ancestral molluscs and echinoderms.

From these soft-bodied creatures arose the Cambrian fauna. Over a period of 20–30 million years, diverse animal groups acquired hard parts. Thus rocks dating from the Cambrian period onwards contain a large number of fossils. Several factors have been put forward to account for the appearance of hard parts, such as the build-up of oxygen, development of new ecological niches and food chains, competition, and the appearance of predators. None of these is satisfactory. The best explanation we have is that it reflects the great change in marine environments that resulted from the break-up and moving apart of continents, and the spread of shallow seas onto their margins.

Cambrian faunas are fascinating not only because they form the first good record of life, but because many groups that appear seem to have been short-lived, 'experimental' animals which failed to survive beyond the period. In the early part of the Cambrian period all life was marine, and most fossils are of small creatures with shells composed of calcium phosphate, including inarticulate brachiopods, gastropods, monoplacophorans, and tiny tube-like shells of unknown affinities. These were superseded by organisms with shells made predominantly of calcium carbonate, including the first tiny bivalves, the arthropod group known as the trilobites, and both stalked and free living echinoderms. By the end of the period, all invertebrate groups with shells or skeletons were present, except the bryozoans, and the earliest colonial hemichordate-like animals known as graptolites appeared.

In the Ordovician period, these groups continued, with brachiopods, bryozoans, gastropods, bivalves, nautiloid cephalopods and trilobites being common, while crinoids and corals were so abundant in some places that their remains later formed limestone rocks. Graptolites were widespread, especially in deeper water. More significant was the appearance, in the early to middle Ordovician period, of the first, primitive marine jawless fish.

Silurian marine benthic faunas were dominated by brachiopods, which filled the role of bivalves in modern communities. The same basic groups were present as in the Ordovician, but corals and crinoids became even more abundant in some places. Giant sea scorpions, or eurypterids, some nearly 3 m long, appeared in the Silurian. New vertebrate groups included the armoured jawless fish, a group related to the lampreys (for example, *Cephalaspis*) and the first jawed fish, known as acanthodians or spiny 'sharks' up to 30 cm long.

The Devonian period saw important changes. In the seas, the first ammonoids, the goniatites, appeared, as did the earliest belemnite-like shells. At this time, animals conquered the land and fresh waters, and in the extensive lake and river sediments of this period (the Old Red Sandstone) are found many cephalaspids and the first bony fish—actinopterygians, dipnoans (lungfish) and rhipidistians. Other groups included the giant arthrodires (up to 8 m long), an armoured carnivorous group. The first amphibians, the labyrinthodonts, appeared in Greenland in the late

Devonian period. They had evolved from rhipidistian fish. The first insects (collembolans) occur in the Devonian rocks of Scotland.

During the Carboniferous period the graptolites became extinct, and many nautiloid, trilobite and brachiopod groups declined. Important new groups included many land animals, such as the first flying insects (some giant dragonfly-like creatures 60 cm across), huge centipedes, millipedes and cockroach-like forms. The earliest reptiles are known from the Carboniferous, and there were diverse amphibians.

The Permian period is the last in the Palaeozoic era, and it saw the final flourish of the goniatites, many brachiopod groups, giant foraminiferans and stalked echinoderms, while on land there were many diverse reptilian groups, including aquatic crocodile-like forms, the first ichthyosaurs, early lizards and those that were eventually to give rise to the mammals.

The close of the Permian period saw a great change in life which marks the boundary of Palaeozoic and Mesozoic eras. Many groups became extinct, including some foraminiferans, many tabulate and all rugose corals, trilobites and some other arthropods, archaic insects, goniatites and many other cephalopods, all Palaeozoic sessile echinoderms, and most brachiopods.

Mass extinctions of this type are one of the enigmas of the fossil record. Explanations are as varied and diverse as the changes themselves, ranging from extraterrestrial causes (radiation or meteorite impacts) to genetic exhaustion, and changing climates and salinities. However, these extinctions did not all occur at the same time. They were spread over millions of years, and replacement by Mesozoic groups occurred over a similar period of time. The changes can probably be attributed to the changing geography of the times. Shelf seas coalesced as continents moved together; marine waters were withdrawn from shallow seas; fluctuations in climate produced large-scale alterations to the environment. These were the same kind of global changes that may have initiated the radiation of animals in the early Cambrian period.

A Devonian rhipidistian. This group of fleshy-finned fish appears in the fossil record from the Devonian to the Permian and its members were probably the ancestors of amphibians.

The Mesozoic era was briefer than the Palaeozoic era, and its fauna was very different. During the Triassic period several new groups appeared in the seas, including modern corals, ceratitic ammonoids, many new groups of bivalves, gastropods and echinoderms. On land, vertebrates were conspicuous, especially herbivorous and carnivorous mammal-like reptiles and thecodonts, with the earliest dinosaurs, and the first tiny mammals. In the Jurassic and Cretaceous periods ammonites were diverse and abundant, as were belemnites, and these groups dominated marine nekton. Rhynchonellid and terebratulid brachiopods were abundant, and many bivalves and gastropod groups diversified. Echinoids diversified into many habitats, and corals, sponges and bryozoans were common. Of the vertebrates, the giant dinosaurs rose to prominence on land, and there were flying reptiles (pterosaurs), together with the first birds, which evolved from the reptiles during the Jurassic period. Giant ichthyosaurs, plesiosaurs and a family of marine lizards, the mosasaurs, dominated the seas.

At the close of the Cretaceous period there was a period of extinctions of a similar scope and scale to that at the end of the Permian period, and this marks the boundary between the Mesozoic and Caenozoic eras. Extinctions affected many groups of vertebrates and invertebrates, both terrestrial and marine. Some bivalves died out, as did many gastropod, foraminiferan and echinoderm groups. All ammonites and belemnites disappeared, and also ichthyosaurs, plesiosaurs, dinosaurs, toothed birds, and much of the ocean plankton.

As with Palaeozoic-Mesozoic faunal changes, reasons for extinction are to be found in global environmental changes, rather than great catastrophes of whatever origin. Almost everywhere in the world, the late Cretaceous was a time when seas retreated, and even in the deep oceans there are traces of this great regression. This major environmental change, which also affected climate, is now believed to be the triggering mechanism for such widespread extinction.

Caenozoic faunas were basically of modern aspect, and many of the creatures present would not look too unfamiliar in a modern forest or shore. Bivalves and gastropods dominated marine faunas in most areas of the world, accompanied by echinoids, corals, crabs, lobsters, barnacles and bryozoans. Although some looked archaic, virtually all are readily recognizable as animals we are familiar with today. Perhaps the one obvious exception is the giant foraminiferan *Nummulites*, which rose to importance during the early Tertiary and built limestones in many areas of the world.

After the extinction of the dinosaurs, the small mammalian groups slowly radiated into terrestrial and aquatic niches, and by the middle of the Tertiary period, giant forms of many groups developed. Marsupials diversified, especially in South America and Australia, and in the late Tertiary of Australia giant kangaroos up to 3 m tall existed. Many rodent groups arose in the Eocene epoch, whilst the first bats are known at this time. Carnivores arose in the Palaeocene. Most fissiped groups (dogs, bears, cats) arose in the Miocene, as did the pinnipeds (seals). Subungulates (aard-

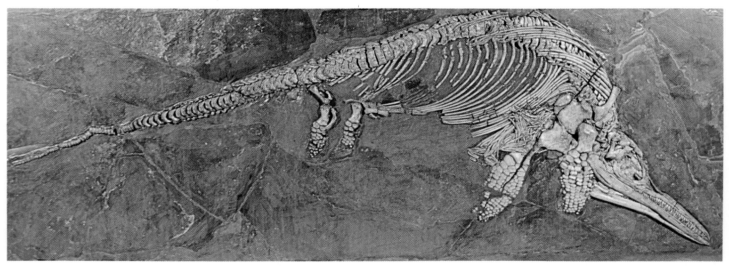

BELOW: a mass of the ammonite *Promicroceras planicostata* from the Lias of Somerset, England. The ammonites (Cephalopoda) are extremely diverse and form the basis for the correlation of the majority of marine Jurassic and Cretaceous rocks.

BELOW: specimens of the giant benthic protozoan *Nummulites*, widely used in the correlation of shallow-water Tertiary rocks. Their remains often occur in sufficient numbers to form limestones, including those of which the Pyramids are built.

ABOVE: a fossil of the fish-like *Ichthyosaurus intermedius* found in the Jurassic Lower Lias of Street in Somerset. It was one of the most familiar of the giant Mesozoic marine reptiles, and it is known to have been an active predator on fish and squid.

varks, elephants, seacows) have their ancestry at the beginning of the Caenozoic era. The evolutionary centre of the elephants was in Africa, and these giant animals have a good fossil record. In the Eocene they began as pig-sized creatures, but by Pliocene times they had evolved into giants, and migrated to Europe, Asia and, somewhat later, North America. Perissodactyls (horses, rhinoceroses) arose in the Palaeocene, and flourished in the Eocene. The largest land animal was *Baluchitherium*—a relative of the rhinoceroses almost 6 m tall—known from Oligocene deposits of Asia. The first horses—the size of a small dog—appeared in the Eocene and the group has been successful ever since. Artiodactyls (cattle, pigs, deer) appeared in the Eocene and included many, long extinct giants, such as the Irish Elk, whose antlers spanned 4 m. Edentates (sloths) known from the Eocene onwards, evolved as giants too, like the South American edentate *Glyptodon*, which was 7 m long.

Primates, the group to which Man belongs, first appeared in the late Cretaceous period, but are rare fossils during much of their history. At least four groups are known from the Palaeocene (mainly from teeth); tarsiers and lemurs appear in the Eocene, whilst from the Oligocene onwards Old and New World primates appear to have evolved separately. The earliest reliable hominids are known from rocks 3.5 million years old, although other creatures with some hominid-like features occur in the Oligocene. Modern man dates back no more than about 250 thousand years.

This glance at the history of life is best put in perspective by the simple analogy of translating it to a single year. On this time scale, the Precambrian ended in late September, animals having been in existence for a few weeks before; the first vertebrates appeared in early October and the first placental mammals in late November; Man comes in late in the evening of December 31st; the 6,000 years of recorded history occupy ninety seconds and a human lifetime lasts a mere second.

This then is the perspective of time against which we must study and understand the animals around us today, and the time interval during which natural selection has produced the diversity of life. Time alone is not the only factor to consider. The pattern of extinctions shows us how important are changes in the world environment for controlling the major events in the evolution of life. We now believe that these great changes in climate, ocean circulation and the distribution of land and sea can all be linked to the constant relative movement of continental blocks as they drift, driven by forces deep in the earth.

More relevant to Recent faunas is the fact that the world is in a time of rapid change. In the last 2.5 million years ice sheets have advanced several times across most of North America, Asia and Europe and as they grew, the seas contracted. In this way migration routes appeared for land animals and disappeared for marine ones, while refrigeration shifted floral zones up to 2,000 kilometres south of their present positions, and animals inevitably moved with them. We are still in an interglacial epoch, and the ice will return, so that patterns of life will change yet again in the next few tens of thousands of years.

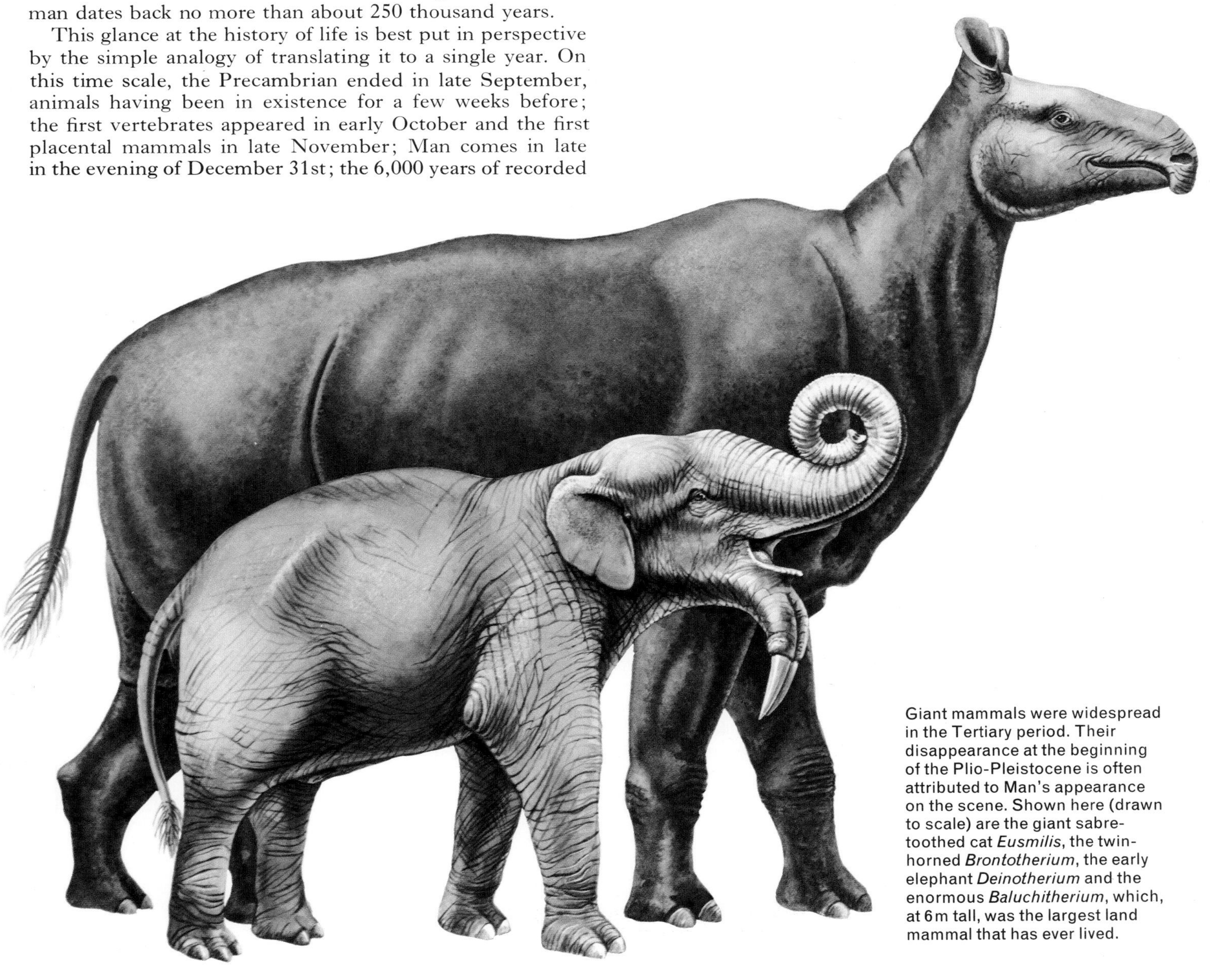

Giant mammals were widespread in the Tertiary period. Their disappearance at the beginning of the Plio-Pleistocene is often attributed to Man's appearance on the scene. Shown here (drawn to scale) are the giant sabre-toothed cat *Eusmilis*, the twin-horned *Brontotherium*, the early elephant *Deinotherium* and the enormous *Baluchitherium*, which, at 6 m tall, was the largest land mammal that has ever lived.

Geographical Distribution

The world, as seen by zoogeographers, is made up of several regions, each characterized by a distinctive fauna. The land masses within the regions are unstable and past continental movement has influenced the distribution of many of today's species.

The study of the distribution of animals on land and in the oceans of the world is known as zoogeography. The subject is a very complex one. The difficulties lie not so much in defining the present distribution of any given species but in explaining how its pattern came about. Finding the answer frequently requires the study and interpretation of data from other fields of science, especially an examination of the fossil record and of past climatic regimes.

For evidence of animal distributions in the past we must turn to the fossil record, and in particular to that of terrestrial vertebrates. Because of their mobility and the ease with which they are contained by physical barriers, a study of fossil reptiles and mammals can reveal valuable information about past land connections and separations between continents. It also becomes obvious that there are many apparent anomalies to be explained concerning past and present geographical distributions of certain animal groups. There are many examples of discontinuous distributions of living animals that are linked by a much wider distribution of fossil remains; tapirs, for instance, are now confined to tropical America and southeast Asia, but fossil forms are known from Europe and North America. There are also cases of past discontinuous distributions not linked by any related forms. A typical example is the early Permian reptile *Mesosaurus* which is found in deposits only in Brazil and South Africa.

One of the better known examples of a discontinuous distribution is that of the prosimian primates. Of these, the galagos are confined to tropical Africa, the lemurs to Madagascar and the tarsiers to southeast Asia, whereas the lorises are found in tropical Africa and Asia. These animals are all adapted to living in warm forested regions and although Africa and Asia are linked by the Sinai peninsula, early zoogeographers supposed that the climate there had always been cold enough to act as a migration barrier. To explain such distributions, therefore, the existence of land bridges was postulated. These were supposed narrow tracts of land which once connected two or more continents across oceans. The prosimian distribution was explained by invoking an ancient land bridge, called Lemuria, which linked Africa, Madagascar and India. Unfortunately, the existence of such land bridges in the south is not supported by the geological facts which have come to light more recently.

We now know that world climate in the geological past was different from that of today and that the warm regions extended much further northwards and southwards. The ancestry of the modern prosimian families can be traced back to two early Tertiary families which were widespread throughout Eurasia when the climate was warmer. Thus the present occurrence of the lorisiforms (galagos and lorises) in separated areas is now interpreted as a relict distribution—

The prosimian primates occur in widely separated parts of the world. The most restricted in range are the lemuriforms, a group which can be found only on Madagascar. This representative of the infraorder is one of the sifakas—*Propithecus verreauxi*

that is, as the climate in the north gradually cooled, these warmth-adapted animals were unable to withstand the change and became confined to more southerly tropical areas which today are separated by the Indian Ocean. The occurrence of lemurs only on Madagascar is now attributed to a chance colonization from the African mainland.

Early workers also postulated similar temporary land bridges connecting, for example, South America and Australia via Antarctica, to account for the presence of closely related animals in continents now widely separated. The marsupials are a classic case: they occur in the Americas and in Australasia but are totally absent from the rest of the world. These supposed land bridge connections between the southern lands seemed to account very nicely for the observed distribution of many animals—including the Permian reptile *Mesosaurus*—but the lack of supporting geological evidence makes this explanation untenable. However, there do seem to have been land connections between North America and Eurasia in the north, one of the best known being the Bering land bridge between Siberia and Alaska.

If many of the postulated land bridges had indeed never existed, then another explanation had to be sought. The unexpected occurrence of, for example, very similar Permian reptiles in both South America and Africa could easily be explained if, at that point in geological time, the continents had been physically joined as one continuous land mass. The evidence of ancient glaciers also suggests that this land mass lay in high southern latitudes. Such a theory would imply that this supercontinent subsequently broke up and its constituent parts then drifted to their present positions. The great difficulty with this revolutionary idea was that, until the early 1950s, nobody could suggest a mechanism by which whole continents could move.

The 17th century philosopher Francis Bacon seems to have been the first to note the remarkable complementary resemblance between the shapes of the west coast of Africa and the east coast of South America. This was in about 1620 but it was not until some 180 years later that Alexander von Humboldt first made the suggestion that this was because the two continents had at one time been joined together. In 1910 the American geologist F. B. Taylor suggested that the movement of continents might have been responsible for the formation of the major mountain ranges of the world, such as the Himalayas and the Alps. The important man of the time, however, was the German scientist Alfred Wegener who, in 1915, published his theory of what came to be known as continental drift.

The idea of continental drift has since been refined and the phenomenon is now more appropriately termed continental displacement. The old concept of continents floating in a sea of more fluid material is no longer favoured but the view that they move relative to one another is now widely accepted by the scientific community.

How they move is best understood by regarding the earth's crust, or lithosphere, as consisting of light rocks, which form the continents, resting on an overall layer of heavier rocks which are exposed at the bottom of the ocean basins. The oceanic crust represents the chilled outer surface of the mantle, the inner part of which (the asthenosphere) is very hot and marks the upper limit of very slow but continuous convection currents which originate from around the earth's core. The crust comprises six major plates and a few sub-plates, most major plates consisting of a continental centre which is partially or wholly surrounded by an extensive oceanic component.

The convection currents bring hot material from the depths of the mantle to the surface, forcing apart the plates along lines of weakness and adding oceanic crustal material to their edges. Such convection currents upwell along mid-oceanic ridges; the Mid-Atlantic Ridge is a well-known example and the effect of the addition of crustal material along this constructive plate margin is gradually to force apart the land masses of Africa and America. Where two plates come together, for example along the margins of the Pacific, the heavier oceanic plate is forced under the lighter continent at trenches, such as the Peru–Chile Trench; these are known as subduction zones or destructive margins. Where plates slide along each other a major crustal wrench

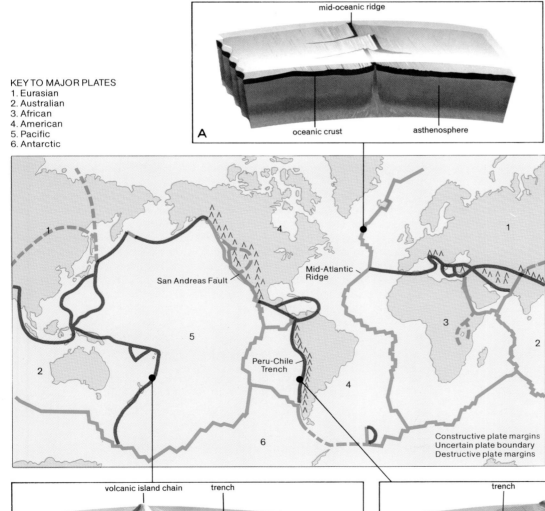

KEY TO MAJOR PLATES
1. Eurasian
2. Australian
3. African
4. American
5. Pacific
6. Antarctic

mid-oceanic ridge
oceanic crust
asthenosphere
A

San Andreas Fault
Mid-Atlantic Ridge
Peru-Chile Trench

Constructive plate margins
Uncertain plate boundary
Destructive plate margins

The distribution of crustal plates and their zones of contact. The three block diagrams show three types of contact. A typical mid-oceanic ridge (A) is virtually symmetrical. As it wells up from below, the plastic asthenosphere gradually solidifies and forms new oceanic crust that may break the surface as an island group, such as the Azores, thus making new land for colonization. At the junction of plates moving towards each other (B, C) the crust tends to be forced down into the mantle. The process at these sub-duction zones, or destructive margins, is usually asymmetrical: one plate sinks below another, normally resulting in a deep trench, and the other plate is lifted up. If this occurs in mid-ocean (B) chains and arcs of islands are formed. Most of the Melanesian islands have been formed in this way and act as a series of stepping-stones, thus allowing colonization of the western Pacific islands. Where an oceanic plate meets the edge of a continent such as South America (C), the continental margin is frequently raised into mountain ranges. Thus, off south Peru, the 6890m deep Peruvian Trench is only 150km away from a 5988m Andean peak, a vertical range of nearly 13km. Subduction zones are characterized by high volcanic and earthquake activity.

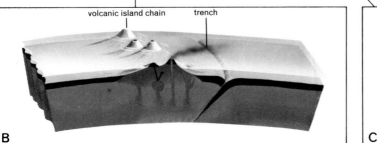

volcanic island chain
trench
B

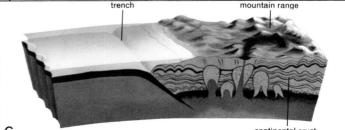

trench
mountain range
continental crust
C

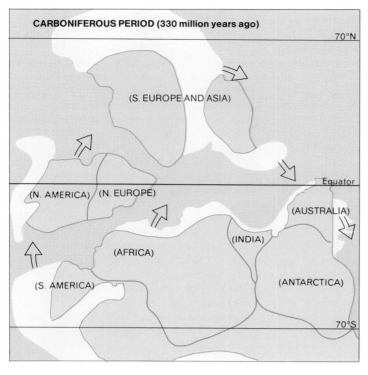

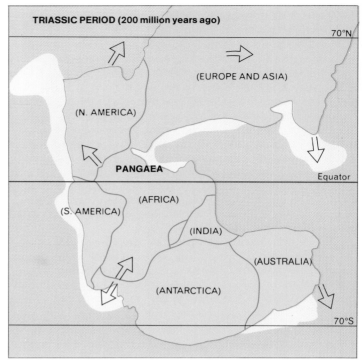

Movement of the continents in the geological past has resulted in an ever-changing picture of land continuity and position.

The distribution of many of today's animals is explained in terms of the movement which has occurred since the Triassic.

fault forms, such as the San Andreas Fault in California. Thus the plates, and the continents which sit on them, are continually moving, very slowly, relative to each other, the power for this movement being derived chiefly from convection currents in the mantle. The dynamics of the phenomenon are known as plate tectonics.

The process of continent formation and break-up, followed by movement and subsequent reunion due to collision, has almost certainly been continuous since the earth began. This has resulted in an ever-changing arrangement of continental shape, continuity and position. There are still differences of opinion regarding the sequence of evolution of the ancient land masses but the basic pattern of events is reasonably clear.

The present phase seems to have started in the Triassic

period about 200 million years ago when all today's continents formed one supercontinent—Pangaea. Before that North America, which formerly lay close to the south pole, drifted northwards and collided with northern Europe about 400 million years ago to form a land mass straddling the equator. In the southern hemisphere there was probably one large continent composed of South America, Africa, Antarctica, Australia and India. All these continents have yielded fossils of a characteristic series of plants, known as the *Glossopteris* flora, which are notably absent from North America, Europe and northern Asia. This distribution is strong evidence for the continuity of today's southern continents and, curiously, India, during Carboniferous times and for their separation from the northern land masses. Northwestwards movement of the southern land mass led to its coalescing with the equatorial and northern continents to form Pangaea. The subsequent break-up of Pangaea into northern Laurasia and southern Gondwanaland began in the Triassic period and the Mediterranean Sea

The zoogeographical regions are designated on the basis of their modern faunas, most of which have evolved since the continents achieved their present positions. The present-day boundaries of the various regions are determined by climate and physical features. For the most part these are the oceans but may be mountains or deserts. The Sahara and Arabian deserts separate the Ethiopian and Palaearctic regions. The Oriental-Palaearctic boundary is the Thar desert in the west, and the Himalayas in the north but is completely climatic in the east. Between the eastern limit of the Oriental region (Wallace's Line) and the Australasian region lies an archipelago (sometimes called Wallacea), whose island faunas form a gradual transition between the two regions. In the New World, the Mexican plateau has a Nearctic fauna, whereas the coastal lowlands are Neotropical.

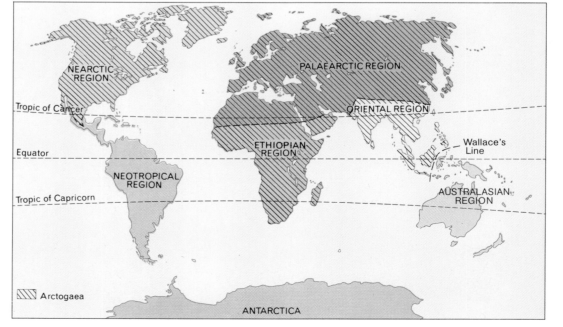

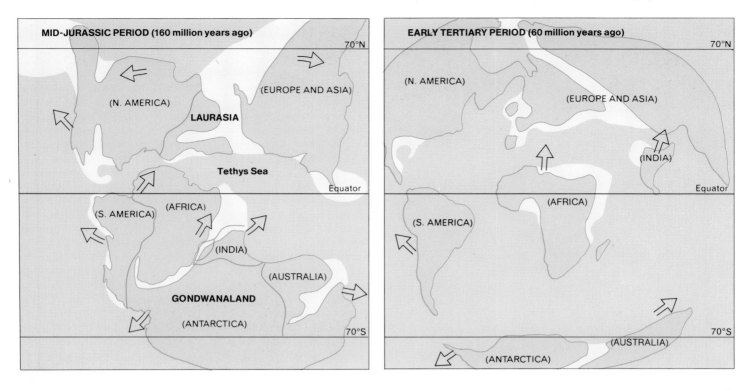

MID-JURASSIC PERIOD (160 million years ago)

70°N

(N. AMERICA)

(EUROPE AND ASIA)

LAURASIA

Tethys Sea

Equator

(S. AMERICA)

(AFRICA)

(INDIA)

(AUSTRALIA)

GONDWANALAND

(ANTARCTICA)

70°S

EARLY TERTIARY PERIOD (60 million years ago)

70°N

(N. AMERICA)

(EUROPE AND ASIA)

(INDIA)

Equator

(S. AMERICA)

(AFRICA)

(AUSTRALIA)

(ANTARCTICA)

70°S

is today's remnant of the Tethys Sea which originally separated these land masses. By the end of the Cretaceous, about 60 million years ago, Gondwanaland had broken up; South America, Africa, Antarctica and Australia had moved far apart, and India was well on its way across the Tethys Sea, ultimately to collide violently with Asia and throw up the Himalayas in the late Tertiary. The North Atlantic began to open up in the Cretaceous and the break between North America and Europe was completed when Greenland separated off in the early Tertiary.

At this time, therefore, most of the present continents were recognizable in shape and positioned more or less as they are now. Asia and North America were connected in the north by the Bering land bridge, and Asia and Australia were linked by an archipelago of islands. The last link to form (in the late Tertiary) was the Panama Isthmus connecting North and South America, although an earlier link existed there in the late Cretaceous period.

So far as the origins of present-day animal distribution are concerned we need go no further back than the early Mesozoic era. Until the Cretaceous period the single land mass of Pangaea had no isolated areas and no effective topographical barriers to the movement of animals. It is not surprising, therefore, that nearly every family of Triassic reptile from North America is also known in Europe, and many occurred in South America, Africa and Asia as well.

The eastern and western hemispheres of the modern world are often called the Old and New Worlds. Africa, Europe, Asia and Australasia are included in the Old World, and the Americas constitute the New World. For the purposes of zoogeographical studies, however, it is customary to divide the modern world into a number of areas whose boundaries are defined by the limits of the range of endemic animal groups (that is, animals peculiar to that area). First there is the broad division into zoogeographical realms which consist of a northern area (Arctogaea) and two southern areas (Neogaea and Notogaea). Arctogaea comprises North America, Europe, Asia and Africa; Neogaea covers South and Central America; and Notogaea includes Australia, Tasmania and New Guinea. These three realms are subdivided into a total of six zoogeographical regions. The Palaearctic, Nearctic, Ethiopian and Oriental regions

lie within the Arctogaean realm; the Neotropical region is Neogaea; and the Australasian region is synonymous with Notogaea. The Palaearctic and Nearctic, because of many marked resemblances in their faunal make-up, are often treated together as forming one large Holarctic region. Although Antarctica has continental status, its fauna is extremely impoverished and is of relatively recent origin. Zoogeographically this large continent is normally regarded as an oceanic island and is therefore not included in the usual scheme of regions.

In discussing geographical distribution the best examples to use are the mammals. Bats, however, are largely unsatisfactory subjects because, like birds and many invertebrates, they are often highly mobile and therefore have great powers of dispersal, even across oceanic barriers.

The Virginian Opossum is a well-known, abundant animal in many parts of North America. Its presence there is remarkable because it is the only living marsupial to occur outside the Neotropical and Australasian zoogeographical regions.

ABOVE: a male Pronghorn. The sole species of the family, Pronghorns are found only in North America—the Nearctic region—where they are the fastest animals (up to 95 kph) on the prairie grasslands.

BELOW: a female hippopotamus (*Hippopotamus amphibius*) and her calf beside the Mara River in Kenya. Hippopotamuses are found only in water habitats in the Ethiopian region.

Arctogaea, the largest of the zoogeographical realms, includes most of the temperate lands of the world and also extends into equatorial latitudes. Though fossil marsupials are known from both Europe and North America, the realm now has no living members of the group with the exception of the Virginian Opossum (*Didelphis marsupialis*), which is common in North America. Neither does it contain any extant running birds other than the African Ostrich (*Struthio camelus*). On the other hand, there are some primitive groups of fish which lack representatives in the other two realms. These include the sturgeons (Acipenseridae), paddlefish (Polyodontidae) and bichirs (Polypteridae). The sturgeons, of which there are about 30 living species, live in the temperate and cold-temperate parts of Eurasia and North America, and are most abundant and diverse in Eastern Europe and Asia. Except for a few, mostly North American species which are confined to freshwaters, sturgeons spend much of their life at sea. Paddlefish are found in China and the eastern United States. Bichirs occur in freshwaters of tropical Africa.

The characteristic fauna of Arctogaea includes a wide variety of ungulates such as the Fallow Deer (*Dama dama*), Red Deer (*Cervus elaphus*) and the Moose (*Alces alces*), and many insectivores such as the Hedgehog (*Erinaceus europaeus*), the Alpine Shrew (*Sorex alpinus*), which occurs only in high mountains, and the Mole (*Talpa europaea*). In contrast, Neogaea has only a few species of ungulates, and the hollow-horned ruminants such as sheep, goats and antelopes, which occur throughout Arctogaea, are not represented at all. Northern deer (subfamily Cervinae) occur throughout tropical and temperate Eurasia but are absent from southern North America and the whole of South America, where they are replaced by a different subfamily (Odocoileinae).

families include the Potamogalidae, containing the two or three species of otter shrew, the golden moles (Chrysochloridae), the giraffes and the hippopotamuses. There are two endemic orders, the aardvarks (Tubulidentata) with a single species and hyraxes (Hyracoidea) with several species, one of which extends as far as Syria. The living fauna of the region most closely resembles that of the Oriental region but some Ethiopian species have Palaearctic relatives and others (particularly some fish, amphibians and reptiles) are distantly related to species in South America.

The Oriental region of Arctogaea includes India, southern China, southeast Asia and the western islands of the Malay archipelago. The fauna contains fewer endemic families than the Ethiopian region and representatives are less widespread. Of 30 mammalian families, excluding the bats, only three are endemic; these are the flying lemurs (Dermoptera), the tree shrews (Scandentia) and the tarsiers (Tarsiidae). Birds, reptiles, frogs and toads are well represented, but only a few salamanders from the Palaearctic have penetrated the northern part of the Oriental region. The eastern limit of the region is defined by Wallace's Line, which runs east of Borneo and Bali along the deep Macassar Strait and separates these islands from Celebes and Lombok. This strait is an ancient geological feature that seems to have acted as a barrier to the eastwards spread of freshwater fish. The abrupt change in the faunal characteristics of the areas on either side of the strait was first pointed out by Alfred Russel Wallace in 1860.

The importance of Arctogaea and its component regions in terms of the past and present history of the distributions of animals is clear. Both marsupials and placental mammals existed in North America in the late Cretaceous period. Marsupials probably appeared for the first time in the Cretaceous and had become moderately widespread by the

ABOVE: the Common Tree Shrew (*Tupaia glis*) in Thailand. The 18 species of tree shrew live on the ground and in the trees of rainforests, but only in the Oriental region.

BELOW RIGHT: Cuvier's Toucan (*Rhamphastos cuvieri*) belongs to the family Rhamphastidae, one of the many families of birds that are found only in South America— the Neotropical region.

Despite the long history of periodic high-latitude land connections between the Nearctic and Palaearctic regions, and the considerable degree of similarity in their faunas, each region has some groups of animals that have never existed in the other. Peculiar to the Nearctic region are the Mountain Beaver (*Aplodontia rufa*), the Pronghorn (*Antilocapra americana*) and pocket gophers (Geomyidae). On the other hand, hedgehogs (Erinaceidae), wild pigs (Suidae) and typical rats and mice (Murinae) occur in the Palaearctic but not in the Nearctic. There are, however, only three families of land mammals that are confined solely to the Palaearctic; these are the jerboas (Dipodidae), the European mole rats (Spalacidae) and the Selevinidae, which contains a single species, *Selevinia paradoxa*, closely related to the dormice. Most of the families of living land mammals now found in Arctogaea occur in both the Nearctic and Palaearctic regions, and some have a much greater range. The squirrels, prairie dogs and marmots (Sciuridae), for example, are distributed in all parts of the world except Australia and Madagascar.

An analysis of the distributions of the familes of terrestrial non-flying mammals presently found in Eurasia and Africa shows that many are endemic to the latter continent, especially south of the Sahara. Furthermore, the present fauna of Africa south of the great desert belt closely resembles that of India south of the Himalayas. The fossil evidence shows that both these areas received much of their present fauna from a source north of the Tropic of Cancer. The extent of this can be judged from the fact that of 16 families now confined to the Ethiopian and Oriental regions, virtually all are represented by fossils in the Palaearctic.

The Ethiopian region (now sometimes called the Afrotropical region) is the most varied of all in terms of its mammalian fauna and, excluding the bats, there are 38 families of which 12 are unique to the region. Endemic

early Tertiary. They colonized Australia, possibly via Antarctica, before these two land masses parted and Antarctica moved southwards, becoming climatically unsuitable for many forms of life. North America was certainly the centre for early placental evolution; the group underwent a period of rapid radiation into many types which spread over the world and replaced the marsupials who were unable to compete with them. Because marsupials survived on different continents for varying periods of time, it would seem that the main evolutionary expansion of the placentals did not take place until the supercontinents had begun the process of disintegration.

Due to the isolation of the Neotropical region for much of the Tertiary period, many unique animals evolved there and, despite the link that later formed between the two American continents, there are still many forms peculiar to South America. The fauna is extensive and varied and, excluding the bats, has 32 families of mammals, of which 16 are unique to the region. The three families of edentates (the anteaters, sloths and armadillos) are confined to the region, except for the Nine-banded Armadillo (*Dasypus novemcinctus*) which has colonized the southern Nearctic. South America is the only great land mass, other than Australia, to have living marsupials but all the families are quite different. Other endemic families include 11 of caviomorph rodents, the New World monkeys (Cebidae) and the marmosets (Callithricidae). To these can be added five endemic families of bats, two orders of birds—the rheas and tinamous—and nearly half of the remaining bird families. The only insectivore on the Neotropical mainland is a species of shrew (*Cryptotis*) in central and extreme northern South America.

The Australasian region is unique in that it has had no land connections with any other region since the break-up of Pangaea. Part of its fauna (for example frogs and marsupials) resembles that of South America, while the terrestrial reptiles, placental mammals and most birds have affinities with those of the Oriental region. The only terrestrial placentals to have reached Australia by natural means, along the stepping-stones of islands from southeast Asia, are the rats and bats.

The vertebrate fauna of Australia is poor compared with that of other regions and, excluding the bats, only nine families of mammals are represented; eight of these, however, are unique to the region. The three living species of monotreme (egg-laying mammals) are confined in their distribution to the Australasian region.

The most striking mammals of the region are the marsupials which, in the almost complete absence of native placentals, have radiated into a great variety of forms, occupying the ecological niches that are elsewhere filled by placentals. Amongst the birds there are ten endemic families of which two are flightless—the emus (Dromaiidae) and the cassowaries (Casuariidae). There are few amphibians, the freshwater fish fauna is poor and there are only two endemic families of reptiles.

For the sake of convenience Polynesia, including New Zealand and the Hawaiian islands, is here included in the Australasian region. With the exception of New Zealand, which has two endemic species of bat, none of these islands has any native terrestrial placental mammals. New Zealand is also the home of the unique lizard-like reptile, the Tuatara (*Sphenodon punctatus*), whose relatives were common in the northern hemisphere during the Mesozoic.

Antarctica, larger than Australia, has one of the most severe environments in the world; the southern ice-cap in winter is the coldest place on earth. There are no native terrestrial mammals but six species of seal breed on land margins and on the pack ice. Of the 43 bird species that breed in Antarctica, all but five are seabirds. Three species of penguin, the Emperor (*Aptenodytes fosteri*), Adelie (*Pygoscelis adeliae*) and Chinstrap (*P. antarctica*), nest exclusively in this icy, inhospitable habitat.

Geographical isolation is a very important factor in the evolution of distinctive faunas, as is well illustrated on oceanic islands throughout the world. In this context it is also significant that at the present time there are only two large land masses that have really distinctive faunas—Australia and South America. Both these continents were

The best known of all marsupials in Australia are the kangaroos. One of the largest species is the Red Kangaroo (*Megaleia rufa*), which inhabits primarily the eastern and southern highlands.

Only a few animals can breed in the frozen Antarctic. Among them is the Chinstrap Penguin, which is found on land or far out in the ocean, usually along the fringes of pack ice.

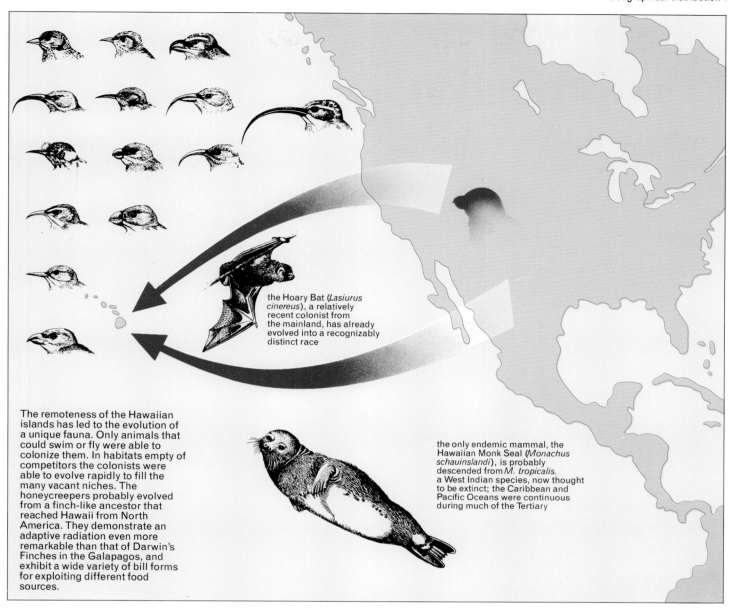

The remoteness of the Hawaiian islands has led to the evolution of a unique fauna. Only animals that could swim or fly were able to colonize them. In habitats empty of competitors the colonists were able to evolve rapidly to fill the many vacant niches. The honeycreepers probably evolved from a finch-like ancestor that reached Hawaii from North America. They demonstrate an adaptive radiation even more remarkable than that of Darwin's Finches in the Galapagos, and exhibit a wide variety of bill forms for exploiting different food sources.

the Hoary Bat (*Lasiurus cinereus*), a relatively recent colonist from the mainland, has already evolved into a recognizably distinct race

the only endemic mammal, the Hawaiian Monk Seal (*Monachus schauinslandi*), is probably descended from *M. tropicalis*. a West Indian species, now thought to be extinct; the Caribbean and Pacific Oceans were continuous during much of the Tertiary

isolated for a considerable length of time in the Tertiary.

A perfect example of an isolated fauna is to be found in the Hawaiian Islands, which are indeed the most remote in the world. They are the home of a unique family of birds, the honeycreepers (Drepanididae). These are believed to have evolved from one common ancestor that reached the islands by chance from the American mainland. In the absence of competition from other species this now-extinct ancestral form radiated to fill various available niches and eventually gave rise to the 23 or so species that today are spread throughout the islands but occur nowhere else.

It is an observable fact that most living species or families of animals have geographical ranges that are restricted to a greater or lesser degree. What are the factors which operate to impose the boundaries? There are topographical barriers that limit animal distributions such as oceans, deserts, high mountain ranges but climatic factors such as temperature and humidity can be equally effective. The extent to which any of these may limit an animal's range will depend on several characteristics of the animal itself; its tolerance of heat or desiccation, for example, will be limited and where these limits are transgressed, there the animal cannot live. The capacity for dispersal also varies and for species that can fly or swim a large stretch of water may not be an obstacle, whereas those that can do neither will be confined

by it. Clearly in the marine environment barriers are less restrictive than those which operate on land, and the dispersal capacity of many of the animals is greater. The oceans occupy almost three-quarters of the surface of the globe and they are connected together. This helps to explain why a world-wide distribution among mammals is confined to some whales, such as the Sperm Whale (*Physeter catodon*) and the Killer Whale (*Orcinus orca*).

Finally, we should remember that all the discussion in this section has been based on the *natural* distributions of animals. Superimposed on this are the artificial distributions which result from Man's activities in introducing a great variety of animals of all kinds to parts of the world which, in the normal course of events, they could not have reached. Accidental or involuntary introductions involve animals such as the Brown Rat (*Rattus norvegicus*) and Black Rat (*R. rattus*) which penetrated to most corners of the world by travelling in the ships of early explorers and traders. Almost without exception they have caused trouble and damage wherever they have arrived. Perhaps the most disastrous introduction in history was that of the Rabbit (*Oryctolagus cuniculus*) into Australia during the latter half of the 19th century. The subsequent devastation of huge areas of crops was an object lesson in what can happen when Man meddles with the delicate balance of nature.

Ecological Distribution

Within each zoogeographical region there is a range of different biomes, such as desert, rain forest and grassland, each typified by a particular climate and vegetation. The nature of a biome and the character of the fauna it supports are closely inter-related.

The zoogeographical realms and their regions sprawl widely across the world, spanning broad latitudinal belts and climatic zones. For example, the Neotropical region stretches from well north of the equator to the subantarctic fringe. But within this region live whole menageries of animals, some widely distributed, others restricted to particular kinds of habitat and locality. It accommodates enormous collections of mammals, birds, reptiles, insects and lesser creatures, all variously specialized for living in its jungles, deserts, grasslands and high mountain environments.

In order to study this great diversity within the realms, ecologists take account of the tendency of plants and animals to live, not merely as individuals, but in communities associated with particular characteristics of climate, soils and other environmental factors. Communities are the units which particularly interest ecologists, and the relationships of plants, animals and environment within them are their chief concern. So they regard the realms as consisting of major communities and minor ones, attending first to the organisms, then to the areas which they occupy.

Major communities of plants and animals, occupying large areas of the earth, are called biomes. These are broadly defined by their vegetation, which is to a large extent controlled by their climate. Tropical rain forest, polar tundra, temperate grasslands and hot deserts are examples of biomes; there is no definitive list, but most ecologists would agree upon about 12 terrestrial biomes of primary importance. To these, rather more secondary biomes can be added, together with a lesser number of marine and freshwater communities. Usually identified by their dominant vegetation, biomes often appear in more than one region. For example, there are tropical rain forests in central and South America, Africa, India and Australasia, and polar desert occurs all around the Arctic basin as well as in Antarctica. Biomes are stable, lasting communities, which have developed over thousands or millions of years and shown themselves able to withstand fire, earthquake, spasmodic flooding and other natural hazards. Because their dominant vegetation depends largely on climate, they yield, shift and change with long-term climatic fluctuations. For example, the tundra is still accommodating to the relatively recent changes wrought by the disappearance of ice from huge tracts of northern lands. They also yield before the sustained impact of man and his domestic animals, which between them can devastate even the greatest forests and turn rich grasslands to desert over a period of a few hundred years. But, generally, the biomes represent the natural world as it was before human interference began.

Within the biomes, plants and animals restrict themselves to the particular home areas or habitats for which they are especially equipped, or 'adapted'. The adaptation

Sheep of the genus *Ovis* occur in both the Nearctic and Palaearctic zoogeographical regions. These are Alaskan Dall Sheep (*O. dalli*) in their typical mountain habitat. Other closely related species occur in eastern Siberia and the American Rockies.

of plants and animals to their environments has been a gradual process, in which the organisms have evolved over many generations to meet the challenges and requirements of the habitats. For some purposes habitats may be thought of as the minor constituents of regions and biomes, the actual places within these larger areas which plants and animals come to inhabit. Again, however, ecologists tend to think in terms of communities of animals and plants rather than places. The minor unit recognized within the biome is the ecosystem, which is defined as a community of interdependent plants and animals together with the environment in which they live. The distinction is not merely a quibble or a game with words, for it recognizes and respects the all-important relationships that exist between any single organism, the other organisms surrounding it, and the environment which contains them all. Yet it is a flexible concept; an ecosystem can be as large as an ocean or as small as a drop of water within it—a whole forest, or a tiny community living under one stone in a far corner of the forest. An ecosystem exists wherever plants and animals stand together in mutual relationship within an environment. Sorting out the relationships, discovering how energy flows and minerals are cycled and recycled through these often highly complex systems, has provided many intriguing and rewarding problems for modern ecologists.

The way in which ecosystems work can be illlustrated by studying the flora and fauna of, for example, a deciduous forest. Each year in spring the trees draw up nutrient minerals in solution from the soil and put out new green leaves from their winter buds. The leaves start to photosynthesize—that is, use some of the radiant energy of sunlight to make sugars and starches. Each year about the same time, myriads of tiny caterpillars hatch from overwintering eggs and begin to feed on the leaves. Their business is essentially to steal the sugars, starches and minerals which the trees have put by, building them into their own bodies or using them as fuel for their own life processes.

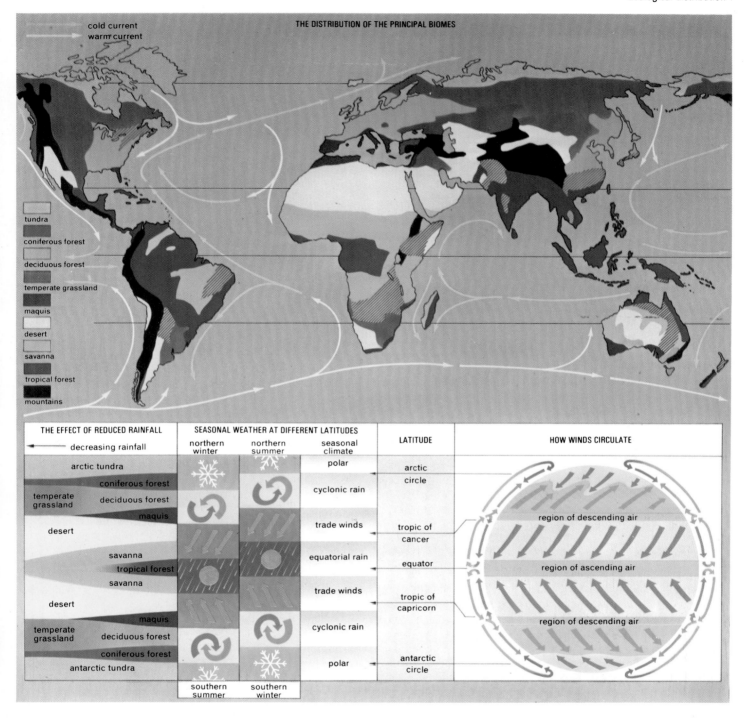

THE DISTRIBUTION OF THE PRINCIPAL BIOMES

cold current
warm current

tundra
coniferous forest
deciduous forest
temperate grassland
maquis
desert
savanna
tropical forest
mountains

THE EFFECT OF REDUCED RAINFALL	SEASONAL WEATHER AT DIFFERENT LATITUDES		LATITUDE	HOW WINDS CIRCULATE
decreasing rainfall	northern winter	northern summer / seasonal climate		

Equatorial regions are warmer than other parts for two reasons: the sun's rays pass through more atmosphere near the poles and are spread over more of the land surface. The warmer air at the Equator rises, cools and sheds its water content as rain. The trade winds moving in to replace this are warm, drying winds. The equatorial air descends to earth at about latitude 35°N and S, and thence it spreads equatorially (as the trades) and polewards (as the westerlies). The earth's spin exerts a twisting influence on the wind direction—clockwise in the north and counterclockwise in the south. The westerlies meet the colder air in the polar regions and produce cyclonic rain and snow (depressions). Because the

earth is tilted on its axis, the climatic zones move north and south with the seasons. Around the Equator, the high, regular rainfall sustains forest growth, but to the north and south rain is seasonal and sustains only grassland (savanna). Where the trades blow all the year round there is little rainfall and only desert vegetation. In temperate latitudes wet conditions support forest, seasonally dry conditions grassland, and very dry areas are maquis or semi-desert. In higher latitudes the rain tends to fall throughout the year, giving rise to boreal forest or, in colder areas, tundra. With increasing distance from the sea, there is less rainfall, and tundra gives way to polar desert.

Titmice—tiny woodland birds with bills like fine forceps—feed themselves and their growing nestlings on the caterpillars, each bird taking several hundred per day. Weasels, owls and other predators in turn feed on the titmice and their young. Each link in the chain represents a transfer of energy and minerals from one organism to another, with the trees as the ultimate providers.

Only the trees and other green plants bring energy and materials into the system through their roots and leaves. All the other organisms depend on them, dividing the plant resources among themselves on a catch-as-catch-can basis. The energy initially stored in the leaves is taken up by animals and gradually dissipated by respiration, returning as heat to the atmosphere, and finally dispersing to outer space. Minerals pass successively from one animal to another, ultimately returning to the soil through the action of bacteria and fungi on droppings or on the dead and decaying bodies of the animals themselves.

Ecologists see these transactions as a chain—a food chain, with each organism a separate link. The green plants are the primary producers, or fixers, and the animals are primary, secondary and tertiary consumers at different trophic (feeding) levels. The fungi and bacteria are decomposers, with the critical role of returning nitrogen and other minerals to the soil and carbon dioxide to the air, making them available for further use. Many such food chains within an environment make up an ecosystem. Food chains seldom involve more than three or four links, because the weight of organisms represented by each successive link falls rapidly; it takes thousands of grams of grass leaves to support a rabbit, and many rabbits are needed to support a single stoat or owl. Equally seldom are food chains found in isolation; this is but one of several dozen cross-linked chains supported by plants, involving many different species of herbivores and a host of predatory carnivores.

A large tree, such as an oak, is in fact a community centre on which three or four hundred other species of plants and animals may depend. The herbivores feed not only on its leaves, but on buds, bark, flowers, sap, roots and acorns. A single leaf may be attacked by a dozen different kinds of insect alone, including caterpillars of several species, greenfly, gall wasps, weevils and leaf-miners. The carnivores feeding on the herbivores may include mites, spiders and half a dozen different kinds of insectivorous birds; the larger carnivores with interests in the community could include rats, hedgehogs, stoats and badgers. These interlinked chains are often described as a food web. Each strand of the web represents a link of dependence on another organism (predator on prey) which results in a one-way transfer of nutrients and energy.

The oak with its community of dependants is a complex ecosystem, but typical in principle of all others. To study an ecosystem, whether great or small, ecologists seek first to establish the role of each organism as producer or primary, secondary or tertiary consumer, recognizing that some organisms have several roles, or may even change roles at different stages in their life cycles. If possible the numbers and bulk (or biomass) of each organism are estimated, and from this the change in the biomass through the annual cycle can be calculated. We can begin to understand how the system works from estimates of the total amount of energy fixed by the producers in a given time, and how this energy is spread and dispersed through the dependent organisms. Only a very few complex ecosystems have so far been budgeted in this way, but the approach is helping us to understand in general how ecosystems work, and is giving us valuable insights into the productivity of particular ecosystems useful to man.

In terms of primary production, tropical ecosystems are by far the most efficient, with rain forest and rich marshlands competing for first place. The wealth of the forest is reflected in the enormous numbers and variety of animals it supports at various trophic levels. Temperate deciduous or mixed forest is only about two thirds as productive as rain forest, coniferous forest only two fifths or less. African savanna compares favourably with the best conifer forests, and is closely matched by temperate agricultural land. Not surprisingly, desert and tundra ecosystems are among the poorest, with only thin populations of animals in evidence.

The richest patches of ocean compare in productivity with good agricultural land, though on average the oceans are little more productive than tundra. Estuaries and mangrove swamps, well endowed with nutrients from the land, can be exceptionally rich and productive. Freshwater ecosystems are generally poor; old-established, shallow lakes with a good accumulation of nutrients can be highly productive, but mountain tarns and fast-flowing streams,

especially in areas of hard, insoluble rocks, are usually deficient in minerals and support only meagre populations of plants and animals.

The major biomes of the world—whether forest, grassland, tundra or desert—are identified mainly by the texture and quality of their dominant vegetation. This makes good sense on several counts; not only is the dominant vegetation easiest to identify, but it also controls the environment and provides particular kinds of habitats both for minor plants and for animals living among it. What determines the kind of vegetation which comes to dominate a region? Soils, drainage, slope of ground and many other factors are involved, but the all-important factor is climate. Sun, wind and rainfall conspire together to decide what kinds of plants

The oak is an important tree in northern deciduous forests. As a primary producer it fixes radiant energy into its tissues. Animals that feed directly on the leaves include the larvae of insects such as the moths and the small, gall-producing wasps. These form an important food for birds like the Blue Tit, which itself is preyed upon by raptorial birds such as the Sparrowhawk. The rich humus of dead leaves is the basic food for large numbers of soil animals including earthworms which, in their turn, are food for shrews, moles, some birds and omnivores like the Badger.

can grow, and what shall grow in greatest abundance to achieve dominance.

On a worldwide scale climate varies primarily with latitude, and changes in broad zones from polar regions to tropics. Vegetation varies with it; school atlas maps of 'world climate' and 'world vegetation' are similar enough in broad outline to make this point clear. There are local anomalies, many of them due to the contours of the land. For example, high mountain chains keep rain from neighbouring plains and turn them into deserts; the desert lands of Arizona and Nevada, in the rainshadow of the southern Rocky Mountains, are a good example. On a smaller scale ridges and valleys, lowlands and uplands, lake basins and peaks all over the world provide local climatic conditions which plants and animals are quick to exploit. But the overall, global pattern of matching climatic and vegetational zones is clear; if anything, the anomalies reinforce the point that climate, more than any other single factor, controls the growth of vegetation and determines the nature of biomes throughout the realms.

Within each biome, and under the general cover of the dominant vegetation, subdominant plants help to provide and shape the habitats on which individual animals depend. Many factors are involved—local climate, exposure, cover and protection against enemies, availability of food, presence or absence of competitors and predators, and a host of others. To all of these, plants make a significant contribution. They are eaten by herbivorous animals which in turn may be eaten by carnivorous and omnivorous animals. All animals use plants for shelter, protection or cover when hunting and foraging. The general character of vegetation is thus important in determining where animals may live, and many different kinds of animals are specialized for living in association with particular kinds of vegetation and even with

Looking across the high savanna of Kenya from the Aberdares with Mount Kenya in the background. Although close to the equator, the East African savanna does not receive enough rain to support tropical forest. The rainfall is highly seasonal and only drought-resistant grasses and scattered, thorny shrubs can survive.

The elements of climate chiefly implicated are rainfall and temperature; between them they provide the groundwater and the atmospheric moisture or humidity to which plants are particularly responsive. Wind too has an important role, though usually a more local one. In broad terms, where rainfall is abundant throughout the year, forests are dominant. In hot climates they are the lush, evergreen jungles called tropical rain forest, grading into monsoon forest where precipitation becomes more seasonal. Temperate climates produce a range of deciduous, evergreen or mixed forests of broadleaved trees and conifers. These may be rich or sparse according to the amount of rainfall, its seasonality, and other environmental factors. Colder conditions with abundant rainfall give rise to the vast coniferous forests which extend like a dark green mantle to the fringe of the Arctic. Here, temperatures and rates of evaporation are lower, so less rain is needed to support forest growth.

Where rainfall is slight or markedly seasonal, forests give way to a variety of grasslands and moorland vegetation, from the savannas of tropical Africa to the steppes of Asia, from the pampas of Argentina to the alpine meadows of central Europe. Sometimes the grasslands are dotted with trees, forming warm forest-savanna or colder taiga; sometimes the trees grow only in sheltered valleys, while the plains themselves are treeless. Always the grasslands are grazed by a variety of highly specialized animals, which are natural managers and essential to their maintenance.

Drier conditions still produce sage-brush, chaparral and maquis vegetation, dominated by plants which grow sparsely, rigorously conserving the small quantities of water available to them. Often they are fire resistant, and may even depend for their continuing existence on frequent devastation by fire. Dry heath and tundra are the equivalent vegetation in cold climates. Where rainfall is meagre and groundwater absent for most of the year, desert conditions prevail. The hot deserts of North Africa and central Australia, and the cold deserts of polar regions and mountain tops are alike in providing some of the harshest environmental conditions on earth. Only a few adaptable animals and plants can survive their rigours.

ABOVE: a Corsican hillside with typical maquis vegetation. In the foreground are tree-heaths; the scrub behind is mainly of strawberry trees. These shrubs bear needle-like or leathery leaves to resist the hot, dry Mediterranean summer. Much of the new growth is in the spring after the ample winter rain.

BELOW: a valley in the southern Nevada desert. The rainfall here is very infrequent and averages only 20 cm each year. The tree-like plants are Joshua trees, large members of the lily family that are adapted to arid conditions; like the cactus they store water in their fleshy stems to carry them over times of drought.

particular species of plants. If the vegetation is destroyed or altered, the animals too disappear, as man has demonstrated only too often in his heavy-handed progress about the continents.

Trees over 30 metres tall, buttressed and laced with lianes and epiphytes, make up the tropical rain forests. Richest of all the biomes, rain forests grow in tropical regions where rainfall exceeds 200 cm spread evenly through the year. Established for tens of millions of years, rain forests have had time to accumulate an immense diversity of species; the Amazon forest alone includes over 4,000 different kinds of trees, and more species of birds than the whole of North America. Rain forest grows continuously in a climate without seasons. Flowering and fruiting are spread throughout the

Tundra occupies an extensive area bordering onto the Arctic Ocean. The severe winters and poor soil prevent the growth of trees other than dwarf willow and birch. Much of the tundra is low-lying bog, flecked with the white flowers of cotton-sedge. The higher and dryer ground is covered with the lichen upon which Caribou feed.

year, so there is always food for herbivorous mammals and birds. Insects, lizards and snakes abound. Ground-living mammals (for example the Bongos, Okapi, forest pigs and big forest Gorillas of African rain forests) must be strong enough to push through dense undergrowth. Smaller mammals of the rain forests live permanently in the canopy far above ground; Kinkajous, porcupines, sloths, Collared Anteaters and many primates, including the long-tailed monkeys, are almost entirely arboreal.

Temperate deciduous and evergreen forests require over 100 cm of rain spread throughout the year, boreal forests rather less. In the sharp cold of the far north needle-leaved species—fir, spruce and pines—predominate; further south larch, aspen and poplar give way to familiar broadleaved trees of warmer climates—oaks, maples, elm, ash, beech and chestnut. The southern edge of the forest zone has frost-free winters and drier summers, favouring broadleaved evergreens which grade into scrub and grasslands. The temperate forests, with their rich leaf litter, fertile soils and multi-layered shrubs and herbs, support a wealth of animal life. Many species of deer, including Red, Roe and Fallow Deer of Europe and White-tailed and Black-tailed Deer of North America, graze and browse among the clearings, paths and dense undergrowth. Most of the smaller woodland mammals—European Hedgehogs, Badgers, Red Foxes and Weasels, for example, and North American skunks, Raccoons and Opossums—are carnivores or omnivores, taking insects, birds and other small creatures from the understorey and leaf litter.

North of the boreal forest lies the Arctic tundra, a broad zone of low shrubs, grasses, sedges and moss. Frozen and snow-covered for three quarters of the year, the tundra awakens for a brief but brilliant summer of eight to ten weeks' intensive growth, culminating in a riot of flowers, berries and colourful leaves. Sedges and marsh plants fill the hollows, which are waterlogged because of frozen subsoil. These are the summer haunt of migrant birds from the south, which flock in by the thousand to take advantage of the rich spring feed. Especially prominent among them are the ducks, geese and waders, which breed, moult and fatten for their return in a hectic race against time.

ABOVE: tropical rain forest by the side of a river in Colombia. The multi-layered structure of rain forest is most clearly seen on river banks. The low scrubby understorey can be seen in the foreground with the tallest canopy trees towering behind. The South American rain forest is the most extensive in the world.

BELOW: coniferous forest fringing a lake in the Banff National Park in Canada. The dark mantle of the evergreen conifers covers much of northern North America and Eurasia. The ample rainfall produces the numerous rivers and lakes which supply the water for a large variety of deciduous trees such as poplars and birches.

Woody shrubs, grasses, mosses and lichens carpet the drier ground of the tundra, providing cover and food for many small herbivorous mammals—lemmings, voles, Arctic Hares and ground squirrels—and a few resident species of birds, including Ravens, Ptarmigans and Snow Buntings. Moose, Musk Oxen and Caribou are the larger grazing and browsing mammals. Insects abound during the brief summer; the tundra is noted for its butterflies, and for biting mosquitos and blackflies which breed in the shallow ponds and streams. Stoats and Arctic Foxes (which are brown in summer and white in winter), Wolves, skuas and Snowy Owls are the main predators. In winter the voles, lemmings and ground squirrels remain active in tunnels beneath the snow, where they are warm and relatively safe from predators.

shrubs of the plains. Lions, Cheetahs, hyaenas and hunting dogs are the main predators of the large game mammals; Servals, Caracals, Meercats and several kinds of venomous and constricting snakes follow the lesser game. Dry grasslands are also the home of many species of ants, termites, locusts and other insects, which support a range of bird, mammal and reptilian predators.

Deserts occur in hot or cold climates where annual rainfall is less than about 25 cm. Few deserts are entirely without rainfall or vegetation. A typical hot desert supports a patchy though permanent covering of thick-stemmed cactuses and tough, almost leafless shrubs. Grasses and ephemeral herbs grow quickly and flower after the transient rainstorms. Strong sunshine by day raises surface temperatures to 50°C

Flying can be difficult for large birds in dense vegetation, and the Hoatzin (*Opisthocomus hoatzin*) is adapted for an arboreal life. It inhabits trees along the banks of rivers in the Amazon basin and can fly only weakly, usually only gliding from tree to tree.

The coniferous forests of Alaska are the home of the Moose (*Alces alces*). Largest of all the deer, they browse deciduous trees and are very fond of aquatic plants. In similar forests in Eurasia this niche is occupied by its smaller conspecific, the Elk.

Drier and colder regions of the Arctic further north are bare, stony deserts, with only a meagre flora of mosses, lichens and algae, and very few resident birds or mammals.

Sometimes regarded as mere transitional zones between forest and desert, grasslands are rich and varied biomes in their own right. They grow where rainfall is seasonal or irregular—enough to promote growing seasons once or twice yearly, but not enough to meet the moisture demands of close-packed trees. Dry grasslands—the Argentine pampas and huge areas of the prairies and steppes—are virtually treeless. Those with good seasonal rains—for example, much of the high African tablelands—are often well peppered with trees and shrubs, especially in valleys and hollows where the water table is high. Grasses grow with a wide range of form and adaptation, from the dense bamboo thickets and elephant grasses of monsoon regions to the thin, sparse annual grasses of near-deserts. Their seasonal growth supports large herds of nomadic grazing animals and flocks of seed-eating birds, which migrate over the plains in regular annual circuits.

The many species of African game mammals are complementary rather than competitive in their feeding. Buffalos prefer coarse grasses, leaving finer growth for zebras and wildebeests; hippos and elephants take the long-stemmed marsh grasses; Impala, Bushbuck and kudus feed mainly in bushy and semi-wooded areas. Eland eat both grasses and shrubs, dik-diks and some other antelopes browse almost exclusively on shrubs and low trees, while giraffes use their long legs and neck to browse far above any possible competitors. Many smaller mammals, including hares, baboons and rodents, also feed on the grasses and

and more; clear night skies allow most of the heat to escape, bringing dew and even frost at ground level. Winds whip the unprotected soil, raising dust storms and gathering sand into shifting dunes. These harsh environments nevertheless provide livings for many species of insects, reptiles, birds and mammals.

Most small desert animals escape extremes of temperature by living in burrows or under stones—even burying themselves in sand—by day, emerging to feed in the evenings when the air is cool and humidity high. Desert lizards and rodents are long legged, often running on hind legs to keep their bodies off the hot sand; desert snakes include many sidewinders, which raise themselves off the sand in coils. Large grazing animals which cannot burrow or find shade— for example, camels, oryxes and desert gazelles—feed mainly at night when the vegetation is most succulent, and stay inactive during the day. Camels lose very little water through sweating or excretion; they allow their body temperature to rise far above normal during the day, and shed the excess heat after dark. Desert foxes have large ears, which help them to dissipate heat and also to hear well at night; so have the rodents and jackrabbits which are their main prey.

Mountains bring great ecological diversity to their realms, compressing a wide range of habitats into a relatively small area. Above the foothills of a high mountain there is a rapid succession of sharply-defined zones of increasing cold, humidity and wind, which end in a region of thinner air and brighter sunshine, surrounded by plants and animals quite different from those below. For example, at the base of the southern Rocky Mountains of Colorado there is a desert landscape of saguaro cactus and sage-brush. Not far above this there is dry mountain forest of pinon pines and junipers, thronged with a rich fauna of seed-eating rodents and birds, with Mule Deer and other larger mammals in the damp gullies. Between 2,000 and 3,000m above sea level grasslands and stands of Douglas firs replace the pines, with aspens and other northern species giving the forest a distinctly Canadian look; Black Bears, Coyotes, Cougars and many other boreal animals reach their southern limits at this elevation. Above 3,000m stands spruce forest, not un-

like the dark forests of the far north, thinning to a semi-desert alpine tundra of dwarf shrubs and coarse grasses, snow-covered for most of the year.

Even more striking is Mount Kenya, which rises to 5,200m from the grass-covered plains of equatorial Africa. The lower slopes of its southern flank, damp and humid, support a rich tropical rain forest, dominated by cedars, hardwood and dense bamboo thickets. Above 3,000m is a zone of mixed forest and parkland, topped by a strange moorland of giant heathers, lobelias and groundsels. Each vegetational zone has its own distinctive community of insects, birds and mammals. Mount Kenya is capped by a desert of permanent snow and ice, only a day's walk from the equatorial plains and little more from the Equator itself.

During the dry season in East Africa the savanna grasses dry out to hay where they stand. Though they still have food value, they provide no water, and grazing animals like these Burchell's Zebra (*Equus burchelli*) must go to waterholes to drink.

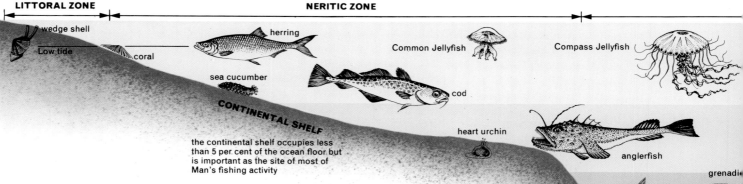

LITTORAL ZONE **NERITIC ZONE**

wedge shell

Low tide coral

herring

sea cucumber

Common Jellyfish

Compass Jellyfish

cod

CONTINENTAL SHELF

the continental shelf occupies less than 5 per cent of the ocean floor but is important as the site of most of Man's fishing activity

heart urchin

anglerfish

grenadier

CONTINENTAL SLOPE

sea lily

The oceans cover almost three-quarters of the earth's surface. The offshore waters can be divided into a neritic zone overlying the continental shelf and an oceanic zone over the deeper regions. The continental shelf is simply an extension of the landmass down to about 200 m with relatively gentle slopes. Beyond this, the floor slopes more steeply down to the abyssal plain, which lies below 3,000 m and covers around half of the earth's surface. The ocean deeps are thus the largest environment on earth, both from the point of view of surface area and volume. All plant life and the greatest diversity of animal life are found in the sunlit upper waters known as the photic or epipelagic zone. As the depth increases, so the light cannot penetrate and plants disappear. Little light reaches below 500 m; this is the bathypelagic zone, extending to the abyss. There is less diversity of animal life here—cephalopods, crustaceans and fish are dominant. This deep-sea fauna is remarkably similar throughout the world, reflecting the essential uniformity of the environment. The floor of the abyss is populated by a wider variety of animals, which feed either on the ooze itself or on the detritus that rains down from above. Separated into basins by shallower ridges, the abyssal faunas show some degree of zoo-geographical differentiation.

The waters of the earth circulate constantly among themselves, recycling periodically through the atmosphere as water vapour, cloud and rain. In bulk they cover well over half the globe, as streams, rivers, lakes, shallow seas, oceans, and the great polar and alpine ice sheets. Ecologists distinguish half a dozen or more aquatic biomes, each with a wide range of habitats and a lively selection of plants and animals within each one.

Life began in water, and aquatic biomes still contain far more living creatures, in greater variety, than all the terrestrial biomes put together. There are many advantages to be gained from aquatic living. Water is denser than air, giving external support that reduces the need for internal skeletal tissues. It provides massive environmental stability; few habitats are more constant/than ocean habitats. Because water is an effective solvent it is seldom pure, and its impurities are life-giving. In falling as rain through the atmosphere it absorbs oxygen and carbon dioxide, and in percolating through soils and rocks it takes up a variety of useful nutrient ions. Plants and animals living in water are bathed in a nutrient soup, which feeds and supports them, brings oxygen, and carries away waste products.

Fresh spring water and melt-water from snowfields contain only a few parts per million of dissolved minerals and support only a meagre flora and fauna; hence the clarity and faunal poverty of mountain streams and tarns. Rivers and lakes are richer, often with hundreds of mineral parts per million in solution, including a reservoir of life-supporting chemicals in the form of suspended organic materials. Oceans are richer still, with mineral concentrations averaging 35 parts per thousand. The muddy waters of tropical rivers, swamps and estuaries and the cloudy green waters of polar oceans in summer include some of the richest habitats to be found anywhere on earth.

Aquatic ecosystems, no less than terrestrial ones, depend on the mainspring of solar energy, captured by photosynthesis and distributed through the various trophic levels of food chains. Curiously, the larger water plants are not especially important in aquatic food webs. Most of the energy in aquatic ecosystems is captured by microscopic, mainly unicellular algae. In spring and early summer we sometimes see them as a green, red or brownish scum in the water, and a thin film covering the weeds and rocks. Insignificant as they may seem, most of the rest of life in the water ultimately depends on them for a living.

On rocky shores and in the rich, well-stirred waters of coral reefs, floating algal cells provide much of the food of filter-feeding worms, sponges, bivalve molluscs, polyps and other tiny creatures living in corners and crevices. Limpets, snails, chitons, and nibbling fish graze the algal film, rasping it from rocks and weeds almost as fast as it grows. Large masses of algal cells floating in the open sea are called phytoplankton. They inhabit only the surface hundred-or-so metres, thinning out rapidly in deeper layers where light is dim and photosynthesis restricted.

BELOW: the water scorpions (*Nepa*) are predatory insects common in north temperate lakes. They may play host, as here, to parasitic water mites (*Hydrachna*). The insects play an important part in freshwater ecosystems.

BELOW RIGHT: a school of mixed coral fishes on the Great Barrier Reef. In these shallow, warm, sunlit waters, productivity is very high and the diversity of animal life is probably the highest in the world

OCEANIC ZONE

krill

plankton

sunfish

dolphin

siphonophores

PELAGIC (PHOTIC) ZONE

— 150

deep sea prawns

lanternfish

dragon fish

Spirula

— 500

deep sea squid

deep sea jellyfish

— 1500

bristlemouths

BATHYPELAGIC ZONE

vampire squid

gulper eel

deep sea anglers

deep sea amphipod

— 3000

Neopilina

tripod fish

— average ocean depth —

— 5000

brittle star

beard worm

ABYSSAL ZONE

ABYSSAL FLOOR

the abyssal floor is covered by soft
oozes consisting of fine muds, bacteria
and skeletal remains of pelagic
animals and plants

— 10000
depth in metres

TRENCH

trenches, great gouges in the sea
bottom, may be over 10000 metres deep

cucumbers

Wherever they occur, these tiny, vital plant cells of surface waters have an enormous capacity for reproduction, especially in spring when the surrounding water is richest in nutrients. Stimulated by the slowly warming water and increasing light penetration, they grow and reproduce several times daily, producing enormous populations that completely change the colour of surface waters.

In fresh waters and oceans alike, phytoplankton is grazed continuously by tiny filter-feeding animals that float at the surface, collectively called zooplankton. Many of these are crustaceans, such as copepods and euphausiid shrimps, and other specialized invertebrates which live all their lives in the upper layers of water. But zooplankton, especially that of the sea, also includes young fish, and the eggs and larvae of many animals that, as adults, live sessile lives on the sand, mud and rocks far below. Most zooplankton animals are filter-feeders, with a variety of intriguing ways of drawing water toward themselves and extracting plant cells from it. Some of the larger forms are carnivores, which prey on their smaller neighbours.

This dense concentration of energy and living material is attacked from all sides by larger, more active predators. Marine zooplankton especially supports an enormous range of micro-feeding carnivores—whalebone whales, seabirds with specially adapted bills, and dozens of species of surface-feeding fish, including many, such as Herring, Mackerel and Whiting, that are caught commercially. These in turn become the prey of the ocean's top predators—large, fast-moving fish (such as sharks, Barracuda and Bonito), seals, toothed whales (including dolphins), squid, and fish-eating seabirds. Many strangely-shaped mid-water fish swim up to the surface at night to feed on the plankton and its attendant carnivores. Ultimately the animals of the deep sea too depend on the plankton, for they are supported by the rain of debris which sinks down to them constantly from the surface. Deep ocean currents pick up the minerals from this steady organic rainfall, eventually returning it to surface waters for use by later generations of phytoplankton.

Animal Names
and
Classification

Throughout this book there are many phrases such as 'the monoplacophoran *Neopilina*' or 'the Tarpon (Elopiformes)'. There are also seemingly incomprehensible charts—classifications—which introduce the different animal groups. Why has such apparently indigestible material been included, what use is it and how was it devised? This section explains how the system of grouping animals has been established, how it works and why it is necessary.

The practice of grouping objects into an orderly array is inherent in our lives and is called classification. Goods in a department store are classified into categories so that they can easily be located. Librarians classify books, usually into two large sections—fiction and non-fiction; the fiction books are further grouped alphabetically and the non-fiction books are usually grouped into travelogues, bibliographies and so on. In other words we classify to bring order to a myriad of data around us. In the animal world there is certainly no shortage of data. Zoologists have so far described about one million species, with body forms ranging from amoeba to Man, and perhaps another three million remain to be described. Clearly some sort of ordering is necessary and the system of classification that we use must be such that newly discovered animals can easily be incorporated into it. Within this classification groups of animals and the individual animals themselves must be labelled according to a precise system of nomenclature. This science of naming and classifying animals is called systematics or taxonomy.

Early Man must have used some form of classification. In all probability he would have grouped animals that were good to eat as distinct from those that were distasteful, or separated those which provided serviceable hides from those that did not. Such classification would have been based on practicality. But more ideologically inclined cultures, such as that of the Greeks, turned their attention to animal classification as part of their search for an understanding of the 'true nature' of the Universe. They, like all others after them, were trying to produce a 'natural' classification. Aristotle believed that animals could be grouped into a series of 'types' according to their 'essences'. Thus the 'essence' of a mammal was the possession of lungs, warm bloodedness and the capacity to bear young which were miniatures of the adult form, not eggs. Once the various 'essences' had been discovered it was easy to fit the animals into their proper groups.

This system of grouping animals typologically, that is by conformity to a recognized type, continued more or less unchallenged well into the 18th century and is still with us today. Even Carolus Linnaeus (1707–1778), the great Swedish naturalist—hailed as the father of modern systematics—worked within a framework of Aristotelian logic. He devised a hierarchical system of classification, rather like a library system, in which he viewed the orderliness of nature as groups within groups. The most fundamental of his groups was the species. Several species would be incorporated into a larger group according to some common attribute. Groups assembled in this way would be included in a still larger group according to some more universal attribute, and so on. The hierarchy which Linnaeus used contained six levels or ranks which graded from the species to the empire (the entire natural world); the intermediate ranks were genus (plural, genera), order, class and kingdom. He did, in fact, recognize a further rank, the variety, which he considered to be smaller than the species, though he did not understand its biological importance; to Linnaeus, varieties were merely accidents in the Divine Creation of the fundamental unit—the species. In modern taxonomy we do not recognize the variety as a formal rank but nonetheless appreciate it as being of evolutionary significance; varieties are now recognized as part of the natural variability within species.

Within the animal kingdom Linnaeus distinguished six classes: mammals, birds, amphibians (which also included animals that are now called reptiles), fish, insects and vermes (worms and worm-like animals). He regarded these as separate rungs on the ladder of perfection—the so-called *scala naturae*—beginning with the worms and rising to the highest of animals—Man.

The 19th century heralded a radical change in our ideas about animal classification. Jean Baptiste Lamark (1744–1829), Alfred Russel Wallace (1823–1913) and especially Charles Darwin (1809–1882) provided the springboard for this when they formulated their ideas on the evolution of living organisms. At last the meaning of the word 'natural' was clear; the orderliness of animals could be expressed in terms of their descent, with modification, from a common ancestor. Clearly, if classifications were to be natural they had to incorporate contemporary ideas about the evolution of animal groups; that is, they had to display evolutionary relationships. Darwin pointed out, however, that there were at least two kinds of such relationships. On the one hand, there was the strict descent of one animal or group of animals from another, the course of which is usually called phylogeny. On the other hand, evolution is also characterized by a morphological divergence of animal types. Modern taxonomists are divided as to which of the evolutionary relationships—phylogeny or divergence—should be stressed in the construction of a classification.

Present classification is based primarily on gross morphology but, increasingly, the taxonomist will have to take note of new criteria which are now being evaluated by related biological disciplines. Biochemists, for example, are beginning to postulate evolutionary pathways of organic molecules, and ethology (the study of comparative animal behaviour), is also displaying its potential value. This tremendous increase in the pool of data will undoubtedly refine our ideas but with it will come the problem of how to cope with so much information. Clearly, the role of the computer will be as paramount in classifying animals as it will be in our everyday lives.

Patience, experience and a good eye are qualities needed by a taxonomist, whose basic work involves detailed examination of specimens and comparison with those already described.

Modern Taxonomy

Internationally recognized names are given to animal species and the hierarchy of groups in which they are classified. Classification attempts to catalogue the characteristics of an animal and to reflect ideas of evolutionary relationship to other organisms.

The task of the modern taxonomist is to classify animals according to their evolutionary relationships. But we must first ask the question—what are we classifying? The basic unit of classification is the species which is defined as a group of interbreeding organisms which is reproductively isolated from other such groups. For the taxonomist this definition is singularly unhelpful since it is very difficult to recognize a species on this criterion. In practice, most animal species are recognized on a purely morphological basis, that is from their external appearance. The limits of a species are determined at points where morphological variation is discontinuous. For instance, a zoologist confronted with a collection of similar looking fishes may count the number of fin rays in the dorsal fin of each. He may find that the individuals can be grouped into those with 12–14 rays and those with 19–21 rays. If further investigation of other characters, such as the number of scales or body proportions, matched the original grouping then separate species would be recognized.

Because this purely typological approach is not logically related to the definition of a species, some people have suggested that the taxonomic species is an artificial grouping and that the basic unit of classification should be the individual. In reality this does not get us very far since individuals will still be grouped together into a higher rank on the basis of morphology. Following the individualists' line we would then have to group caterpillars of several species together as distinct from the adults into which they metamorphose. It is clear that species are real. They are not less real just because we may have difficulty in recognizing them. In practice, the recognition of species on morphological criteria has rarely posed a problem since, in the many cases where interbreeding experiments have been carried out using morphologically recognized species, the results have demonstrated that there is overwhelming support for the validity of such criteria.

The next task of the taxonomist is to arrange the species into larger, more embracing groups, according to their evolutionary relationships. These groups will then be arranged into still more embracing groups and so on. To do this we use a hierarchy of ranks which is derived from the work of Linnaeus. There are seven basic ranks but, because some groups are larger (for example, the insects) or have been more intensively studied (for example, the mammals) than others, it is often convenient to use intermediate ranks to demonstrate fully ideas about the evolution of the group.

The two basic criteria used to group animals in modern classification are phylogeny and morphological divergence. The consequences for classification of using either of these may be rather different. A strict phylogenetic classification will group together animals which, we believe, shared a unique common ancestor irrespective of the degree of

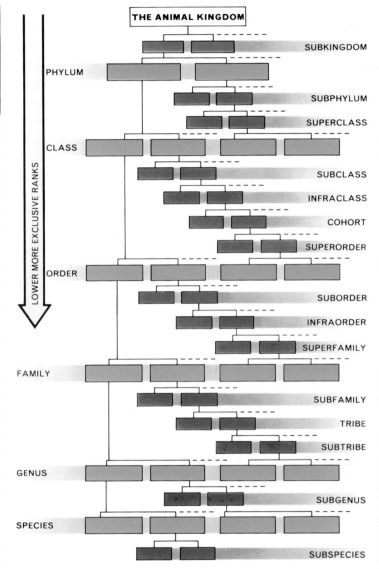

Zoological classification uses a hierarchical system of ranks, ranging from the most inclusive above to the most exclusive below. The basic unit is the species, to which all animals must be referred; other ranks are only recommended but those shown here in blue are used in most classifications. Similar species are grouped into a genus which, with other genera, is grouped into a family, and so on. Intermediate ranks, shown here in orange, are available for large or diverse groups but some of these are used only rarely.

morphological divergence. A classification based on morphological divergence considers the phylogeny to be of lesser importance. For example, we know that among living reptiles (which include turtles, snakes, lizards and crocodiles) the crocodiles shared a common ancestor with birds. Lizards, snakes and turtles are more distantly related, but our current classification does not reflect this. Turtles, lizards, snakes and crocodiles are grouped together as the class Reptilia, and birds are grouped as a separate class—Aves. This scheme of classification stresses the extreme morphological (and

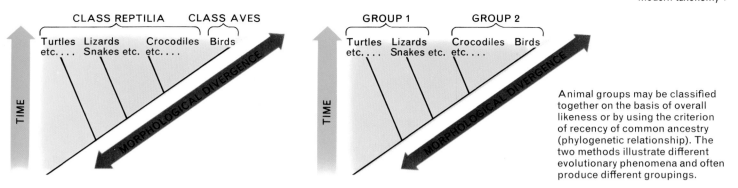

CLASS REPTILIA   CLASS AVES

Turtles   Lizards   Crocodiles   Birds
etc....   Snakes etc.   etc....

TIME

MORPHOLOGICAL DIVERGENCE

GROUP 1   GROUP 2

Turtles   Lizards   Crocodiles   Birds
etc....   Snakes etc.   etc....

TIME

MORPHOLOGICAL DIVERGENCE

Animal groups may be classified together on the basis of overall likeness or by using the criterion of recency of common ancestry (phylogenetic relationship). The two methods illustrate different evolutionary phenomena and often produce different groupings.

physiological) divergence of birds. In effect, grouping crocodiles with lizards, snakes and turtles is a legacy from Linnaeus's time. Unfortunately, a great part of our current animal classification is of this typological nature with the result that the student often has to assimilate a great deal of supplementary information in order to discover the phylogenetic relationships.

Animal names and the names of taxonomic ranks are always expressed in scientific language. Many of these names will be unfamiliar and may be difficult even to pronounce, let alone understand. Why therefore do we need scientific names? Are not common names, such as lion, sufficient? Why does such an animal need to have two parts to its scientific name—*Panthera leo*?

The answer to these questions is that the name of an animal should be a symbol which unambiguously refers to it alone. Common or vernacular names do not serve this purpose. For example, 'gopher' is the common name of a rodent in western Canada and also of a turtle in Florida.

some books will adopt a convention, as this one does, that if a common name is intended to relate to a single species it should begin with a capital letter. Thus 'Short-tailed Albatross' is *Diomedea albatrus*, but 'short-tailed albatross' can refer to any species in the genus *Diomedea* which has a relatively short tail. Unfortunately, such conventions and checklists are by no means universally followed, with the result that the unqualified use of vernacular names frequently leads to confusion.

To facilitate unambiguous communication across a variety of languages animals are therefore given formal Latin names which are governed by a set of recommendations, known as the Code, drawn up by the International Commission for Zoological Nomenclature. The Code specifies those rules with which scientists are requested to comply when they erect a name for an animal. The ultimate objective of the Code is to provide an animal species with a unique name.

Latin or 'latinized' words have been chosen for use in scientific names for two reasons: Latin was the international

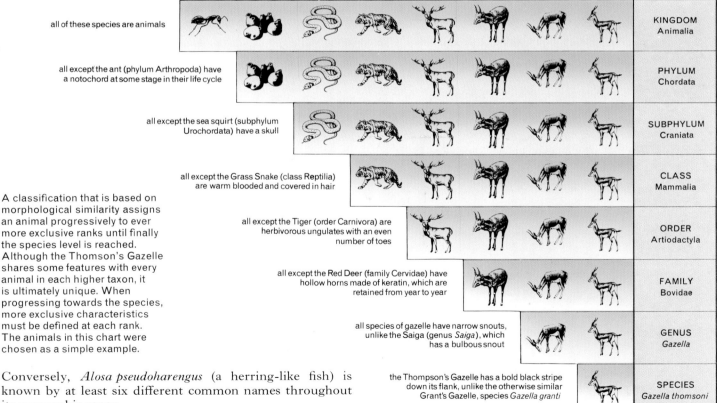

all of these species are animals	KINGDOM Animalia
all except the ant (phylum Arthropoda) have a notochord at some stage in their life cycle	PHYLUM Chordata
all except the sea squirt (subphylum Urochordata) have a skull	SUBPHYLUM Craniata
all except the Grass Snake (class Reptilia) are warm blooded and covered in hair	CLASS Mammalia
all except the Tiger (order Carnivora) are herbivorous ungulates with an even number of toes	ORDER Artiodactyla
all except the Red Deer (family Cervidae) have hollow horns made of keratin, which are retained from year to year	FAMILY Bovidae
all species of gazelle have narrow snouts, unlike the Saiga (genus *Saiga*), which has a bulbous snout	GENUS *Gazella*
the Thompson's Gazelle has a bold black stripe down its flank, unlike the otherwise similar Grant's Gazelle, species *Gazella granti*	SPECIES *Gazella thomsoni*

A classification that is based on morphological similarity assigns an animal progressively to ever more exclusive ranks until finally the species level is reached. Although the Thomson's Gazelle shares some features with every animal in each higher taxon, it is ultimately unique. When progressing towards the species, more exclusive characteristics must be defined at each rank. The animals in this chart were chosen as a simple example.

Conversely, *Alosa pseudoharengus* (a herring-like fish) is known by at least six different common names throughout its geographic range.

Although common names are therefore generally unsuitable for world-wide communication, they are nonetheless often used, and for this reason certain conventions have arisen to deal with them. First, checklists are published which contain the recommended common names for certain groups of animals living in particular places, for example, American freshwater fish or West Palaearctic birds. Second,

language of scientific communication well into Linnaeus's time and it is a stable language, untainted by dialect or fashion. It should be pointed out, however, that biological Latin is rather different from classical Latin. It has a much simpler grammar and employs such letters as j, k and w, which are not present in the classical Latin alphabet.

The rationale behind the scientific names of animal species is derived from Linnaeus, who successfully introduced order into chaos. Up to Linnaeus's time, animals were given long names which were, in effect, phrases that described the morphology. The Honeybee was known as *Apis pubescens, thoracae subgriseo, abdominae fusco, pedibus posticis glabris utrinque marginae ciliatis* (a hairy bee with a greyish thorax, dark-coloured abdomen and hairless hind legs which have bristles on both sides). Linnaeus decided that such a cumbersome name was unnecessary and introduced what is widely known as the binomial system whereby an animal has two parts to its name—for example, *Apis mellifera* (Honeybee), *Ostrea edulis* (Edible Oyster).

When written, names in the binomial system are always italicized, for example *Equus zebra*. The first word denotes the genus to which the animal belongs; it is always written with an initial capital letter and is treated as a Latin noun in the nominative singular. The second word denotes the species; it never begins with a capital letter and can be treated as a noun in apposition, a noun in the genitive or an adjective. In all cases it must be singular and agree with the generic name in gender. The function of the specific name is to act as a qualifying description of the generic name in order to distinguish that particular breeding population from others in the same genus. Thus *Rhinoceros unicornis* is the Indian One-horned Rhinoceros and *Rhinoceros sondaicus* is the Javan Rhinoceros. A specific name cannot stand alone because it only describes an attribute.

The Code allows almost any derivation of animal names as long as the words are 'latinized' by the addition of a Latin ending. Generic names tend to be the most variable. Many are derived directly from Latin, such as *Bos* (cattle) and *Homo* (men). Others are latinized Greek words, such as *Astacus* (crayfish). Still others are formed from a variety of languages, for example *Tupaia* (tree shrews) comes from the Malayan language and *Lama* (llamas) is derived from a Peruvian Indian word. Some are simply made up words, such as *Tatera* (gerbils). Sometimes there is an attempt to construct names which describe attributes of the animals: for example, *Osteoglossum* means 'bony tongue' and alludes to a very distinctive feature of this fish genus. There are unfortunately many examples of generic names being formed from people's names (for example *Moythomasia*—a fish genus named after J. A. Moy-Thomas) or from places (for example *Singida* from a province of that name in Tanzania), and in consequence it is not so easy to trace their derivation.

COMMON ROOT WORDS USED IN ANIMAL NAMES

The following list presents some common examples of the many Latin (*L*) and Greek (*Gr*) root words used in the names of animals and their taxonomic groups. Many of them are almost invariably used as prefixes (e.g. **amphi-**bia); others, shown here preceded by a hyphen, generally appear as suffixes (e.g. nemato-**morpha**). Root words shown preceded by a tilde (~) appear as both prefixes and suffixes (e.g. **Aspi-do-**gaster, Cephal-**aspi-s**). A root word shown both preceded and followed by a tilde indicates that, when used as a suffix, the root word is complete in itself (e.g. Micr-**aster**). In all other cases the letters that most commonly follow the root word are indicated; word endings are shown by letters without hyphens.

a- (*Gr*) without
ab- (*L*) from, away
~**acanth-, o-, a** (*Gr*) spine
-acis (*Gr*) point, barb
~**acti-, ni-, no-, s** (*Gr*) ray
ad- (*L*) to, toward
alb-, i-, o- (*L*) white
allo (*Gr*) other, different
amphi- (*Gr*) on both sides, double
an- (*Gr*) without, not
ancyl-, o- (*Gr*) crooked
arch-, aeo, eo- (*Gr*) ancient
arch-, e-, i- (*Gr*) first, beginning
arthr-, o- (*Gr*) joint
~**aspi-, do-, s** (*Gr*) shield
~**aster~, i-, o-** (*Gr*) star

bath-, o-, y- (*Gr*) deep
bi- (*L*) two, double
~**brachi-, o-, a** (*Gr*) arm
brachy- (*Gr*) short
branch-, i-, o- (*Gr*) gill, fin

caec- (*L*) blind
capit-, i-, o- (*L*) head
~**caud-, a, ata** (*L*) tail
~**cephal-, o-, a** (*Gr*) head
~**chaet-, o-, a** (*Gr*) bristle
chel-, i-, o- (*Gr*) claw
chlamy- (*Gr*) cloak
~**cirr-, i-, us** (*L*) curl of hair
~**cnid-, a, o** (*Gr*) nettle
-coelia (*Gr*) abdominal cavity
~**coel-, o-, a** (*Gr*) hollow

crassi- (*L*) thick
crypt-, o- (*Gr*) hidden
cycl-, o- (*Gr*) circle, wheel
~**cyst-, o-, is** (*Gr*) bladder, bag

~**dactyl-, o-, a, us** (*Gr*) finger, toe
deca- (*Gr*) ten
dendr-, o- (*Gr*) tree
dent-, i-, o- (*L*) tooth
~**derm-, o-, a, ata** (*Gr*) skin
di- (*Gr*) two, separate
dicr-, o- (*Gr*) forked
dipl-, o- (*Gr*) double
~**disc-, i-, o-, us** (*Gr*) round plate
-dont-, a, ia (*Gr*) tooth

echin-, o-, us (*Gr*) spiny; sea urchin; hedgehog
ecto- (*Gr*) outside
-ell-, a, um, us (*L*) small
endo- (*Gr*) within, inner
ento- (*Gr*) within, interior
enter-, o- (*Gr*) intestine, gut
eo- (*Gr*) early
epi- (*Gr*) upon, over, beside
eu- (*Gr*) good, well

fil-, i-, o-, um (*L*) thread
-form-, a, es, is (*L*) form, shape

~**gast-, ero, r-, ro-, er** (*Gr*) stomach, belly
giga-, n- (*Gr*) giant

~**gloss-, o-, a, us** (*Gr*) tongue
~**gnath-, o-, a, us** (*Gr*) jaw
gon-, a-, o- (*Gr*) seed, generation, offspring
gon-, i-, io-, ia (*Gr*) angle
gorg-, o- (*Gr*) grim, fierce
gracil-, i- (*L*) slender
~**grise-, o-, a** (*L, medieval*) grey
gymn-, o- (*Gr*) naked

haem-, a-, ato-, o- (*Gr*) blood
heter-, o- (*Gr*) other, different
~**hipp-, o-, us** (*Gr*) horse
hydr-, o-, a (*Gr*) water
hypo- (*Gr*) under, beneath

~**ichthy-, o-, s** (*Gr*) fish
iso- (*Gr*) equal

lamell-, i- (*L*) plate
~**lat-, i-, us** (*L*) broad, wide
~**lepi-, do-, s-, s** (*Gr*) scale
lept-, ale, ino-, o- (*Gr*) fine, slender
~**leuc-, o-, a** (*Gr*) white
lingu-, l-, a (*L*) tongue
~**loph-, i-, o-, us** (*Gr*) crest, tuft
~**luc-, i-, us** (*L*) clear, shining

macro- (*Gr*) large
meg-, a-, alo- (*Gr*) large, great
melano- (*Gr*) black
meso- (*Gr*) middle
meta- (*Gr*) between
mono- (*Gr*) one, single
-morph-, a (*Gr*) form

neo- (*Gr*) new, recent
nephr-, i-, o- (*Gr*) kidney
~**neur-, o-, a** (*Gr*) nerve cord
noct-, i- (*L*) night
~**not-, a-, o-, um** (*Gr*) back

olig-, a-, o- (*Gr*) few; scant; small
~**onch-, o-, a, us** (*Gr*) barb, hook
omni- (*L*) all
~**ophi-, d-, o-, s** (*Gr*) snake
~**ophthalm-, i-, o-, us** (*Gr*) eye
opisth-, i-, o- (*Gr*) behind, hind

~**orni-, th-, tho-, s** (*Gr*) bird
~**oste-, o-, us** (*Gr*) bone
~**ostrac-, o-, um** (*Gr*) shell

pachy- (*Gr*) thick
palae-, o- (*Gr*) ancient
~**pect-, en-, in-, o-** (*L*) comb
~**pedi-, a** (*L*) foot
peri- (*Gr*) around
-phaga (*Gr*) eat
-philo (*Gr*) loving
-phor-, a, ea (*Gr*) carry, bear
~**phyll-, o-, a, ia** (*Gr*) leaf
phyl-, et-, o- (*Gr*) tribe
phyla-, ct (*Gr*) guard
phyto- (*Gr*) plant
~**pithec-, an-, o-, us** (*Gr*) ape
placo- (*Gr*) plate; flat
plagi-, o- (*Gr*) oblique, slanting
plesio- (*Gr*) near
~**pleur-, o-, a** (*Gr*) side
-pod-, a (*Gr*) foot
poly- (*Gr*) many, much
~**por-, i-, a, us** (*L*) pore, small opening
pro- (*Gr*) before, in front of
pseud-, o- (*Gr*) false
~**pter-, o-, ygo-, a, yx** (*Gr*) wing; feather; fin

-ramia (*L*) branch
~**rhynch-, o-, us** (*Gr*) snout, beak
ruf-, esc-, i-, o- (*L*) red

~**saur-, o-, us** (*Gr*) lizard
siphono- (*Gr*) tube
-soma (*Gr*) body
spiro- (*L*) spiral, coil
~**spor-, o-, a, us** (*Gr*) seed
steno- (*Gr*) narrow
~**stom-, ato-, a** (*Gr*) mouth
syn- (*Gr*) together

~**thec-, o-, a** (*Gr*) case, box; cup
tricho- (*Gr*) hair

uni- (*L*) single
~**ur-, o-, a** (*Gr*) tail

~**zo-, o-, a** (*Gr*) animal

The derivation of specific names is simpler since in most cases they express a definite attribute, such as colour—*Picus viridis* (Green Woodpecker), size—*Arapaima gigas* (Giant Arapaima), shape—*Diphyllobothrium latum* (Broad Tapeworm of Man), occurrence—*Alces americanum* (American Moose), habitat—*Apodemus sylvaticus* (Wood Mouse). Alternatively species are often named in honour of someone, for example *Artibeus watsoni*—Watson's Leaf-nosed Bat.

The scientific names of ranks higher than genus, such as class Anthozoa or order Galliformes, always consist of only one word and are treated as plural nouns. These names are never italicized but should begin with a capital letter. For two ranks the Code demands that specified endings shall be used: a subfamily name ends with -inae and a family name must end with -idae; thus Talpinae is a subfamily of the Talpidae—a family of moles. The Code also recommends suffixes for names of the ranks of tribes (-ini) and superfamily (-oidea). Additionally in certain groups such as birds and fishes a convention recognizes the standard ending -formes for an order, for example Pelicaniformes and Salmoniformes. In such cases the rank may easily be inferred by looking at the ending of the word.

Above the rank of genus there are basically two ways in which the names are formed. They may be derived directly from a generic name: for example family Anguillidae, suborder Anguilloidei and order Anguilliformes are all derivatives of the genus name *Anguilla* (eels). Family names are most often derived in this manner. The second method involves constructing a name which describes an attribute (often the one which is used to group the animals at that rank), for example phylum Arthropoda (jointed-foot), class Bivalvia (shell with two valves).

From time to time a taxonomist will find a collection of individual specimens which, he feels, represent a new species, that is one which was hitherto unknown and unnamed. Before he can name it, however, he must go through certain procedures. Once he has searched the literature and has assured himself that the specimens do not conform to any recorded species, he must describe them and make it perfectly clear in what ways his species differs from any other. He must then choose one specimen and designate it as the so-called type specimen or holotype. The holotype is very important since it is the name bearer and the one to which all other taxonomists may refer for future questions of identity. Quite often a series of other specimens is chosen which displays the range of morphological variation. These may be designated as paratypes.

The taxonomist will then devise a name and he must check to see if that name is available; that is, he must assure himself that the specific name has not already been used for other species of the same genus or, if he is describing a new genus as well, that the generic name has not been used for any other animal. The name, the designation of the holotype and the description of the new species must then be published in a source to which a specific date can be attached and to which other scientists have access; this is usually a scientific journal. Once this procedure is completed the name may be considered valid. Strictly, for completeness, the author's name and the date of original description should appear after the scientific name; for example, the Pacific Barracuda is *Sphyraena argentea* Girard, 1854. This convention, however, is not always followed. In cases where an animal is later transferred to a new genus the original author's name should appear in brackets after the new name; for example, the Atlantic Tunny is *Thunnus thynnus* (Linnaeus). In this case it was Linnaeus who originally named the fish *Scomber thynnus* but a later researcher decided that it was too unlike other members of the genus *Scomber* (the mackerels) to be included with them.

ABOVE: the scientific names of animals often reflect obvious characteristics or geographical range. The name of these hover-flies, *Helophilus pendulus*, translates to 'hanging sun-lover', which is probably a reference to their habit of hovering, as though suspended on a thread, on bright, sunny days.

BELOW: the scientific name of the Giant Anteater, *Myrmecophaga tridactyla*, makes reference to its diet and the three enlarged digging claws on each foreleg. The generic name is derived from *myrmeco-* (Gr.) = ant, and *phag-* (Gr.) = eat; the specific name comes from *tri-* (L.) = three, and *dactyl-* (Gr.) = finger.

BELOW: *Recurvirostra americana* is the American Avocet, shown here on its nest. The generic name combines *recurvi-* (L.) and *rostri-* (L.) = recurved beak, a characteristic common to avocet species. This is qualified by the species name *americana*, which indicates the country of origin of this particular species.

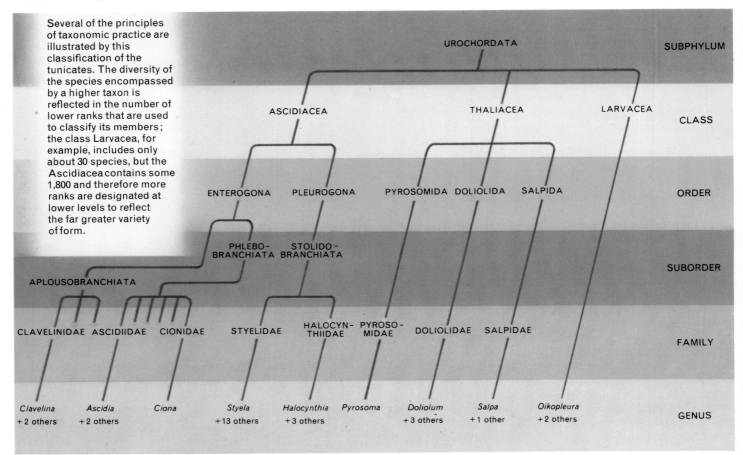

Several of the principles of taxonomic practice are illustrated by this classification of the tunicates. The diversity of the species encompassed by a higher taxon is reflected in the number of lower ranks that are used to classify its members; the class Larvacea, for example, includes only about 30 species, but the Ascidiacea contains some 1,800 and therefore more ranks are designated at lower levels to reflect the far greater variety of form.

The starting point for the availability of names is taken as 1st January 1758, the year in which Linnaeus published the tenth edition of his famous book *Systema Naturae*. Any name used before that date is technically available. Of course during the last two centuries mistakes have been made. Sometimes the same animal has been described by different scientists independently. For instance, in 1836 Yarrell described and named the lancelet *Amphioxus lanceolatus*. However, it was later discovered that Costa had described the same animal in 1834 and had given it the generic name *Branchiostoma*. By the rule of priority *Branchiostoma* stands as the official name of the genus and *Amphioxus* becomes a junior synonym which, incidentally, has been retained as a common name and is therefore no longer italicized.

Synonymy may also occur in higher ranks but the Code makes no insistence about priority in this case. For instance, mayflies were first placed in the order Odontata in 1806, then in an order Ephemerida in 1817 and lastly in an order Ephemeroptera in 1896. It is this latest name which remains in current usage. Some names, although strictly synonyms, have received equal usage by different scientists. Thus, the class name for the bivalve molluscs is accepted as Pelecypoda by Americans and as Bivalvia or Lamellibranchiata by European zoologists.

The best way of illustrating the ways in which names are derived and classifications constructed is by examining a classification in some detail. Here we have chosen that of the subphylum Urochordata (= Tunicata). This subphylum, commonly known as the tunicates (a vernacular derived from its now less widely accepted synonym), is an exclusively marine group which includes sea-squirts, salps, and the peculiar larvaceans. There are about 78 genera (the number varies slightly according to different authorities) and for many centuries they were regarded as molluscs. In the last century, however, examination of the larvae suggested that tunicates should be regarded as primitive chordates.

Like most animal classifications, that of the Urochordata employs the Linnaean hierarchical system with more universal attributes being expressed at each successively higher rank. Thus the most universal attribute is the possession of a notochord restricted to the tail region of at least one stage in the life cycle (*uro* = tail, *chorda* = notochord). Another universal attribute is the presence of a protective coat or tunic around the animal's body wall. This feature gave rise to the alternative name of the subphylum—Tunicata.

The initial subdivision of the tunicates into three classes is based primarily on habitat and life cycle. Thus members of the class Ascidiacea (named after one of the best known genera) are sessile in the adult stage and have motile larvae. Members of the class Thaliacea (named after a now disused group called Thalia by Aristotle) are pelagic when adult and have motile larvae. The class Larvacea is so named because members of this group resemble the larvae of the other two classes throughout their lives.

Below the rank of class there are several features worthy of note. For example, not all of the ranks are used within each class. This is because the three classes contain very different numbers of species. The class Larvacea contains only three genera which are morphologically very similar to one another. It is therefore considered unnecessary to place them in separate families (though some authorities do). There is little point in using the ranks of suborder and order because each would only contain one category of the next lower rank. In other words, they would be coextensive, as each would contain the same complement of species.

As an aside it may be mentioned that an extreme example of the omission of ranks is to be found in the classification of the horseshoe worms (phylum Phoronida). These strange animals are thought to be so very different from other invertebrates that they have been accorded phylum status; but there are only two known genera, and in their classification all the ranks between phylum and genus are omitted.

In the tunicates, the class Thaliacea contains seven genera which fall into three morphological 'types'. Each of these types is characterized by a genus which gives its name to a family and to an order; for example, *Pyrosoma* is the type-genus of the family Pyrosomidae which is, in turn, the type-family of the order Pyrosomida. It can be seen that the families and orders of thaliaceans are coextensive and, strictly, we do not need to use the order rank.

The class Ascidiacea contains approximately 68 genera and, as one would expect, the members between them show a considerable range of body form. Each of the twelve families is recognized as conforming to a type-genus from which it takes its name, for example, *Ciona* is the type genus of the family Cionidae. The families may be grouped into three suborders according to the condition of the branchial sac or pharynx: in the Aplousobranchiata (which means simple gills) the branchial sac does not have folds or inner longitudinal bars; in the Phlebobranchiata (vein-like gills) the branchial sac does not have folds but does have inner longitudinal bars; and in the Stolidobranchiata (solid or immovable gills) the branchial sac has both folds and inner longitudinal bars. The first two of the suborders is united by a more universal attribute, the position of the gonad. In both the Aplousobranchiata and the Phlebobranchiata the gonads develop within a loop of the intestine. This is the basis for grouping them as the order Enterogona (*entero* = gut, *gona* = gonad). The Stolidobranchiata have gonads which develop at the side, in the mantle wall—hence the name of the order, Pleurogona (*pleuro* = side). Once again we notice that the suborder is coextensive with the order and is technically unnecessary but here it is helpful to include it as it implies that there are three conditions of the branchial sac.

The classification of the tunicates is essentially typological in nature and does not necessarily reflect the phylogenetic relationships within the group. There is, for instance, no consensus as to whether all members of the class Ascidiacea are more closely related to one another than any are to members of the Thaliacea or Larvacea. This uncertainty has arisen primarily because scientists cannot agree on which type of life cycle is primitive. Earlier ideas held that the tadpole-like larvaceans represented the primitive condition and that the form such as that seen in the ascidiacean adult was the derived condition. A classification reflecting this idea would place together thaliaceans and ascidiaceans in a higher category. Other workers have suggested that the tadpole-like animal evolved from an ascidiacean-like animal (by prolonging the larval stage), implying that the larvaceans are derived. But the question remains—to which ascidiacean or thaliacean are the larvaceans most closely related? In this case therefore the typological classification of the tunicates stands in default of agreement about the course of phylogeny within the group.

Because classification is designed to bring evolutionary order to an astonishing variety of animals, it is to be expected that increasing knowledge and changing ideas should be reflected in periodic revisions. At the same time, because much animal taxonomy is typological in nature, classifications of the same group may differ according to the choice of characters used to link its members at various ranks. For these two reasons animal classification is subject to both change and variation at any time. Despite this constant state of flux, the fundamental need for orderliness outweighs the inconvenience of continual modification.

Though sea squirts (members of the class Ascidiacea) show a great range of colour and body form, such features can be very variable, even within a species, and are therefore of limited taxonomic value. Specimens are assigned to orders within the class using more fundamental characters. *Clavelina* (below) is placed in the Enterogona because it has a single gonad within the loop of the intestine. *Halocynthia* (left) and the stylelid genus *Botrylloides* (far below) have paired gonads in the body wall and are therefore placed in the Pleurogona; these genera are referred to different families using other internal characters.

The
Animal
Kingdom

A comprehensive survey of the groups

There are over a million described species of animals living today and perhaps 300,000 more are known as fossils. It is the task of zoologists to study this bewildering variety in such a way that its complexity can be understood in terms of a more manageable number of principles and concepts. One of the first steps in achieving this is to attempt to classify the members of the kingdom—a procedure which can be approached in several different ways. Animals may be grouped according to the habitats they exploit, the ways they obtain food, how they reproduce, or on other single criteria, but the value of such classifications is always limited. This is because during evolution animals have repeatedly explored different ways of living; parasitic animals, for example, are found to have a wide range of body forms modified in a variety of ways to ensure survival. Consequently, although special classifications are of use, much more general—or natural—classifications are preferred. Such natural classification, which relies on the use of many features, is found to reflect strongly two aspects of animal life: the basic structural organization or body plans, and the ways in which the body plans are constructed during development.

Experience over many decades has shown that a natural classification, which is reflected in the phylum-by-phylum treatment in this book, embraces all animals, allows us to discern the progression from simple to complex forms and, in conjunction with the fossil record, gives us deep insights into past evolution.

Despite their variety of forms and functions, however, all animals face the same basic problems. The progression from simple to more complex body plans reflects the ways which animals have explored in their attempts to overcome these problems more and more effectively. The simplest of animals—protozoans—show a wide diversity of lifestyles and considerable complexity of internal structures, but the range of habitats which they can exploit is limited by their small size and their extreme vulnerability to large changes in the external environment. Clearly, there is a limit to the size of a unicellular organism that can be maintained efficiently as an integrated unit.

All other animals are multicellular. A group of cells, acting together, allows the isolation of some cells within an internal (but extracellular) environment bounded by other cells. Whereas in a protozoan the single cell must carry out all the functions of feeding, reproducing and moving, as well as conserving the internal environment, in a multicellular animal a continuous outer envelope of cells can concentrate its energies upon protecting the inner cells from environmental changes. Freed from this burden, inner groups of cells can specialize to perform with greater efficiency such activities as feeding, digestion, reproduction and movement. This principle is demonstrated simply by the minute, parasitic, motile members of the phylum Mesozoa, whose multicellular bodies are little more than an envelope of cells surrounding a mass of reproductive cells. With this low level of cellular

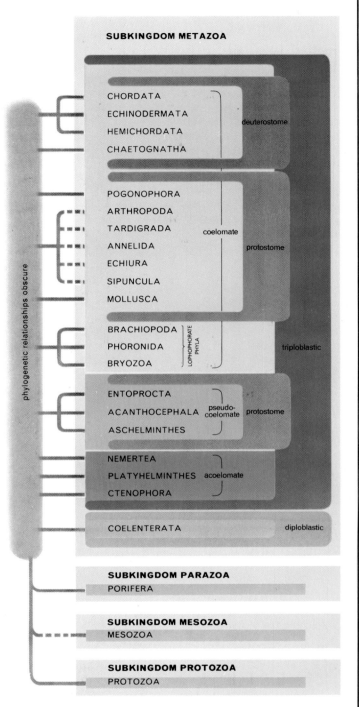

Classifying animals enables us to recognize a range of groups, or phyla, whose members have certain common characteristics. Using more fundamental characteristics, such as the number of body layers, the form of the body cavities and the pattern of embryo development, the phyla can be assigned to a further set of more embracing categories. On the basis of these criteria phyla may be ranked in order of increasing complexity; this sequence conforms to the picture of evolutionary history suggested by the fossil record. Common ancestry can be postulated with some confidence for phyla that share combinations of basic characters but tracing earlier phylogenetic relationships involves much greater speculation. Not all the phyla fall neatly into every category; recent work, for example, has shown that the three lophophorate phyla have a mixture of protostome and deuterostome features and cannot be put into either group. Future work will probably cast further doubts on the value of these concepts and the generalizations which are based on them.

specialization, however, mesozoans are still very susceptible to external environmental changes.

For an active life, animals need a steady supply of nutritive chemicals, and the source of such chemicals is external. At the same time, cells produce toxic waste products and the only safe dumping ground for these also lies outside the body. For these reasons the relationship between the surface area and the volume of an animal is crucial. A multi-cellular animal can perform certain activities more efficiently if it is large and it can only be efficient and multicellular if covered in a protective envelope. However, the larger it becomes and the more efficient is its envelope the greater is the danger to the cells deep inside. They may become starved of oxygen and food and poisoned by waste products.

One escape from this difficulty lies in throwing the body surface into many folds, so that no cell lies further than a millimetre from the animal's surface. This approach has been adopted by the sponges (subkingdom Parazoa) whose bodies are riddled with tortuous cavities filled with the water of the environment. However, so vast is the surface area of a sponge, while so low is the protective efficiency of the enveloping cells, that sponges have little control over their internal environments and so are severely restricted in their habitats.

A second escape from the difficulty lies in having a strongly protective envelope of cells while developing internal transporting systems for moving nourishment, oxygen and waste products safely and rapidly within the body. All the remainder of the multicellular animals—the subkingdom Metazoa—have adopted this plan of construction, developing one or more of the three systems of food transport (alimentary system), nitrogenous waste transport (excretory system), and oxygen transport (blood in a vascular system).

To see how the various metazoan groups have developed these systems, it is helpful to look at the process by which a single-celled egg begins its development. All the phyla within the Metazoa are characterized by the fact that the fertilized egg develops first into a solid ball of cells (morula) and then into a hollow ball of cells (blastula). There then follows a variety of infoldings and migrations of cells—collectively termed gastrulation. These ensure the separation of groups of cells which can specialize for a variety of functions within an outer envelope of special cells (the ectoderm).

After the process of gastrulation, the animal consists of a hollow ball with an opening (the blastopore) connecting the cavity (the coelenteron) with the exterior. In the coelenterates and platy-helminths this remains as the only opening and serves as both mouth and anus. In most other phyla a second opening is formed some distance from the blastopore. In most cases this second opening becomes the anus and the blastopore becomes the mouth; this is known as the protostome condition. In chaetognaths, hemichordates, echinoderms and chordates the reverse is true: the second opening forms the mouth and the blastopore becomes the anus—the deuterostome condition.

Small, unicellular protozoans, living freely in water, represent the lowest level of animal organization; simple diffusion is sufficient to transport food and waste around the body, and the relatively stable aquatic conditions remove the need for complex systems to regulate the animals' internal environment. Fish also rely on the constant temperature and chemical composition of large water bodies for maintaining internal balance, but are larger and more complex; internal transport systems distribute food and oxygen to the tissues and carry away waste products. Adaptations of the more complex morphology fit the animals to various specialized life styles. The physiology of mammals has developed to free them from the need to remain in water; body temperature is held constant, fertilization is internal and water can be efficiently conserved. Such independence enables mammals to exploit a great variety of habitats.

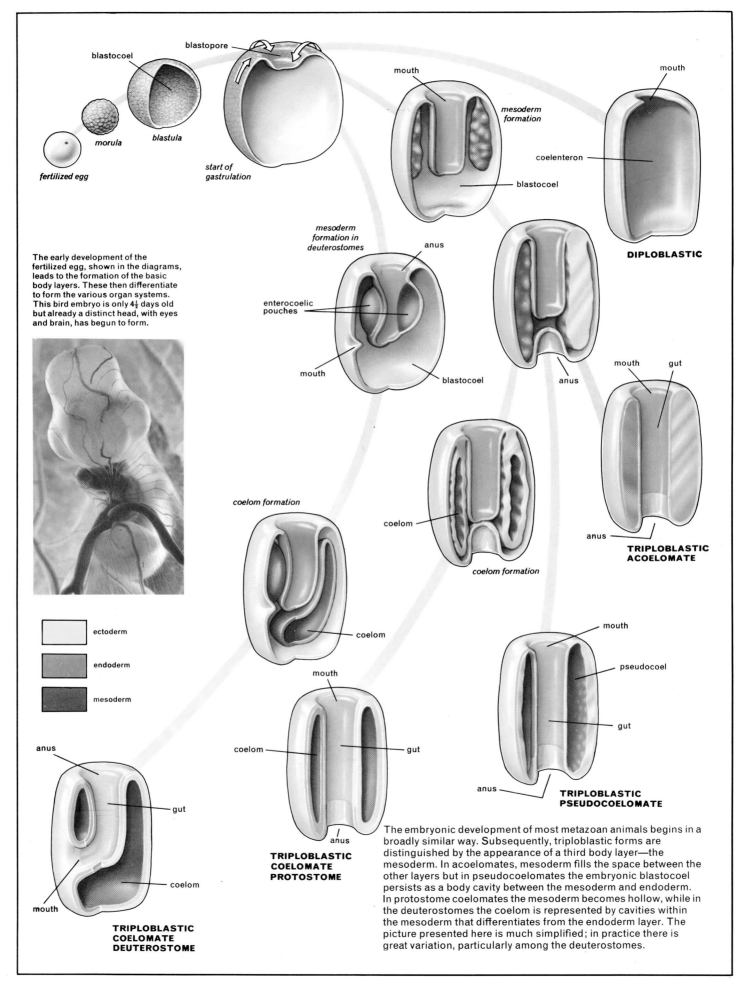

The early development of the fertilized egg, shown in the diagrams, leads to the formation of the basic body layers. These then differentiate to form the various organ systems. This bird embryo is only 4½ days old but already a distinct head, with eyes and brain, has begun to form.

ectoderm

endoderm

mesoderm

blastocoel

morula

blastula

fertilized egg

start of gastrulation

blastopore

mouth

mesoderm formation

coelenteron

blastocoel

DIPLOBLASTIC

mouth

mesoderm formation in deuterostomes

anus

enterocoelic pouches

mouth

blastocoel

anus

mouth   gut

anus

TRIPLOBLASTIC ACOELOMATE

coelom formation

coelom

coelom formation

coelom

coelom

mouth

coelom   gut

anus

TRIPLOBLASTIC COELOMATE PROTOSTOME

mouth

pseudocoel

gut

anus

TRIPLOBLASTIC PSEUDOCOELOMATE

anus

gut

coelom

mouth

TRIPLOBLASTIC COELOMATE DEUTEROSTOME

The embryonic development of most metazoan animals begins in a broadly similar way. Subsequently, triploblastic forms are distinguished by the appearance of a third body layer—the mesoderm. In acoelomates, mesoderm fills the space between the other layers but in pseudocoelomates the embryonic blastocoel persists as a body cavity between the mesoderm and endoderm. In protostome coelomates the mesoderm becomes hollow, while in the deuterostomes the coelom is represented by cavities within the mesoderm that differentiates from the endoderm layer. The picture presented here is much simplified; in practice there is great variation, particularly among the deuterostomes.

In the phylum Coelenterata only one transport system, the gut, is developed and the body wall consists of only two layers of cells (ectoderm and endoderm). In this so-called diploblastic condition gaseous exchange and excretion take place by diffusion through the thin layers directly into the external medium which bathes them both. In all other Metazoa, however, a third, middle tissue layer (mesoderm) arises during gastrulation producing the triploblastic, or three-layered condition. Either this mesoderm must remain thin and the animal small, or else it must be drained and supplied by excretory and vascular transport systems. In animals described as acoelomates the mesoderm thickens only to a certain extent, is loosely packed, and is drained by excretory organs and tubes, while in the pseudocoelomates the blastocoel (the space within the hollow blastula) persists in the adult as a so-called pseudocoelom between the mesoderm and endoderm. This fluid-filled space may contain loose cells (mesenchyme) and its contents are also flushed out by an excretory system. The approach adopted by all other animals, which are therefore grouped together as the coelomates, is to develop large, regular, fluid-filled cavities (coelomic pouches) within the mesoderm, effectively splitting the tissue into an inner layer forming the gut muscles and an outer layer providing muscles of the body wall.

This coelomate condition has been further modified in four ways. In one the coelomic cavity forms as a series of pouches along the body length giving a condition called metameric segmentation, as in the phylum Annelida. A second modification involves great reduction of the coelom, but retention of superficial segmentation, hardening of the body wall and development of paired limbs (phylum Arthropoda). A third involves development of a hard external skeleton, reduction of the coelomic cavity, loss of metameric segmentation and expansion of the vascular system within a soft body wall as the hydrostatic part of the locomotory apparatus (phylum Mollusca). The fourth modification is the formation of the coelom as only two or three pairs of pouches (which may coalesce to a single cavity) and slight development of segmentation (the chordates and other deuterostomes, and the lophophorate phyla).

In many animals the early stages of development of the individual (embryogenesis) produce animals which are minute replicas of their parents, subsequent development involving mostly growth and the achievement of sexual maturity with little reorganization of body plan. But this is not the only kind of development. In many other animals, particularly marine forms, embryogenesis results in a small individual very different from the adult in appearance and habits, and with regard to the habitats it can occupy. Such a developmental stage is called a larva, which must undergo very rapid and radical changes in organization and lifestyle to become adult. These changes are termed metamorphosis, a phenomenon well known in the butterflies whose larvae are the familiar caterpillars. Metamorphosis is common within the animal kingdom and many animals go through one or more larval stages before becoming adults. Examples include the trochophore larvae of molluscs, which develop into veliger larvae; the nauplius larvae of crustaceans, which may give rise to several other different larval types; and the tornaria larvae of hemichordates, which metamorphose directly into adults.

Some features of animal organization are given special emphasis even though they do not reflect faithfully modern ideas of natural classification. Thus the term 'invertebrate' is used to refer to all multicellular animals which lack an internal skeletal backbone, whereas the remainder are called vertebrates. The phylum Chordata, however, contains representatives of both these divisions. Although the two terms reflect outmoded zoological ideas of fundamental relationships they are still retained for the convenience of dividing the kingdom into two groups.

All but one of the 24 phyla described in this book have only invertebrate members. This disparity may seem surprising, especially as to many of us the vertebrates, which include fish, amphibians, reptiles, birds and mammals, are the most familiar of all animals. But it reflects the infinitely greater variety of basic form found among the invertebrates. It may also seem strange that many of the invertebrates are worm-like animals, but close examination reveals such enormous differences between them that it is impossible to present them as closely related and so they are placed in separate phyla. The reason for this wealth of variation is that most of these groups emerged some 3,500 million years ago and they demonstrate the evolutionary exploration that occurred at that time. Initially, marine environments alone were sufficiently hospitable to permit evolutionary exploration of forms and functions, and to this day the marine fauna is by far the most varied. Only a few annelids and molluscs, many arthropods and the tetrapod vertebrates have adapted to life on land. They alone have evolved the devices and strategies which enable them to be sufficiently independent of a copious, external water supply.

Although classifying animals by comparative anatomy, embryology and basic physiological requirements is extremely useful, it provides only limited information. Thus, to say that a shrimp is a member of the class Crustacea in the phylum Arthropoda tells us several facts about the development and body organization of shrimps. But many questions remain: Are shrimps all marine? Are they parasitic? Do they have separate sexes? What do they feed on? How do they move about? Do they have larval stages? Conventional classification will not provide us with the answers.

This section describes more fully the individual phyla and their component classes, and reveals a variation of forms, lifestyles and activities that is astonishingly rich. Nevertheless, it is both humbling and exhilarating to know that, as yet, we understand only a fraction of this animal world.

The Animal Kingdom

This chart, which serves as a pictorial contents list, shows the entire kingdom divided into the groups which are individually presented on the following pages. For the most diverse taxa general introductions precede the detailed accounts of the lower groupings they contain.

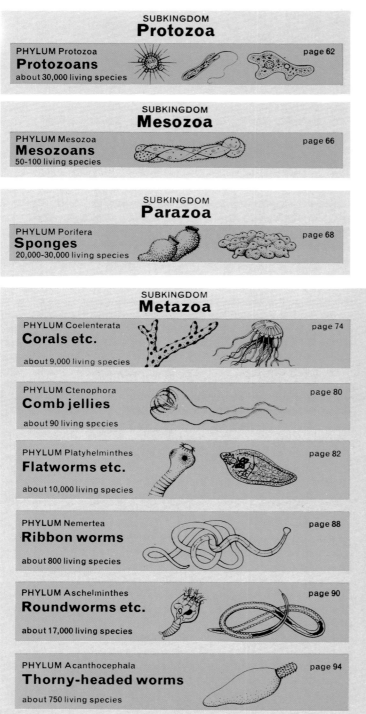

SUBKINGDOM Protozoa

PHYLUM Protozoa
Protozoans
about 30,000 living species
page 62

SUBKINGDOM Mesozoa

PHYLUM Mesozoa
Mesozoans
50-100 living species
page 66

SUBKINGDOM Parazoa

PHYLUM Porifera
Sponges
20,000-30,000 living species
page 68

SUBKINGDOM Metazoa

PHYLUM Coelenterata
Corals etc.
about 9,000 living species
page 74

PHYLUM Ctenophora
Comb jellies
about 90 living species
page 80

PHYLUM Platyhelminthes
Flatworms etc.
about 10,000 living species
page 82

PHYLUM Nemertea
Ribbon worms
about 800 living species
page 88

PHYLUM Aschelminthes
Roundworms etc.
about 17,000 living species
page 90

PHYLUM Acanthocephala
Thorny-headed worms
about 750 living species
page 94

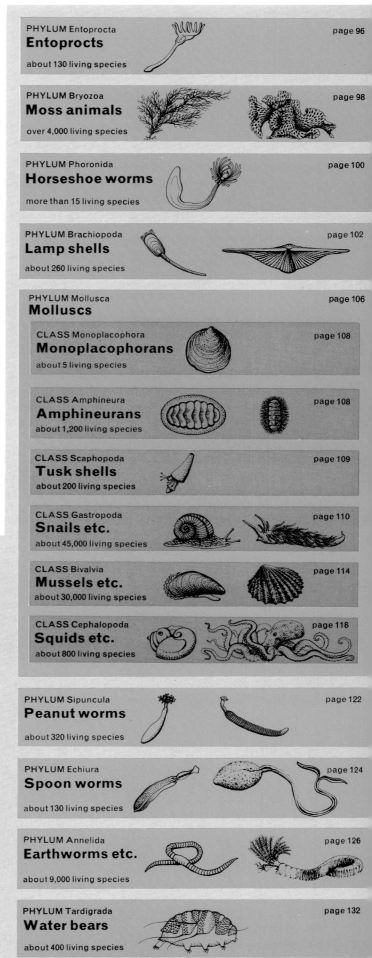

PHYLUM Entoprocta
Entoprocts
about 130 living species
page 96

PHYLUM Bryozoa
Moss animals
over 4,000 living species
page 98

PHYLUM Phoronida
Horseshoe worms
more than 15 living species
page 100

PHYLUM Brachiopoda
Lamp shells
about 260 living species
page 102

PHYLUM Mollusca
Molluscs
page 106

CLASS Monoplacophora
Monoplacophorans
about 5 living species
page 108

CLASS Amphineura
Amphineurans
about 1,200 living species
page 108

CLASS Scaphopoda
Tusk shells
about 200 living species
page 109

CLASS Gastropoda
Snails etc.
about 45,000 living species
page 110

CLASS Bivalvia
Mussels etc.
about 30,000 living species
page 114

CLASS Cephalopoda
Squids etc.
about 800 living species
page 118

PHYLUM Sipuncula
Peanut worms
about 320 living species
page 122

PHYLUM Echiura
Spoon worms
about 130 living species
page 124

PHYLUM Annelida
Earthworms etc.
about 9,000 living species
page 126

PHYLUM Tardigrada
Water bears
about 400 living species
page 132

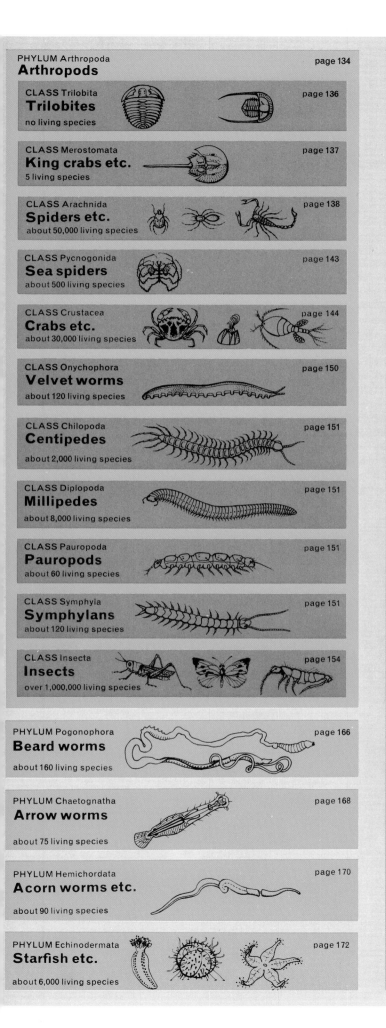

PHYLUM Arthropoda
Arthropods
page 134

CLASS Trilobita
Trilobites
no living species
page 136

CLASS Merostomata
King crabs etc.
5 living species
page 137

CLASS Arachnida
Spiders etc.
about 50,000 living species
page 138

CLASS Pycnogonida
Sea spiders
about 500 living species
page 143

CLASS Crustacea
Crabs etc.
about 30,000 living species
page 144

CLASS Onychophora
Velvet worms
about 120 living species
page 150

CLASS Chilopoda
Centipedes
about 2,000 living species
page 151

CLASS Diplopoda
Millipedes
about 8,000 living species
page 151

CLASS Pauropoda
Pauropods
about 60 living species
page 151

CLASS Symphyla
Symphylans
about 120 living species
page 151

CLASS Insecta
Insects
over 1,000,000 living species
page 154

PHYLUM Pogonophora
Beard worms
about 160 living species
page 166

PHYLUM Chaetognatha
Arrow worms
about 75 living species
page 168

PHYLUM Hemichordata
Acorn worms etc.
about 90 living species
page 170

PHYLUM Echinodermata
Starfish etc.
about 6,000 living species
page 172

PHYLUM Chordata
Chordates
page 180

SUBPHYLUM Urochordata
Sea squirts etc.
over 2,000 living species
page 184

SUBPHYLUM Cephalochordata
Lancelets
14 living species
page 188

SUBPHYLUM Craniata
Craniates
page 190

CLASS Myxini
Hags
32 living species
page 192

CLASS Heterostraci
Heterostracans
no living species
page 192

CLASS Cephalaspidomorphi
Lampreys etc.
31 living species
page 193

CLASS Elasmobranchiomorphi
Sharks etc.
about 630 living species
page 194

CLASS Acanthodii
Spiny sharks
no living species
page 198

CLASS Osteichthyes
Bony fish
over 18,000 living species
page 198

CLASS Amphibia
Amphibians
about 2,900 living species
page 206

CLASS Reptilia
Reptiles
about 6,300 living species
page 216

CLASS Aves
Birds
over 8,600 living species
page 230

CLASS Mammalia
Mammals
page 254

SUBCLASS Prototheria
Prototherians
6 living species
page 256

SUBCLASS Metatheria
Marsupials
about 240 living species
page 258

SUBCLASS Eutheria
Placentals
about 3,800 living species
page 262

SUBKINGDOM Protozoa

PHYLUM Protozoa

Protozoans

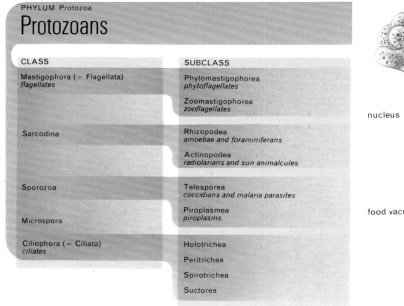

CLASS	SUBCLASS
Mastigophora (= Flagellata) *flagellates*	Phytomastigophorea *phytoflagellates*
	Zoomastigophorea *zooflagellates*
Sarcodina	Rhizopodea *amoebae and foraminiferans*
	Actinopodea *radiolarians and sun animalcules*
Sporozoa	Telosporea *coccidians and malaria parasites*
	Piroplasmea *piroplasms*
Microspora	
Ciliophora (= Ciliata) *ciliates*	Holotrichea
	Peritrichea
	Spirotrichea
	Suctorea

- NUMBER OF LIVING SPECIES: about 30,000
- DIAGNOSTIC FEATURES: body not divided into cells
- LIFESTYLE AND HABITAT: free living in aquatic environments and parasitic or symbiotic in a wide range of plants and animals
- SIZE: mostly less than 0.5 mm in diameter
- FOSSIL RECORD: Precambrian to Recent but particularly important as rock formers in the Cretaceous; at least 30,000 species

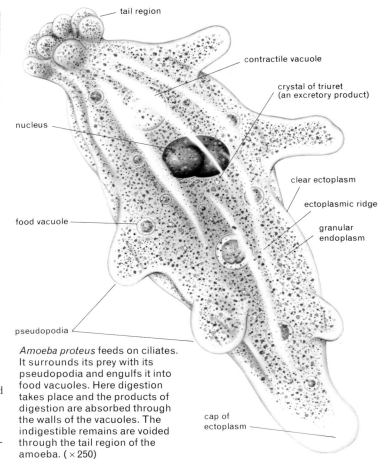

Amoeba proteus feeds on ciliates. It surrounds its prey with its pseudopodia and engulfs it into food vacuoles. Here digestion takes place and the products of digestion are absorbed through the walls of the vacuoles. The indigestible remains are voided through the tail region of the amoeba. (×250)

The basic building unit of all living organisms is the cell. A cell has a central nucleus containing the chromosomes which bear the hereditary instructions for basic life processes, cytoplasm surrounding the nucleus and containing the organelles which carry out these instructions, and a bounding plasma membrane which controls the entry and exit of various substances. Most animals are multicellular, the activities of their many different kinds of cells being integrated through communication systems involving either direct contact between cells or transport of messenger molecules between them. The simplest animals, however, can be regarded as those that consist of a single cell which is also an organism capable of independent existence. Such unicellular animals compose the subkingdom and phylum Protozoa.

Protozoans are usually microscopic in size—some flagellates as small as 0.002 mm across—but certain fossil foraminiferans (*Nummilites* species) may reach 15 cm in diameter. Their world distribution is limited by habitat rather than geography. Some of these habitats may be fleeting—drops of dew on leaves, 'cuckoospit', the water film around soil particles, and the surface of snow. Among the parasitic habitats are the latex vessels of plants, the blood, guts, muscle and brain of animals, and inside living cells of all kinds.

Several protozoans exhibit extraordinary complexity within the confines of a single cell—often through elaboration of the mechanisms whereby the organisms propel or feed themselves. Thus in the Mastigophora which move by the beating of flagella, these structures may be present in thou-

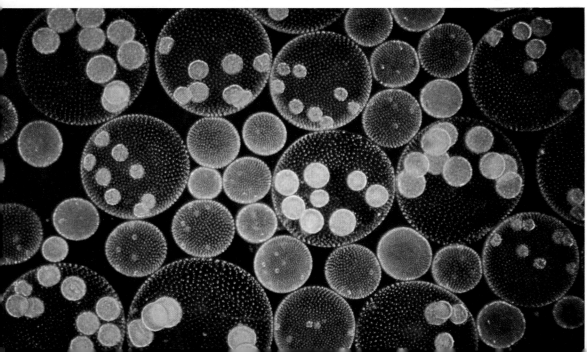

LEFT: the globular colonies of the phytomastigophorean *Volvox* are usually quite easy to see with the naked eye. Each sphere consists of thousands of phytoflagellates (seen here as tiny glowing dots) linked to one another by fine threads of cytoplasm. 'Daughter colonies' are visible inside the larger 'parent' colonies. (×100)

RIGHT: *Glenodinium* is a phytomastigophoran from freshwater plankton. It belongs to the order Dinoflagellida, which has two flagella: one lies in an equatorial groove (visible here), while the other trails behind. (×950)

FAR RIGHT: *Trypanosoma brucei* is shown here in a stained blood smear. The flagellum is attached along the length of the body and its beating gives rise to the 'undulating membrane'. (×2,500)

sands on the surface of the larger parasitic forms such as *Trichonympha*. In amoebae and other Sarcodina, movement is effected through flowing extensions of the cytoplasm called pseudopodia—seen at their most elaborate in the lace-like food-catching nets of the foraminiferans. In the Ciliophora cilia may be fused in rows or bunches to form more powerful beating blades for use in swimming or feeding.

Most protozoans are heterotrophic, that is feed in animal-like fashion, taking in particles of food by phagocytosis ('cell eating') or food molecules by pinocytosis ('cell drinking').

and so they have a special organelle, the contractile vacuole, to pump water out of the cell; the vacuole fills with water in a bladder-like fashion, then suddenly discharges its contents to the exterior.

The most common method of reproduction in protozoans is asexual, by binary or multiple fission. Sometimes division of the cytoplasm is incomplete and whole 'colonies' of protozoans are formed as a result. Sexual reproduction is common in Sporozoa and Ciliophora but found rarely in the other classes.

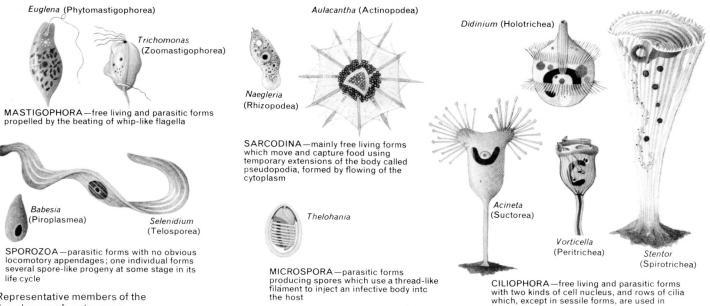

Euglena (Phytomastigophorea)

Trichomonas (Zoomastigophorea)

MASTIGOPHORA—free living and parasitic forms propelled by the beating of whip-like flagella

Babesia (Piroplasmea)

Selenidium (Telosporea)

SPOROZOA—parasitic forms with no obvious locomotory appendages; one individual forms several spore-like progeny at some stage in its life cycle

Representative members of the five classes of protozoans.

Aulacantha (Actinopodea)

Naegleria (Rhizopodea)

SARCODINA—mainly free living forms which move and capture food using temporary extensions of the body called pseudopodia, formed by flowing of the cytoplasm

Thelohania

MICROSPORA—parasitic forms producing spores which use a thread-like filament to inject an infective body into the host

Didinium (Holotrichea)

Acineta (Suctorea)

Vorticella (Peritrichea)

Stentor (Spirotrichea)

CILIOPHORA—free living and parasitic forms with two kinds of cell nucleus, and rows of cilia which, except in sessile forms, are used in locomotion

Many have a special cytostome (cell mouth) through which food is taken into vacuoles in which digestion takes place. In the phytoflagellates, feeding is partly or wholly autotrophic, that is plant-like, for these organisms usually possess chloroplasts with green chlorophyll pigments that enable them to trap the radiant energy of the sun and utilize this energy to build complex foodstuffs from carbon dioxide, mineral salts and water through the process of photosynthesis.

Protozoans have no special organelles for the exchange of gases comparable to gills or lungs in higher animals, nor do they have special excretory systems for getting rid of wastes. These processes simply take place through the plasma membrane by diffusion. Protozoans living in freshwater face the problem of water continually entering the body by osmosis

The simplest protozoans are to be found among the Mastigophora. This class illustrates the artificiality of the boundary between the animal and plant kingdoms, for some of the phytoflagellates can feed autotrophically in the light and heterotrophically in the dark. Genetic mutations can result in loss of the ability to synthesize chlorophyll or loss of the entire chloroplast, so that the flagellate stock becomes permanently colourless and heterotrophic. Photosynthesizing phytoflagellates are important primary producers in plankton. Sometimes their numbers are sufficient to colour the water as in the notorious 'red tides' caused by species of *Gymnodinium* which secrete substances poisonous to fish and man. Abundance of the minute phytoflagellates of the order Coccolithophorida ('white water') in the sea, however, is a

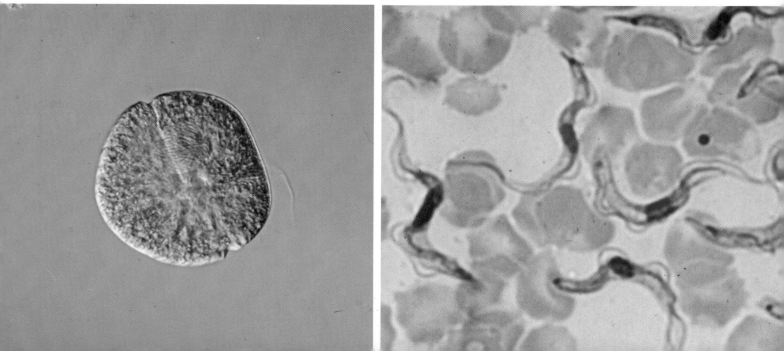

good omen for herring fishermen. The coccolithophorids had their heyday in the Cretaceous period of geological time and chalk rock is composed largely of their tiny surface scales.

Zooflagellates are best known as important parasites. Species of *Trypanosoma* in African game animals do not seem to do their hosts much harm but when transmitted by the tsetse fly to domestic cattle they cause a severe wasting disease, nagana, which is one of the principal drawbacks to stock rearing in Africa. Some varieties of *Trypanosoma brucei* can be transmitted from game (or domestic cattle) to man, causing sleeping sickness. In South America, another try-panosome, *T. cruzi*, transmitted by biting bugs (for example *Rhodnius*) causes the widespread Chagas' disease in man. A related zooflagellate, *Leishmania*, has several species responsible for disfiguring skin diseases or fatal anaemia (kala azar) in people of the tropics. The intestinal flagellate *Giardia* is the commonest cosmopolitan protozoan parasite of mankind, occasionally causing severe diarrhoea. Zooflagel-lates indirectly affecting man are those such as *Trichonympha* which live in termites and digest for their hosts the wood which these damaging insects consume.

The class Sarcodina includes the amoebae. *Amoeba proteus*, much studied in the classroom, is one of the larger species and is found in permanent bodies of water; unlike the smaller and more ubiquitous soil amoebae, it probably never forms a cyst. *Naegleria* abounds in the surface scum of stagnant water as well as in the soil; should this amoeba invade the lining of the nose during swimming it can cause fatal meningitis in man. The human dysentery amoeba, *Entamoeba histolytica*, is normally a harmless bacteria-eater in the colon but once it starts feeding on the gut wall it causes severe intestinal dis-order and loss of blood. Its cysts pass out in the faeces and infect another person through contamination of food or drinking water.

Testate amoebae secrete a shell or construct one from particles in the soil in which they live—and can avoid desiccation by retracting into the shell. The related marine foraminiferans form shells of calcium carbonate, often many chambered. Most foraminiferans feed at the bottom of the sea but some, such as *Globigerina*, float in the surface plankton. Their discarded shells sink to form the 'Globigerina ooze' which is believed to cover nearly one-third of the ocean floor. Extinct foraminiferans are among the most studied of fossils for they are much used in identifying rock strata—especially in the oil industry. Some foraminiferan shells are the main constituents of limestone rocks, perhaps the most famous being those used to build the pyramids and sphinx of Egypt which are almost wholly composed of shells of the large *Nummulites*.

The radiolarians are floating marine sarcodinans which secrete intricate skeletons of silica; they have radiating spine-like pseudopodia (axopodia) and frothy cytoplasm, both of which contribute to their buoyancy. 'Radiolarian ooze' is a marine deposit composed of their skeletons. Closely related are the sun animalcules (heliozoids) which are common in freshwater.

The all-parasitic Sporozoa have elaborate life cycles usually involving an alternation of multiple fission (schizogony) with sexual reproduction in which reproductive cells (gametes) fuse in pairs (fertilization) and the resulting zygote undergoes several fissions (sporogony) to produce infective sporozoites. In the malaria parasite, *Plasmodium*, schizogony occurs in the vertebrate host, fertilization and sporogony in the vector. *Plasmodium falciparum* is mankind's greatest killer. Malaria is controlled by killing off the *Anopheles* mosquito vector or preventing it from breeding, by people at risk taking prophy-lactic drugs (which prevent infection) or by treating patients with curative drugs such as chloroquine.

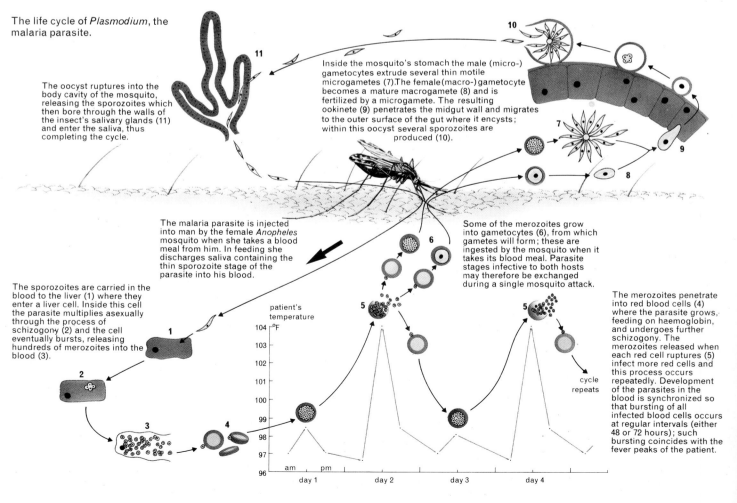

The life cycle of *Plasmodium*, the malaria parasite.

11 The oocyst ruptures into the body cavity of the mosquito, releasing the sporozoites which then bore through the walls of the insect's salivary glands (11) and enter the saliva, thus completing the cycle.

Inside the mosquito's stomach the male (micro-) gametocytes extrude several thin motile microgametes (7).The female (macro-) gametocyte becomes a mature macrogamete (8) and is fertilized by a microgamete. The resulting ookinete (9) penetrates the midgut wall and migrates to the outer surface of the gut where it encysts; within this oocyst several sporozoites are produced (10).

The malaria parasite is injected into man by the female *Anopheles* mosquito when she takes a blood meal from him. In feeding she discharges saliva containing the thin sporozoite stage of the parasite into his blood.

Some of the merozoites grow into gametocytes (6), from which gametes will form; these are ingested by the mosquito when it takes its blood meal. Parasite stages infective to both hosts may therefore be exchanged during a single mosquito attack.

The sporozoites are carried in the blood to the liver (1) where they enter a liver cell. Inside this cell the parasite multiplies asexually through the process of schizogony (2) and the cell eventually bursts, releasing hundreds of merozoites into the blood (3).

The merozoites penetrate into red blood cells (4) where the parasite grows, feeding on haemoglobin and undergoes further schizogony. The merozoites released when each red cell ruptures (5) infect more red cells and this process occurs repeatedly. Development of the parasites in the blood is synchronized so that bursting of all infected blood cells occurs at regular intervals (either 48 or 72 hours); such bursting coincides with the fever peaks of the patient.

patient's temperature

°F
104
103
102
101
100
99
98
97
96

cycle repeats

am   pm

day 1   day 2   day 3   day 4

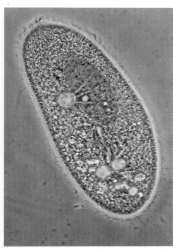

ABOVE LEFT: shells of *Elphidium* and other related foraminiferans (subclass Rhizopodea). The many chambers of the shell are formed progressively in a spiral. (×30)

ABOVE RIGHT: *Acanthometra*, in the subclass Actinopodea, has a skeleton that is made of strontium sulphate, arranged as spicules radiating from the body. (×400)

LEFT: *Paramecium caudatum*, one of the best known ciliates. (×200)

RIGHT: *Dendrocometes paradoxus* (subclass Suctorea) lives attached to appendages of the crustacean *Gammarus* and feeds on moving ciliates caught with its branched tentacles. (×300)

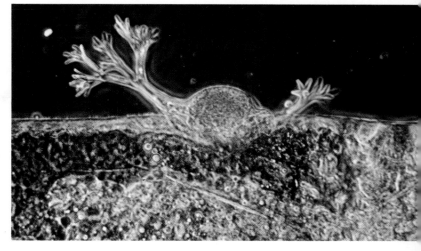

In the coccidians, for example *Eimeria*, the causative agents of coccidiosis in domestic animals, asexual and sexual reproduction take place in the gut or liver of the one host. The oocyst which forms from the zygote acquires a tough wall and passes out with the faeces. It releases its sporozoites when it arrives in the gut of another host animal. *Toxoplasma gondii* has a typical coccidian cycle in the cat, but should its sporozoites get into man or other non-feline animal, riotous asexual reproduction results in the disease toxoplasmosis. Piroplasms are sporozoans whose life cycle alternates between two hosts—vertebrate blood cells and the gut and tissues of an arthropod vector, usually a tick. The disease they cause, piroplasmosis, is an important killer of domestic stock in the tropics. The microsporans were once included in the Sporozoa but are now given a separate class; they are important pathogens of fishes and insects causing, for example, the pebrine disease of silkworms which was studied by Pasteur.

The Ciliophora—ciliates—are widespread as both free-living and parasitic protozoans. Ciliates have two nuclei—the macronucleus which is responsible for directing the synthetic activities of the individual organism and the micronucleus which generates gamete nuclei during the elaborate sexual processes induced by starvation. The commonest sexual process, conjugation, involves exchange of gamete nuclei between two apposed ciliates (conjugants), followed by formation of a zygote nucleus in each conjugant and regeneration of both types of nuclei from the zygote nucleus. Such regeneration causes both of the ex-conjugants to begin a new series of binary fissions.

Many ciliates, for example the holotrich *Paramecium*, feed entirely on bacteria; some holotrichs such as *Nassula* eat algal filaments, while others, for example *Didinium*, are voracious carnivores. Several ciliates are sedentary, for example *Vorticella* species (in the subclass Peritrichea) and produce motile 'swarmers' which serve to disperse them. The suctorians are not recognizable as ciliates in their sedentary phase for they lack cilia and feed by capturing other ciliates on their club-shaped tentacles then sucking out the contents of their prey. When they divide, however, they produce typical ciliated swarmers which swim away and settle down to produce a new adult. The large and beautiful spirotrich ciliate *Stentor* has blue, green and amethyst coloured species which can feed as attached organisms or detach and swim away to find new feeding grounds. Ciliates abound in the rumen of cud-chewing ungulates and in the colon of horses and other perissodactyls; the relationship is believed to be a symbiotic one but not so obligatory as that which exists between termites and their flagellates.

Protozoans can serve as useful indicator organisms in assessing water purity or pollution, the presence of certain species being indicative of whether oxidation or reduction of organic material is in progress. Ciliates in particular have an important role in sewage treatment, their grazing on bacteria being responsible for the production of a clear effluent. The significance of protozoan diseases in human health can be gauged from the fact that of the six most important tropical diseases listed on the World Health Organization as subjects of an intensive control programme, no less than three—malaria, trypanosomiasis and leishmaniasis—are caused by parasitic protozoans.

SUBKINDGOM Mesozoa

PHYLUM Mesozoa

Mesozoans

ORDER
Dicyemida
Orthonectida

- NUMBER OF LIVING SPECIES: between 50 and 100

- DIAGNOSTIC FEATURES: body consists of one or more central cells surrounded by a single layer of ciliated cells and lacks a gut or other body cavity

- LIFESTYLE AND HABITAT: exclusively parasitic in a range of marine invertebrates

- SIZE: generally less than 5 mm long

- FOSSIL RECORD: none

The Mesozoa is an enigmatic phylum of obscure parasites: some are rare and all are small, being generally less than 5 mm long. The name Mesozoa was coined about a century ago by van Beneden to reflect his view that the Dicyemida might represent an evolutionary stage of development between the unicellular Protozoa and the complex and multicellular Metazoa. From time to time other animal groups have been assigned to the Mesozoa, but today only two groups—the Dicyemida and the Orthonectida—with perhaps only seven genera between them are generally held to belong to the phylum. Even the rank of phylum is in some doubt, for a number of authorities consider the Mesozoa to be merely degenerate flatworms.

Dicyemids are found fairly commonly in the kidneys (nephridia) of cephalopod molluscs such as octopuses: the immature stages swim in the fluid of the host's tissues whereas the adults are attached to the spongy tissue of the kidney. An adult generally consists of only 20 to 30 cells, depending on the species, one central cell (the axial cell) being surrounded by a layer of ciliated cells sometimes known as jacket cells. At the anterior end eight or nine cells carry more numerous cilia and form the calotte or polar cap. The axial cell is concerned with reproduction and is therefore metabolically active when producing the reproductive cells, known as axoblasts. There being no gut or other body cavity, basic nutrients are presumably passed to the axial cell from the outer cells. Perhaps to aid such transport numerous microvilli extend out from the surface of the axial cell into pits formed by infoldings of the membranes of the outer cells. The electron microscope has also revealed, within the axial cell, a complicated arrangement of membranes which invests the axoblasts and the young larvae which can develop from them.

An adult dicyemid, known as a vermiform, can reproduce either asexually or sexually. During asexual development an axoblast first divides into two cells of unequal size. The smaller daughter cell divides to produce the ciliated jacket cells of the new young adult, the larger cell becoming the new axial cell. The young vermiform eventually forces its way out of the parental axial cell and completes its development in the kidney of its cephalopod host.

Sexual reproduction is more complex. In this process an axoblast undergoes a complex series of divisions to produce a gonad which has both spermatazoa and oocytes. The spermatozoa are unusual in being amoeboid, not flagellated. Spermatozoa fuse with the oocytes, usually from the same gonad, and so effect self-fertilization. Sperm can sometimes escape from a gonad, however, and can be seen lying freely within the parental axial cell and even between the jacket cells. Sperm may therefore on occasion fertilize oocytes from other gonads within the parent and even, perhaps, leave the adult

altogether and cross-fertilize oocytes of other individuals. A fertilized oocyte divides several times and develops into a small 28-cell larva known as a dispersal larva or infusoriform. These larvae escape from the parent and, leaving the cephalopod host by way of the urine, enter the sea. It was once thought that the larvae entered an intermediate host, but this now seems unlikely or, at least, unnecessary. Exactly how the larvae enter new cephalopod hosts is unknown, but they probably sink to the sea floor and await the arrival of a passing octopus or squid, whichever is the appropriate host for them. The route taken through the octopus to its kidney is also uncertain, but once there a larva eventually develops into a vermiform and so completes its life cycle.

These reproductive processes of the dicyemids do not represent a strict 'alternation of generations', for asexual reproduction does not necessarily alternate with sexual

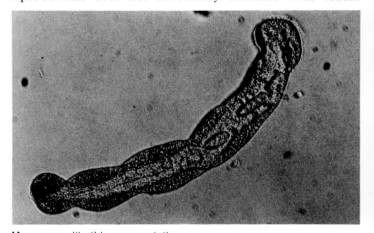

Mesozoans, like this representative of the order Dicyemida, have a very simple design. When viewed under the microscope (above) their mobile bodies seem featureless except for the many cilia. These cover all the body surface, as the drawing (right) shows. There is no trace of a gut or any other body cavity.

calotte

cilia

jacket cells

axial cell

axial cell nucleus

axoblast

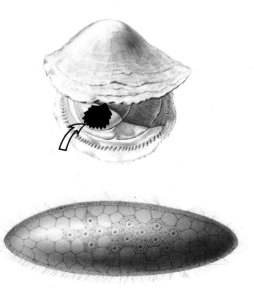

Orthonectids parasitize bivalve molluscs (above) and other hosts, either within the body cavity or in the gonad. Dicyemids, however, are more host specific; they are restricted to cephalopod molluscs such as octopuses (right), where they live within the kidneys.

reproduction in the life cycle. Recent studies indicate that in a sparse population of dicyemids asexual reproduction predominates and this results eventually in a very dense infection in the cephalopod kidney; sexual reproduction then occurs, producing larvae which leave the host to pioneer new infections in other hosts. So it would seem that any individual adult can reproduce either asexually or sexually.

The orthonectid mesozoans differ from the dicyemids in several ways. Their body tends to have a ringed appearance and they parasitize chiefly flatworms, polychaete annelids, bivalve molluscs and brittle stars, where they are found in the body cavity and in tissues such as the gonads. Their structure has recently been shown to be more complicated than was once thought, with the result that some authorities would no longer consider them to be members of the Mesozoa. During reproduction adult males and females leave the host. Spermatozoa released from the males penetrate the females and fertilize their oocytes. The resulting ciliated larvae infect new hosts and ultimately reproduce asexually by forming multinucleate plasmodia, from which a new generation of males and females develops.

Speculation has always been rife as to the evolutionary significance, if any, of the Mesozoa. Are they merely degenerate flatworms, which by right should be included in the phylum Platyhelminthes? There are certainly similarities between mesozoans and the larval stages of some species of flukes, but among parasites similarities may readily arise through parallel adaptation to similar conditions rather than indicate close relationships. One authority, nevertheless, ranks the Mesozoa as a class of the Platyhelminthes. Alternatively, the Mesozoa may have evolved directly from ancestors of the modern Protozoa, by developing a simple multicellular form.

If from protozoan stock, are the Mesozoa merely a blind evolutionary alley, like the sponges (Porifera), or do they represent an important intermediate grade of cellular organization between the unicellular Protozoa and the simpler of the multicellular Metazoa (the coelenterates and platyhelminths)? Esoteric zoological arguments have been brought to bear on this point but opinions still differ. The evolutionary history of the mysterious but fascinating Mesozoa remains somewhat obscure.

BELOW: the complicated life cycle of a dicyemid has several features characteristic of parasites in general: high fecundity, devices to ensure that sperm and oocytes come together, and a free living phase that moves from one host to another. Even if present in large numbers, successful parasites do not totally debilitate their hosts, for this would impair their own chances of survival.

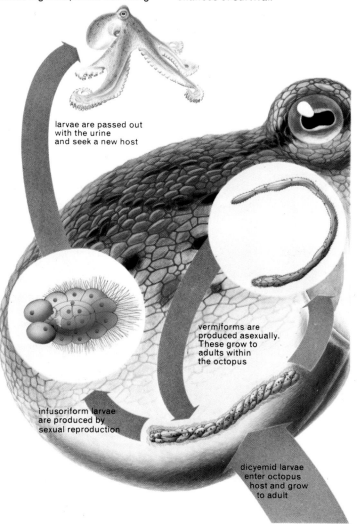

larvae are passed out with the urine and seek a new host

vermiforms are produced asexually. These grow to adults within the octopus

infusoriform larvae are produced by sexual reproduction

dicyemid larvae enter octopus host and grow to adult

SUBKINGDOM Parazoa

PHYLUM Porifera

Sponges

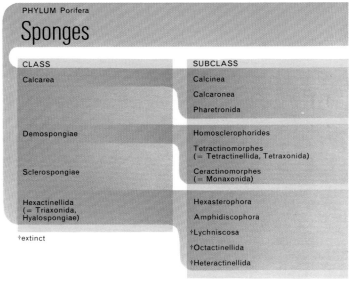

CLASS	SUBCLASS
Calcarea	Calcinea
	Calcaronea
	Pharetronida
Demospongiae	Homosclerophorides
	Tetractinomorphes (= Tetractinellida, Tetraxonida)
Sclerospongiae	Ceractinomorphes (= Monaxonida)
Hexactinellida (= Triaxonida, Hyalospongiae)	Hexasterophora
	Amphidiscophora
	†Lychniscosa
	†Octactinellida
	†Heteractinellida

†extinct

- NUMBER OF LIVING SPECIES: probably 20,000 to 30,000

- DIAGNOSTIC FEATURES: multicellular with no discrete tissues and organs but with a skeleton composed of collagen (with or without mineral elements), and a tubular water filtration system driven by specialized uniflagellate cells (choanocytes)

- LIFESTYLE AND HABITAT: free living sessile habit (stalked, encrusting or boring), mainly marine at all depths and latitudes but some occur in freshwater

- SIZE: from sheets or irregular mats of tubes of $0.25\,cm^3$ to arborescent, vase or barrel shapes of 100 litres in volume

- FOSSIL RECORD: Precambrian to Recent; dominant as reef builders within the Palaeozoic era; about 150 genera

BELOW: between the tidemarks many sponges grow as thin, encrusting mats, from which arise oscular mounds and chimneys. This green British specimen of the world-wide species *Halichondria panicea* (class Demospongiae) spreads across some 10 cm of rock. The green colour is given by single-celled algae which live amongst the sponge cells.

BELOW RIGHT: the exhalant vent or osculum of a sponge. Below the rim or 'chimney' of the osculum, major exhalant canals can be seen opening into a common atrium. By changing the shape of the oscular chimney the sponge can alter the velocity and discharge distance of the filtered, deoxygenated water. The smaller, inhalant ostia perforate the entire surface.

It is misleading to think of the sponges merely as those absorbant natural products which, until their replacement by synthetic substitutes, were so widely used industrially and domestically. Throughout the world's oceans and in lakes and rivers, ranging in form and volume from delicate sheets and tubes of less than $0.25\,cm^3$ to squat tubs of 100 litres in volume, and coloured white, orange, yellow, red, purple, blue, brown or black, live 20,000 to 30,000 species of Porifera. Of these, only about ten species provide the horny skeletons which we refer to as 'bath sponges'. In common with some other invertebrates—for example bivalve molluscs—sponges form living filter-beds. Always immobile in their adult forms and spreading themselves only by budding or growth and by minute creeping or swimming larvae, the sponges are structurally and physiologically organized for pumping water continually through themselves. From this water they extract oxygen and the finely particulate and dissolved organic material for their nourishment. By the currents that they exhale are removed carbon dioxide and other waste products.

Because sponges exhibit no discernible movement and respond only a little and slowly to even strong stimuli, they were not accepted into the animal kingdom until after 1825, when careful observation and the perfection of the light microscope revealed their all-pervading water currents and the nature of the cells which move the water and exploit its flow. The sponges, however, do not fit well within the higher multicellular animals—the subkingdom Metazoa—for their cells are not clearly arranged into layers to form permanent tissues or organs and they organize themselves from the cells of their larvae in an unique manner. For these reasons they are usually placed in a separate subkingdom—the Parazoa ('sort of animals').

Sponges have no special filtration organs. Instead the whole mass of the sponge is permeated by a three-dimensional network of water spaces and fine tubes, the aquiferous system, whose walls are formed of flattened cells, the pinacocytes. A similar single layer of flattened cells covers the outer surface of the sponge and, where this makes contact with a solid substratum, these same cells secrete, as a cement, the iodine-rich protein spongin, which is unique to the phylum. These single layers of pinacocytes have no basement membrane, which is important, for it leaves the cells free to wander as necessary. The sheets and tubes that the pinacocytes form are called pinacoderm.

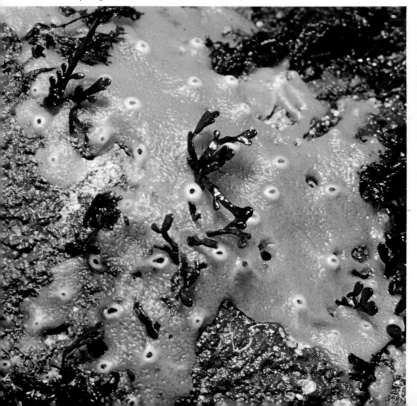

A diagram of a very small portion of the outer part of a demosponge. Water enters the sponge through the ostia and then passes into the choanocyte chambers—shown here cut in half. Between the layers of pinacocyte cells lining the aquiferous system are the many collencyte and archaeocyte cells and the spongin and silica spicules of the skeleton. Archaeocytes and their daughter-cells wander in the mesohyl, dividing, receiving food from a choanocyte, secreting and enlarging spicules, laying down spongin threads to bind spicules together, and breaking up to liberate mucilages.

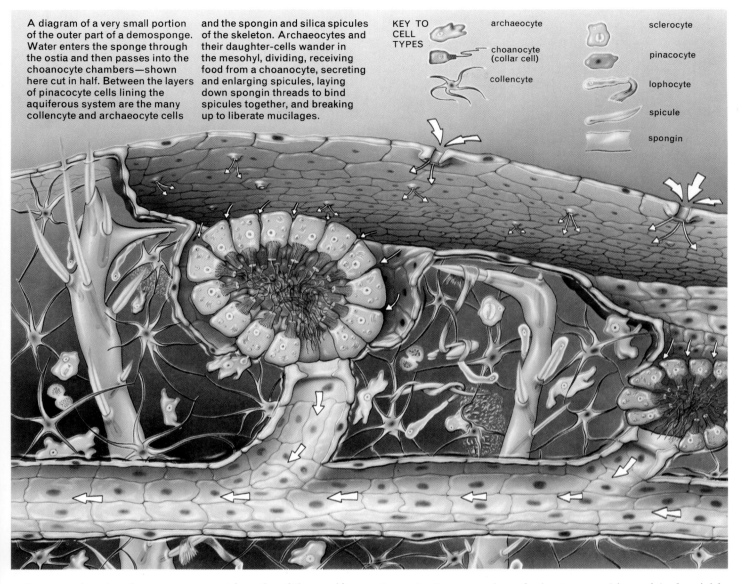

KEY TO CELL TYPES

archaeocyte

choanocyte (collar cell)

collencyte

sclerocyte

pinacocyte

lophocyte

spicule

spongin

Interrupting the often very tortuous lengths of the aquiferous tubes or canals lie the pumps—choanocyte chambers—which move the water. With cross-sectional diameters of 20 to 60 μm, these structures are ovoid, spherical, or thimble-shaped expansions of the aquiferous canals and their walls are formed of a single layer of cells, the choanocytes or collar cells. Not only do these cells drive the water currents, but they also capture with their collars of microvilli, and then ingest, the food particles of about 1 μm diameter and the large dissolved organic molecules drawn in with the inhalant currents. Although collar cells are known to occur in scattered groups in some other animals, for example, protochordates, they are best known elsewhere as the cells of some flagellated protozoans, a fact which for many years encouraged the idea that sponges are little better organized than loose groups of Protozoa.

The choanocyte chambers act as pressure pumps. The simultaneous but asynchronous beating of the single flagellum on every choanocyte tends to raise the water pressure in the chamber. As a result, water flows out of the chamber but, because one of the openings in the chamber wall is of much larger diameter than any of the others, the different frictional resistances to water flow ensure that water escapes only through the larger opening into the exhalant canals. The escaping water is replaced by unfiltered water from the inhalant system seeping through the many smaller apertures. Thus, flow through the chambers is always one-directional.

Water is drawn into the aquiferous system from the outside through a huge number of minute pores (the ostia) of variable aperture in the surface pinacoderm. Passing inward down narrowing inhalant canals it enters the choanocyte chambers, where it is filtered, and then escapes under pressure into the exhalant canals. These merge and increase in diameter in their passage from the pumping chambers to the few, large exhalant vents, the oscula. Because of this changing geometry, the walls of the exhalant canals offer little frictional resistance. The water can, however, be 'stopped down' at the rim of each osculum by an 'iris diaphragm' of cells. In this way potential energy is converted to kinetic energy and the exhaled jet transports the stale water fast and far away from the sponge. This is most important, for sponges are in continuous danger of re-inhaling their recently exhaled water and the great range of their shapes is due to different strategems for throwing stale water to a safe distance.

Small as the choanocyte chambers are, the quantities of water which can be pumped by a sponge are prodigious. One small common marine sponge, of a volume of about 4 cm³, can filter nearly 0.5 litres of water in an hour. This feat is performed by about 2.5 million choanocyte chambers, each pumping at very low pressure, but coupled into the aquiferous system in parallel. Some sponges grow to very large volumes and a specimen of the subtropical ceractinomorph genus *Verongia* of 10 litres volume is estimated to pump 360 litres of seawater per hour. The human kidneys, often cited as high-speed biological filters, can manage a filtration rate little better than 7.5 litres per hour.

From the point of view of the biology of an individual sponge and when considering the role of sponges in their environments, it is helpful to regard these animals as pumping filters. But the cells and their products which fill the spaces between the pinacoderm sheets and tubes are most unusual. To begin with, no tissues are discernible in this mesohyl, as it is termed to distinguish it from the middle body layers of other multicellular animals. Instead, there is a 'mixed community' of cells within which two major types of cells and their products can be recognized. The first type of cell—the collencytes—bear long, fine processes of cytoplasm which radiate to make contact with one another and so form a three-dimensional network. Through this network move the amoeboid cells of the other main type—the archaeocytes. As the archaeocytes move they come up against the bases of the choanocytes, from which they receive undigested or partly digested food, which they will use themselves or pass on to other cells in the mesohyl. Moving deeper within the mesohyl the archaeocytes may divide. Some of them may change into additional collencytes and, by increasing the extent of the meshwork, cause growth. Others may transform themselves into pinacocytes and become part of the pinacoderm. Others will transform themselves into sclerocytes and lay down mineral elements of the supporting skeleton. Yet others will transform into lophocytes and secrete and lay down fine trails of spongin fibres, binding together the mineral elements of the skeleton. If the need arises, an archaeocyte will divide and the daughter cells become modified into the choanocytes of a choanocyte chamber.

Pumping and filtering activities do not complete the versatility of the choanocytes. Following division, choanocytes transform to spermatozoa and are wafted out of the sponge in the exhalant current. If fortunate, they will be inhaled by another sponge, in which case they will be captured by a choanocyte in a chamber. This capturing choanocyte, after absorbing its flagellum and collar and holding the sperm in its cytoplasm, will migrate into the mesohyl, there to be engulfed by a ripe egg—itself derived from a choanocyte and swollen by the food transported to it by archaeocytes. Nourished by archaeocytes, the dividing cells of the resulting zygote form a solid or hollow multicellular larva with an outer layer of ciliated cells enclosing archaeocytes and collencytes. This larva may sometimes achieve its final form by inverting itself completely. The beating of the cilia helps the larva to be expelled in the parent's exhalant currents. After drifting or creeping for a few hours or days, the larva settles; its special organization then fails and after a brief period of apparent cellular anarchy the archaeocyte–collencyte system gains control and a tiny filtering sponge is organized. The organizing powers of the archaeocytes and collencytes are so great that if a piece of adult sponge is crushed through fine bolting silk, a small perfectly functional sponge will reorganize itself from the cell debris within a few days.

The skeletons of sponges are designed primarily to support the aquiferous system from collapsing under gravitational, tidal current and other forces. The skeleton must also maintain the form of the sponge against the small, but persistent pressures generated by the choanocyte chambers. In fact, the sponge skeleton is proof against large tension and compression forces. In most sponges it consists of mineral elements, the spicules, bound together with spongin. In this respect, sponges exploit the physical properties used by civil engineers in 'two-phase' systems such as reinforced concrete.

Rooted in the abyssal mud by tufts of long spicules, a particularly fine specimen of *Pheronema grayi* (class Hexactinellida) filters the slow ocean currents at a depth of 1,400 m.

The proportions of mineral and spongin in the skeleton vary enormously. The bath sponges and some related forms in the Dictyoceratida (an order within the Ceractinomorphes and sometimes called the 'keratose' sponges) have spongin only. By contrast, some of the Desmophorida (a tetractino-morph order) have little spongin, and cement their mineral spicules together with more mineral. Some of the Dictyo-ceratida have reverted to a 'two-phase' system during their evolution and pick up sand grains and the spicules of other,

BELOW: in Greece a batch of dictyo-ceratid bath sponges, *Euspongia officinalis*, is prepared for final drying after the cellular material has been removed by decomposi-tion, beating and washing.

RIGHT: two cleaned whole skele-tons of the Venus Flower Basket, (*Euplectella aspergillum*). In this and related hexactinellid sponges the glass spicules are fused into a rigid, frilled meshwork.

dead sponges and bind them together into thick tracts of spongin. The Dendroceratida, one of the ceractinomorph orders, do not even have spongin organized into large tracts, but only a fine network of spongin fibrils, so that they have the consistency of firm jelly.

Frequently the spicules are arranged in a three-dimensional network with dense spongin cementing the spicule ends and fine spongin fibrils running from the spicule surfaces into the mesohyl ground substance. The spicule meshwork may be highly regular, with sides of 50 to 300 µm, the length of the individual spicules. The spicules may occur in several differ-ent sizes and shapes. This range of spicules may be divisible into two size categories—the macroscleres and the micro-scleres. The microscleres, together with spongin fibrils, may then make up a smaller-scale skeleton within the coarser meshwork of the macrosclere skeleton.

The spicules consist either of crystalline calcium carbo-nate or of colloidal silica. Only in one small group of modern sponges, the Sclerospongiae, are calcareous and silicious mineral elements present in the same animal. Spicules vary enormously in size, from 4 to 5 µm up to 1 m or more in length. The calcareous spicules rarely exceed 1 mm in length and most are very much smaller. They are most often single-rayed or three- or four-rayed, are truly crystalline and, no matter how complex their shape, each appears as if carved from a single crystal. They are formed inside cavities bounded by two or six sclerocytes, elongating and thickening as these cells move apart. By contrast, silicious spicules arise inside cells, where the silica is laid down, in a vacuole, around a protein thread. Their construction may start in one place in the mesohyl and the enlarging spicules be trans-ported by cells to their final places in the skeleton.

A small selection of spicules from a variety of sponges. Some are simple geometric shapes that articulate neatly and are bound together with spongin or extra mineral to form large meshworks. Other, smaller and more irregular forms lie scattered within the larger meshwork and provide another skeleton, or else form a protective rind. Spicule types provide much information about relationships amongst sponges.

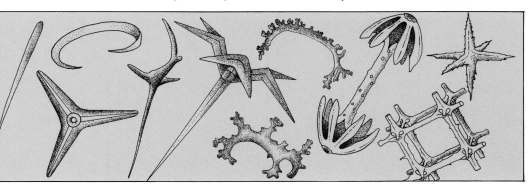

ABOVE: a group of small, cream demosponges growing amongst sea squirts in tropical waters. This dense growth reveals that, as for other sedentary animals, sponges must compete hard for substrate upon which to grow.

LEFT: a small calcareous sponge growing among bryozoans in shallow water off the New Zealand coast. Each branch bears a single osculum. The water-filtration systems of all the branches interconnect near the base.

BELOW LEFT: a demoiselle fish gaining protection from an erect, branching specimen of the demosponge *Aplysilla rosea*.

The chemical composition of the spicules of the skeleton is used to divide the Porifera into four classes. Within these classes the forms of the spicules, their patterns of arrangement into the skeleton, and the relative amount of spongin are important aids to further classification.

The class Calcarea, whose members construct only calcareous spicules, is a relatively small group of species. Because of taxonomic uncertainties, estimates of the numbers of the species range from about 60 to about 600—a fact which reflects the extraordinary morphological plasticity of sponges. The calcareous sponges are usually small and delicate, often with relatively simple aquiferous 'ducting' and are confined to the seashore and the shallower ocean waters. Their fossils first appear in the Carboniferous period.

Those sponges which never produce calcareous spicules comprise the classes Hexactinellida and Demospongiae. The Hexactinellida, whose members have essentially six-rayed silica spicules, are confined to deep, cool ocean water and to the shallower cold water near the poles. Often their larger spicules are fused into rigid basketwork of such elegance that the cleaned dried skeletons of, for example, the Venus' Flower Basket (*Euplectella*) were prized as decorative objects in Japan and in Victorian and Edwardian Europe. Their anatomy is somewhat different from that of all other sponges in that their aquiferous systems consist of perforated, folded sheets of cells rather than of discrete ducting. Often they rely on persistent ocean currents for their supply of food and oxygen and pump only at low pressure. The earliest fossil examples come from the early Cambrian period. Because of their inaccessibility, far less is known of the biology of this class than of other kinds of sponge.

The members of the class Demospongiae are the most commonly occurring of all the sponges and their species probably comprise more than 95 per cent of all extant forms. The earliest known members of the class occur as fossils in the Cambrian period. The variety in the form of their spicules provides the basis of classification within the class. The horny sponges, including the bath sponges, are demosponges, as also are the desmophorids. The species of demosponge vary in size from thin films to huge, vase-shaped forms which may exceed 100 litres in volume. They occur at all depths from the seashore to the bottoms of the deepest ocean trenches and it is they alone which have invaded freshwater successfully. They have done this several times and there is no continent (except, possibly, Antarctica) whose rivers and lakes do not have their complement of demosponge species.

Only during the past 15 years have the living members of the fourth class, the Sclerospongiae, been understood as sponges. Their skeletons consist of silica spicules lying above a thick base of calcium carbonate. Nowadays they are known only from the deeper recesses of coral reefs, but their skeletons bear strong resemblances to fossil Stromatoporoidea—a group once numerous and previously thought to have died out in the Palaeozoic era. Some authorities think that some of the stromatoporoids were ancient sclerosponges, which failed to compete successfully with the demosponges and were driven into their present-day habitats. In this they resemble a small group of the Calcarea, the subclass Pharetronida, once common but today confined largely to the harsh environments of underwater caves.

It is only very recently that the significance of sponges in the economy of the marine environment has been understood at all well. Best understood of all, perhaps, is their complex role in coral reefs. Here occur many of the demosponges of the family Clionidae (in the subclass Tetractinomorphes) whose members bore chemically into calcareous substrata, destroying dead coral skeletons and mollusc shells, and so contributing largely to the formation of calcareous sediments. Other sponges in the reef system help to bind together loose sediment, while the sclerosponges consolidate the deeper parts of the reef. All over the reef occur myriads of sponges, not only filtering the water, but also providing food for browsing molluscs, annelid worms and fishes, while all but the most poisonous species provide, in their irregular surfaces and aquiferous systems, refuges for a host of small organisms. In the deeper parts of the southern oceans and parts of the Arctic Basin sponges occur in vast masses, apparently providing a similar range of foods and dwelling places. Here demosponges and hexactinellids predominate, removing vast quantities of silica from the water into their skeletons. The spicules themselves, falling from dying sponges, form thick dense mats, which provide a special habitat for many small invertebrates. Photographs of the often barren-looking abyssal ocean plains frequently show the bottom littered with delicate sponges which belong to the ceractinomorph order Xenophyphophora, and from amongst them rise the delicate and stalked or chunky masses of hexactinellids.

In the earliest fossil-bearing rocks of the Precambrian period occur traces of strange creatures, the Archaeocyathida, whose affinities remain obscure though they most closely resemble the Calcarea. Apparently, they were organized for large-scale water filtering, and it is thought that here lie some of the earliest known ancestors of all the sponges. It seems almost as if, as long as there have been cells, so have there been sponges. From the earliest times it was a successful strategy to filter water through very flexibly organized masses of cells and it is a successful strategy now. In the past, biologists have sought in the sponges patterns of organization foreshadowing the Metazoa. Perhaps the Porifera are one of the earliest successful strategies for living—and not the less successful because conservatism was part of the strategy.

LEFT: a large goblet-shaped demosponge growing on a deep coral shelf off the Cayman Islands. Its inhalant ostia lie outside the cup while the exhalant oscula lie on the inside.

BELOW: only the tips of the inhalant and oscular papillae of the boring sponge *Cliona celata* protrude from cavities excavated within the shell of this bivalve (*Cyprina islandica*).

SUBKINGDOM Metazoa

PHYLUM Coelenterata (= Cnidaria)

Corals and other coelenterates

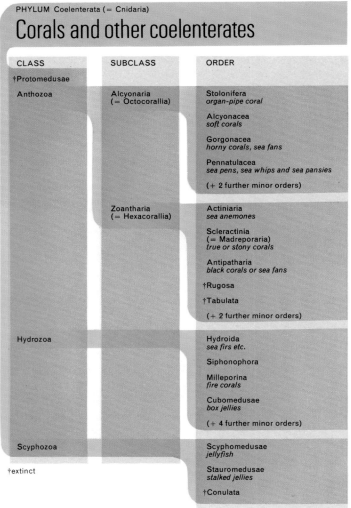

CLASS	SUBCLASS	ORDER
†Protomedusae		
Anthozoa	Alcyonaria (= Octocorallia)	Stolonifera *organ-pipe coral*
		Alcyonacea *soft corals*
		Gorgonacea *horny corals, sea fans*
		Pennatulacea *sea pens, sea whips and sea pansies*
		(+ 2 further minor orders)
	Zoantharia (= Hexacorallia)	Actiniaria *sea anemones*
		Scleractinia (= Madreporaria) *true or stony corals*
		Antipatharia *black corals or sea fans*
		†Rugosa
		†Tabulata
		(+ 2 further minor orders)
Hydrozoa		Hydroida *sea firs etc.*
		Siphonophora
		Milleporina *fire corals*
		Cubomedusae *box jellies*
		(+ 4 further minor orders)
Scyphozoa		Scyphomedusae *jellyfish*
		Stauromedusae *stalked jellies*
		†Conulata

†extinct

- NUMBER OF LIVING SPECIES: about 9,000

- DIAGNOSTIC FEATURES: acoelomate with body wall divided into two layers (diploblastic) separated by jelly-like mesoglea; primarily radially symmetrical; no anus; mouth usually surrounded by tentacles bearing stinging nematocysts

- LIFESTYLE AND HABITAT: mostly marine and free living, often sessile; a few forms are parasitic and some inhabit freshwater

- SIZE: 1 mm to 25 m but mostly 2 to 20 cm, though coral and sea fan colonies may regularly exceed 2 m

- FOSSIL RECORD: Precambrian to Recent; coral groups important as rock-formers from mid-Ordovician to Recent; about 4,000 species

The predominantly marine phylum Coelenterata contains such diverse and at first sight unrelated animals as jellyfish, hydroids (sea firs), corals, sea anemones, the Portuguese-man-of-war, sea fans and the common freshwater *Hydra*. The group is widely distributed and in the tropics corals are of particular significance as they form reefs and atolls which reach their maximum development in Indonesia, the Great Barrier Reef off the east coast of Australia and in the central west Pacific Ocean. Coelenterates comprise a very old group and include some of the earliest fossils known. Well developed jellyfish are known from the Precambrian—about 1,000 million years ago. Undoubtedly the wide range of form into which the group has diversified can be related to the great length of time over which it has evolved. Coelenterates are the simplest known animals to possess true tissues and, compared with other simple phyla such as Porifera, Ctenophora and Platyhelminthes, their success in invading a wide variety of

marine habitats is outstanding. They have not, however, made a success of either the freshwater or the parasitic ways of life, though examples of each are found within the phylum. Because some of the freshwater forms, such as *Hydra*, are highly successful, it remains a puzzle to coelenterate biologists why more species are not found in the rivers and ponds of the world.

The most remarkable feature of the phylum is that it includes both sedentary forms, called polyps, and the very different free swimming, gelatinous medusae or jellyfish. All coelenterates can, in fact, be related to one or other of these two grades of organization. In many species, both grades are represented in the life cycle—a phenomenon known as alternation of generations—but in some only the polyp grade is represented and in others only the medusa. The main advantages conferred on a species having a free swimming medusa stage are greater exploitation of planktonic food resources than can be achieved by the polyp stage, and wider dispersion of the species.

Polyp and medusa stages share certain features which characterize the phylum: they have a central mouth, which is usually surrounded by tentacles, but the gut into which it leads lacks an anus, waste products being evacuated through the mouth; the body is made up of two tissue layers—ectoderm and endoderm—which are separated by a gelatinous layer known as the mesoglea (this diploblastic arrangement of the body wall sets the coelenterates apart from all higher animals which are triploblastic, having a third tissue layer—the mesoderm—between the other two); special surface cells called cnidoblasts are present; these contain nematocysts which are used in connection with food capture and also for defence.

In those coelenterates that exhibit an alternation of generations it is the medusa generation which represents the sexual phase and the polyp stage is purely vegetative. In some hydroids the medusa remains attached to the polyp on which it develops and in a few is so degenerate that it is scarcely recognizable as a medusa, appearing more like an aberrant polyp. The polyp stage in many species can reproduce asexually by budding off miniature individuals and this is the usual method of colony extension.

Diadumene neozelanica, a typical sea anemone, seen from above the mouth. The regular arrangement of the tentacles is obscured by their close packing. Like many anemones it is brightly coloured.

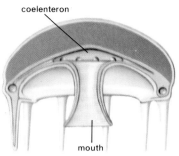

coelenteron

mouth

MEDUSA STRUCTURE

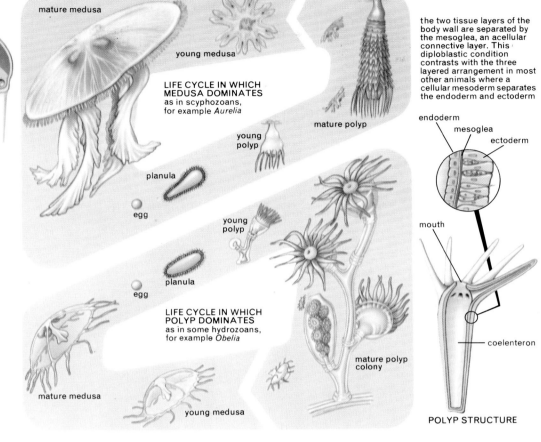

mature medusa

young medusa

LIFE CYCLE IN WHICH
MEDUSA DOMINATES
as in scyphozoans,
for example *Aurelia*

planula

egg

young
polyp

mature polyp

the two tissue layers of the
body wall are separated by
the mesoglea, an acellular
connective layer. This
diploblastic condition
contrasts with the three
layered arrangement in most
other animals where a
cellular mesoderm separates
the endoderm and ectoderm

endoderm

mesoglea

ectoderm

mouth

young
polyp

planula

egg

LIFE CYCLE IN WHICH
POLYP DOMINATES
as in some hydrozoans,
for example *Obelia*

mature polyp
colony

mature medusa

young medusa

coelenteron

POLYP STRUCTURE

Many coelenterate life cycles
show alternation of generations.
A sexual medusa gives rise to
asexual polyps which in turn
produce medusae. The two stages
are equally important, but one
or other is usually dominant in
a particular group. Both polyp
and medusa have the diploblastic
body plan. Medusae resemble
inverted polyps in which the
mesoglea has expanded to assist
in flotation. Although medusae
are soft, their fossil traces have
been found in Precambrian rocks.
The earliest known fossil polyps
come from early Cambrian
deposits but it remains unclear
which of the two body forms
arose first.

Sexual reproduction involves the production of eggs and
sperm by the medusa stage in those forms which have one,
and by the polyp stage in those in which the medusa stage is
lacking (for example all the Anthozoa). Some species are
dioecious while a few are hermaphrodite and species vary
as to whether the gametes are retained or released into the
sea. The development of the fertilized egg almost invariably
results in a small, free-swimming planula larva which typically settles, 'nose-first', and in most species develops into a
sedentary polyp. In most scyphozoans, however, the planula
metamorphoses into an attached polyp-like larva called a
scyphistoma, at the free end of which miniature medusae
form asexually and are released serially when their development is completed.

Most coelenterates are predatory carnivores and among
both sedentary and free-swimming forms there are those
which feed on quite large prey. Others, however, feed on
microscopic plankton which is sifted from the water by the
tentacles. Prey is not captured by the tentacles themselves
but by minute organelles called nematocysts which lie
embedded in the tentacular surface. Nematocysts are produced by special cells known as cnidoblasts and take a
variety of forms depending on their specific purpose. Most
types contain a long hollow thread which is shot out when
the nematocyst is stimulated by the proximity of prey. In
some types the thread has minute barbs and simply entangles
the prey, while in others it is capable of piercing the victim
and paralysing toxins are then pumped through the thread.
Once the prey is immobilized the tentacles bend over or
contract and transfer the food to the mouth.

Movement in coelenterates is restricted to simple muscular contractions. In polyps both the column and the
tentacles are capable of limited movement which either
serves to draw the stimulated region away from danger or
plays a part in feeding. Sea anemones (Actiniaria), though
sedentary, can creep slowly over the substrate. Movement
in medusae involves the tentacles but also takes the form of
jerky contractions of the whole bell, each followed by a slower
recovery stroke. The contractions propel the animal through
the water in a direction away from the mouth. All such movements are under the control of a primitive form of nervous
system, known as a nerve net. A coelenterate lacks a brain or
any other coordinating centre and the nerve net only functions
in response to localized stimuli.

A section of the tentacle of a
coelenterate showing how the
unique cnidoblasts are arranged.

Each cnidoblast contains a single
nematocyst armed with a hollow
thread. Over 30 types are known
some with threads 20 times the
capsule's length. Stimulation of
the cnidocil opens the operculum
allowing water into the thread.
The resulting changes in pressure
cause the thread to fly out,
turning inside-out along its whole
length. The cnidoblast on the
right is simplified for clarity.

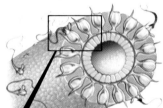

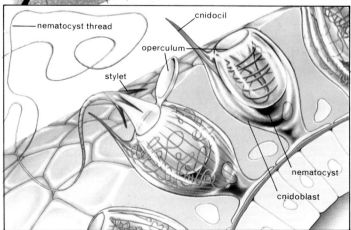

nematocyst thread

cnidocil

operculum

stylet

nematocyst

cnidoblast

Although opinions vary as to how many classes the 30 or so orders of living coelenterates represent, the most widely recognized scheme involves three—the Anthozoa, Hydrozoa and Scyphozoa. A fourth class is wholly fossil. Anthozoans are the most primitive of the major groups. All have polyps but none has a medusa. In some—the subclass Zoantharia (sea anemones, corals and related forms)—the tentacles and internal vertical partitions or mesenteries are arranged in multiples of six: in the other subclass, the Alcyonaria or Octocorallia, they are in eights. The anemones, order Actiniaria, have a more basic polyp plan than do most coelenterates. The 1,000 or so species vary but little from the familiar seashore anemones of intertidal rockpools. Some species enter an association with free living animals of different groups. Thus the clownfishes (family Pomacentridae) and anemone prawns (*Periclemenes*) find protection among anemone tentacles, from the stings of which they seem immune, and hermit crabs carry anemones on the backs of the shells they inhabit.

In general terms the polyps of the true or stony corals (order Scleractinia) each resemble an anemone, but differ in that each produces a cup-shaped skeleton of calcium carbonate to which it is fixed. Primitively, corals were probably

ABOVE: *Cerianthus* is a member of a minor zoantharian order—the Ceriantharia. Unlike true sea anemones it lacks a pedal disc and has two kinds of tentacles.

BELOW: *Clava squamata*, an athecate hydroid found intertidally on weed and rocks. Another small hydroid, *Dynamena pumila*, is also growing on the main stem of this seaweed.

solitary and this condition is retained in forms such as the cup coral, *Caryophyllia*, but in most living species many polyps aggregate to form colonies which can produce massive skeletons. When numerous—in warmer and for the most part shallow, tropical waters—these can form impressive reefs and atolls.

An organization of the polyps into conspicuous colonies has been arrived at independently by two other, at first sight similar but in fact unrelated, anthozoan groups—the orders Gorgonacea and Antipatharia. Both groups are known as sea fans, though antipatharians are also called black corals after the colour of their axial skeleton. The skeleton in both

ABOVE: a siphonophore, *Physophora hydrostatica*, from the Sargasso Sea. Several kinds of zooid make up the free living colony. (×4)

BELOW: an alcyonacean coral *Dendronephthya* from the Indian Ocean. Numerous calcareous spicules add to the rigidity of the skeleton. It is shown here about life size.

groups is chitinous and in the tissues of gorgonians there are also numerous minute calcareous spicules. Another difference is that antipatharian polyps are usually octoradiate, whereas those of gorgonians are basically hexaradiate.

The other living anthozoan groups contain fewer species than those already mentioned. The order Stolonifera includes dissimilar but all simple colonial genera such as *Tubipora* (the organ-pipe coral) which has a red skeleton and inhabits tropical reefs. Its deep red skeleton is composed of myriads of minute calcareous spicules cemented together. The familiar 'dead man's fingers' (*Alcyonium*) typifies the order Alcyonacea, the soft corals. The colony is a fleshy mass and embedded spicules support the polyps. Completely different are the sea pens, sea whips and sea pansies in the order Pennatulacea. Their colonies consist of a chitinous skeletal rod or plate along which is borne a thin tissue layer and from which project relatively fleshy polyps. In erect forms the elongated basal region is soft and worm-like and anchors the colony in muddy sea beds.

The largely shallow-water class Hydrozoa includes both fewer orders and fewer species than does the Anthozoa but is undoubtedly more varied. The major difference is the introduction in hydrozoans of a medusa stage in the life cycle—they show an alternation of generations. In most hydrozoans the medusa is the most conspicuous stage but in some the sessile polyp (hydroid) phase is equal in importance. In others the medusa has been entirely lost, while a few have lost the hydroid stage and retained only the medusa. The basic plan can be seen in the sea firs, order Hydroida. An asexual, vegetative polyp stage typically buds off single medusae (hydromedusae) from one or more special places. These free-swimming sexual medusae give rise to larvae which eventually settle in places remote from the parent and develop into new colonies or individuals.

In all hydroid colonies there are at least two forms of individual polyps or zooids with different structures and functions. The most numerous are the gastrozooids which have a mouth and tentacles, and capture and ingest prey on behalf of the whole colony. The zooids with a reproductive function—the gonozooids or gonophores—in their turn produce the medusae. In addition, some colonial hydrozoans may also have special defensive zooids—known as nematophores or dactylozooids depending on the group—located near the gastrozooids and armed with nematocysts. These different forms of zooid reach extreme specialization in the order Siphonophora.

In siphonophores the medusae are rarely released by the parent colony but remain attached and are modified for reproductive and other functions. In, for example, the infamous Portuguese man-of-war (*Physalia*) a medusa forms the bluish air bladder, up to 30 cm long, which rides high in the water and keeps the colony afloat. Beneath the float hang numerous fine tentacles which may reach 25 m in length and are armed with millions of nematocysts which can cause painful stings. *Physalia* is found in all warm seas, and where conditions are right can form huge shoals which sometimes stretch for hundreds of kilometres.

The Milleporina of warm and shallow water are like corals in general appearance but differ in that their massive calcareous skeleton is constructed of numerous closely-compacted tubules, and in that they release medusae. In the box jellies, order Cubomedusae (sometimes given full class status, as the Cubozoa), each polyp metamorphoses into a single large, square-looking medusa: there is no budding off of medusae as in most other hydrozoans. Though most box jellies are harmless, two genera, *Chironex* and *Chiropsalmus*, are the dreaded and deadly sea wasps of tropical Australian waters: a sting from only a moderate contact can cause death in two minutes.

The true jellyfish and the related stalked jellies comprise the third and smallest living class—the Scyphozoa. Most jellyfish (order Scyphomedusae) have a life cycle in which a sessile, polyp phase (termed a scyphistoma) vegetatively buds off numerous medusae from its free end. The medusae are planktonic, are often large—some species may have a bell diameter of one or more metres—and bear the gametes. Some forms of the open ocean have a modified polyp stage which remains attached to the 'parent' medusa, thus freeing the species from the need to inhabit shallow seas. By contrast, the sessile medusae of the exceptional stalked jellies (order Stauromedusae) are largely intertidal and sublittoral where they attach themselves to rock and weed. They grow to about 8 cm in height and are found in all seas.

BELOW: *Haliclystus auricula*, a stauromedusan, on the green alga *Ulva lactuca*. Though they resemble polyps, stalked jellies are sessile medusae; the polyp stage has been entirely suppressed. (×15)

RIGHT: an open ocean jellyfish, *Pelagia noctiluca*. The sessile scyphistoma stage is missing from the life cycle of this species as an adaptation to a wholly planktonic existence.

BELOW RIGHT: a small group of *Dendrophyllia* polyps. Most scleractinian corals expand only at night. During the day they withdraw completely to escape being eaten by browsing fish. (×15)

Several coelenterate groups are known only as fossils. One of the most ancient and perhaps the simplest is the class Protomedusae (Precambrian to Ordovician) which comprised jellyfish-like forms with a radially lobed body without marginal tentacles or true mouth. One of the best known examples is *Brooksella*. The scyphozoan order Conulata (middle Cambrian to early Triassic) comprised sedentary, supposedly medusoid forms with an unusual chitinous exoskeleton made up of numerous plates, in the shape of an inverted cone with a rectangular cross-section. The upper, widest end of the cone was closed by an operculum. Two other fossil orders, the Rugosa (Ordovician to Permian) and the Tabulata (Ordovician to Permian), have left no descendants although they resembled modern corals. Fossils of both groups form an important constituent of coralline limestone rock.

Modern corals likewise continue to be major rock builders: the coral reefs of today will form the limestone of tomorrow. Reefs occur widely in tropical seas and reach their greatest development in the Great Barrier Reef some 1,500 km long

off the north-east coast of Australia. Although coral reefs are mostly composed of countless coral colonies, the wave-swept seaward parts of the reefs are dominated by calcareous algae related to seaweeds which, like corals, secrete calcareous skeletons. Their contribution to reef growth, and in particular to seaward advancement of the reef front, is considerable.

The cells of many reef corals possess numerous symbiotic single-celled algae (zooxanthellae) which contribute sugar (derived from photosynthesis) and other nutrients to the coral cells. In return, by being in the coral tissue, the algae have a protected place to live and receive carbon dioxide (essential to photosynthesis) from their hosts. During the day most corals contract their tentacles and the symbionts in their tissues can thus better receive sunlight and photosynthesize, but at night the coral tentacles are extended for feeding on passing plankton. Most corals, therefore, have two distinct food sources.

Although no two reefs are the same, most conform to a general plan. On the sheltered side there is usually a reef-flat lagoon, about 5 m in depth, the bottom of which is covered with sand and broken coral fragments. Proceeding seawards one meets in the lagoon first large and then, nearer the outer edge, smaller coral colonies. The tops of the corals are usually at a common level, corresponding roughly to that of low tide. The depth of the sea bed gets less until it nearly breaks the surface at the reef edge by the open sea, but beyond this edge it increases steeply on the exposed side of the reef. Beneath the living colonies the sea bed is chiefly composed of reef limestone consisting of dead and compacted remains of corals, calcareous algae, shells and to a lesser extent the hard skeletal remains of other animals.

Despite this general conformity to a common reef plan the earliest voyagers in tropical seas must have been struck by the diversity of topographical types of coral reefs to be seen. Some coral growth simply extends outwards from the shore and forms a fringing reef. In other places reefs encircle islands, or, as in the barrier reefs, grow offshore roughly parallel with the coast, and form intervening lagoons up to 70 m deep. A third kind of reef, the atoll, is simply a ring of coral reef encompassing a lagoon. At their most impressive they are tens of kilometres across. The relation between these three, at first sight distinct, kinds of reef was first realised by the distinguished English naturalist Charles Darwin (1809–1882) at the age of 27. His theory, eventually published in 1842 and now at last recognized as both ingenious and correct, proposed that atolls typically start out as fringing reefs around a volcanic island. Growth of coral is greatest on the seaward side since here food and water-borne nutrients and oxygen are most abundant. On the landward side rain water runs off from the land and gradually, by reducing the salinity of the lagoon water and by bringing down choking silt, inhibits coral growth. Thus the reef grows outwards to become an encircling barrier reef. Darwin proposed that slow sinking of the central island would result in its eventual disappearance, leaving, as the fringing coral continued its upward and outward growth, an atoll surrounding a circular lagoon. This was an ingenious idea but, though it seemed plausibly to explain these mysterious features, it failed to gain widespread acceptability for a long time.

In 1952 one aspect of Darwin's theory was substantiated when a borehole was drilled down into the reef on Enewatak Atoll in the Marshall Islands group in the Pacific Ocean. At an incredible depth of 1,141 m the borehole passed through the coralline rock and into the volcanic rock (basalt) beneath. This basalt represented the sunken remnants of what had been a volcanic island some 50 million years previously, a time span which indicates that the rate of sinking is about 2 mm every century.

ABOVE: a coral island off the coast of New Guinea. Reef flats spread from it over a wide area, the distant breakers marking the fringes. In time the island may subside—part of the progression in atoll formation. Drifting coconuts often colonize such islets.

BELOW: part of the Great Barrier Reef off the coast of Australia, exposed at a very low spring tide; at normal low tides only the tips can be seen. Coral reefs provide homes for thousands of animal and plant species, and are one of the world's richest habitats.

SUBKINGDOM Metazoa

PHYLUM Ctenophora

Comb jellies

CLASS	ORDER
Tentaculata	Cydippida *sea gooseberries etc.*
	Lobata
	Cestida
	Platyctenea
Nuda	Beroida

- NUMBER OF LIVING SPECIES: about 90

- DIAGNOSTIC FEATURES: acoelomate with triploblastic body wall; basically radially symmetrical but with superimposed bilateral symmetry; lack nematocysts and anus usually present; eight rows of ciliated comb plates used for locomotion

- LIFESTYLE AND HABITAT: all marine and free swimming; most are planktonic but several creeping forms are known

- SIZE: 5 mm to 10 cm with one genus (*Cestum*) reaching 1.5 m

- FOSSIL RECORD: none

Ctenophora, or comb jellies, typically are planktonic animals found in coastal waters and the open ocean. Most are globular and transparent and are between 10 and 100 mm long. Almost all species are voracious carnivores, the prey animals—small crustaceans, small fish, arrow worms and larvae of various animals—typically being trapped by two long, branched tentacles armed with sticky cells called colloblasts. The tentacles can be retracted completely into deep pits when not in use. However, some ctenophores, for example *Beroe*, lack tentacles and engulf their prey by suddenly expanding their sac-like bodies while others, such as *Lampetia*, have adopted a crawling existence and browse.

Although ctenophores resemble coelenterate medusae in both general appearance and their way of life they are not closely related to coelenterates, and constitute a quite distinct phylum. Thus ctenophores, instead of having an alternation of generations between sessile and free swimming stages, simply release gametes into the sea where direct development involving a ciliated larva takes place. Whereas medusae propel themselves by vigorous contractions, ctenophores move by beating eight longitudinal rows of numerous tiny comb plates (from which their name derives) which are under the control of a nerve centre at one end. No ctenophore has nematocysts—the organelles present in all coelenterates —but the non-toxic colloblasts of the tentacles perform a broadly similar function. Furthermore, ctenophores have a three-layered organization—the triploblastic condition (endoderm, mesoderm and ectoderm)—whereas coelenterates have only two layers—the diploblastic condition (endoderm and ectoderm)—separated by a tough and jelly-like mesoglea. Some ctenophores have a fully-functioning anus, unknown in coelenterates, and the embryology of the two groups is quite different. In view of these fundamental differences it is indeed remarkable that the superficial similarities between the two groups should have led to their being classified together so frequently.

The most obvious difference between the two ctenophore classes is the presence or absence of the two or more long tentacles. Among the orders in the larger class (Tentaculata) are the most primitive and simplest of ctenophores, the Cydippida. In most of these the gelatinous, transparent body is spherical, ovoid or pear-shaped and the tentacles are carried prominently in two dorsal pits. Cosmopolitan examples are the planktonic *Pleurobrachia* and *Hormiphora*—the sea

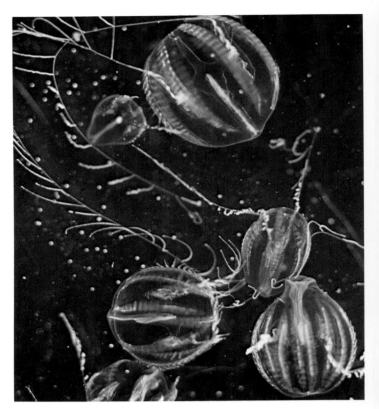

ABOVE: *Pleurobrachia* is common in coastal waters of the North Atlantic where it forms part of the diet of herring.

BELOW: the structure of *Pleurobrachia* and an enlarged view of the apical organ. The extensile tentacles can be fully withdrawn.

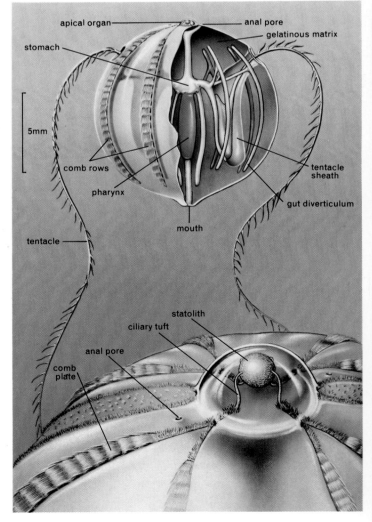

gooseberries. A few cydippids, such as *Lampetia*, have slug-like bodies and have secondarily adopted a crawling habit.

The order Lobata, which probably evolved from the Cydippida, includes *Mnemiopsis* (the sea walnut of North American beachcombers) which is about 8 cm long, and has four long, arching processes arising from the gelatinous body and two similar but shorter processes in between. Like cydippids, lobates have two or more tentacles which, although not protected in sheaths, are similarly used to catch small planktonic animals. Although frail and seemingly unsuited to the open ocean *Mnemiopsis* occurs throughout warmer parts of both Indo-Pacific and Atlantic Oceans. There can in fact be few animals as large which are so numerous.

Venus's Girdle (*Cestum veneris*) is the most widely known example of the order Cestida. Its body is absurdly wide—being expanded laterally up to 1.5 m—but is only about 1.5 cm long. The animal is therefore ribbon-shaped. The rows of comb plates run along the edges of the ribbon and, unusually among swimming ctenophores, play little part in movement, which is by muscular sinusoidal undulations of the body. The animal in fact moves perpetually sideways. Venus's Girdle is found in all warm seas and has even been reported occasionally from the Arctic.

Most ctenophores are grouped together in the order Platyctenea. Some of these have adopted a slug-like, creeping existence (although a few have retained limited swimming powers) and are compressed dorsoventrally, that is in the equatorial plane, the tentacles remaining on the upper, aboral surface and the mouth below. Like cydippids they have two tentacles which can be retracted into sheaths, but the rows of comb plates disappear once the larval stage is passed. Examples are *Coeloplana*, which hangs from and creeps along

LEFT: *Beroe* has no tentacles but is still an efficient predator, feeding by sudden expansion of its bell-shaped body and drawing in the prey by suction. There are several species in the Atlantic.

ABOVE: at a depth of 15 m, a diver in the Sargasso Sea watches the peculiar undulating motion of a Venus's Girdle. The delicate ribbon can withstand rough seas and the species is widespread.

the water surface, catching any food in its tentacles, and *Tjalfiella* and *Vallicula*, which are thought to browse.

The remaining ctenophore order, the Beroida, is sufficiently distinct to be removed to a separate class, Nuda, distinguished by lacking tentacles. A typical and worldwide example, *Beroe*, with a length of up to 5 cm, is marrow-shaped with a large mouth at one end and feeds on smaller animals which even include other planktonic ctenophores such as *Pleurobrachia*.

Almost all the free swimming ctenophores propel themselves by the rows of comb plates, which are special structures not found anywhere else in the animal kingdom (although they are paralleled by some protozoan organelles). The comb rows of *Pleurobrachia* are probably fairly typical, and are among the best known. The eight rows consist of several dozen transverse flapping plates, arranged one behind the other, each of which is composed of 100,000 or more unusually long cilia (about 0.1 mm) joined together in a close and regular arrangement. Each plate is stimulated to beat by the movement of its neighbours so that waves of beating pass down each row. The comb rows seem to be under the control of the apical sense organ situated at the end of the animal opposite to the mouth, and are connected to it by a primitive nervous system. The apical organ includes a small, solid (probably calcareous) body called a statolith which is supported in *Pleurobrachia* by four tufts of fused cilia. As the animal is pitched around in the sea the statolith is attracted by gravity first to one side of the animal and then to another and by appropriate stimulation of the comb rows the animal apparently repostures itself semi-automatically. Jellyfish, in contrast, are unable to posture themselves in this way.

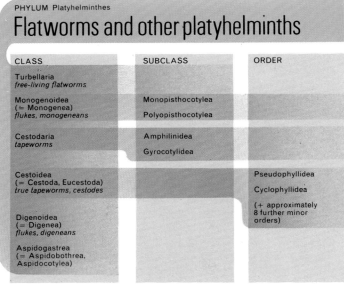

SUBKINGDOM Metazoa

PHYLUM Platyhelminthes

Flatworms and other platyhelminths

CLASS	SUBCLASS	ORDER
Turbellaria *free-living flatworms*		
Monogenoidea (≡ Monogenea) *flukes, monogeneans*	Monopisthocotylea Polyopisthocotylea	
Cestodaria *tapeworms*	Amphilinidea Gyrocotylidea	
Cestoidea (≡ Cestoda, Eucestoda) *true tapeworms, cestodes*		Pseudophyllidea Cyclophyllidea (+ approximately 8 further minor orders)
Digenoidea (≡ Digenea) *flukes, digeneans*		
Aspidogastrea (≡ Aspidobothrea, Aspidocotylea)		

- NUMBER OF LIVING SPECIES: about 10,000

- DIAGNOSTIC FEATURES: dorsoventrally flattened, simple triploblastic body lacking a coelom, anus and blood vascular system

- LIFESTYLE AND HABITAT: some free living in marine and freshwater habitats but mostly parasitic in a wide range of invertebrate and vertebrate hosts

- SIZE: usually 1 to 10 mm but some tapeworms may exceed 12 m

- FOSSIL RECORD: none substantiated

The Platyhelminthes represent the simplest grade of organization of the triploblastic Metazoa. Most platyhelminths are small and flattened, their form being determined by the ability of gases and nutrients to diffuse through the tissues since they have no circulatory system. Characteristically, the internal organs are surrounded by a packing tissue, the parenchyma, and osmoregulation is carried out by flame cells or protonephridia. Almost all platyhelminths are hermaphrodite with complex reproductive organs; the female system is divided into an ovarium, producing gametes, and a vitellarium producing yolk. Elsewhere in the animal kingdom the yolk is an integral part of the egg cytoplasm.

An idea of the ancestral turbellarian form may be reconstructed from primitive features found throughout the group today. This hypothetical type was probably a small creeping marine organism with a ciliated epidermis, epidermal muscle strands and a simple nerve net; a ventral mouth opened into a mass of nutritive cells or into a primitive digestive cavity and the body interior was packed with mesenchymal cells; gametes were formed from amoeboid cells and simply discharged through the mouth or body wall. This form has many similarities to the planula larva found in coelenterates and may perhaps have given rise to evolutionary lines to all triploblastic metazoan groups.

Present day members of the class Turbellaria are principally free living and aquatic. Freshwater forms can be found attached to weed or to the underside of stones in ponds and rivers but most turbellarians are marine. They glide over the substratum by means of their ciliated epidermis and this delicate layer also functions in respiration, so the few terrestrial forms—the land planarians such as *Bipalium kewense*—occur in damp environments.

Most turbellarians are predatory; they have a well-developed pharynx which, in many forms, can be everted through the mouth. This serves for capturing smaller organisms (protozoans and coelenterates) but equally enables some turbellarians to suck out the body contents of large animals including crustaceans and even oysters.

Turbellarians have considerable powers of regeneration: detached fragments can develop into complete individuals. This capacity also enables some forms to reproduce asexually; these simply divide by fission and each half regenerates the missing parts. The new tissue originates chiefly from undifferentiated mesenchymal cells. Similar processes enable them to withstand prolonged starvation, the worm utilizing its mesodermal tissue during starvation and regenerating it when feeding is resumed.

BELOW: most of the turbellarians are marine. They measure a few millimetres in length and feed on the smaller invertebrates which live on the sea bed.

RIGHT: terrestrial turbellarians occur naturally in humid, tropical vegetation but may be found also in similar environments such as greenhouses and botanical gardens.

At metamorphosis the tadpole loses its gills and the tiny parasites migrate over the skin of the host and enter the bladder via the cloaca.

Once established inside the bladder the parasite waits three years until the first spawning of the frog before laying eggs itself.

A swimming larva hatches and infects a host tadpole, becoming attached to its gills.

The eggs of both parasite and host are laid at the same time each spring.

The adult parasite is attached to the frog's bladder by the suckers and hooks of its posterior disc.

Although turbellarians are predominantly free living, some live in close association with other animals. Some, for example, are commensal on the bodies of echinoderms, molluscs and crustaceans. Their prey includes fellow commensals and water-borne organisms such as protozoans which are particularly abundant on the surface of the partner's body, but this diet is otherwise little different from that of free living members of the class. One group of turbellarians, the temnocephalans, are all commensals, principally on freshwater crayfish and prawns, and are found in South America, Australia and New Zealand, Madagascar and the Balkans. These isolated regions were originally linked in the supercontinent of Pangaea during the early Mesozoic era when the ancestral crustaceans were evolving. The present geographical distribution may therefore represent evidence of a very ancient association between flatworm and crustacean partners which must already have been established before separation of the land masses by continental drift. Yet despite its long history, the commensal association has not, in

ABOVE: the life histories of the monogenean *Polystoma integerrimum* and its frog host are exactly synchronized. Both mature only when three years old and then lay their eggs during the same few weeks each spring. This ensures that the parasite larvae hatch soon after the tadpoles emerge from the spawn. The larvae attach to the gills of the tadpoles and eventually migrate to the urinary bladder of the adult frog host.

RIGHT: *Temnocephala brenesi* lives ectocommensally on the gills of freshwater shrimps in Central America. Temnocephalans have a posterior disc for attachment to the crustacean partner and an anterior ring of tentacles used to capture free living prey and force it into the mouth. (×10)

regions communicating directly with the exterior. Life cycles are typically simple, involving only one host, and transmission is generally achieved by an actively swimming larva, the oncomiracidium. This mode of infection restricts potential hosts to those which are aquatic or semiaquatic. Monogeneans are characterized by an elaborate attachment organ at the hind end—the opisthaptor—which is variously equipped with suckers, clamps, hooks and anchors to maintain the parasites on their exposed habitats. Subdivision of the class reflects the principal modes of life: those with a relatively simple opisthaptor (the monopisthocotyleans) generally live on the skin of fish and have a diet of host epidermal cells and mucus; the polyopisthoctyleans feed on blood and have a complicated haptor specially designed for more enclosed but no less precarious sites—the gills of fish, the urinary bladder of frogs, toads and aquatic tortoises.

The polyopisthocotylean *Polystoma integerrimum*, which occurs in the urinary bladder of the European Frog *Rana temporaria*, provides a most remarkable illustration of the close ecological and physiological adaptation of a parasite to its host. The entire reproductive processes of *Polystoma* and *Rana* are synchronized, which results in the parasite laying its eggs only when the host spawns. So close is the coordination that it has been suggested that the parasite's reproduction is controlled by the sex hormones of the frog.

this group, developed into parasitism. In some other turbellarian groups, some species have become totally parasitic but few show real structural modification towards this way of life. Turbellarian ancestors, however, are considered to have given rise, in the distant past, to all the other platyhelminth classes which are represented exclusively by parasites.

The Monogenoidea are parasitic principally on coldblooded vertebrates and occur on external surfaces or in

The Cestodaria and Cestoda totally lack an alimentary canal and absorption of food takes place over the general body surface; physiologically they may be considered to be turned inside out, for their external surface functions in digestive uptake rather like the intestine of other animals. Only soluble foods can be absorbed and the parasites usually occur in nutrient-rich regions of the host's body, principally the alimentary canal.

Cestodarians are an isolated group, parasitic in primitive fish; their elongated bodies are not divided into segments, as in the cestodes, and there is only one set of reproductive organs. *Amphilina* lives in the body cavity of sturgeons (*Acipenser*) and is equipped with a powerful anterior proboscis with which it can bore through the host's tissues. *Gyrocotyle* occurs in the intestine of rabbit fish (chimaeras) and attaches by a much-folded posterior organ, the rosette. These cestodarians are thought to be survivors of a once-flourishing group which parasitized ancient fishes, perhaps in the early Mesozoic era. In several respects cestodarians seem to be intermediate between monogeneans and cestodes.

The Cestoda, or true tapeworms, are generally parasitic as adults in vertebrates. Almost exclusively their life cycles require one or sometimes two intermediate hosts and transmission tends to be passive: the egg or larva being eaten by the intermediate host which is, in turn, eaten by the final host in which the parasite matures. The tapeworm attaches to the host's intestinal wall with a head or scolex equipped with a range of sucker-like structures and sometimes hooks. In almost all cestodes, the tape-like body (strobila) is composed of a series of segments or proglottids; these are budded off from a region behind the head and each contains a complete set of reproductive organs; as they are displaced backwards first the male and then the female systems become functional and finally the terminal segments become packed with eggs. This replication of reproductive units leads to an enormous increase in biotic potential. The rat tapeworm *Hymenolepis diminuta* (order Cyclophyllidea), for example, may produce 250,000 eggs per day for as long as its host lives, a potential total production of 100 million individuals. If all these succeeded in infecting hosts and reached maturity, they would amount to more than 20 metric tons of tapeworm strobilae. It is quite clear, however, that most of the offspring perish before achieving their aim, and this reflects the hazards of the parasitic way of life.

The two principal cestode groups, Pseudophyllidea and Cyclophyllidea, are adapted respectively for transmission in aquatic and in terrestrial ecotypes. Species of the pseudophyllidean *Diphyllobothrium* infect fish-eating birds and mammals, including man. The eggs are released from gravid proglottids and pass out with the host's faeces and they hatch in water to release a small ciliated larva, the coracidium. This lives for about 12 hours and if, in this time, it is eaten by a copepod crustacean the larva bores into the body cavity. When a copepod containing an infective larva (the procercoid) is eaten by a fish, the parasite is released, penetrates into this host's tissues and settles down to await transfer to a predator of the fish. The parasite (now a plerocercoid) is released in the bird or mammal's intestine as its previous host is digested and it attaches to the gut wall and grows to maturity.

BELOW LEFT: the head, or scolex, of *Taenia solium* is armed with four muscular suckers and also a double ring of long and short hooks for firm attachment to the intestine of its host—Man. This species is prevalent wherever undercooked pork is eaten, but is absent from Jewish and Moslem societies. (prepared specimen × 60)

BELOW: the 'fish' tapeworm, *Diphyllobothrium latum*, can grow to 12 m in length. It infects Man in fishing communities in the Arctic, the Baltic and the Great Lakes of North America. Several animals, including cats, dogs and bears, act as reservoirs for the disease, which may then be transferred to Man via fish hosts.

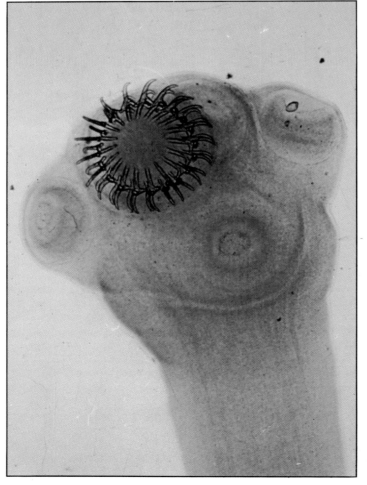

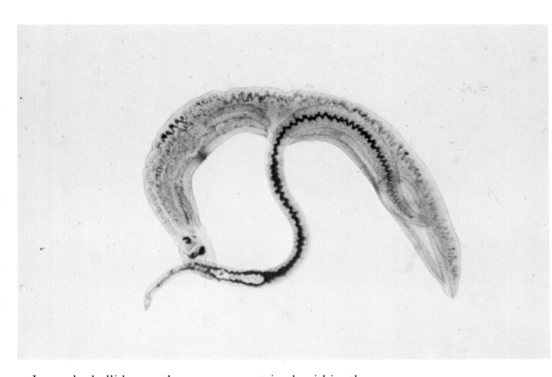

LEFT: the body form of *Schistosoma mansoni* is almost cylindrical; this is an adaptation to its very specialized tubular habitat—the blood vessels. The schistosomes are also exceptional among platyhelminths in that they have separate sexes; the mature female lives permanently within a groove in the ventral surface of the male's body. In this microscope preparation the female has become partially separated from the male. This species is found in Africa and parts of the New World tropics. (× 18)

BELOW: *Clonorchis sinensis* (class Digenoidea), the Oriental Liver Fluke of Man, is largely confined to the Orient and until recently the infection rate in some areas of Japan was 35 per cent of the population. Particular sources of infection are traditional dishes that include raw fish dipped in vinegar or sauce. (prepared specimen, stained to show its internal structure, × 16)

In cyclophyllideans, the eggs are retained within the gravid proglottid which is passed out intact with the host's faeces; additional protection for the eggs is provided by tough embryonal membranes. In the case of the human 'beef' tapeworm *Taenia saginata*, eggs ingested by cattle hatch in the duodenum, the larvae (oncospheres) bore through the gut wall and enter the blood circulation; they are then distributed around the body and settle in cysts within the muscles. If insufficiently-cooked infected beef is eaten by man, the larva (now a cysticercus) emerges in the intestine, attaches and develops to maturity.

Some of the most complicated platyhelminth life cycles are found in the Digenoidea with up to six larval stages intervening between successive adult generations. There may be several intermediate hosts but the first of these is almost always a mollusc, and most digeneans mature in a vertebrate final host. Two free-swimming penetrative stages may occur —a miracidium which infects the mollusc and a cercaria which is released from it—but there are many variations including completely terrestrial life cycles.

Most adult digeneans occur in the alimentary canal and associated organs of their vertebrate hosts. Attachment is typically by two simple muscular suckers, usually one surrounding the anterior mouth and the other situated on the ventral surface. The diet of digeneans is varied and includes epithelial tissue of the host, mucus, blood, and even protozoans which also inhabit the host's gut. There are three groups of digeneans, however, which live actually inside the host's blood vessels. The best known of these are the schistosomes, agents of the serious disease bilharzia (schistosomiasis) in man. In addition to feeding on blood cells, *Schistosoma* can absorb soluble nutrients from the blood circulation through the surface of its body. Bilharzia has an ancient history: mummies taken from the pyramids of the twentieth Egyptian dynasty (3,000 years ago) have been found infected with it and the disease is still prevalent throughout the region to this day.

The Aspidogastrea is a small but distinct group, with only about 30 species which have some affinities with the Digenoidea. They are characterized by an enormous attachment organ composed of a series of sucker-like alveoli which covers the greater part of the ventral surface. Some are ectoparasites, but chiefly the group is endoparasitic in a range of hosts, including molluscs, fish and turtles.

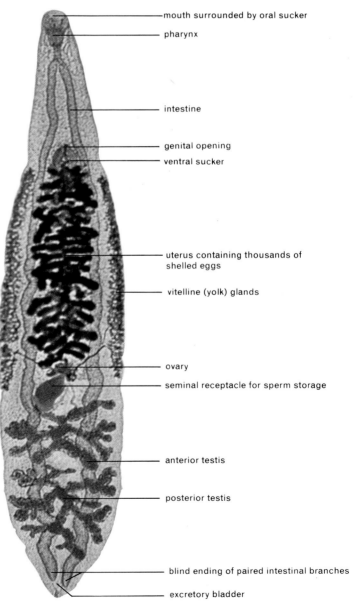

mouth surrounded by oral sucker

pharynx

intestine

genital opening

ventral sucker

uterus containing thousands of shelled eggs

vitelline (yolk) glands

ovary

seminal receptacle for sperm storage

anterior testis

posterior testis

blind ending of paired intestinal branches

excretory bladder

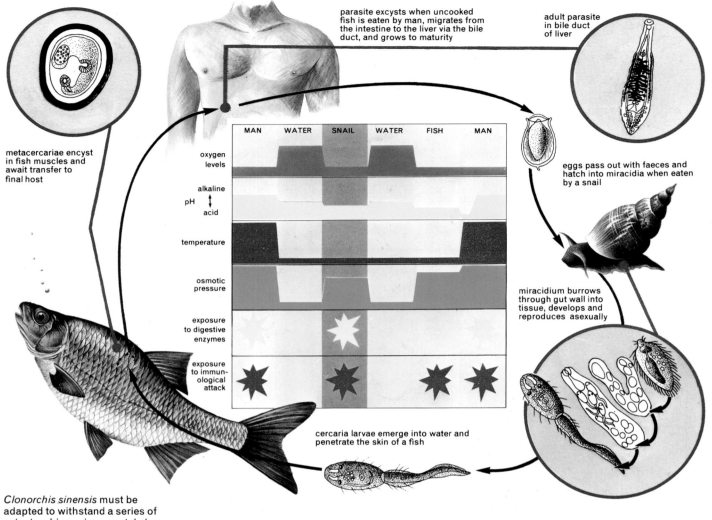

parasite excysts when uncooked fish is eaten by man, migrates from the intestine to the liver via the bile duct, and grows to maturity

adult parasite in bile duct of liver

metacercariae encyst in fish muscles and await transfer to final host

eggs pass out with faeces and hatch into miracidia when eaten by a snail

miracidium burrows through gut wall into tissue, develops and reproduces asexually

cercaria larvae emerge into water and penetrate the skin of a fish

MAN	WATER	SNAIL	WATER	FISH	MAN

oxygen levels

alkaline
pH
acid

temperature

osmotic pressure

exposure to digestive enzymes

exposure to immunological attack

Clonorchis sinensis must be adapted to withstand a series of catastrophic environmental changes in the course of its life cycle.

The Platyhelminthes are delicate soft-bodied animals and have left no fossil record, so assessment of their interrelationships is based largely on comparative morphology, life cycles and larval characteristics. The parasitic groups are generally considered to have originated from turbellarian ancestors. Older classifications linked monogeneans and digeneans together as the class Trematoda, but it is now recognized that there is a closer relationship between monogeneans, cestodarians and cestodes than between either of these and the digeneans. Monogeneans are almost exclusively vertebrate parasites; the tapeworms too were primitively parasites of vertebrates and incorporated arthropod intermediate hosts into their life cycles as a later, secondary feature. On the other hand, the ancestral digeneans were probably parasites of molluscs, and the vertebrate hosts, in which the parasites now mature, were a later addition to facilitate completion of the life cycle. The adoption of the parasitic way of life has, however, involved such profound modifications that these evolutionary relationships are largely a matter for speculation.

It was once common for parasites to be considered degenerate animals, but in fact the reverse is true: the ability to live on or in the body of another animal requires considerable specialization and adaptation. Parasitic platyhelminths with complicated life cycles must pass through a sequence of profoundly different environments, and the change from one to the next is often accompanied by what would be for free living animals a series of catastrophic shocks. For instance, the Oriental Liver Fluke, *Clonorchis sinensis*, inhabits the tissues successively of a mammal, a snail and a fish,

and two intervening stages are adapted for a free swimming life in water. These environments have vastly different conditions of temperature, acidity, salt concentration and so on, and the parasite is exposed to attack by digestive enzymes and by the host's immunological defences.

For some parasites, major problems in their lives are easily identified. Monogeneans, principally ectoparasites of fish, are continuously subjected to powerful water currents and their adaptations involve a bewildering array of attachment organs. For some digeneans, the problems are more subtle but just as ingeniously countered. The bilharzia parasite, *Schistosoma mansoni*, lives in the blood vessels which carry dissolved nutrients from the intestine to the liver. Here it is supplied with all its physiological needs. It is bathed in an inexhaustible supply of food, oxygen is provided and all its wastes are transported away by the circulating blood. But, being in the blood stream, the parasite is in one of the most hostile environments imaginable: fully exposed to attack by the host's immune system. Normally any foreign body in the blood is destroyed by macrophage cells and by antibodies. However, schistosome parasites are known to live and reproduce for more than 20 years in the human bloodstream. Recent research has shown that within a few weeks of entering the mammalian body, *Schistosoma mansoni* coats itself with a layer of proteins which are indistinguishable from those of the host itself. In man, the parasite has a disguise of human proteins; in a monkey, a disguise of monkey proteins. The host then regards the digenean as its own tissue and the parasite lives happily in this state of deception.

Once safely in or on its host's body, the parasite generally has an unlimited food supply and an environment more or less regulated for it by the host. The most fundamental problem for the parasite is finding that safe habitat—the host—in the first place. The enormous difficulties of successful transmission are reflected in the vast reproductive output of most parasites. Some parasites use a sequence of hosts, each one eaten by the next in the line and, at the same time, transferring the parasite. Some parasite stages, however, are free living and must actively locate, recognize and penetrate or attach to a potential host. Most platyhelminth larvae are small, only 0.1 to 0.5 mm long, and short-lived, swimming in water for only 12 to 18 hours. They could never hope to locate and overtake a moving host and so infection might be restricted to chance collisions.

A moment's reflection would suggest that locating one particular species of fish host in the open sea would be an almost impossible task for a monogenean larva. Recent research has shown that the larva of *Entobdella soleae*, a monogenean parasitic on the skin of the Sole (*Solea solea*), has an internal clock and this regulates the hatching of the eggs to the first three hours of each period of daylight. The Sole is nocturnal and rests on the sea bed during the day, so the parasite has the maximum period in which to locate the host while it is stationary. The related *E. hippoglossi* infects the Halibut (*Hippoglossus hippoglossus*) which is diurnally active and rests on the seabed at night. The clock of this species is set so that its larvae hatch at the start of darkness. Location and identification of the host depend on chemical recognition of the specific mucus of the fish by the swimming larva. One other monogenean, *Acanthocotyle lobianchi*, has dispensed with a swimming larval stage and the difficult task of actively searching for the fish hosts (rays, *Raja* spp.); its non-ciliated larva hatches from the egg capsule only in response to the chemical stimulus of the particular host sitting more or less on top of it. Within seconds of recognizing the mucus, the larvae emerge and directly attach. These larvae have great patience—and food reserves—and can wait in a fully-prepared state for three months for the right fish to land nearby.

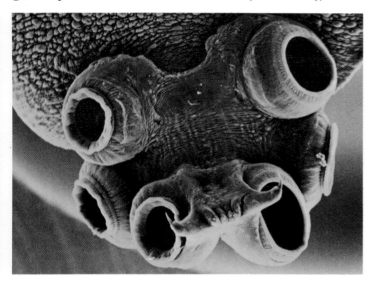

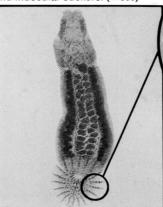

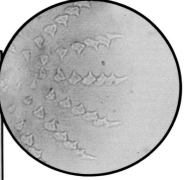

LEFT: monogeneans inhabiting soft, flat surfaces, such as *Protopolystoma xenopodis* in the bladder of the toad *Xenopus*, attach by hooks and muscular suckers. (×800)

LEFT: suckers cannot adhere readily to the rough skin of cartilaginous fish; the haptor of *Acanthocotyle lobianchi* carries radial rows of spines which engage on the sharp scales of its host (a ray). (×20)

RIGHT: monogeneans living on fish gill filaments employ pincer-like clamps for attachment. In *Diclidophora merlangi*, the gill fluke of Whiting, there are four pairs of clamps borne on peduncles. (×100)

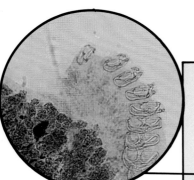

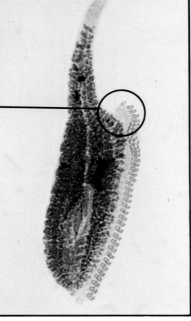

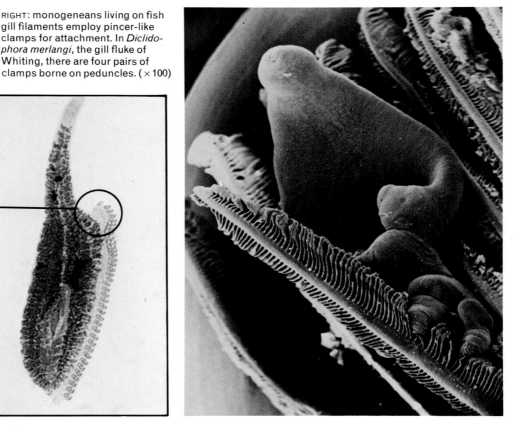

RIGHT: gill flukes demonstrate a close adaptation to the conditions of their microhabitat. They are subjected to a continuous unilateral current of water flowing over the gill lamellae and some have become permanently asymmetrical. *Gastrocotyle trachuri* anchors firmly to the gills of its host with clamps developed along one side of its body—that facing upstream; the rest of its body lies downstream, offering least resistance to the water current. (×30)

SUBKINGDOM Metazoa

PHYLUM Nemertea (= Rhynchocoela)

Ribbon worms

CLASS	ORDER	SUBORDER
Anopla	Palaeonemertea	
	Heteronemertea	
Enopla	Hoplonemertea	Monostilifera
		Polystilifera
	Bdellonemertea	

- NUMBER OF LIVING SPECIES: about 800

- DIAGNOSTIC FEATURES: unsegmented and acoelomate, with a through gut, a closed blood vascular system and an eversible muscular proboscis situated in an enclosed fluid-filled chamber —the rhynchocoel

- LIFESTYLE AND HABITAT: mostly marine but a few freshwater and terrestrial forms; predominantly free living but a few are parasitic or commensal in invertebrates

- SIZE: a few millimetres to 30 m, but mostly less than 50 cm

- FOSSIL RECORD: unsubstantiated—Cambrian and Silurian

The popular names of bootlace worms or ribbon worms aptly describe the general shape of most members of the phylum Nemertea. Apart from the true deep-water pelagic species and unusual forms such as the commensal bdellonemertean *Malacobdella*, nemerteans are typically benthic marine animals with soft slender bodies capable of extreme contraction and elongation. This makes them ideally suited for squeezing beneath embedded boulders and stones or into rock crevices, burrowing into soft mud, sand or gravel or creeping among the holdfasts and fronds of algae. The size of individual species is quite variable; they range from a few millimetres in length up to several metres, although most are probably less than about 50 cm long and only 3 or 4 mm wide. The largest known species is the heteronemertean *Lineus longissimus*, which may attain a length of 30 m or more even though it is no thicker than a pencil. Some species live in tubes, lining them with their own mucus, or inhabit the abandoned burrows of other invertebrates.

A medium-sized specimen of *Lineus longissimus*—an intertidal and sub-littoral European species, common on rocky shores. When uncovered by a receding tide it has this typical tangled appearance.

Nemerteans are characterized as unsegmented, bilaterally symmetrical, acoelomate invertebrates, with a gut which has a separate mouth and anus, a closed blood vascular system, and an eversible muscular proboscis situated above the gut in an enclosed tubular and fluid-filled chamber, the rhynchocoel. It is the possession of a through gut and blood vascular system which sets the nemerteans apart from the turbellarian platyhelminths to which they are clearly related.

The Bdellonemertea and the polystiliferous Hoplonemertea have flattened bodies quite unlike those of the more conventional benthic species. The bathypelagic Polystilifera live entirely in the oceanic depths and individuals have been found in hauls made from as deep as 4,000 m. *Malacobdella* is leech-like in form and has a posterior ventral sucker for attaching itself to the mantle lining of its bivalve host.

Nemerteans are found all over the world. Most forms are free living in marine coastal or littoral localities and a few species, such as the heteronemerteans *Baseodiscus delineatus* and *Lineus ruber*, have a quite cosmopolitan distribution. Other nemerteans, however, have a much more restricted geographical range and may be known from only single localities. Some unusual habitats are occupied, particularly

A species of the palaeonemertean genus *Tubulanus* in shallow water off the Australian coast. The banded pattern is characteristic not only of many tubulanid forms but also of many other species.

The proboscis is used to catch living prey. In this lineid nemertean the everted proboscis has ensnared a polychaete worm which will then be swallowed whole.

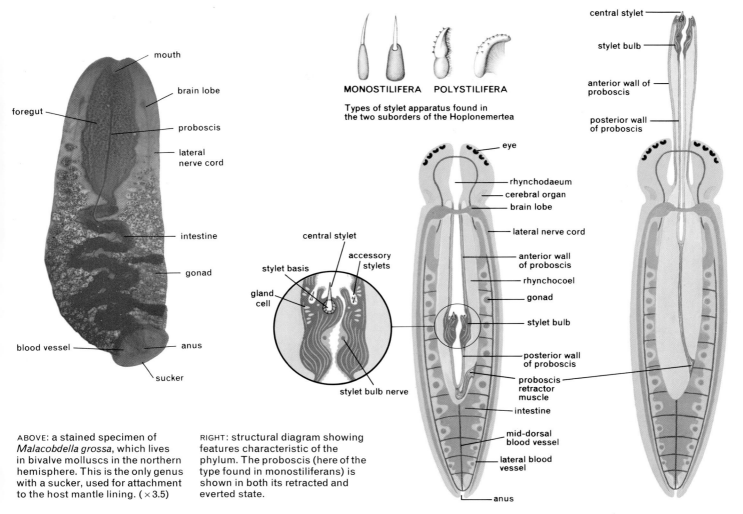

MONOSTILIFERA POLYSTILIFERA

Types of stylet apparatus found in
the two suborders of the Hoplonemertea

ABOVE: a stained specimen of
Malacobdella grossa, which lives
in bivalve molluscs in the northern
hemisphere. This is the only genus
with a sucker, used for attachment
to the host mantle lining. (×3.5)

RIGHT: structural diagram showing
features characteristic of the
phylum. The proboscis (here of the
type found in monostiliferans) is
shown in both its retracted and
everted state.

by the monostiliferous Hoplonemertea; some genera, such as *Potamonemertes* from New Zealand, inhabit freshwater and species of *Geonemertes* are found on land, up trees or above high water mark on Atlantic, Pacific and Indian Ocean islands; *Carcinonemertes* parasitizes crabs and *Gononemertes* lives within the atrium of sea squirts and is known only from Scandinavia and Australia.

Many nemertean species are more or less uniformly coloured in various shades—several of the pelagic forms are brilliantly coloured in red, orange or yellow; paler individuals often appear almost translucent. Other species are strikingly patterned with stripes, bands, speckles or contrasting colours arranged into precise geometric patterns.

Movement in nemerteans involves the use of both the epidermal cilia and the body-wall musculature. Ciliary locomotion is used by larvae, juveniles and small-sized adults. Larger nemerteans use muscular activity and during locomotion the body shows great variation in its diameter and shape. A few heteronemertean species can swim actively in an eel-like fashion. Bathypelagic hoplonemerteans generally move only sluggishly, being able to float freely with little expenditure of muscular effort because of their low specific gravity and broad, flattened shape.

The classification of Nemertea is based on internal morphology and can only be determined by histological studies. The major taxa are distinguished by the position of their nervous system, within the body wall layers (Anopla) or internal to the main musculature (Enopla), and the organization of their proboscis. The arrangement of the proboscis apparatus, as it appears in the Nemertea, is not found anywhere else in the animal kingdom. All members of the phylum possess a proboscis, although in some species its construction is simplified. In anoplans it is uniformly constructed

and either unarmed or provided with large numbers of small barbs, whereas in the enoplans three regions are distinguishable, the middle portion housing the stylet apparatus bearing either a single large (Monostilifera) or many small (Polystilifera) sharp, needle-like stylets. Mostly the proboscis is undivided, but in three heteronemertean genera (*Gorgonorhynchus*, *Panorhynchus*, *Polybrachiorhynchus*) it is complexly branched. The proboscis may be very short but in most species it is longer than the body and, when retracted, lies loosely coiled up in the rhynchocoel.

The proboscis is used principally for capturing live prey but may, as in some monostiliferan hoplonemerteans, be used for rapid locomotion. Secretions produced by the proboscis may be sticky or poisonous; sticky substances help the organ to grip, whereas toxins are injected through wounds inflicted by the stylet apparatus to paralyse or kill the prey. After being captured, the food is released by the proboscis and swallowed for digestion. Nemerteans are mostly carnivorous or scavenging and feed on a wide variety of invertebrates, often ingesting the prey whole.

Most nemerteans have separate sexes. Fertilization is usually external, but some species are hermaphrodite and a few are ovoviviparous with internal fertilization. In palaeo-, hoplo- and bdellonemerteans development of the fertilized egg is direct, with a small worm emerging from the egg, but in heteronemerteans development is indirect and involves an intermediate larval stage that in some species is retained within the egg membrane and is never pelagic but in others is, for a while, free swimming. In both cases the larva eventually metamorphoses to form a miniature adult.

A few nemertean-like fossils are known from as far back as the Cambrian period but their identification is far from certain and beyond this the phylum has no fossil record.

SUBKINGDOM Metazoa

PHYLUM Aschelminthes

Roundworms and other aschelminths

CLASS	SUBCLASS	ORDER
Rotifera (= Rotatoria) *wheel animalcules*		Seisonacea (= Seisonidea)
		Bdelloidea
		Monogononta
Gastrotricha		Macrodaysyoidea
		Chaetonotoidea
Kinorhyncha (= Echinodera)		
Priapulida		
Nematoda *roundworms or eelworms*	Adenophorea (= Aphasmidia)	Trichosyringida
		Enoplida
		(+ 12 further orders)
	Secernentea (= Phasmidia)	Strongylida
		Ascaridida
		Spirudida
		(+ 3 further orders)
Nematomorpha (= Gordiacea) *gordian or horsehair worms*		Gordioidea
		Nectonematoidea

- NUMBER OF LIVING SPECIES: about 17,000

- DIAGNOSTIC FEATURES: mostly worm-like and unsegmented, with an external cuticle, a pseudocoelom between body wall and gut, an anus and complex nervous system but no circulatory system

- LIFESTYLE AND HABITAT: mostly free living bottom dwellers in marine and freshwater environments and in the soil; many nematodes are parasitic on plants and animals

- SIZE: 0.04mm to 1m but mostly between 0.1 and 2mm

- FOSSIL RECORD: sparse, but some nematodes with their insect hosts are known from Eocene and early Oligocene; probably fewer than 10 species

The phylum Aschelminthes contains several miscellaneous groups of animals, the classification of which has always been controversial. The name is derived from the Greek *ascos* (cavity) and *helminth* (worm), and refers to the presence of a body cavity in these worm-like animals. The body cavity lies between the body wall and the digestive tract, has no inner mesodermal lining and embryologically is not a true coelom. It is called a pseudocoelom and is peculiar to the phylum. Most aschelminths are unsegmented, have a body wall consisting of a non-chitinous cuticle, an epidermis, and a layer of muscle which varies in amount and structure from group to group. They also, with few exceptions, have a complete, straight, digestive system consisting of mouth, pharynx or oesophagus, intestine, rectum and anus. The excretory system consists of protonephridia in all groups except the nematodes and nematomorphs. There is no circulatory system, and specialized respiratory organs are found only in the priapulids. The nervous system consists of a circumpharyngeal ring from which longitudinal nerves extend along the body wall.

At first sight, the six classes of the Aschelminthes appear to be completely unrelated, and many scientists prefer to regard them as separate phyla. This is especially so for the nematodes which are, numerically, the largest and the most important class of aschelminths. Some trichosyringid nematodes have been found associated with their insect hosts in Rhine lignite (Eocene) and Baltic amber and some free living nematodes have also been found in Baltic amber (early

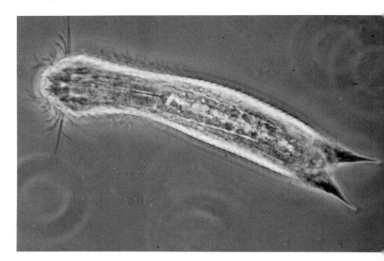

Lepidoderma (Chaetonotoidea) is a gastrotrich which lives in water. In size and habits it resembles some ciliate protozoans. Adhesive tubes in the forked tail are used for attachment. (× 600)

Oligocene), but otherwise there is no fossil record of any members of the phylum.

The class Rotifera comprises transparent, microscopic, aquatic animals (0.04 to 2mm long), which are often mistaken for protozoans. They are attractive little animals which abound throughout the world in freshwater habitats such as lakes, ponds, puddles, rainwater gutters, and the leaf axils of mosses; a few species live in salt water. Most of the 1,500 or so species are free living and solitary, but some form colonies and others have become parasitic.

A typical rotifer, such as *Philodina roseola* in the order Bdelloidea, has an anterior head, an elongated trunk, and a narrow, tail-like posterior foot, which is usually moveable and often ends in two toes. The head bears a crown of cilia on a retractable disc; this is called the corona and is double in some species. The beating cilia give the appearance of a wheel spinning, and give rise to the popular name of wheel animalcules. The cilia draw food, such as microorganisms, towards the mouth, which passes it into a muscular pharynx —the mastax—that contains cuticular teeth and jaws. Certain predatory rotifers project long teeth through the mouth to grasp protozoans, rotifers, and other microscopic animals. The beating cilia on the corona also bring about locomotion. Red eye spots are present in some species.

In a typical life history, females give rise to other females for most of the year by parthenogenetically producing unfertilized eggs. During the sexual season, or when the environment deteriorates, some females produce a smaller egg that develops into a male while other females produce an egg that must be fertilized. The males inseminate these females by injecting sperm through the body wall. The fertilized eggs do not hatch until conditions are favourable. They can withstand desiccation and are readily dispersed by wind. Only females hatch from these eggs. Adults in habitats liable to dry up can also withstand desiccation. They do this by contracting into a small, shrivelled ball which loses most of its moisture and enters a state of cryptobiosis ('hidden life'), when their metabolism almost ceases. They can remain in this state of suspended animation for years, but rapidly imbibe water when it appears and resume their normal activities.

Gastrotricha are often found in the same habitats as rotifers and resemble them in many ways. They are microscopic (0.1 to 1.5mm long), worm-like, and are common bottom dwellers among debris in fresh and salt water. The 200 or so species have a worldwide distribution. A typical gastrotrich, such as *Chaetonotus maximus* (in the order Chaetonotoidea) has a rounded head, an elongated neck and, usually, a forked tail containing adhesive tubes. Adhesive tubes are a characteristic feature of gastrotrichs—some species have as many as 250 tubes projecting from their body surface. The ventral surface of the animal bears bands of cilia which bring about a gliding motion. The arched, upper surface is usually covered with spines or scales. Gastrotrichs feed on single-celled algae, bacteria, protozoans and other microscopic animals, by means of a sucking action of the pharynx.

The Macrodasyoidea are hermaphrodite, whereas the Chaetonotoidea reproduce parthenogenetically, males being unknown. The eggs hatch to release young that resemble the adult, and maturity is reached in about three days. The Chaetonotoidea are mainly found in freshwater and can produce two types of egg, one of which hatches immediately, the other being dormant and able to withstand desiccation.

Kinorhyncha are yellowish, microscopic (less than 1 mm long), marine worms that live in mud or sand, where they feed on microorganisms. The spiny body is superficially divided into 13 or 14 'segments' or zonites. They creep about by digging their head spines into the mud, contracting the body towards the head, then digging in the tail spines to prevent backward slip. The body is then extended forwards and the cycle repeated. The sexes are separate but similar in appearance; sexual reproduction occurs throughout the year, and the eggs hatch into larvae that moult several times before reaching the adult stage. Most of the 100 species have been found on European coasts, but they probably have a worldwide distribution.

Priapulida are yellow or brown marine worms, about 8 cm long. There are two genera, *Priapulus* and *Halicryptus*, both of which inhabit mud or sand in shallow coastal waters of the more northern and more southern countries of the world. Their spined, eversible proboscis is used to seize prey such as polychaetes. They are unusual in the Aschelminthes in possessing one or two branched respiratory appendages at their posterior end. The sexes are separate, eggs and sperm being shed into the sea, where fertilization occurs. The egg hatches to produce a rotifer-like larva that eventually moults into a juvenile; this continues to grow and moult until the adult is produced.

Representative members of the six aschelminth classes. There is great diversity of form and size, not only within the phylum but also in each class, as shown by the two rotifers illustrated. Most nematomorphs lack the bristles shown on *Nectonema* (in the order Nectonematoidea) and are easily mistaken for nematodes.

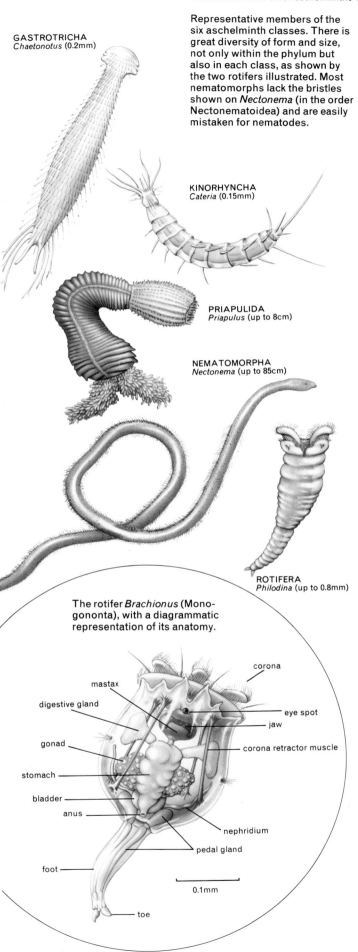

GASTROTRICHA
Chaetonotus (0.2mm)

KINORHYNCHA
Cateria (0.15mm)

PRIAPULIDA
Priapulus (up to 8cm)

NEMATOMORPHA
Nectonema (up to 85cm)

ROTIFERA
Philodina (up to 0.8mm)

NEMATODA
Strongyloides (2mm)

The rotifer *Brachionus* (Monogononta), with a diagrammatic representation of its anatomy.

corona

mastax

digestive gland

eye spot

jaw

gonad

corona retractor muscle

stomach

bladder

anus

nephridium

pedal gland

foot

0.1mm

toe

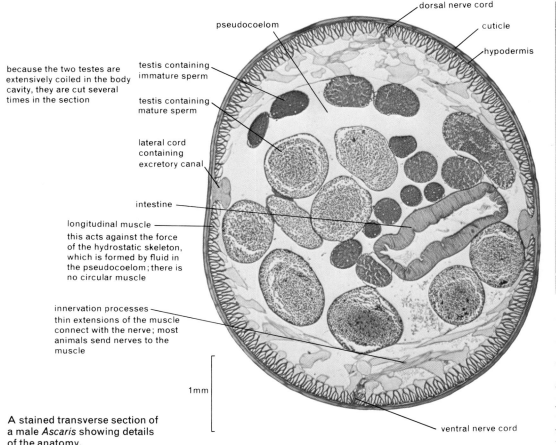

pseudocoelom

because the two testes are extensively coiled in the body cavity, they are cut several times in the section

testis containing immature sperm

testis containing mature sperm

lateral cord containing excretory canal

intestine

longitudinal muscle

this acts against the force of the hydrostatic skeleton, which is formed by fluid in the pseudocoelom; there is no circular muscle

innervation processes

thin extensions of the muscle connect with the nerve; most animals send nerves to the muscle

dorsal nerve cord

cuticle

hypodermis

1mm

ventral nerve cord

A stained transverse section of a male *Ascaris* showing details of the anatomy.

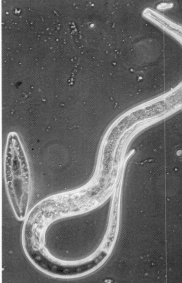

ABOVE: *Rhabdolaimus* is an adenophorean nematode which lives in soil and freshwater. Free living nematodes are found in nearly all environments; they occur in soil, marine mud, the bottom of lakes and rivers, in deserts and in hot springs. Some are able to survive for years in a desiccated state. The boat-shaped object touching the worm is a diatom—a simple form of plant. (× 150)

Nematoda (known variously as roundworms, eelworms, threadworms and pinworms), are spindle-shaped and unsegmented, mostly microscopic in size, but some animal-parasitic species can be large, for example, the Guinea Worm (*Dracunculus*) in the order Spirudida can reach 1 m long. This group of animals rivals the insects in their distribution and in the numbers of individuals—there are thousands of species. Enormous numbers of free living nematodes inhabit marine and freshwater mud and soil, several million being found per square metre in the top few centimetres of sublittoral mud. Numerous species are parasitic on plants and animals and they are one of the most important groups of animals that are parasitic on man, his crops and his livestock.

Nematodes move by undulatory propulsion, the longitudinal muscles of the body wall acting against the cuticle and the hydrostatic skeleton in the pseudocoelom. The muscles are unusual in that they send innervation processes to the nerve. Nematodes lack true cilia, and have amoeboid sperm. Fertilization is internal, and the female usually lays shelled eggs, the shell being very resistant in some ascaridids (for example *Ascaris*). The egg hatches and releases a larva (juvenile) which must undergo four moults before reaching maturity. Hatching may occur immediately or may be delayed until suitable conditions appear, one or more moults occurring in the egg. In many animal-parasitic species the egg must be eaten by a suitable host before it will hatch. In some groups the free living, third-stage larva retains the cuticle of the previous stage as a protective sheath, this being lost when the larva is transferred to a suitable environment, such as fresh dung for some free living species, or the alimentary tract of a suitable host for certain animal-parasitic species.

Most free living nematodes are found in the subclass Adenophorea but some animal-parasitic and plant-parasitic species belong to this group, notably the pork trichina worm (*Trichinella spiralis* in the order Enoplida) which infects man and other mammals if they eat raw or undercooked meat containing the larva of this nematode. Most of the parasitic species belong to the subclass Secernentea. Free living species feed upon a wide range of materials, such as bacteria, in which case they have a relatively simple buccal cavity, or on prey such as other nematodes or rotifers, when they have an enlarged buccal cavity containing teeth or a stylet. Animal-parasitic species may feed on gut contents, tissue fluids or tissues, and their buccal cavity is appropriately modified. The strongylid hookworms *Necator* and *Ancylostoma*, for example, feed upon tissues and blood and have large buccal cavities containing teeth and jaws, whereas fluid-feeding species have simple mouthparts. Plant-parasitic species possess a mouth stylet which is used to pierce plant cells and to withdraw the cell contents. They cause enormous economic losses to crops throughout the world. Some species transmit virus diseases to plants. Nematodes also attack insects, and some species are being developed as biological control agents.

The life cycles of animal-parasitic nematodes are often complex and may involve the use of an intermediate host or an insect vector, as is the case with the spirurids that cause river blindness (*Onchocerca*) or elephantiasis (*Wuchereria*) in man. Human infections are commonest in warm climates, and several species cause major diseases (for example, hookworm) in these regions. The large roundworm of man (*Ascaris*) infects about a quarter of the world's population, but its success is not surprising because a single female can lay as many as 200,000 eggs a day, and these eggs can remain viable in soil for years.

The 80 species of Nematomorpha, or gordian worms, are closely related to nematodes. They can reach 100 cm in length and are often found coiled and tied in knots. They are also called horsehair worms because they are often seen wriggling in horsetroughs, ponds and streams, giving rise to the old belief that they were horse hairs which had come to life in the water. The Nectonematoidea are marine, and their

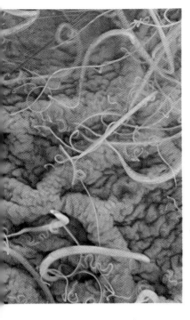

ABOVE: whipworms, *Trichuris trichuira* (Enoplida), attacking the intestinal lining of the host. The slender portion of each worm is the anterior end which bores into the tissues.

RIGHT: an adult nematomorph, *Gordius* (Gordioidea), emerging from its dead host—a spider. The larva is parasitic but the adult is free living.

larvae are parasitic in crustaceans. Adults belonging to the Gordioidea lay their eggs in water, usually attached to plants, and these hatch to release a larva if they are eaten by an insect such as a grasshopper or certain beetles. The larva feeds on

the body fluids and tissues of the living insect, taking weeks or months to develop to the adult stage. The adults emerge when the weakened insect falls into the water and they do not feed.

The Aschelminthes are a diverse phylum of invertebrates which display many differences in morphology and habit. None of the groups is clearly derived from another. They have all, however, reached a similar level of organization in having a body cavity in the form of a pseudocoelom and a through gut, and for this reason the phylum can be considered a 'grouping of convenience'.

Scanning electron micrographs of the heads of two nematodes with widely differing diets. On the left is the strongylid hookworm *Ancylostoma*, a widespread parasite of man in the tropics; the mouth bears large, curved teeth for feeding on the gut wall of its host (× 1,000). The soil-dwelling adenophorean *Rhabditis* bears a smaller, six-lipped mouth for feeding on bacteria. (× 10,000)

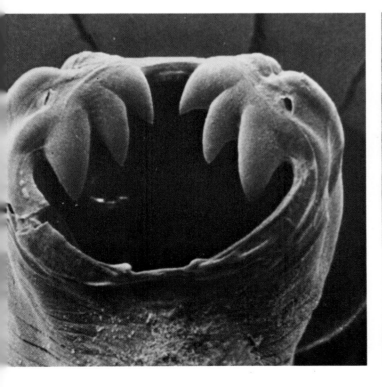

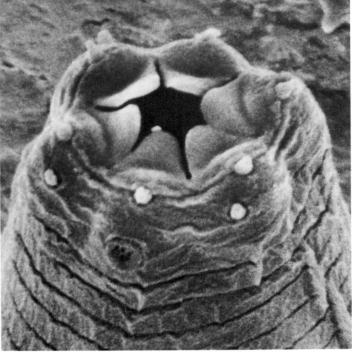

SUBKINGDOM Metazoa

PHYLUM Acanthocephala

Thorny-headed worms

CLASS	ORDER
Palaeacanthocephala	Echinorhynchida
	Polymorphida
Archiacanthocephala	Oligacanthorhynchida
	Gigantorhynchida
Eoacanthocephala	Gyracanthocephala
	Neoacanthocephala

- NUMBER OF LIVING SPECIES: about 750

- DIAGNOSTIC FEATURES: pseudocoelomate worms, which lack a mouth and gut and possess an anterior eversible proboscis armed with hooks

- LIFESTYLE AND HABITATS: cosmopolitan gut parasites of vertebrates utilizing arthropods as intermediate hosts

- SIZE: 1 mm to 1 m, but mostly 5 to 20 mm

- FOSSIL RECORD: one species from the middle Cambrian

ABOVE: acanthocephalans have an eversible proboscis that is armed with numerous backward-pointing hooks. This is the feature from which the name thorny-headed worms is derived. The number and arrangement of the hooks varies between the species. (prepared specimen, ×80)

BELOW: a prepared specimen of *Filicollis anatis*. The presence of ovarian balls shows that this is a young female. The female of this species is unusual in that it has a swollen proboscis and a long neck which passes through the muscle layers of the host's gut wall. (×8)

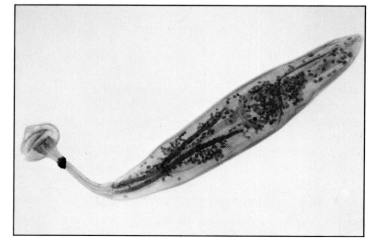

The phylum Acanthocephala—thorny-headed worms—is composed exclusively of parasites which inhabit the gut of all the major vertebrate groups (with the exception of the cartilaginous fish) but are especially common in bony fish and birds. Acanthocephalans are described as pseudocoelomate—their body cavity is not a true coelom—are usually less than 20 mm long and are characterized by the presence of a short anterior proboscis and by the absence of a gut. The proboscis is armed with hooks, hence the name Acanthocephala, which means 'thorny head'.

The proboscis is eversible—it can be turned inside out—and for purposes of attachment it buries itself deep into the gut wall of the host. The retraction of the proboscis into the anterior end of the body is controlled by a proboscis sac to which are attached retractor muscles. In the absence of a gut, food is absorbed through the body wall, which is a syncytium composed of several layers, the outer one being adapted to absorb nutrients from the gut contents of the host. In the body wall are interconnecting canals, the lacunar system, filled with fluid. Respiration seems to be anaerobic, and excretion and limited osmoregulation probably occur through the body wall, although one small group of acanthocephalans (the Oligacanthorhynchida) possesses primitive excretory organs called protonephridia. The body cavity contains little but the reproductive system.

The Acanthocephala are not closely related to any living group, but are generally considered to be very primitive and distantly related to the Aschelminthes. Their early origin is demonstrated by mid-Cambrian fossil evidence—that of a worm called *Ottoia* from British Columbia—which indicates that forms, possibly free living, occurred before the evolution of vertebrates.

Acanthocephalans are dioecious, the female normally being larger than the male. The gonads develop in a membranous structure called a ligament sac (sometimes there are two) which links the proboscis sac with the terminal genital pore. The most obvious parts of the male reproductive system are the two testes, the cement glands and the copulatory bursa. During mating the eversible copulatory bursa grips the female and after transfer of sperm the cement glands produce a cement which plugs the vagina. In young females the ovary fragments into ovarian balls which float in the ligament sac and produce the eggs.

Within the egg a larva called an acanthor develops, and when the egg, after release from the vertebrate host, is eaten by a suitable arthropod this larva bores through the gut wall of its new host using hooks on its anterior end. Within the arthropod's body cavity the larva begins to absorb food and grow. Now called an acanthella, it continues to develop until it reaches a stage, the cystacanth, at which it can infect the vertebrate host. A cystacanth is somewhat like a miniature adult, although it is often enclosed in a cyst wall. The final host, for example, a fish or a bird, usually becomes infected by feeding directly upon the arthropod host; but in many cases other vertebrates may act as paratenic hosts, that is, hosts in which the parasite can survive, but not develop further.

The classification of the Acanthocephala is a matter of disagreement because of different interpretation of the taxonomic value of certain morphological characters. Using features, such as the arrangement of the longitudinal vessels of the lacunar system in the body wall and the nature of the cement glands, the group has, however, generally been divided into three classes. At a lower taxonomic level, the number and arrangement of hooks on the proboscis are the major features used to distinguish the species.

The Palaeacanthocephala are of variable size and usually parasitic in aquatic vertebrates. Their principal features are that the main longitudinal lacunar canals are lateral, the nuclei in the body wall are fragmented and numerous, and the cement glands (usually six or less) are separate and have fragmented nuclei. The echinorhynchidans, such as *Echinorhynchus* from marine and freshwater bony fish, occur in cold-blooded aquatic vertebrates and use crustaceans as intermediate hosts: the polymorphidans, such as *Polymorphus* from ducks, occur mainly in aquatic birds and mammals, and use crustaceans or insects as intermediate hosts.

The Archiacanthocephala are usually large forms which parasitize terrestrial birds and mammals, and use insects or myriapods as intermediate hosts. Their principal features include dorsal or dorsal and ventral main longitudinal lacunar canals, few nuclei in the body wall and separate cement glands (usually eight) which possess giant nuclei. Examples of this group are the gigantorhynchidan *Moniliformis*, which is especially common in insectivorous mammals, and the oligacanthorhynchidan *Oligacanthorhynchus*, which is parasitic in birds, especially birds of prey, and often utilizes snakes as paratenic hosts.

The Eoacanthocephala are small forms which live in cold-blooded aquatic vertebrates, especially bony fish, and use crustaceans as intermediate hosts. Their principal features are that the main longitudinal lacunar canals are dorsal and ventral, there are a few giant nuclei in the body wall and the cement glands are a syncytium with several nuclei.

Few species of acanthocephalans have any veterinary importance although the polymorphids *Polymorphus* and *Filicollis*, parasites of ducks, often caused emaciation and sometimes death, especially in cases where worms rupture the host's gut wall. As humans are not generally in the habit of eating raw arthropods, or even raw vertebrates which may serve as paratenic hosts, acanthocephalans have little medical importance. Several worms, such as the gigantorhynchidans *Moniliformis* and *Macracanthorhynchus* are, however, occasionally recorded from man. They are usually acquired by accidentally eating beetles. It is worth noting, however, that acanthocephalan eggs, possibly *Moniliformis*, have been found in fossilized human faeces from about 1800 BC in Utah, USA. In the nineteenth century an Italian scientist bravely infected himself with *Moniliformis* and suffered great pain, delirium, diarrhoea and a 'ringing in the ears'!

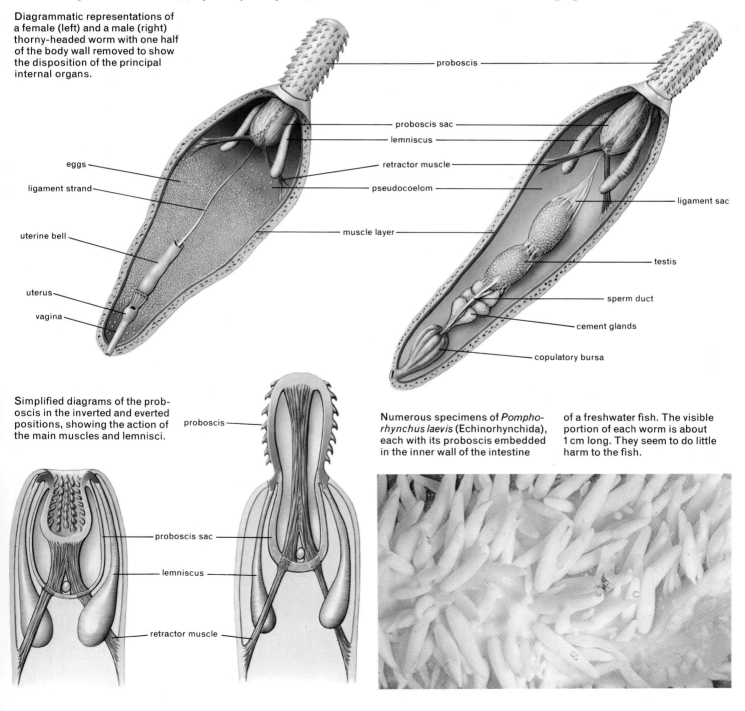

Diagrammatic representations of a female (left) and a male (right) thorny-headed worm with one half of the body wall removed to show the disposition of the principal internal organs.

proboscis
proboscis sac
lemniscus
retractor muscle
pseudocoelom
muscle layer
ligament sac
testis
sperm duct
cement glands
copulatory bursa

eggs
ligament strand
uterine bell
uterus
vagina

Simplified diagrams of the proboscis in the inverted and everted positions, showing the action of the main muscles and lemnisci.

proboscis
proboscis sac
lemniscus
retractor muscle

Numerous specimens of *Pomphorhynchus laevis* (Echinorhynchida), each with its proboscis embedded in the inner wall of the intestine of a freshwater fish. The visible portion of each worm is about 1 cm long. They seem to do little harm to the fish.

SUBKINGDOM Metazoa

PHYLUM Entoprocta

Entoprocts

FAMILY	GENUS
Pedicellinidae	Pedicellina
	Myosoma
	Barentsia
	(+ 3 further minor genera)
Urnatellidae	Urnatella
Loxosomatidae	Loxosoma
	Loxocalyx
	Loxosomella
	Loxomespilon
	Loxostemma

- NUMBER OF LIVING SPECIES: about 130

- DIAGNOSTIC FEATURES: stalked, pseudocoelomate zooids with a calyx bearing a ring of retractile tentacles, a U-shaped gut with both mouth and anus opening within the tentacular circle, paired nephridia, but no vascular system

- LIFESTYLE AND HABITATS: mostly marine but one freshwater genus; all sessile, attached to hard substrates or commensal on other marine invertebrates

- SIZE: individual zooids 0.5 to 4 mm high, but most less than 1 mm

- FOSSIL RECORD: none

The phylum Entoprocta consists of inconspicuous sessile animals, mostly less than 1 mm long, which are rarely found without careful searching. With the exception of one genus they are all marine and inhabit shallow, inshore waters of most seas. Although most species are known from Europe and the North Atlantic, this simply reflects the lack of study they have received elsewhere in the world, and without doubt many tropical and subtropical species have yet to be

ABOVE: *Loxosomella* reproduces asexually by budding new zooids directly from the calyx. At an advanced stage of development the buds detach and grow separately.

ABOVE: the growth form of three entoproct genera. These varied forms are adapted to different environments, in particular, the local density of food particles.

discovered. Individual entoprocts—zooïds—may be solitary, though aggregated in a mass, or they may be colonial, with all the zooids linked by a slender, encrusting stolon. Two colonial genera, *Pedicellina* and *Barentsia*, are often found between the tidemarks attached to hydroid coelenterates, bryozoans and sponges, or encrusting shells and stones; but most entoprocts are commensal and are found only by examining suitable invertebrate host animals.

The Entoprocta were formerly classified with the Bryozoa, which they resemble superficially, in a group called the Polyzoa. For convenience, entoprocts are now usually grouped with other pseudocoelomate phyla such as Aschelminthes but seem to have little in common with any of them. No fossil entoprocts are known, but the group is thought to have diverged long ago from the common stock of some other invertebrate phyla (known collectively as the protostomes) which have a similar embryological development.

The body wall of an entoproct zooid is soft and flexible and consists of a thin outer cuticle and an epidermis that are often reinforced by a thin inner layer of muscle which enables it to perform limited nodding movements. Each zooid consists of an erect stalk, or peduncle, which bears a cup-shaped structure known as the calyx at its free end. The calyx bears a rim of 6 to 40 short, thick, ciliated tentacles which are partly retractile and can fold over the top of the calyx so that they lie protected within a muscular flap of the body wall which links their bases. The viscera are contained in the calyx in a cavity or pseudocoelom which is filled with a viscous fluid containing a few mesenchyme cells. The gut is U-shaped, and the tentacle ring encloses both the mouth and the anus, the latter usually on a papilla with a cavity, the atrium, between it and the mouth. It is this position of the anus, within the circle of tentacles, which gives the phylum its name. In this respect, entoprocts differ from bryozoans in which the anus is situated outside the tentacular ring. A double nerve ganglion lies immediately above the stomach. Paired nephridia discharge into the atrium, but there are no special respiratory organs and no circulatory system.

Entoprocts are filter feeders; water currents are drawn between the bases of the tentacles and pass out through the opening of the calyx. Food particles are trapped by lateral cilia and pass down the inner edge of the tentacles, entangled in mucus, to a basal ciliated groove which leads to the mouth. There are no clear rejection paths and unwanted particles are simply flicked into the exhalant current of water.

Barentsia gracilis zooids attached to the elytra of the polychaete worm *Aphrodita*. Each zooid is seldom longer than 1.5 mm.

Reproductive processes are poorly understood, but hermaphrodite and dioecious species are known. The gonads are paired but their ducts join before opening into the atrium. Fertilization is thought to be internal and the eggs, as they develop into larvae within the protection of the atrium, obtain nutrients from the parent. When the larvae are eventually released they are ciliated and resemble trochophores. Colonial species may also reproduce asexually by budding (setting free) new zooids from the stolon, but in solitary forms daughter zooids bud directly from the calyx.

The three families of Entoprocta are distinguished by the form of the zooid and habit. The Pedicellinidae (for example, *Pedicellina* and *Barentsia*) are all colonial and have a calyx which is separated from the peduncle by a diaphragm and is freely shed, either in response to unfavourable conditions or perhaps when preyed on by small, grazing invertebrates. The calyx is, however, renewable and colonies may be found with calyces in all stages of regeneration.

Another family, Urnatellidae, contains the only freshwater genus, *Urnatella*, which is represented by one species in both the eastern United States and western Europe and another in India. The stolon of *Urnatella* is reduced to a small discoid attachment, new zooids frequently bud from peduncles and colonies are often dense and luxuriant. The calyx is shed readily and most colonies show a preponderance of headless peduncles among their numbers.

BELOW: the calyx of *Pedicellina cernua* with (inset) a zooid. The atrial muscles, shown incomplete on the right, close the atrium in which embryos develop, linked to their parent by a placental structure. In monoecious species the posterior gonad on each side is a testis, and the anterior gonad is an ovary.

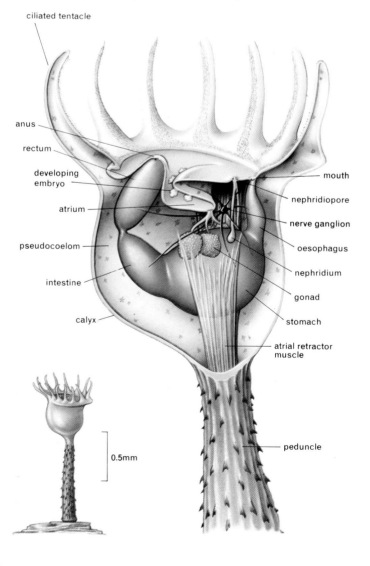

ciliated tentacle

anus

rectum

developing embryo

atrium

pseudocoelom

intestine

calyx

0.5mm

mouth

nephridiopore

nerve ganglion

oesophagus

nephridium

gonad

stomach

atrial retractor muscle

peduncle

ABOVE: a zooid of *Pedicellina cernua* in lateral view. The tentacles are extended for feeding, and the stomach, which forms the main part of the alimentary canal, appears as a dark mass.

The Loxosomatidae (which contains about 100 of the 130 species in the phylum) are all solitary; the peduncle is often short and has a broad base which incorporates a muscular, sucking disc or a cement gland. Forms such as *Loxosoma* with a sucking disc can re-attach themselves if dislodged and may even move. In those with a basal cement gland (for example, *Loxosomella*) the zooid is permanently secured to the substratum. Loxosomes, unlike other entoprocts, have a continuous stalk and calyx, and the calyx is not renewable. The peduncle varies in length and is often indistinct or absent altogether. All loxosomes are commensal on marine invertebrates, especially polychaete annelids. They occur in dense clumps, called loxosome populations, which arise through characteristically rapid asexual reproduction—new zooids are budded from the calyx, break loose and attach themselves close to the parent.

Populations of *Loxosomella phascolosomata* are found on the proximal tip of the sipunculan *Golfingia vulgaris*. Although this association is common, it is not highly specific and the loxosome may live on other sipunculans, or on the bivalve mollusc *Lepton*. *Phascolion strombi*, another sipunculan, may have any of five species of *Loxosomella* attached to it, with as many as three different species on the same animal, but *Loxomespilon perezi* is restricted to the polychaete *Sthenelais*. Loxosomes appear to settle only on particular parts of their hosts, such as the exhalant openings of sponges, the proximal trunk of sipunculans or the parapodia of polychaete annelids. These restricted distributions probably help to reduce interspecific competition, and the loxosome population flourishes where it is able to benefit from the host's respiratory or feeding currents.

SUBKINGDOM Metazoa

PHYLUM Bryozoa (= Polyzoa, Ectoprocta)

Moss animals

CLASS	ORDER
Phylactolaemata	
Stenolaemata	Cyclostomata
	†Cystoporata
	†Trepostomata
	†Cryptostomata
Gymnolaemata	Ctenostomata
	Cheilostomata

†extinct

- NUMBER OF LIVING SPECIES: more than 4,000

- DIAGNOSTIC FEATURES: coelomate and unsegmented zooids, invariably colonial, with a lophophore of ciliated tentacles, a U-shaped gut and an anus outside the lophophore, but without vascular and excretory systems

- LIFESTYLE AND HABITATS: some freshwater species but mostly marine, in all seas from the shore to the abyss but most frequent between 50 and 200 m; sedentary, colonies erect or encrusting

- SIZE: individuals range from 0.25 to 1.5 mm long, with an average of 0.6 mm, but colonies can range from 0.5 cm across to coral-like growths up to 20 cm diameter

- FOSSIL RECORD: Ordovician to Recent; about 15,000 species

The Bryozoa—commonly known as moss animals or sea mats—are mostly sedentary animals which form colonies of individuals or zooids by repeated budding from a primary zooid called an ancestrula. Each zooid is a functionally independent unit, with a crown of ciliated tentacles (the lophophore), a gut, a coelom, reproductive organs and a simple nervous system. Yet, all the zooids of a colony form a closely integrated whole and constitute a single genetic individual.

Most bryozoans are marine; they range from the intertidal zone to abyssal depths, but are most abundant in shallow, continental shelf seas. The colonies are soft, fleshy or gelatinous, or calcified, and are easily recognized. On rocky shores they may be found encrusting the undersides of rocks, or on holdfasts of large seaweeds, and colonies of erect species with tufted or spiralled branches commonly occur in sheltered crevices. Details of zooid structure, however, cannot be seen without magnification and, perhaps for this reason, bryozoans are frequently neglected. Under the microscope they reveal a striking beauty of form.

Although bryozoans have a superficial resemblance to coelenterate polyps, and also to entoprocts (with which they were once grouped), they are considerably higher in the scale of evolution: they are triplobastic and coelomate and probably shared an ancestor with the Phoronida and Brachiopoda, the other two lophophorate phyla. Bryozoans made their first appearance as fossils in the Ordovician period and reached their peak in the late Mesozoic era.

The bryozoan lophophore is everted through an orifice at the distal end of the zooid; when retracted it is enclosed by an introverted portion of the body wall (the tentacle sheath), and the orifice is closed by a hinged operculum, or by muscular contraction. The tentacles form an inverted cone around the mouth, the gut is U-shaped and the anus opens at the base of the lophophore, outside the ring of tentacles—hence the alternative name of the phylum—Ectoprocta. The zooid is enclosed by an outer, non-living cuticle, which contains an opening for the lophophore. This cuticle is secreted by the epidermis and consists largely of chitin, but in a majority of species it is impregnated with calcium carbonate. Bryozoans are filter feeders: food particles are driven towards the mouth by currents generated by the tentacle cilia and rejected particles pass out between the bases of the tentacles in the exhalant current. Such small animals as bryozoans have no need of special organs for respiration and excretion because both these functions are carried out by diffusion of dissolved molecules through the body wall.

Most bryozoans are hermaphrodite but a few species are dioecious and some show marked sexual dimorphism. Simple gonads develop from specialized areas of peritoneum (the membranous lining to the coelom) and liberate their products into the coelom. Sperm pass through pores at the tips of the tentacles and aggregate on the expanded lophophores of neighbouring zooids. Eggs escape through a minute pore at the base of the lophophore, and are fertilized internally or, less commonly, externally. In a few species the egg develops

LEFT: delicate, lacy colonies of cheilostome *Sertella* make up the centrepiece of this attractive group of bryozoans. *Sertella* is a common genus in the warm, shallow seas of the world.

BELOW: the freshwater phylactolaemate *Cristatella mucedo* is the only bryozoan that is capable of locomotion; the colony crawls on a common, flattened sole and can move up to 2 cm a day.

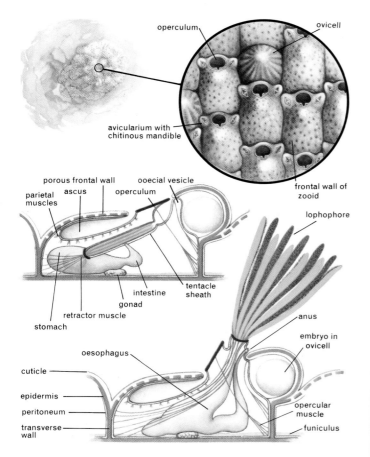

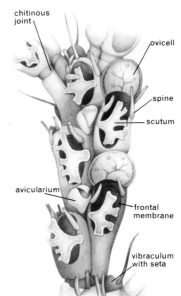

LEFT: a colony of the encrusting, intertidal species *Schizoporella unicornis*. It is a member of the Ascophora, a cheilostome sub-order in which the frontal wall of the zooid is calcified. Embryos are brooded in the special ovicells. The diagrammatic sections show a generalized ascophoran at rest and with the lophophore expanded. Contraction of the parietal muscles lowers the floor of the ascus, which then fills with sea-water. The coelomic pressure of the zooid is thus increased and the lophophore is everted; withdrawal is assisted by retractor muscles. The ooecial vesicle seals the opening of the ovicell during incubation of the embryo. The funiculus links all the zooids of the colony.

RIGHT: part of a branch of the erect bryozoan, *Scrupocellaria reptans* (Cheilostomata). The avicularia, vibracula and scuta are all polymorphic individuals.

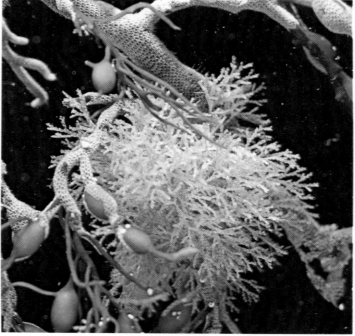

ABOVE: two typical, intertidal cheilostomes growing on a seaweed: lacy colonies of *Electra pilosa* encrusting the midrib, and a large, stiff tuft of *Scrupocellaria scruposa*.

into a shelled larva called the cyphonautes which swims and feeds in the plankton, but in most it is brooded and develops into a bulky larva which settles within hours of release.

The class Phylactolaemata, which includes the few freshwater forms, are gelatinous, uncalcified bryozoans which encrust stones, twigs and water plants. They are characterized principally by a crescent-shaped lophophore. There are perhaps 50 species throughout the world, but they may achieve broad geographical ranges by producing resistant, asexually formed buds called statoblasts. These are small and light, may be sticky or hooked, and are spread by wind or water, or are transported attached to other animals—the phenomenon known as phoresy.

The Cyclostomata is a geologically ancient group which has declined steadily from a peak of abundance and diversity in the Cretaceous period. Cyclostomes have tubular zooids with calcified walls and form erect or encrusting colonies. The zooid orifice is terminal and circular, without an operculum, and embryos are brooded in special, enlarged reproductive zooids called gonozooids.

The Ctenostomata have uncalcified zooids and the colonies may encrust, forming thick, gelatinous sheets, or are delicate and ramifying, with the zooids attached to long, branching stolons. The orifice is terminal or frontal, without an operculum, and embryos are brooded within the tentacle sheath. *Zoobotryon verticillatum*, a stoloniferous ctenostome, is an important fouling species in tropical and subtropical waters.

The earliest known genera of Cheilostomata are from Jurassic rocks and the order began to diversify rapidly in the early Tertiary period. Today, they are the largest group of living bryozoans. The zooids are typically box-shaped with a hinged operculum closing the frontal orifice. Embryos develop in special brood chambers called ovicells, and polymorphism of zooids is well developed. Most forms are calcified and the order contains erect and encrusting genera.

Several encrusting cheilostomes are important fouling organisms on ships' hulls, drilling rigs and harbour installations.

Polymorphism is characteristic of all bryozoans but is most developed in the Cheilostomata. Polymorphic zooids, called heterozooids, are not easily recognized as such and only careful study has revealed their zooid form. Heterozooids are budded in the same way as normal unmodified zooids (autozooids) and can themselves bud both heterozooids and autozooids. In some types (for example, stoloniferous ctenostomes) the orifice is absent and the heterozooid is simply an enclosed, coelomate chamber known as a kenozooid. But in most heterozooids the operculum is highly modified, with a consequent increase in its musculature at the expense of the lophophore. The most common heterozooids are the avicularium, in which the operculum is modified as a snapping mandible, and the vibraculum, in which it forms a whip-like seta.

SUBKINGDOM Metazoa

PHYLUM Phoronida

Horseshoe worms

GENUS
Phoronis
Phoronopsis

- NUMBER OF LIVING SPECIES: 15 (but more known only as larvae)

- DIAGNOSTIC FEATURES: coelomate worms that live in self-secreted chitinous tubes, have a food-gathering lophophore of ciliated tentacles, a U-shaped gut, vascular system and blood containing red corpuscles

- LIFESTYLE AND HABITATS: all marine, in shallow waters of temperate and tropical seas; sessile, bottom living attached to mollusc shells or stones or buried in sand; a few bore into shells

- SIZE: from 6 mm to 20 cm, but mostly less than 10 cm

- FOSSIL RECORD: none substantiated

The Phoronida (horseshoe worms) are a phylum of sessile, coelomate tubicolous animals found in the shallow waters of tropical and temperate seas. They range in size from 6 to 200 mm long, and have an unsegmented body with a slender trunk, a swollen, posterior end bulb and, at their anterior end, a horseshoe-shaped lophophore. Each individual secretes a tube, the major constituent of which is probably chitin. The secretion, at first fluid and sticky, hardens on contact with seawater and so collects an outer coat of shell fragments, pebbles and sand. Most species described are from European, Australian and Japanese waters, and from the Atlantic and Pacific coasts of North America, but the development of marine studies in other regions will almost certainly reveal new species.

Phoronids have left no fossil record, although a few trace fossils from the Palaeozoic era have been interpreted as phoronid tubes, and their phylogeny is largely a matter for speculation. Because they have a lophophore and are coelomate, phoronids are usually grouped with the Bryozoa and Brachiopoda, as the lophophorate phyla; all three display morphological and embryological features of both the deuterostome and protostome divisions of the animal kingdom.

The soft, elongate body of the phoronid is well provided with muscles so it can move in its tube and evert its lophophore. The lophophore consists of two crescentic, parallel ridges, each bearing a single series of slender, ciliated tentacles. The horns of the crescent are morphologically dorsal and may be enrolled in tight spirals. Ciliary currents pass down the tentacles and out between their bases; food particles are entangled in mucus and are carried along the groove between the lophophore ridges, and into the mouth. The gut occupies most of the coelom, expanding as a stomach in the end bulb and looping back towards the lophophore where the anus opens on a small papilla close to the mouth. Beside it are the openings from the paired nephridia. A nerve ring at the base of the lophophore sends fibres into the coelomic cavities of the tentacles, and is also linked to a diffuse nerve net in the body wall. The trunk muscles are innervated by a single, branching giant fibre. The vascular system comprises two longitudinal vessels which are linked posteriorly by a diffuse plexus in the peritoneum of the stomach, and anteriorly by a single vein in each tentacle. The blood has red corpuscles that contain the pigment haemoglobin.

Most phoronids are hermaphrodite and although some dioecious species are known none is sexually dimorphic. Gametes are shed into the coelom and pass to the exterior through the external openings of the nephridia. Fertilization is usually external but can be internal and the egg hatches into a ciliated, planktonic larva.

The fifteen known species of phoronid are divided between the genera *Phoronis* and *Phoronopsis*. *Phoronopsis* is distinguished by an erect fold between the lophophore and the trunk. Although phoronids are not generally common, they may be locally abundant. *Phoronopsis viridis* is found on sandy substrates along the Californian coast; with up to 300 tentacles it is the largest species known and can grow up to 20 cm long. The tube of this species is cylindrical, with its outer surface thickly covered with fine sand grains, and is buried vertically in sand. In other species the tubes are closely entangled and are camouflaged with pieces of shell or stone, so resembling those of certain polychaete annelids. All species are able to re-burrow and secrete new tubes if dislodged from the first. *Phoronis ovalis* is the smallest species, being only 6 mm long with about 25 tentacles. It burrows in mollusc shells or other organic carbonates and has been recorded from Japan, New Zealand, Brazil and Europe. The burrows are shallow, lying just under the inner surface of the shell and open to the exterior by a sharp, right-angled bend. Populations of this phoronid are dense, with up to 150 per square centimetre of shell. *Phoronis australis*, which lives within the fabric of the tube of the anthozoan coelenterate *Cerianthus*, has the most specialized habitat.

Most phoronids have a protracted summer breeding period. Eggs are shed into the sea to develop or are brooded, supported by the innermost tentacles, within the concavity of the lophophore. *Phoronis albomaculata* attaches its eggs to the inner surface of its tube, or to adjacent rocks. At least one species, *P. ovalis*, reproduces asexually and gives rise to crowded aggregations which can probably be regarded as colonies. In some species the lophophore is periodically shed and a new one develops rapidly. More remarkably, the shed lophophore may itself regenerate as a new individual. *Phoronis hippocrepia* degenerates completely each winter, the animal developing anew in the spring from small fragments remaining in the tube.

Phoronids are probably more familiar as larvae than as adults. The free swimming, oval actinotroch larva, 1 to 5 mm long, is a common constituent of inshore summer plankton in many seas. The domed anterior end is formed by a broad extension of the body wall, the preoral lobe, and the larva shows similarities to the trochophore type found in other

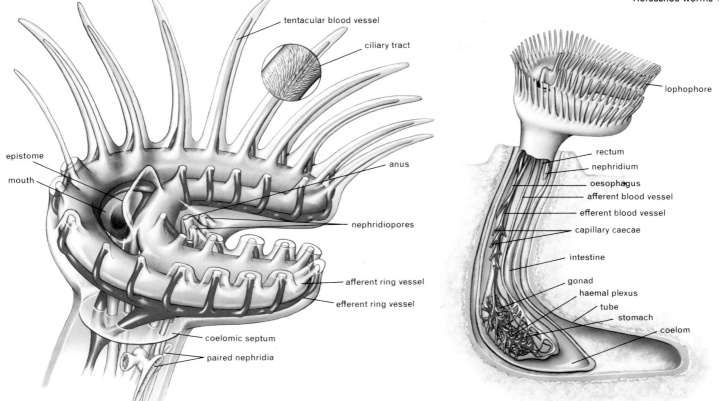

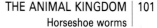

ABOVE: a diagram of *Phoronis* in its chitinous tube in the sand. The lophophore has a large number of food collecting tentacles (here reduced in number for clarity). Each tentacle contains a single blood vessel which receives blood from the efferent ring vessel and delivers it to the afferent ring vessel. The nephridia collect waste matter through the ciliated funnels and the nephridiopores open near the anus.

BELOW LEFT: the tubes of *Phoronis australis* are closely associated with the sea anemone *Cerianthus*.

In this species the horns of the lophophore are enrolled in tight spirals.

BELOW CENTRE: *Phoronis hippocrepia* normally bores into calcareous rocks. The lophophore is a simple crescent. In these feeding worms the dorsal blood vessel is visible.

BELOW RIGHT: an actinotroch larva. The gut and red clumps of blood corpuscles are visible through the body wall. The domed preoral lobe indicates the anterior end. The ciliated band at the hind end is the main propulsive organ. (× 40)

invertebrates such as annelids. A median, ciliated girdle becomes lobed and develops into a ring of ciliated larval tentacles. These are lost before settlement but the rudiments of the adult tentacles are seen at their bases. The actinotroch swims and feeds in the plankton for some weeks before it finally settles on the sea floor and undergoes a profound metamorphosis, emerging as a juvenile phoronid.

The first known phoronid was described as a planktonic organism in 1847. It was 10 years before an adult was discovered and a further 10 years before it was realised that the planktonic '*Actinotrocha*' was a larval stage of the benthic *Phoronis*. Today, more species are known as larvae than as adults—some authorities estimate that more than 30 larval forms may be recognized.

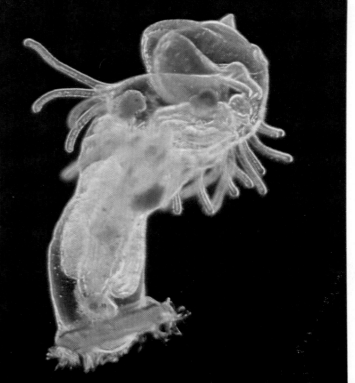

SUBKINGDOM Metazoa

PHYLUM Brachiopoda

Lamp shells

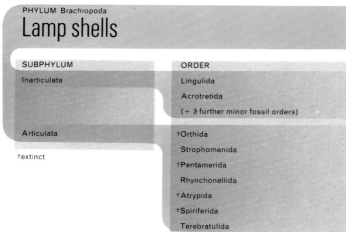

SUBPHYLUM	ORDER
Inarticulata	Lingulida
	Acrotretida
	(+ 3 further minor fossil orders)
Articulata	†Orthida
	Strophomenida
	†Pentamerida
	Rhynchonellida
	†Atrypida
	†Spiriferida
	Terebratulida

†extinct

● NUMBER OF LIVING SPECIES: about 260

● DIAGNOSTIC FEATURES: coelomates whose body is enclosed in a shell composed of a dorsal and ventral valve, and which have a loop-shaped tentaculate lophophore and usually a pedicle (stalk) for attachment

● LIFESTYLE AND HABITAT: all marine at all depths and latitudes, usually attached to hard substrates or anchored in soft sediment

● SIZE: less than 1 mm to 4 cm; some extinct forms reached 30 cm

● FOSSIL RECORD: early Cambrian to Recent with peak abundance in late Palaeozoic; about 2,000 genera

BELOW: the Recent inarticulate *Glottidea*, a burrowing lingulid. In life the animals live in deep burrows. Bristle-like setae form inhalant and exhalant apertures.

RIGHT: a terebratulid brachiopod is a typical representative of the Articulata. Note the similarity to a Roman oil lamp, from which the common name is derived.

Brachiopods (commonly known as lampshells because of their resemblance to Roman oil lamps) comprise a wholly marine, benthic and largely bottom living group of minor significance today. In the past, however, they were far more prominent and there can be no doubt that the phylum has declined markedly since its peak of abundance in mid-Palaeozoic times. Extinct forms show a much wider range of morphology than their living counterparts; some were free living, others occurred in dense reef-like masses, and yet others may have possessed limited swimming abilities. In the Palaeozoic era at least, brachiopods dominated communities on flat-bottomed continental shelves and were often rock builders. The living forms are perhaps most familiar to the naturalist as providing, in the inarticulate genus *Lingula*, one of the classic examples of a 'living fossil', for that genus has survived unchanged in habits and morphology for almost 500 million years.

The distribution of living brachiopods is sporadic, and they tend to occur in dense clusters (presumably to facilitate fertilization of eggs when released). In spite of this, individual species have wide ranges; some *Terebratulina* occur over the whole of the northern hemisphere. The group as a whole ranges from the poles to the tropics, the articulates and calcareous inarticulates being most abundant in temperate waters, and the lingulids the most typical denizens of warmer conditions. The depth range of brachiopods is considerable: some live intertidally, a *Pelagodiscus* (a lingulid) is known from abyssal depths, but most forms inhabit continental shelf environments in waters whose depths range between tens and hundreds of metres.

Bivalve molluscs and brachiopods are often confused by the expert as well as the layman, as they both have paired calcareous shell valves. This evolutionary convergence, however, is a response to the need to have an openable shell when filter feeding. The two groups can usually be differentiated in that most bivalves have left and right mirror image valves, a ligament joining the two valves together, widely separated adductor muscle scars and extensive gills. Brachiopods have dorsal and ventral valves with the plane of symmetry bisecting them, a pedicle, a lophophore but no gills, and a distinctive shell structure. Phylogenetically, the groups are widely separate, bivalve origins lying close to the annelids whereas brachiopods relate to bryozoans and phoronids, the other so-called lophophorate phyla.

The ventral or pedicle valve is usually the larger of the two and is extended behind as a beak or umbo. A fleshy 'stalk' or pedicle extends posteriorly out of the shell, and serves to attach the animal to the substrate, usually a hard surface such as rocks or shells. In a few living brachiopods, the end of the pedicle is divided into filaments which are anchored in soft bottom sediments (*Terebratulina*); in

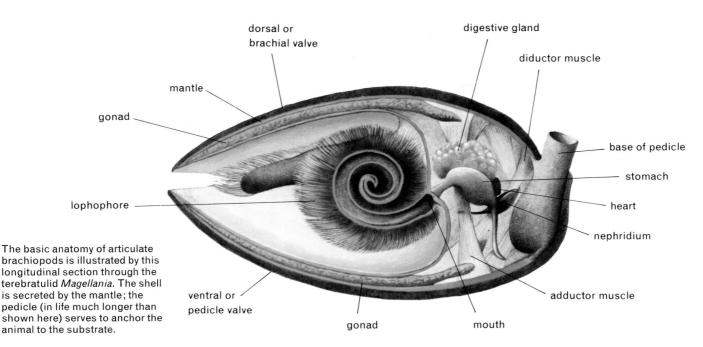

dorsal or
brachial valve

mantle

gonad

lophophore

digestive gland

diductor muscle

base of pedicle

stomach

heart

nephridium

ventral or
pedicle valve

gonad

mouth

adductor muscle

The basic anatomy of articulate brachiopods is illustrated by this longitudinal section through the terebratulid *Magellania*. The shell is secreted by the mantle; the pedicle (in life much longer than shown here) serves to anchor the animal to the substrate.

others it is modified into a burrowing organ (*Lingula*) where the animals live in soft mud or sand; others encrust hard surfaces (the inarticulate *Crania*).

The valves are opened by the diductor muscles and closed by the adductors. The shell is strengthened by a variety of ribs and spines, and its margin may be folded to separate inhalant and exhalant water currents. Each valve is lined and secreted by a thin sheet of tissue, the mantle, between which is a large mantle cavity, filled by seawater and open to the exterior. Housed in the cavity is a loop-like structure, the lophophore, which consists of a pair of variously contorted arms or brachia. These usually have internal calcareous supports, the brachidium, in addition to a hydrostatic skeleton (early workers believed that these arms extended beyond the shell and were perhaps used in locomotion, hence the name of the phylum which means, literally, 'arm footed'). The lophophore bears ciliated filaments which generate circulatory currents of water into and out of the mantle cavity. Microplankton and particulate organic matter are filtered out by the filaments and transferred to the mouth along the ciliated food grooves on the lophophore. In living inarticulates, the gut extends from the mouth through the coelom to an anus but in living articulates the gut is blind, and faeces are ejected through the mouth.

Sexes are separate in most living brachiopods but a few hermaphrodite forms are known. Gametes are released directly into seawater in all but a few forms in which eggs are fertilized in a brood pouch, where the larvae spend their early developmental stages. Most larvae are planktonic, usually for a matter of hours or days, and this is the main means of dispersal of the groups.

Among the Inarticulata, there are two main orders, the Lingulida and Acrotretida, both of which range from the Cambrian to Recent times. All lingulids have shells built of alternating layers of organic material and calcium phosphate (termed chitinophosphate). The shell is typically elongate and elliptical in outline, with flattened, convex valves ornamented only by concentric growth lines; the vermiform muscular pedicle, generally two or three times longer than the shell, emerges posteriorly between the valve edges. Lingulids range from intertidal mudflats to the outer margin of the continental shelf and most of them burrow deeply in mud or sand to which they are anchored vertically by the pedicle. When feeding the front end of the shell lies at the surface, and water currents bearing food particles and oxygen are sucked into the mantle cavity by way of tubular siphon-like structures produced by bunches of elongate setae on the mantle margin. When alarmed, the animal draws down into its burrow through contraction of the pedicle.

The lophophore serves as a feeding as well as a respiratory organ and is a feature common to all lamp shells. In *Terebratulina caput-serpentis* (left) it consists of a spiral of ciliated filaments that draw in water and food particles (indicated by stippled arrows in the diagram, which shows the valves gaping to expose the lophophore); food is filtered off and the water expelled (clear arrows). The setae extending from the valve margins of the upper specimen are sensory and if touched trigger closure of the shell.

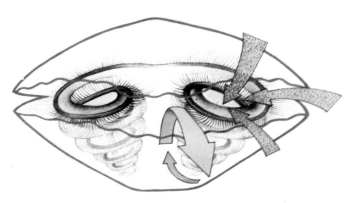

In the Acrotretida, the shell form is quite distinct: it is circular with a conical dorsal valve and a flattened disc-shaped ventral valve; the pedicle emerges through a hole (foramen) in the ventral valve. The shell is chitinophosphatic, and ornamented only by concentric growth lines. Typically, acrotretids live firmly attached to rock surfaces and shells (a few fossil forms may have lived attached to algae). One aberrant group, the Craniacea (Ordovician to Recent), have calcitic rather than chitinophosphatic shells and are the only group of cemented inarticulates. They have no pedicle; instead the ventral valve is moulded by cement to shells or rock surfaces.

Most living articulate brachiopods are referred to the two groups Rhynchonellida and Terebratulida. Rhynchonellids have calcitic shells which are strongly biconvex (as have all articulates) with no apparent hinge line. The lophophore supports are in the form of a pair of calcareous spikes (crura). These brachiopods first appeared in the early Ordovician and had peaks of abundance during the Silurian and Jurassic periods. The pedicle, which emerges through a foramen in the umbo of the ventral valve, is functional, and in living representatives serves to hold individuals together in clumps. The shell bears fine concentric growth lines, and strong radial ribs, which give a zig-zag contact to the valve margins. The margins are commonly further modified by a distinct anterior fold or folds which serve to separate the inhalant and exhalant currents.

At present rhynchonellids are subsidiary in numbers to the terebratulids, which first appeared at the beginning of

A colony of the red terebratulid *Waltonia inconspicua*. Most of the individuals are feeding, with valves slightly gaping. White tubes of serpulid polychaetes are visible on the shells.

the Devonian period and reached their peak during the period from the later Mesozoic era to modern times. The most obvious superficial difference between the two groups is the absence of strong radial ribs in most terebratulids though, paradoxically, if living forms alone are considered, there are more ribbed terebratulids than rhynchonellids. Assignment of a species to the relevant order may be

ABOVE: the Devonian spiriferid *Mucrospirifer mucronata* from Canada. An elongate hinge-line and wing-like extensions of the shell typify this group, which flourished in the Palaeozoic.

The aberrant Permian productid *Echiaurus* (Strophomenida) had a bowl-shaped lower valve which is characteristic of the group. In life the long spines were used to support and anchor the shell in the sediment, the animals having lost a functional pedicle.

BELOW: the strophomenid *Leptaena depressa* from Silurian Wenlock limestone of the Welsh borders. Flattened shells, with a roughly semi-circular outline are typical of the group.

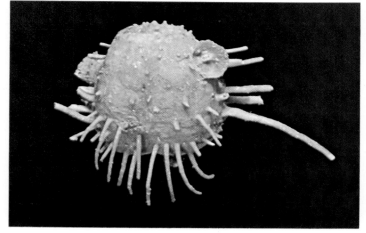

confirmed by microscopic inspection of the shell, for the terebratulid shell is punctate, that is, it is perforated by a series of minute cylindrical holes which extend perpendicularly through almost the entire thickness of the shell. In life, these holes house slender extensions of the mantle surface called caecae, of uncertain function, though they may serve as food stores and as 'anti-fouling' devices. Most terebratulids live in clusters attached to hard bottoms, although some (for example, *Terebratulina*) have branching pedicles, rooted in soft sediment.

The articulate brachiopods first appeared during the early Cambrian period, but were small (1 cm or less), inconspicuous members of the shallow water benthic communities. By the Ordovician they became common. The root stock of all later articulates were the Orthida—simple, usually gently biconvex shells, which rarely exceed 2 cm long and are ornamented by fine radial ridges. There is a long hinge line which gives an approximately semicircular outline to the shell. Typically the animals were attached by the pedicle and lived in shallow water environments. Appearing in the Cambrian period, orthids survived to the close of the Permian.

From the Orthida arose the Strophomenida, a typically Palaeozoic group, which was particularly common in the Ordovician and Carboniferous, although they were most abundant in the Permian; a few survived to the Jurassic and two genera have persisted through to the present day. Early representatives had a bowl-like, semicircular shell with a long hinge and very narrow body cavity. Ornamented by fine radial ribs, these forms had lost a functional pedicle and lay loose on the sea floor. Some strophomenids perhaps swam by episodic clapping of the valves. From these modest origins however, arose the largest and most aberrant brachiopods, the productids, many of which lived partially buried in sediment, sometimes supported or attached by enormous tubular spines. *Gigantoproductus* with a 20 to 30 cm hinge, is the largest known brachiopod.

A second group to arise from the orthids were the Pentamerida—strongly complex, typically smooth shells with prominent interior platforms for the attachment of muscles. They ranged from mid-Cambrian times to the late Devonian with a peak during the early Ordovician. These animals probably lay loose on the sea floor, maintaining their umbo-down position by virtue of extensive thickenings of the posterior parts of the shell.

The final major group of extinct brachiopods were the Spiriferida which ranged from the Ordovician to the early Jurassic, although they are a predominantly Palaeozoic group and were most abundant in the Devonian.

Atrypida

Terebratulida

Rhynchonellida

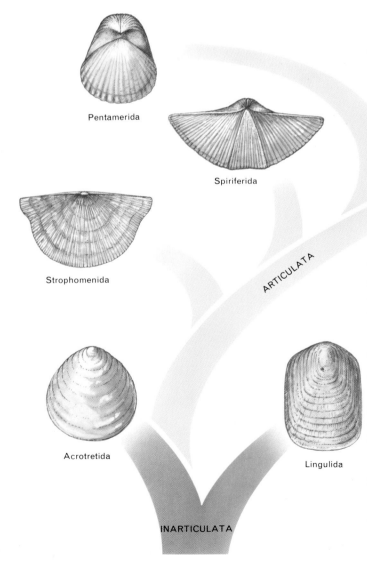

Pentamerida

Spiriferida

Strophomenida

Acrotretida

Lingulida

ARTICULATA

INARTICULATA

Orthida

The evolutionary relationships of the brachiopods. The first articulates, the Orthida, gave rise directly to three groups, the Strophomenida, Spiriferida and the Pentamerida. The remaining articulate orders diversified from the Pentamerida. It is of interest that, while most of the brachiopods are now extinct, two of the five orders that persist are the ancestral inarticulates.

A typical spiriferid (for example, *Spirifer*) has a strongly biconvex shell with the hinge produced into elongate 'wings', strong anterior folds and an ornament of radial ribs. This group represented the peak of feeding and respiratory efficiency within the brachiopods, for each branch of the lophophore was coiled into a spire that was directed dorsally or inwards. In fossils its position is shown by the similarly coiled calcareous supports or spiralia. Some spiriferids were attached by a pedicle, but many were free living on the sea bed, lying on the broad flattened area between the umbos.

The final extinct group, the Atrypida, were rather like rhynchonellids, from which they probably evolved, but differ in having a complex brachidium like that of the spiriferids. They lived from the mid-Ordovician to Triassic.

Brachiopods can thus be seen as a group whose inconspicuousness in modern faunas belies their major role in ancient marine faunas. Abundant in the Palaeozoic and Mesozoic, surviving several phases of near extinction with renewed evolutionary radiations, their place in Tertiary and Recent faunas has been taken by bottom living bivalves, which are as common in shallow marine environments today as brachiopods were in the past.

SUBKINGDOM Metazoa

PHYLUM Mollusca

Molluscs

CLASS

Monoplacophora

Amphineura *chitons and solenogasters*

Scaphopoda *tusk shells*

Gastropoda *snails and slugs etc.*

Bivalvia *mussels and oysters etc.*

Cephalopoda *squids and octopuses etc.*

- NUMBER OF LIVING SPECIES: about 80,000.

- DIAGNOSTIC FEATURES: fundamentally bilaterally symmetrical coelomates with segmentation almost or completely absent; muscular foot, typically ventral; head usually with concentrated sense organs; dorsal mantle of tissue folded to form a cavity which usually houses gills; well developed blood vascular system and usually an internal or external calcareous shell.

- LIFESTYLE AND HABITAT: usually free living, attached or unattached, in marine, freshwater or terrestrial environments; rarely parasitic and symbiotic.

- SIZE: less than 1 mm to 18 m, usually 2 to 20 cm.

- FOSSIL RECORD: early Cambrian to Recent, especially common in Mesozoic and Tertiary; probably more than 60,000 species.

The phylum Mollusca, encompassing the monoplacophorans, chitons, tusk shells, bivalves, gastropods, squids, cuttlefish and octopuses, *Nautilus*, and the extinct belemnites and ammonites, is second in number and diversity only to the Arthropoda. Alike in their fundamental morphology, molluscs have undergone a remarkable adaptive radiation to produce, within the cephalopods, the most intelligent (*Octopus*) and the largest (the squid *Architeuthis*) of invertebrates. Many molluscs produce shells of aragonite and/or calcite which form some of the most beautiful and sought after of natural history objects, and which have also served as decoration, money, food, cooking utensils, tools, and even musical instruments for mankind. They have a fossil record extending back nearly 570 million years, to the beginning of the Cambrian period, and are among the first organisms with hard parts to appear in the fossil record. At many times in geological history molluscs have occurred in such numbers as to be rock formers, and one fossil group, the ammonites, has attracted man's curiosity since antiquity.

The phylum as a whole is very clearly characterized as comprising coelomate, bilaterally symmetrical animals (although sometimes secondarily modified to an asymmetrical condition) distinguished by a nearly complete or complete absence of segmentation. By virtue of the great diversity of the group, it is perhaps easiest to discuss basic anatomy in terms of a hypothetical archetypal ancestor, from which the other groups may be derived.

The soft body of this hypothetical creature is protected by a dorsal calcareous shell covering the visceral mass. The shell is attached to the body at various sites of muscle insertion, and at the shell margin. Over wide areas, there is a fluid-filled cavity between the mantle (the sheet of tissue that secretes the shell) and the shell proper. Within this fluid-filled cavity, deposition of the calcareous shell takes place, the organic and inorganic components being secreted into a liquid and thence precipitated as the hard shell material.

There is a ventral, muscular contractile foot, on which the animal crawls; an anterior head bearing a terminal mouth; a pharynx with the toothed, rasp-like radula—a tongue-like structure with which the animal feeds; and eyes, tentacles and other sense organs. At the posterior end there is a fold

Mollusc shells are highly diverse in form. The calcareous shells of the gastropods, cephalopods and bivalves shown here have exquisite shapes and patterns, and colours which may be muted or bold. They not only reflect the variety of habitat and way of life among the molluscs, but also make beautiful ornaments and tools.

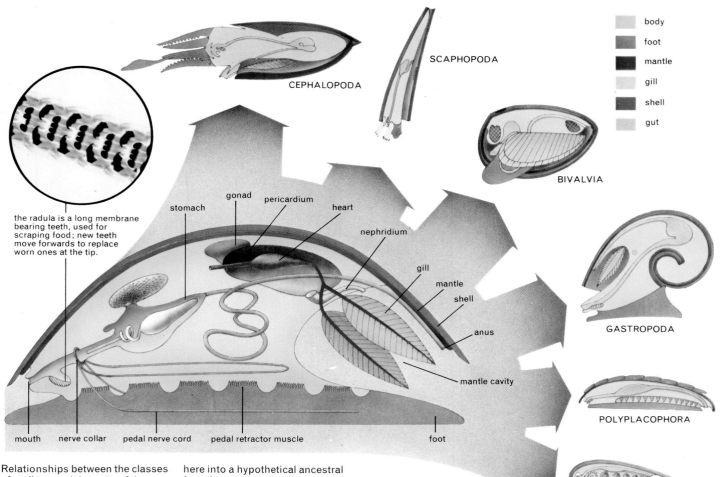

the radula is a long membrane bearing teeth, used for scraping food; new teeth move forwards to replace worn ones at the tip.

CEPHALOPODA

SCAPHOPODA

BIVALVIA

GASTROPODA

POLYPLACOPHORA

MONOPLACOPHORA

body
foot
mantle
gill
shell
gut

stomach · gonad · pericardium · heart · nephridium · gill · mantle · shell · anus · mantle cavity

mouth · nerve collar · pedal nerve cord · pedal retractor muscle · foot

Relationships between the classes of mollusc and the unity of the phylum are indicated by common features such as the foot, head, mantle cavity and gills—combined here into a hypothetical ancestral form from which the six living classes can be derived. Inset is a detail of the structure of a typical molluscan radula.

of body wall, leaving a large mantle cavity between the mantle and visceral mass. This cavity houses the gills, and into it discharge the digestive, excretory and reproductive products. Within the main mass of the body there is the coiled intestine, a heart with two auricles, an open circulatory system, and the gonads.

Visceral mass, mantle, head, foot and mantle cavity are common to most molluscs, but each has been modified in varying ways as a response to the selective pressures which have led to the diversification of the group. Closest to the archetype (and perhaps the rootstock of the molluscan radiation) are the diminutive, cap-shaped Monoplacophora, a group characterized by serial repetition of organs. In the Polyplacophora or chitons, the head is distinct, but lacks eyes and tentacles, and the shell is represented by a longitudinal series of eight plates. The related Aplacophora or soleno-gasters are worm-like, with a poorly differentiated head, and a shell reduced to a coat of calcareous spicules.

In the Scaphopoda, the tusk shells, the shell is tusk-shaped, and open at both ends. The head is poorly differentiated, and the gills have been lost. The Gastropoda—snails—have a well developed head, and their univalved shell is usually coiled into a spiral. At an early stage in embryological development, many snails undergo a process known as torsion—the mantle/shell with enclosed visceral mass moves through 180° in a counterclockwise direction in relation to the head and foot, so that the originally posterior mantle cavity comes to occupy an anterior position. Bivalvia lack a head and radula, and are enclosed in a dorsally hinged shell which has two lateral valves. The mantle cavity is greatly enlarged, and the gills are generally used for feeding as well as being respiratory organs.

The most divergent molluscs are the Cephalopoda. Here, the foot has migrated anteriorly, surrounds the head, and is modified into a crown of tentacles. The shell in forms such as *Nautilus* and the extinct ammonites is external; it is divided into a series of gas-filled chambers, and functions as a buoyancy apparatus. It serves a similar function, but is internal, in cuttlefish. By contrast, octopods have lost the shell completely. In many cephalopods the mantle cavity has a restricted opening and a well developed muscular wall which enable the animal to swim rapidly by jet propulsion.

Molluscs have conquered almost all possible environments save the air. Although their evolutionary origins lie in the sea, both bivalves and gastropods successfully invaded freshwaters during the Palaeozoic era, and possible terrestrial gastropods also appeared at this time, although undisputed land snails are best known only from the Jurassic period onwards. In the seas, molluscs are known from the abyss to the intertidal zone, although most are littoral. The great majority are bottom dwellers. Some are surface crawlers, such as the chitons, monoplacophorans and most gastropods, while many bivalves burrow into soft sediment, as do some gastropods, the aplacophorans and scaphopods. A boring habit is seen in some bivalves and a few gastropods. By contrast other bivalves are sessile, encrusting forms (for example, oysters), or are attached by strong, horny fibres called byssal threads (for example, mussels). Swimming is best developed in the cephalopods, although some gastropods and bivalves can also move in this way. Planktonic cephalopods and gastropods are known. On land, there are gastropods that burrow into soil, live in caves or even in the canopy levels of tropical rainforests, and can withstand prolonged dought and even freezing.

SUBKINGDOM Metazoa
PHYLUM Mollusca

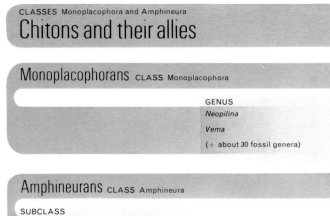

CLASSES Monoplacophora and Amphineura

Chitons and their allies

Monoplacophorans CLASS Monoplacophora

GENUS

Neopilina

Vema

(+ about 30 fossil genera)

Amphineurans CLASS Amphineura

SUBCLASS

Polyplacophora *chitons or coat-of-mail shells*

Aplacophora *solenogasters*

A typical amphineuran, the Lined Chiton (*Tonicella lineata*). The jointed shell, which is embedded in the marginal girdle, enables the animal, if detached from its rock, to roll up like a woodlouse.

○ NUMBER OF LIVING SPECIES: Monoplacophora: about 5. Amphineura: about 1,200.

○ DIAGNOSTIC FEATURES: Monoplacophora: untorted molluscs with saucer-like shell, radula and serially repeated muscles, nephridia and other organs. Amphineura: typically elongated molluscs, the majority with an external eight-plated shell; radula well developed; numerous paired gills present.

○ LIFESTYLE AND HABITAT: Monoplacophora: free living bottom dwellers on deep ocean floor. Amphineura: mainly free living and exclusively marine bottom dwellers at all depths; some parasitic on hydroid coelenterates.

○ SIZE: Monoplacophora: 2 mm to 3 cm. Amphineura: 2 mm to 30 cm, mostly less than 5 cm.

○ FOSSIL RECORD: Monoplacophora: early Cambrian to Recent, with peak in early Palaeozoic, about 30 genera. Amphineura: late Cambrian to Recent; about 60 genera (350 species).

The class Monoplacophora forms a suitable introduction to the Mollusca because there is widespread agreement that its members lie close to, if not actually form, the rootstock of the molluscan radiation. Diminutive (2 mm to 3 cm) cap-shaped shells with several symmetrical pairs of dorsal muscle scars had long been known as fossils from the early part of the Cambrian period. Then in 1950–1952 the Danish Gala-thea Expedition dredged a living species (named *Neopilina galatheae*) from a depth of 5,000 m off the Pacific coast of Mexico. This and four other species subsequently dredged from deep waters off Peru, California, and in the Red Sea constitute a fascinating group of 'living fossils'.

Anatomically, the animals are remarkable. Bilaterally symmetrical, they have a flat, saucer-shaped shell which has not undergone torsion, a ventral foot, an anterior head which lacks eyes, and a posterior anus. The mantle is separated from the foot by a groove in which are situated five pairs of gills. Internally, five pairs of retractor muscles and six pairs of nephridia are present. This serial repetition of organs led to the conclusion that *Neopilina* was in fact a segmented mollusc, supporting the long-held view (based on the striking similarity between molluscan and polychaete annelid embryology) that annelids and molluscs arose from some common coelomate stock; latterly, however, it has been regarded as more probably secondary in origin.

The Amphineura is a wholly marine class of mollusc, the most familiar representatives of which comprise the subclass Polyplacophora, the chitons or coat-of-mail shells—common inhabitants of rocky shores, although known from depths of up to 4,200 m. These animals retain many features reminiscent of the basic molluscan archetype. The body is flat-tened and elongated, with a large muscular foot, an anterior head and a posterior anus. The head lacks both eyes and tentacles but within the mouth is a well developed radula. Along both sides of the foot, and separating it from a marginal band of muscular tissue (termed the girdle) is a long groove opening into the mantle cavity, which houses the numerous paired gills. The most striking feature of the chitons is the eight-plated shell, which occupies the mid-region of the dorsal surface. Chitons eat a variety of foods. Some are wholly herbivorous and feed on marine algae; others eat bryozoans, hydroid coelenterates and barnacles.

The subclass Aplacophora is a small worldwide group of only 100 or so species which is sometimes accorded independent class status. The animals are all worm-like and rarely more than 2–3 cm long. They lack a shell but the mantle surface, which invests the body, bears numerous calcareous spicules. The aplacophorans all live in deep water. Several forms burrow into sand and mud, where they feed on annelids and other small invertebrates; some live free on the bottom or entangled with alcyonarian corals and seaweeds, and a few are parasitic on hydroids.

A ventral view of the monoplaco-phoran *Neopilina galatheae*, a living fossil which represents the closest surviving form to the probable ancestors of the main molluscan classes. The shell is up to 4 cm long.

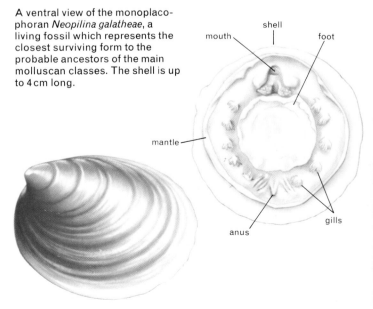

mouth

shell

foot

mantle

anus

gills

SUBKINGDOM Metazoa
PHYLUM Mollusca

CLASS Scaphopoda
Tusk shells

GENUS

Cadulus

Dentalium

Siphonodentalium

Entalina

- NUMBER OF LIVING SPECIES: about 200.

- DIAGNOSTIC FEATURES: molluscs with an elongated body enclosed in a tusk-shaped shell; thread-like tentacles (captacula) used in feeding; radula present and gills absent.

- LIFESTYLE AND HABITAT: exclusively marine burrowers at all depths.

- SIZE: 2 to 15 cm with some fossils up to 30 cm.

- FOSSIL RECORD: Devonian to Recent but rare before Tertiary; about 100 species.

The Scaphopoda is a small but distinctive molluscan class of perhaps 200 known species. These are wholly marine, burrowing animals commonly known as tusk shells, the name deriving from the shape of the shell which, in many forms, is an elongated cylindrical tube, curved and tapering like an elephant's tusk. The shell of adults varies from a few millimetres to 15 cm long in living forms; fossil representatives up to 30 cm long are known. The shell colour is usually white, ivory, yellow or buff, but brown and bright green species are also known.

The earliest undisputed fossil representatives of the group occur in the Devonian period, and there are doubtful records from the Ordovician of the USSR. These are rare fossils in late Palaeozoic and early Mesozoic rocks; modern forms do not appear until the Cretaceous period, and the maximum diversity and abundance of the group seem to have been reached only in geologically recent times.

The body of the scaphopod is greatly elongated. The mantle lines the shell, and encloses the remainder of the body other than the head and foot, which protrude at the larger,

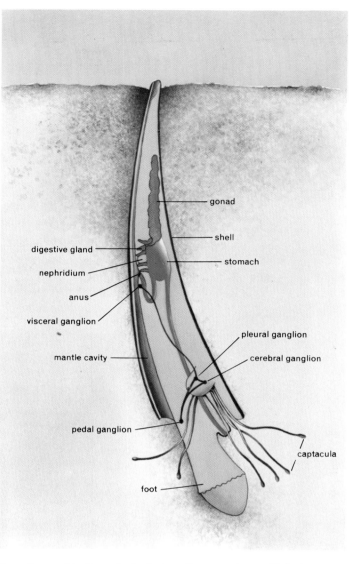

A median section through a typical scaphopod *Dentalium*. These animals live buried upside down in sand, taking foraminiferans and other organisms with their captacula. There are no gills; respiration takes place across the wall of the mantle cavity.

A live *Dentalium* that has been disinterred from its burrow. The surface of the tusk-like shell bears fine growth ridges. At the mouth of the shell, the foot and captacula are visible.

anterior end. The head is simplified, being reduced to a proboscis-like extension, and lacks eyes and tentacles. Within the mouth is a massive, strong radula. Two lobes on either side of the head bear numerous thread-like tentacles, called captacula, which are used in feeding. The foot is cone-shaped in many species, and is used in burrowing.

Sexes are separate in scaphopods, and the large, unpaired gonad fills much of the narrower, posterior end of the body. Eggs and sperms are discharged through the right nephridium into the mantle cavity, and so pass out into the open sea, fertilization being external. The subsequent development of scaphopods closely parallels that of bivalves, a free swimming trochophore larva being succeeded by a bilaterally symmetrical veliger. The larval mantle and shell are initially bilobed, but then the mantle lobes fuse ventrally, so producing a continuous cylindrical mantle (and thus shell). The body then gradually elongates.

A few scaphopods live in shallow water muds, but most inhabit sands at depths of between 6 and 1,000 m, with maximum abundances in the neritic and bathyal zones. Species are known from all latitudes, but are rarely observed live; they are, however, often washed up in such numbers onto beaches as to suggest local abundance in offshore waters. All scaphopods are carnivorous. The captacula on either side of the head extend out through the surrounding sediment, searching out foraminiferans, bivalve spat and other small organisms which are trapped and drawn back into the proboscis in large numbers. The lateral teeth of the radula grasp the particles in turn and remove them to the buccal cavity, where they are shredded and are then passed to the stomach for digestion.

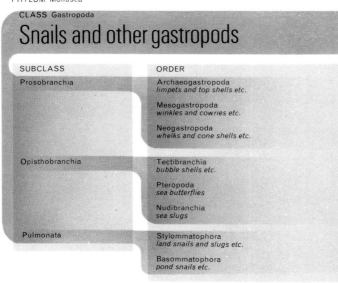

SUBKINGDOM Metazoa
PHYLUM Mollusca

CLASS Gastropoda

Snails and other gastropods

SUBCLASS	ORDER
Prosobranchia	Archaeogastropoda *limpets and top shells etc.*
	Mesogastropoda *winkles and cowries etc.*
	Neogastropoda *whelks and cone shells etc.*
Opisthobranchia	Tectibranchia *bubble shells etc.*
	Pteropoda *sea butterflies*
	Nudibranchia *sea slugs*
Pulmonata	Stylommatophora *land snails and slugs etc.*
	Basommatophora *pond snails etc.*

● NUMBER OF LIVING SPECIES: about 45,000.

● DIAGNOSTIC FEATURES: typically asymmetrical torted molluscs, usually with a shell and with a well-developed radula; mantle cavity sometimes modified into a lung.

● LIFESTYLE AND HABITAT: mostly free living in freshwater, marine and terrestrial environments, with crawling, burrowing, swimming, and planktonic representatives; a few parasitic on echinoderms or commensal.

● SIZE: less than 1 mm to 60 cm (*Aplysia* reaches 1 m) but mostly 1 to 10 cm.

● FOSSIL RECORD: early Cambrian to Recent but most abundant from late Tertiary to Recent; probably more than 20,000 species.

The Gastropoda form the largest class of extant molluscs, and have a fossil record extending back to the early Cambrian period. From their first appearance, they have been persistent members of marine faunas, but also invaded freshwater during mid-Palaeozoic times. They are at their peak diversity today, and show a wide range of morphological variation and life habits. Some forms have even converted the mantle cavity into a lung and have thereby invaded the terrestrial environment.

Most gastropods bear a single, conical, variously coiled shell, often closed by a calcareous or organic lid, the operculum. In some, such as nudibranchs and slugs, the shell is greatly reduced or even lost, and the animal is naked. Anatomically, gastropods retain many of the basic features of the molluscan archetype, albeit somewhat modified. The viscera are thus concentrated together; there is a head with well developed sensory organs, and a mouth with a specialized rasping radula; the foot is flattened and muscular and is used in locomotion (crawling, burrowing or swimming); and the mantle cavity houses the gills in all except the pulmonates, where these are lost.

Gastropods seem to have evolved from monoplacophorans by heightening and coiling of the shell (there are coiled fossil monoplacophorans), eliminating serial repetition of organs and, most important, undergoing torsion. This is one of the most fundamental features of the group: the mantle/shell, and the enclosed visceral mass, has moved through 180° in a counterclockwise direction in relation to the head and foot, so that the mantle cavity comes to occupy an anterior position. Torsion occurs during larval life, as a simple result of asymmetrical muscle development. The principal advantages of this change seem to be related to early life, as it allows retraction of the larval swimming organ (the velum) into the shell, thus protecting it: in untorted forms this manoeuvre would be blocked by the foot. There can be little doubt, however, that torsion has created serious problems for adult gastropods and many of the adaptations exhibited by the group represent attempts to adjust to these difficulties, in particular fouling, because waste from the anus and kidneys would otherwise be dumped over the head. Primitive gastropods referred to the order Archaeogastropoda solved the fouling problem by developing a slit in the shell and mantle, and withdrawing the anus from the mantle/shell margin to discharge at the base of this slit.

With the evolution of a compact helically coiled conical shell, the gastropods faced additional problems. A spiral cone cannot be carried vertically like a flat or planospiral, since all the weight falls to one side. In these forms there has thus been a further distortion so that the shell is carried obliquely to the long axis of the body. This effectively restricted the mantle cavity to one side of the body, the other being pressed against the shell, resulting in a reduction or loss of one gill, heart auricle and kidney, to produce a markedly asymmetrical anatomy. Problems of sanitation were now solved variously by retaining a slit or perforation in the shell, or by developing a long inhalant siphon to the side of the head or, finally, by a process of detorsion.

The basic division of the gastropods is thus made on the basis of anatomical differences of a fundamental nature, and three subclasses are recognized: (1) the Prosobranchia, which

BELOW AND RIGHT the edible Common Whelk (*Buccinum undatum*) shows many of the typical gastropod characteristics. Features include the coiled shell, torted neural loop, muscular, crawling foot and well-developed head bearing eyes and tentacles.

shell
kidney
heart
osphradium
hypobranchial (mucous) gland
gill (ctenidium)
digestive gland
mantle cavity
rectum
anus
gonad
mantle siphon
stomach
visceral ganglion
operculum
retractile proboscis
foot
gland of Leiblein
cerebral ganglion
salivary gland

RIGHT: torsion occurs during larval development and is a striking feature of the Gastropoda. The mantle, shell and enclosed visceral mass rotate through 180° relative to the foot; the anus moves round and then discharges over the head.

BELOW: main features of typical untorted and torted anatomy.

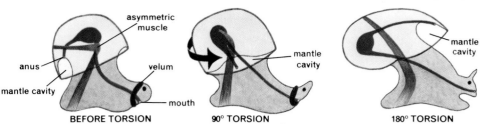

BEFORE TORSION

90° TORSION

180° TORSION

cerebral ganglion
pleural ganglia
stomach
aorta
pallial ganglion
ventricle
auricle
visceral ganglion
gill (ctenidium)
anus

NO TORSION

180° TORSION

have an anterior mantle cavity, gills and anus and have therefore undergone torsion; (2) the Opisthobranchia, which show varying degrees of detorsion, secondary bilateral symmetry and a reduced shell; and, finally, (3) the Pulmonata, which are torted but have lost their gills, the mantle cavity being modified to a lung.

Gastropod ecology is as variable as their morphology. Some are pelagic, and most have pelagic or planktonic larvae. Most benthic forms creep on hard to soft substrates, at depths ranging from the intertidal zone to the abyss. Other gastropods burrow, or are pelagic as are the heteropods which swim with a highly modified foot; some live on floating algae (for example, the mesogastropod *Litiopa*) and the truly planktonic *Janthina* produces a float of air bubbles. A few forms bore, encrust, or are parasitic.

Most gastropods graze on vegetation or encrusting animals such as sponges, bryozoans or ascidians by rasping away at them with their horny radula. Others have evolved a long, everted proboscis and become active predators (*Murex*) or, like *Conus*, have developed a poisonous tooth or stylet with

which they stab and poison their prey (chiefly fish). Some are parasitic on bivalves, echinoderms or corals, while others feed on carrion or on faeces.

Three groups of Prosobranchia are recognized. The order Archaeogastropoda contains primitive forms which have retained paired gills, heart auricles and kidneys, and a marginal slit or notch in the shell is prominent. These are mostly marine animals such as the top shells (family Trochidae), limpets (*Patella*) and keyhole limpets (*Emarginula*). An important, chiefly fossil group is the suborder Pleurotomarina. The Mesogastropoda have only one gill, heart auricle and kidney, those of the right side having disappeared. These forms are mainly marine and include periwinkles (*Littorina*), worm shells (*Vermetus*), slipper limpets (*Crepidula*), the pelagic *Janthina* and the swimming heteropods. In the order Neogastropoda, the parts of the nervous system are concentrated together and a specialized, gill-like sense organ, the osphradium, is highly developed. The shell often bears a long canal housing the inhalant siphon, which acts as a mobile

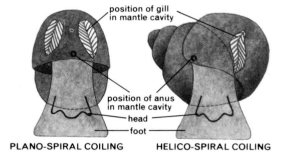

position of gill in mantle cavity
position of anus in mantle cavity
head
foot

PLANO-SPIRAL COILING

HELICO-SPIRAL COILING

With the evolution of a compact helico-spiral shell anatomical modification is necessary to carry the shell obliquely to the body, notably the loss of one gill and reduction of other organs.

'nose', with the osphradium at the base, constantly sensing the incoming water. Neogastropods are usually carnivorous, and have a proboscis to house the radula, and sometimes a poison gland. Typical representatives are *Buccinum*, *Murex*, *Voluta*, *Terebra*, and cone shells (*Conus*).

BELOW: the Ormer (*Haliotis tuberculata*) is a typical member of the Archaeogastropoda. The holes in the shell are outlets from the mantle cavity. Here the animal's head is to the right.

RIGHT: when not feeding on its prey (the siphonophore *Velella*) the pelagic *Janthina janthina* floats upside-down from a raft of tough, transparent bubbles secreted by the foot.

ABOVE: the nudibranchs or sea slugs are opisthobranchs that have lost their shells. The respiratory cerata on the middle of the back are a prominent feature of many representatives of the order.

BELOW: the slug *Arion rufer* is a terrestrial pulmonate in the order Stylommatophora—characterized by two pairs of tentacles. The hole behind the head leads into the mantle cavity which acts as a lung.

The Opisthobranchia is composed of three orders. In the Tectibranchia, the shell, although reduced (and covered by mantle in some), is still present. There is one gill, although some forms are secondarily bilaterally symmetrical. Included in this group are the carnivorous bubble shells (*Actaeon*, *Bulla*) and the vegetarian sea hares such as *Aplysia*. The order Pteropoda comprises the sea butterflies in which the foot is expanded to form a swimming web and the shell is reduced to a mere cone or is missing altogether. Some pteropods (for example *Cavolina*) are carnivorous; others, such as *Limacina*, trap suspended food particles on special mucous tracts in the mantle cavity and the strings of mucus are transported by ciliary action to the mouth. The brightly coloured sea slugs comprise the third order, the Nudibranchia, whose members include *Doris*, *Glaucus*, *Fiona*, and *Flabellina*; these are secondarily symmetrical, naked gastropods in which both mantle cavity and true gills are absent, respiration is through the general body surface, or through gill-like structures called cerata on the back. Most nudibranchs are carnivorous and feed on sponges, alcyonarian corals, and hydroid coelenterates, but a few catch plankton.

Members of the pulmonate order Stylommatophora are characterized by having two pairs of tentacles and eyes located at the tip of the posterior pair. Wholly terrestrial, these gastropods include the land snails such as *Helix*, *Bulmius*, *Partula*, and the giant *Achantia*. Most of these pulmonates are herbivorous and eat vegetation, fruits and vegetables, although they may also take carrion. Some forms are active carnivores and hunt earthworms. Many land snails possess extraordinary powers of resistance to drought, and can survive the loss through desiccation of up to 80 per cent of body weight for periods of up to a year. The common reaction to such conditions is aestivation: the snails retire deep into their shells and secrete a mucous cover or epiphragm across the aperture. By these means they may survive for up to five years. Temperate species hibernate in a similar manner. The land slugs (for example, *Limax*, *Testacella*, *Arion*), like land snails, are mainly herbivorous; many are nocturnal, and may burrow, especially when exposed to desiccation.

Pulmonates with only one pair of tentacles with eyes located at their base are classified in the order Basommatophora. These forms are primarily freshwater and typically air-breathing, and include the pond and river snails (for example,

Planorbis corneus, the Great Ramshorn Snail, is a typical planospirally coiled, freshwater member of the Basommatophora. The single pair of tentacles is characteristic of the order.

ABOVE: many pulmonates, such as the Roman Snail (*Helix pomatia*), have elaborate courtship rituals. They are hermaphrodite and cross-fertilize each other by exchanging spermatophores.

BELOW: *Conus textilis* with its egg mass. Localizing the eggs in this way increases the efficiency of external fertilization. Eggs shed loosely into the sea suffer much greater wastage.

Lymnaea and *Planorbis*) and the freshwater limpets (for example, *Ancylus*). Some forms have developed secondary gills, and take water into the mantle cavity. Several families (for example, Siphonariidae) have a limpet-like habit and live on rocky marine or brackish shores. Both the marine and freshwater Basommatophora graze on algae. Several freshwater forms are of great economic importance as they are intermediate hosts to schistosome flukes (Platyhelminthes), including those that infect man. The African *Bulinus truncatus*, for example, carries the parasite *Schistosoma hematobium* which causes the human disease bilharzia or schistosomiasis.

Perhaps the most fascinating evolutionary adaptations shown by gastropods are the reproductive strategies which have enabled them to invade freshwater and terrestrial environments. In primitive archaeogastropods, eggs and sperms are simply shed into seawater, and fertilization is external, by chance, and is accompanied by great wastage. As an advance on this, some forms such as *Conus* produce gelatinous egg ribbons, but fertilization is still external, the only economy being the evolution of triggered spawning, induced by the presence of another individual of the appropriate sex. The eggs and sperm are thus brought together in the same place and at the same time.

More advanced gastropods have evolved special genital ducts, first, to allow internal fertilization and, second, to provide nutritive and protective layers to the egg. The latter allows the shortening, or even suppression of a pelagic larval life, and thus eliminates the associated dangers. At its simplest, the genital duct is merely a groove which leads from the posterior opening to the anterior end of the mantle cavity. In females this may be developed into an enclosed tube which in some forms serves as a uterus for egg development; in males a penis may be developed. Sexes are separate in primitive gastropods, but opisthobranchs and pulmonates are hermaphrodite. A few species, such as the slipper limpet *Crepidula*, begin life as a male, pass through a hermaphrodite phase and finish life as a female.

In hermaphrodite gastropods, fertilization is generally mutual, but terrestrial forms have developed additional reproductive strategies. Relatively few, large-yolked eggs are produced, with a tough, calcified shell (in the land snail *Achantia* these are as large as a thrush's egg), and distinctive patterns of courtship occur. In the garden snail *Helix*, there is a special dart sac associated with the vagina, in which a crystalline calcareous spicule is secreted. Almost a centimetre long, this 'lover's dart' is driven deep into the body of the partner and serves to stimulate copulation. Sperms are enclosed in a chitinous envelope, the spermatophore, which the partners exchange during copulation.

In some slugs, the dart sac has evolved into a stabbing organ, the sarcobelum, with which mating individuals stimulate each other before copulation. Some of the larger slugs, such as *Limax*, climb trees or walls before mating, and after several hours of spiralling courtship, entwine themselves and leap, head downwards, into the air, where they hang suspended on a thick mucous strand up to 25 cm long. The giant (up to 10 cm) penes are unrolled, their free ends entwine to form a knot, and a sperm mass is exchanged. Penile retraction follows, the slugs ascend their mucous cord (devouring it as they go) and part. A particularly impressive species, *Limax redii*, is reported to possess a penis 85 cm long!

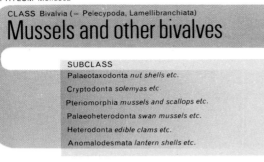

SUBKINGDOM Metazoa
PHYLUM Mollusca

CLASS Bivalvia (= Pelecypoda, Lamellibranchiata)

Mussels and other bivalves

SUBCLASS

Palaeotaxodonta *nut shells etc.*

Cryptodonta *solemyas etc.*

Pteriomorphia *mussels and scallops etc.*

Palaeoheterodonta *swan mussels etc.*

Heterodonta *edible clams etc.*

Anomalodesmata *lantern shells etc.*

- NUMBER OF LIVING SPECIES: about 30,000.

- DIAGNOSTIC FEATURES: molluscs with a shell of two valves hinged together dorsally and joined by a ligament; head reduced, lacking sense organs and radula; large posterior gills usually adapted for feeding as well as respiration.

- LIFESTYLE AND HABITAT: mostly free living in marine environments at all depths with crawling, burrowing, boring, attached or swimming forms; some in freshwater and a few parasitic.

- SIZE: under 1 mm to 1.5 m but mostly less than 20 cm.

- FOSSIL RECORD: middle Cambrian to Recent with a Tertiary peak; probably more than 20,000 species.

Clusters of the Edible Mussel (*Mytilus edulis*) are anchored by bunches of byssal fibres. The animals show the marked anterior reduction of the shell common in byssate bivalves.

The Bivalvia (also known as Pelecypoda or Lamellibranchiata)—clams, oysters, cockles, and mussels—are among the most widely encountered molluscs. They are a wholly aquatic group being found in fresh, brackish and marine waters. They inhabit all latitudes and extend from abyssal depths of 11,000 m to the intertidal zone and to temporary freshwater pools where they can survive long periods of desiccation. The earliest bivalves appeared in the middle part of the Cambrian period some 500 million years ago, and the group reached rock forming proportions during the late Palaeozoic, Mesozoic and Tertiary eras. It was during the Tertiary that they reached their maximum peak of diversity.

Bivalves show marked divergence from the basic molluscan archetype. The mantle has expanded greatly, and is modified into voluminous folds which hang down on either side of the visceral mass. The body has become compressed laterally, and the head has been reduced and lost as a reflection of the animal's sessile, passive mode of life. Typically,

the foot is modified into a hatchet-shaped burrowing organ although cemented, free bottom-living and byssate forms are known. Most bivalves are suspension feeders and filter seawater for suspended organic detritus and microorganisms. The gill, primarily a respiratory organ, has evolved a secondary filtration function, and is relatively greatly enlarged in comparison with that of other molluscs. The shell, which consists of left and right valves, is joined and hinged dorsally by a partially calcified organic ligament which tends constantly to pull the valves apart against the tension of the adductor muscles, which extend from valve to valve and bring about closure when they contract.

The whole of the bivalve body is enclosed by the mantle, which is typically fused around the margins except for limited regions, for example, where the foot extrudes. Within the mantle cavity thus produced, lie the visceral mass, the ventral foot, and the gills or ctenidia. Water currents flow in and out of specific openings in the mantle and bathe the gills. In

The anatomy of a bivalve. The left shell valve, gill and labial palps have been removed and the foot partly cut away to show the internal organs. Bivalves differ markedly from the archetypal mollusc, being laterally compressed and having no head.

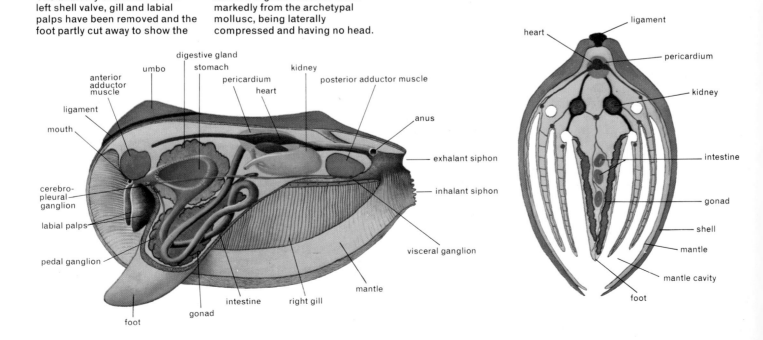

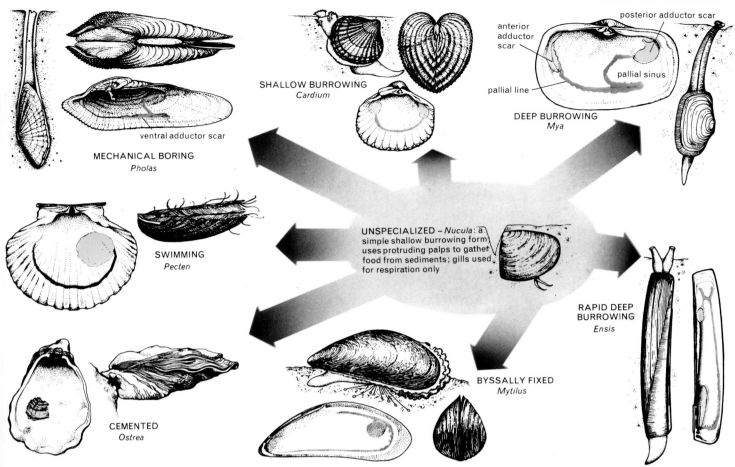

MECHANICAL BORING
Pholas

ventral adductor scar

SHALLOW BURROWING
Cardium

DEEP BURROWING
Mya

anterior adductor scar

posterior adductor scar

pallial sinus

pallial line

SWIMMING
Pecten

UNSPECIALIZED – *Nucula*: a simple shallow burrowing form uses protruding palps to gather food from sediments; gills used for respiration only

RAPID DEEP BURROWING
Ensis

CEMENTED
Ostrea

BYSSALLY FIXED
Mytilus

The bivalves have radiated from unspecialized forms like *Nucula* to fill a variety of niches. The habits of a species can often be deduced by examining the muscle scars inside its shell. Usually two adductor muscle scars are joined by the pallial line—the scar left by the mantle muscles. The anterior body parts of many swimming and fixed species are reduced and the anterior adductor muscle is small or absent. In *Ostrea* the adductor scar and pallial line coincide. Deep burrowing species have long siphons which protrude through gapes in the shell; a sinus in the pallial line marks the cavity into which the siphons can be at least partially withdrawn.

many bivalves fusion of the mantle has produced extensile tubular inhalant and exhalant siphons at the anterior end of the shell. Associated with the gills and situated on either side of the mouth are the flap-like labial palps—triangulate extensions of the lips. In early bivalves where the gills serve only as organs of respiration, these palps extend (as palp probiscides) through the margins of the valves, and pick up organic particles from the surrounding sediment; in suspension feeders, they aid in the sorting of filtered material and its transfer to the mouth. The anus discharges into the mantle cavity at the posterior end of the animal, and faeces are swept away in the exhalant current.

The foot is primarily an organ of locomotion and is used in burrowing, creeping and jumping. Sometimes associated with the foot is a byssus, a cluster of hair-like threads secreted by the byssal gland, and by which the animal may attach itself to the substrate. All bivalves are byssate during early life, and in some (for example, *Mytilus*) this method of attachment is retained to maturity. Sexes are separate in most bivalves, and fertilization is generally external. Some forms, for example, scallops and cockles, are hermaphrodite. Oysters change sex throughout their life cycle, and self fertilization occurs in some hermaphrodite freshwater pea mussels (Sphaeridae). Larval development typically consists of a free swimming trochophore, succeeded by a veliger larva, but direct development occurs in the pea mussels, and some swan mussels have a veliger stage which parasitises some species of fish.

The bivalve shell is the most conspicuous, often brightly coloured or strongly ornamented, part of the animal. The two valves are secreted by the mantle, and consist of very shallow assymetrical cones. External ornament and shape are markedly adapted to the diverse life habits adopted by the group, and the life style of both living and fossil bivalves is easily inferred from the shell morphology. Also, internal features such as the structure of the hinge plate which articulates the valves and lies below the beak-like umbo, the position and nature of the adductor and other muscle scars (byssal, pedal) together with shell microstructure are all used in assessing relationships and life habits within the group.

Modern burrowing bivalves typically have two equal, usually elongated valves. The forms that live in shifting sands, or burrow rapidly have streamlined, elongate, cyclindrical shells, as in the razor shell (*Ensis*). In deep burrowers, like the gaper *Mya*, anterior and posterior ends of the shell may gape permanently, and neither the foot, nor the long siphons connecting the animal to the sediment/water interface can be withdrawn. Other burrowers have developed sharp ribs with which to cut through sediment during burrowing; and some have ribs, flares and spines which help to anchor and stabilize them in sediment after burial. Typical shallow burrowing genera are the cockles (*Cardium*), *Mercenaria*, and *Venus* (commonly eaten as clams).

Boring bivalves inhabit hard substrates such as rock, shell and wood. These animals face essentially the same problems as do deep burrowers so far as feeding and respiration are concerned and in consequence have a similar body form— long siphons, a vertical life orientation, and permanent gapes. Many borers (for example, *Pholas*) have shells with short external spines at the anterior end which are used in rotational boring; others have fluted grooves which pick up sediment particles that are then used as tools. Another form, the date mussel, *Lithophaga*, bores chemically by secreting a compound which dissolves away calcareous substrates.

Byssal fixation is seen best in bivalves such as *Mytilus* and the pearl oyster *Pinctada*. In forms which lie on one valve (for example, scallops and *Pteria*), the two valves of the shell are of different shapes, and ear-like expansions develop at the hinge, in part to prevent the flipping over of the shell by water currents. Many byssate bivalves show a marked reduction of the anterior part of the body (most obviously reflected in a small anterior adductor muscle scar). This is an evolutionary change which results in the elevation of the posterior siphons well clear of the substrate and the migration of the byssal retractor muscles to a point directly above the byssus, thus increasing their efficiency in holding the shell fast.

Cementation, as displayed by many oysters, represents a far more permanent mode of attachment. The shell is thick and massive, and may bear spines—all features of protective advantage in exposed habitats. The lower, attached valve is larger, and the smaller upper valve is usually lid-like—an asymmetry which serves to lift the valve margin clear of the bottom and up into oxygen- and nutrient-rich waters. Swimming is best developed in scallops (family Pectinidae) and file shells (*Lima*), although even such unlikely animals as the razor shell (*Ensis*) can move in this way. In general, swimming is only an escape mechanism, perhaps evolved from the clapping of the valves to expel clogging sediment. Features associated with rapid swimmers are thin, reduced shells, marginal gapes through which water is expelled, and in forms such as *Pecten*, symmetrical ears and a large umbonal angle. Bivalves which swim vertically do so with limited efficiency since their shape offers little resistance to sinking.

The subclass Palaeotaxodonta is one of the first groups of bivalves to appear in the early Palaeozoic era. Living representatives are the nut shells—genera such as *Nucula* and *Nuculana*—in which the gills are typically used only for respiration, and the palp proboscides sift sediment for food. These animals live in organic-rich muds and are characterized by a comb-like hinge of many teeth, equal adductor muscle scars, and, usually, a nacreous (mother of pearl) shell on the interior. The subclass Cryptodonta is an essentially Palaeozoic group represented in living faunas by the genus *Solemya* which has survived unchanged since the Devonian period. This form has very simple gills used for both respiration and feeding and lives buried deep in sediment. The shell is brittle and has a very reduced hinge.

A much larger and more important group is the subclass Pteriomorphia, a diverse group today, but closely linked phylogenetically since its origin in the Ordovician period. Representative forms include ark shells (*Arca*) and dog cockles (*Glycimeris*)—solid, porcellaneous shells with many small teeth in each valve and a distinct separation of umbos. These range from the intertidal zone downwards and either burrow (*Glycimeris*), nestle in depressions or are byssally fixed (*Arca*). Better known are the many byssate and cemented forms such as the mussels (*Mytilus*), horse mussels (*Modiolus*), pearl mussels (*Pinctada*), fan shells (*Pinna*), oysters (for example, *Ostrea*), scallops (Pectinidae) and file shells (*Lima*). Most permanently attached bivalves fall into this group, which also includes some boring (for example, *Lithophaga*) and swimming (some pectinids) forms, and the largest

known bivalve, the Cretaceous *Inoceramus* which reached more than 1 m across. Many pteriomorphs have a nacreous inner shell and are the source of sea pearls.

The Palaeoheterodonta is a subclass characterized by the frequent differentiation of the hinge teeth into shorter cardinal teeth below the umbo and elongate lateral teeth on either side, and by the presence of a nacreous inner shell. The oldest known bivalves belong to this group, and they are particularly important in the Palaeozoic. Living representatives are the shallow marine family Trigonidae, a once important group of shallow burrowers in Mesozoic times, but now restricted to the genus *Neotrigonia* which lives off the shores of Australia, and the freshwater family Unionidae (swan mussels). The unionids first appeared in the Devonian, and are common fossils in Carboniferous sequences. Today, they have undergone a remarkable diversification in the lakes and rivers of the world, producing elongate, *Mytilus*-like forms, species with ears reminiscent of the marine pearl mussel, *Pinctada*, and even the oyster-like encrusting family Etheridae which occurs widely in the tropics. Swan mussels are the source of freshwater pearls.

The subclass Heterodonta differs from the Palaeoheterodonta in having non-nacreous shells, and a mantle margin fused into long siphons. This adaptation to a burrowing life became widespread in the early part of the Mesozoic era, and led to a great diversification of the group, which continues to the present day. Most are burrowers and inhabit all depths and environments, as, for example, do the genera *Venus*, *Ensis*, *Mercenaria*, *Mya*, *Tellina*, and *Cardium*. A lesser

LEFT: a swimming scallop, *Pecten jacobeus*, showing the sensory tentacles and eyes studding the mantle margin. Locomotion is by forceful expulsion of water on either side of the umbo.

RIGHT: *Tridacna gigas* is a giant clam which attaches itself to the substrate by means of a byssus. The permanently gaping shell is lined with thick mantle tissue, crowded with symbiotic algae.

BELOW: the Sand Gaper or Soft-shell Clam (*Mya arenaria*) is a deep-burrowing bivalve. Water is circulated around the mantle cavity through the long siphons which protrude from the burrow.

BELOW RIGHT: the Common Piddock (*Pholas dactylus*) can bore into quite hard rock using the sharp external ridges on its shell. Here the rock has been broken open to show the animal in its borehole.

number have adopted a byssate habit (*Cardita*), are borers (*Pholas*), or show an evolutionary convergence with oysters (for example, *Chama*). Other heterodonts have conquered freshwater, such as the mussel-like pea mussels (Sphaeridae). The giant clam *Tridacna* has evolved a symbiotic relationship with single-celled algae which live within its tissues.

The final subclass, the Anomalodesmata, is a further group of siphonate burrowers with nacreous shell interiors, and a reduced hinge. It first appears in the Ordovician period, and is represented today by genera such as *Pholadomya* and *Pandora*. One of the more remarkable features of some members of the group is the modification of the gill into a septum which divides the mantle cavity into two chambers one above the other. Food particles are strained off as water is pumped through the gill from the lower chamber into the upper exhalant chamber.

SUBKINGDOM Metazoa
PHYLUM Mollusca

CLASS Cephalopoda

Squids and other cephalopods

SUBCLASS	ORDER
Nautiloidea	
†Endoceratoidea	
†Actinoceratoidea	
†Bactritoidea	
†Ammonoidea *ammonites, ceratites and goniatites*	
Coleoidea	†Belemnitida *belemnites*
†extinct	†Phragmoteuthida
	Sepiida *squids and cuttlefish*
	Octopodida *octopuses*

- NUMBER OF LIVING SPECIES: about 800.

- DIAGNOSTIC FEATURES: molluscs in which the foot surrounds the head and is modified into tentacles; radula present; sense organs, particularly eyes, well-developed; highly muscular mantle cavity used in jet propulsion; shell external, reduced and internal, or absent.

- LIFESTYLE AND HABITAT: exclusively free living in marine environments at all depths.

- SIZE: 1 cm to 18 m but mostly 10 to 80 cm.

- FOSSIL RECORD: late Cambrian to Recent; especially common in Ordovician, Jurassic and Cretaceous; probably more than 20,000 species.

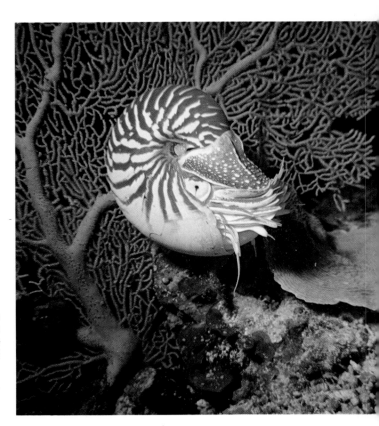

The name Cephalopoda means 'head-foot', and describes one of the diagnostic features of the class: the foot has migrated from the ventral position to surround the head, where it is modified into a crown of tentacles. Included in the class are the living squids (*Loligo*), cuttlefish (*Sepia*), octopuses (*Octopus*), argonauts (*Argonauta*), *Nautilus* and the fossil nautiloids, ammonoids and belemnites. The cephalopods have attained the largest size of any known invertebrates, and although most forms are from a few centimetres to a metre long, the North Atlantic giant squid *Architeuthis* has been reported to reach a length of 18 m with a body circumference of 4 m, and the Pacific octopus *Octopus punctatus* has arms up to 5 m long.

Of the two extant subclasses, the Nautiloidea are represented by *Nautilus*, sole surviving genus of the once dominant group of externally shelled cephalopods. Although widespread in the past, *Nautilus* is now confined to the southwestern Pacific and off southern Australia, with scattered records in the Bay of Bengal and off Madagascar. The shell is planospiraled: some 20–25 cm in diameter in adults, white in colour, and with irregular brown radial stripes on the early whorls and upper half of the mature shell. When sectioned, the shell is seen to consist of two regions: all but the last third of the outer whorl is divided into chambers by concave, saucer-like septa, perforated by a central siphuncular tube. This part of the shell (the phragmocone) is gas-filled, and acts as a buoyancy apparatus. The undivided, terminal part of the shell contains the body which extends back through the phragmocone as the siphuncular cord.

The nautiloid body consists of a visceral mass and mantle cavity. In the mantle cavity are two pairs of gills, and into it the gonads and anus discharge. The cavity opens externally below the head in the form of a tubular spout, the hyponome, through which water can be forcibly expelled by muscular contraction. The head is surrounded by a crown of about 90 tentacles, devoid of hooks or suckers. There is a pair of

large eyes, which lack a lens and function on the pinhole camera principle. The mouth is equipped with horny jaws rather like a parrot's beak, and there is a radula which aids in swallowing. Above the head is a tough fleshy fold, the hood, which closes off the shell when the animal withdraws.

Little is known of the biology of *Nautilus*. The animals are noctural, and are known from depths down to 600 m Food —crustaceans, fish and carrion—is grasped by the tentacles, and cut up by the jaws before being ingested. Locomotion is by means of forcible expulsion of water from the hyponome, the tentacles being used for temporary attachment and perhaps in limited movement. The shell gives the animal neutral buoyancy, and chambers are added throughout life, although the mechanism of growth is not well understood. The sexes are separate, and large, attached eggs are laid on hard objects. The young hatch as miniature adults.

LEFT: *Nautilus macromphalus* seen in the sea off New Caledonia. The shell, dorsal hood and crown of tentacles (a modified foot) are all cryptically coloured. The vertical section of the shell shows the gas-filled chambers that give balance and buoyancy, and the septal holes through which the siphuncle (an extension of the body) passes.

ABOVE: the squid *Loligo vulgaris* is a fast-swimming member of the class which occurs in large numbers in shallow waters. It is an active and efficient predator.

RIGHT: the diminutive, benthic cuttlefish *Sepiola* is 2–5 cm long. It hides itself in the sand to wait for its prey of small crustaceans.

Loligo, the Atlantic squid (order Sepiida), is a typical living member of subclass Coleoidea. Usually 20–30 cm long, the animal is streamlined, and has a long, tapering body, a rounded head with large eyes (equipped with a lens as in the vertebrate eye) and a crown of ten suckered arms. Two arms, the tentacles, are longer than the others, have flattened ends and are used for seizing prey. There is no external shell and the body instead is enclosed in muscular mantle tissues. The visceral mass is dorsal and the ventral mantle cavity houses the one pair of gills. The mantle cavity has a muscular wall, and when water is expelled through the forward-pointing hyponome the animal shoots backwards in a rapid jet propelled motion, steered by a pair of muscular posterior fins. The hard parts of *Loligo* are reduced to a long quill-like organic 'pen' above the visceral mass.

Squids are active swimmers and move around in huge shoals. Many have such powers of locomotion that they can leap out of the water. All have the ability to change colour to match their background or mood, and all possess an ink sac. When alarmed, this sac discharges a cloud of brownish sepia to confuse pursuers while the animal escapes. Sexes are separate in squids and complex courtship rituals have evolved. Squids show great morphological diversity, and although they are usually inhabitants of surface and coastal waters, some extend to the abyss. Deep sea forms such as *Chiroteuthis* have cylindrical bodies, large eyes, and incredibly long, whip-like arms up to six times their body length.

Cuttlefish (for example, *Sepia*) are also members of the order Sepiida but differ from squids in that they are generally dorsoventrally flattened, with a continuous fin along the sides of the body. Internally, they differ further in having a much larger, dorsal shell, the familiar 'cuttlebone'. This structure is derived from the external chambered shell of the nautiloid-like ancestors of the group and the soft, porous tissues form a part gas- and part liquid-filled float which gives the animal neutral buoyancy. Variation in the degree of flooding of the chambers allows the animals to migrate vertically through the water during their diurnal wanderings in search of food. Cuttlefish are common inhabitants of coastal waters and are mainly nocturnal scavengers and active predators.

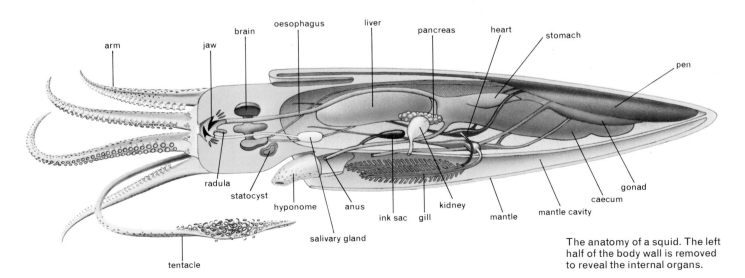

The anatomy of a squid. The left half of the body wall is removed to reveal the internal organs.

The other common group of living cephalopods is the Octopodida, distinguished most obviously from other coleoids by the presence of eight rather than ten arms. Octopods have lost all trace of a shell (except for the minutest vestige seen in the first—and only—fossil octopus, *Palaeoctopus newboldi*, from the late Cretaceous of the Lebanon), and the arms form a crown, linked together by a web which surrounds the mouth. Most octopods are benthic, and the sac-like body lacks fins in genera such as *Octopus* itself, although rapid jet propulsion is possible. The suckered arms are used to capture the prey and also serve as sense organs. The animal

LEFT: a female Paper Nautilus (*Argonauta nodosa*) resting inside her 'shell'; this is in fact an egg-case secreted by the tentacles and should not be confused with the ammonite or nautiloid shell.

BELOW: *Octopus vulgaris* at rest. The funnel-shaped structure at the front of the sac-like body is the hyponome. The eyes of octopuses are developed far beyond those of any other invertebrate group.

Mesozoic and Tertiary rocks, and even these are rarely found in any numbers. Only a handful of species in the genus *Nautilus* have survived through to the present day—true 'living fossil' representatives of this once major group.

Coleoids and ammonoids are better known descendants of the Palaeozoic nautiloid stock. The ammonoids evolved by way of a group of straight and curved nautiloid-like cephalopods known as bactritoids. Early forms, commonly termed goniatites, are widespread in Palaeozoic rocks, and are typically globose, smooth or feebly ornamented shells best distinguished by the fact that their septa were much more strongly folded (a mechanism to support the shell wall), so that the suture—the trace where the septa meet the shell, visible on moulds when the shell is removed—is markedly angular. The heyday of ammonoids was in the Mesozoic era, however, and the beautiful ammonite shells, with intricately frilled suture lines, are common fossils in many marine sediments.

Occurring in great numbers and variety—some 2,000 genera are known—ammonites probably occupied the ecological niches now filled by squids and octopods: as benthic browsers and scavengers, nektonic carnivores and plankton feeders. Ammonite shells vary enormously in size and coiling: some were adult at less than 1 cm in diameter, and the giant *Parapuzosia seppenradensis*, from the later part of the Cretaceous, reached more than 2 m across. Planospirally coiled shells are the most frequent, but at various times, in particular during the Cretaceous, the bizarrely uncoiled and sometimes straight ammonites became common. Sexual dimorphism with small decorated males (microconchs) was

Arcestes, a Triassic ammonite, preserved as a calcitic internal mould of the shell. The intricate pattern is a trace of the suture lines which mark the junctions between the septa and shell wall.

commonly hides in crevices or builds a house of pebbles and shells in which it lurks, waiting for any unsuspecting potential food supply to pass by.

Not all octopods live in this way. Members of the superfamily Argonautacea are pelagic inhabitants of the surface layers of the oceans. The paper nautilus *Argonauta* itself is best known from the coiled 'paper-thin 'shell'—actually an egg-case secreted by the arms of the female—which floats by virtue of entrapped air. The shell is carried about and serves both as a brood chamber for the eggs and as a retreat for the female, whose posterior end usually remains within it. The much smaller male (about a tenth of the size of his mate) does not secrete a shell and often lives in the female's shell. Deep sea octopods show perhaps the most remarkable anatomical modifications. In genera such as *Cirroteuthis*, the web reaches almost to the tips of the arms, and the suckers are modified into filaments to trap food particles. These animals swim slowly, opening and closing the web to feed; they have lost virtually all their muscle tissues and are transparent, and as fragile as a jellyfish. Octopods are the most intelligent invertebrates, with highly developed nervous systems, powerful vision, and some ability to learn from experience and to store knowledge.

The evolutionary origins of the Cephalopoda lie in the Monoplacophora sometime during the Cambrian period. The changes involved the heightening of the shell, the division of the apical part into chambers, and the development of a siphuncle which was able to control the gas/fluid proportions in the chambers, enabling the shell to take on a buoyancy function in addition to its early supporting and protective role. The earliest cephalopods from the latter part of the Cambrian period are tiny curved or cylindrical nautiloid-like shells, and this group evolved, in the Ordovician, into a range of loosely to tightly coiled, straight, curved or even helical, gastropod-like shells. These diverse nautiloid types survived into the later parts of the Palaeozoic; some of the straight forms reached great sizes—more than 9 m for the shell alone—and they were an important component of marine nekton. Nautiloid types as a whole declined dramatically towards the end of the Palaeozoic era; only a few genera occur in

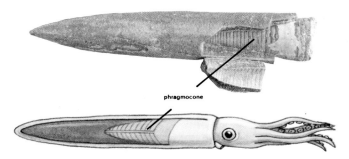

phragmocone

The internal skeleton, or guard, of a belemnite and a reconstruction showing its position in the animal. The front portion of the fossil has been broken open to show the chambered phragmocone. Belemnites were most abundant in the Jurassic and Cretaceous.

widespread in ammonites, suggesting that complex courtship rituals like those seen in recent coleoids had evolved.

A further important fossil group, especially in the later part of the Mesozoic, are the belemnites. These, like the ammonoids, arose from the bactritids towards the end of the Devonian, but here the tissues enveloped the shell. The phragmocone remained as a float, but was enclosed in a sheath of calcium carbonate—the familiar bullet-shaped belemnite guard—which acted as ballast and helped maintain neutral buoyancy and a horizontal position. The body chamber of the ancestral bactritid was reduced to a ventral protection for the viscera, and the external appearance of the animal was very much like a living squid. Although they were almost extinct by the close of the Mesozoic, the belemnites probably gave rise (by reduction of the internal shell and guard) to the squids (first known from the Carboniferous) and other modern coleoids.

The fossil record of the cephalopods can be seen as almost the complete opposite of that of the living forms. Today, only one externally shelled form, *Nautilus*, survives, and coleoids are dominant. In the past, the reverse was the case for most of Palaeozoic and Mesozoic times, when the shelled nautiloids and ammonoids were at their peak.

SUBKINGDOM Metazoa

PHYLUM Sipuncula

Peanut worms

FAMILY
Sipunculidae
Golfingiidae
Phascolosomatidae
Aspidosiphonidae

- NUMBER OF LIVING SPECIES: about 320

- DIAGNOSTIC FEATURES: unsegmented, coelomate worms, with body divided into trunk and a protrusible introvert bearing terminal tentacles; a long U-shaped gut with anus anterior and dorsal

- LIFESTYLE AND HABITATS: all marine, free living burrowers at all depths

- SIZE: extended length (trunk plus introvert) up to 65 cm, but most species between 3 and 15 cm

- FOSSIL RECORD: no certain remains, but deformities in mid-Devonian corals possibly caused by sipunculans

Sipunculus robustus shows the skin pattern characteristic of its family. The dorsal anus is clearly visible and at the posterior end two commensal bivalves can also be seen.

The Sipuncula are unsegmented, worm-like animals, readily recognized by their extensile introvert—a slender proboscis-like structure which is regularly protruded from, and withdrawn into, the main part of the body called the trunk, the latter being cylindrical, flask-like or even globular in shape. As a group, they are exclusively marine and are found in all seas, a few species, such as *Sipunculus nudus*, being almost cosmopolitan in distribution. Many forms occur between the tidemarks whereas others are found only in deep water and may extend to depths of 5,000 m or more. All sipunculans are defenceless, sedentary creatures that generally burrow into mud, sand or gravel, but some can bore into harder substrates such as coral rock. Most, if not all, are detritus or deposit feeders, ingesting organic material by swallowing the sediment in which they live using ciliated tentacles to transport particles to the mouth which is situated at the tip of the protruded introvert.

In feeding, the introvert is protruded hydraulically, fluid within the coelom being pushed into it on contraction of the muscles in the trunk wall. Withdrawal of the introvert is effected by the contraction of special retractor muscles, one to four in number, extending across the coelom from the introvert tip to the middle of the trunk. Food passes through a U-shaped gut to the anus situated on the anterior part of the trunk. The gut fills most of the body cavity and is usually coiled upon itself into a double spiral. The gonads are attached at the bases of the ventral retractor muscles.

In all sipunculans the sexes are externally alike and, with one exception, are separate. Gametes are released from the body cavity through the sac-like excretory organs (nephridia) which open to the exterior by small pores at the same level as the anus. Fertilization is external and the embryo normally develops to a floating trochophore larva before metamorphosing to become a bottom-dwelling adult. The pelagic phase is

The anatomy of *Golfingia* seen through the left side. Here the animal is shown in a partially extended state and the intestine does not occupy the full length of the body cavity.

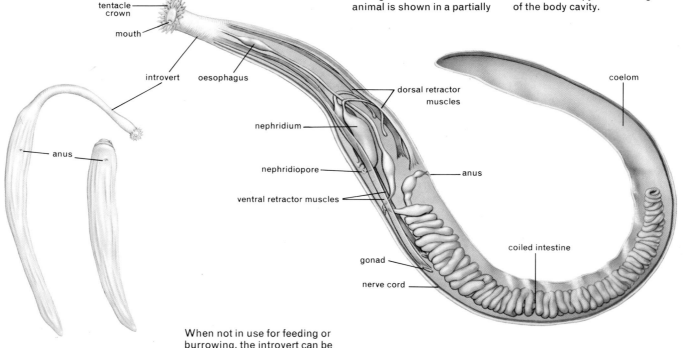

tentacle crown

mouth

introvert   oesophagus

anus

dorsal retractor muscles

nephridium

coelom

nephridiopore

anus

ventral retractor muscles

gonad

nerve cord

coiled intestine

When not in use for feeding or burrowing, the introvert can be fully withdrawn into the trunk.

often short but some species have a long planktonic development involving a characteristic larval stage known as a pelagosphaera.

As a group the Sipuncula show great variation in structure and not all forms can be easily accommodated in a scheme of classification. At present, four families are recognized, the chief characters separating them being the arrangement of the tentacles, the structure of the body wall and the presence or absence of areas of thickened skin, called shields, on the trunk. In the arrangement of the tentacles, two main types can be recognized; in one type, the tentacles are arranged in a circle around the mouth (families Golfingiidae and Sipunculidae) and, in the other, they are arranged in a crescent dorsal

The tip of the introvert of a typical golfingiid bears a ring of tentacles which completely encircles the crescent-shaped mouth. The arrangement in sipunculids is on a similar plan.

In phascolosomatids the tentacles form an incomplete ring with the nuchal organ lying in the gap. As it is ventral to the tentacles, the mouth here is obscured behind them.

to the mouth and enclosing the nuchal organ, a cushion-like sensory area (families Phascolosomatidae and Aspidosiphonidae). The body wall contains outer circular and inner longitudinal layers of muscle. In the golfingiids both muscle layers are continuous sheets, giving the body a smooth outline but in sipunculids both layers are in bands and often give the trunk surface a distinctive 'tessellated' aspect; in most species of the other two families only the longitudinal muscle is in bands, so giving these species a 'striped' appearance. Shields are a characteristic feature of aspidosiphonids and these thickened skin structures usually appear as darker caps on the anterior end, and sometimes also on the posterior end of the trunk.

The Golfingiidae, considered to be the most primitive family, contains the largest genus in the phylum—*Golfingia* —which has about 100 species; its curious name, a dedication to the game of golf, was invented by an eminent professor of zoology who studied a specimen while on a golfing holiday at St Andrews in Scotland in 1884! The Sipunculidae (for

Phascolosoma nigrescens removed from its burrow. It is common in warmer waters where it bores into coral rock. Usually only the introvert (shown here extended on the left) can be seen protruding from the burrow to gather the detritus which accumulates in crevices.

example, *Sipunculus* and *Siphonosoma*) are further characterized by the presence of longitudinal canals or sacs within the body wall; the family contains the largest-sized members of the phylum, several of which grow to more than 0.5 m long. The Phascolosomatidae include one large genus, *Phascolosoma*, which is distinguished by the numerous rings of flat, recurved hooks on the introvert behind the tentacles. The Aspidosiphonidae includes smallish species, most of which belong to the genus *Aspidosiphon* and are borers.

As with many sedentary burrowers, sipunculans act as hosts to a variety of animals, both commensals and parasites. On European shores, a careful search of the burrows of the common species *Golfingia elongata* or *G. vulgaris* will often reveal small bivalve molluscs, such as *Mysella bidentata*, lying alongside, but unattached to, the worm; other bivalves attach themselves by byssal threads to sipunculans, for example the *Erycina*-like species found in the tropics on *Sipunculus robustus*. Parasites are less well known but one copepod (*Akessonia*) effectively castrates its host *Golfingia minuta* while living in the coelom.

Many sipunculans, such as *Phascolion strombi* found in the Atlantic Ocean, occupy discarded gastropod, scaphopod and foraminiferan shells, thus gaining protection while remaining near or on the surface of the sediment. A tropical form, *Aspidosiphon jukesi*, shows greater specialization in this direction in that the young stage occupies a small shell onto which a solitary coral settles and subsequently completely overgrows the shell, thus providing an enlarging shelter for the growing sipunculan. On coral reefs, sipunculans are a major destructive element since they bore into the coralline rock, thus considerably weakening its structure. The mechanism by which boring is achieved is poorly understood but it is likely that chemical secretions alter the calcareous framework of the rock thus permitting its abrasion by hard skin structures, such as the shields.

Borings and growth deformities, such as those produced by *Aspidosiphon*, provide the only evidence of sipunculans in the geological record, no plausible fossilized remains of soft parts having been described. Without such clues, the affinities of the Sipuncula to other invertebrate groups are unclear. The similarity in the pattern of development indicates a close relationship to the Annelida, but the total lack of segmentation is regarded as such a basic difference that the group is given the status of a separate phylum.

SUBKINGDOM Metazoa

PHYLUM Echiura

Spoon worms

ORDER	FAMILY
Echiuroinea	Bonelliidae
	Echiuridae
Xenopneusta	Urechidae
Heteromyota	Ikedaidae

- NUMBER OF LIVING SPECIES: about 130

- DIAGNOSTIC FEATURES: unsegmented, coelomate worms, with a non-retractable muscular proboscis and a ventral pair of spines behind the mouth.

- LIFESTYLE AND HABITAT: mostly marine, but a few estuarine; free living, mostly in burrows

- SIZE: body ranges up to 75 cm long, but most species between 3 and 15 cm long; proboscis extensile and of variable length

- FOSSIL RECORD: none substantiated

The Echiura are called spoon worms because of their scoop-like proboscis which originates from the anterior end of a sausage- or cigar-shaped body or trunk. This proboscis is highly muscular and often capable of great extension but, un-like the introvert of sipunculans, cannot be withdrawn into the trunk. Externally, echiurans have few features; most have a pair of spines below the mouth and some also have one or more rings of spines around the posterior end of the trunk, but for the most part, the different genera and species are recognized by their internal anatomy. Nearly all echiurans are marine, but a few are known to live in brackish water. As a group, they are worldwide in distribution and occur from the littoral zone to the abyssal regions, some species extending to depths of 6,000 to 10,000 m.

Echiurans are soft-bodied and helpless creatures; they gain a measure of protection from their enemies by construct-ing more or less permanent burrows in mud and sand or in rock crevices and similar niches. Most of them feed on detri-tus, eating organic material mixed with sediment, and the tip of the proboscis is used to select particles which are then transported by cilia along the length of the proboscis to the mouth. In *Bonellia*, for example, the proboscis may be extended to a length of up to 3 m in searching for food, but in other species, such as *Echiurus*, the scoop-like proboscis is

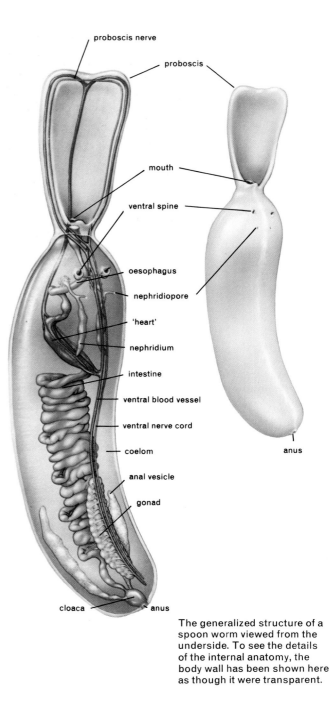

The generalized structure of a spoon worm viewed from the underside. To see the details of the internal anatomy, the body wall has been shown here as though it were transparent.

The greatly extended proboscis of *Bonellia* in action, sweeping over the sea bed in its search for food. It is thought to creep along by the action of cilia on the underside.

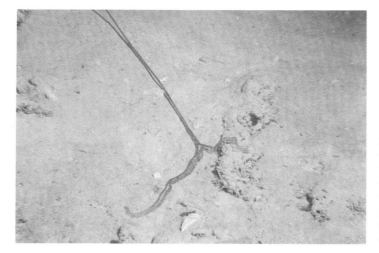

not so extensible and is used in a less selective manner to gather large quantities of mud. A more specialized form of feeding has evolved in another species, *Urechis*, which secretes a slime funnel within its burrow and this funnel acts as a sieve, straining out particles suspended in the water pumped through the burrow; after a period of time, the funnel with its trapped food material is eaten and a new funnel is formed.

Internally, the body cavity of echiurans is mostly occupied by a long gut; often this is three times the length of the trunk, and closely associated with its middle section there is a tubular structure known as the siphon, the function of which is obscure. Before opening at the anus, the gut expands to form a cloaca and in some species this is used for respiration, water being exchanged through the anus by pumping. Two voluminous sacs empty into the cloaca; these are the anal vesicles which have an excretory function. A closed blood vascular system may or may not be present. In the anterior region of the body cavity there are usually one to four pairs of thin-walled sacs (or nephridia) which act as storage organs for mature eggs and sperm.

ABOVE: *Bonellia viridis*, a green Mediterranean species, removed from its burrow. The soft, sac-like body is about 5cm long. The bifid proboscis, here much contracted, can be clearly seen.

RIGHT: the Innkeeper (*Urechis caupo*) with its commensals.

Clevelandia ios

slime funnel

proboscis

Hesperonoë adventor

Cryptomya californica

Scleroplax granulata

The sexes are separate and mostly they are indistinguishable externally, although in the Bonelliidae the two sexes are very different. The gonads lie above the nerve cord or close to the cloaca; the sex cells or gametes develop in the body cavity and are stored in the nephridia before release to the exterior where fertilization takes place. Larval development involves a planktonic trochophore stage which metamorphoses to a bottom-dwelling adult.

Of the three orders of Echiura one, the Heteromyota, contains only one obscure species; the chief differences between the other two orders, Xenopneusta and Echiuroinea, lie in the vascular system and the cloaca. In the Urechidae (the only family in the Xenopneusta) there are no blood vessels and the cloaca acts as a respiratory organ; whereas in the Echiuroinea, blood vessels are present and the cloaca is not modified for respiration. All four species of the Urechidae belong to the genus *Urechis* and have a short proboscis; the best known example is *Urechis caupo*, a large species up to 5 cm in diameter and 45 cm long, found on the coast of California. In the Echiuroinea the Bonelliidae are distinguished from the Echiuridae by having males which are usually very small and parasitic on the female. The proboscis of bonelliids usually has a bifid tip and is very extensible, often to 2 or 3 m; many of the 50 or so species in this family live in abyssal regions. The Echiuridae—the largest family with some 80 species—typically have a short, thick, scoop-like proboscis, which is not bifid, a greenish or reddish coloured trunk, and are commonly found in shallow water.

Among echiurans, *Urechis caupo* from California is one of the most interesting because it acts as a host to a variety of commensal animals which utilize its burrow as a refuge. This 'hospitability' is indicated by its specific name, *caupo* being Latin for 'inn keeper'. Among its usual 'guests' are a scale worm, *Hesperonoe*, the crabs *Scleroplax* and *Pinnixa*, a goby fish *Clevelandia* and a small clam, *Cryptomya*. Apart from protection, all these animals benefit from living with *Urechis* in that the current of water pumped through the burrow by the spoon worm brings them a supply of both food and oxygen.

A remarkable feature of echiurans is the size difference between the male and female of bonelliids—the female *Bonellia* may be 20,000 times larger than the male, probably

the greatest difference known in the animal kingdom. The bonelliid male is small, often only a few millimetres long, and is degenerate in that it lacks a proboscis and vascular system, since both are unnecessary in its parasitic mode of life on or in the female. It resides in the gut or in the nephridium where the eggs are stored, and in some instances up to ten males may be found living on one female. The manner in which the sex of an individual is determined is equally curious. The larvae swim close to the sea bottom and if before settling a larva does not make contact with a female worm, it develops into a female; if it does encounter a female, however, it settles and develops into a parasitic male, at first living on the proboscis and later entering the female trunk through the mouth. Evidence suggests that a chemical secreted by the female proboscis is responsible for inducing the development of the males.

Echiurans have no fossil record and the affinities of these animals are not readily apparent. As with sipunculans, they are similar to the Annelida in their development but are quite distinct from that phylum in that they lack completely any signs of segmentation.

SUBKINGDOM Metazoa

PHYLUM Annelida

Earthworms and other annelids

CLASS	SUBCLASS	ORDER	SUBORDER
Aclitellata	Polychaeta *marine worms*	Orbiniida	
		Cossurida	
		Spionida	
		Capitellida *lugworms etc.*	
		Opheliida	
		Phyllodocida	Phyllodocina *paddle worms*
		Spintherida	Aphroditina *scale worms and sea mice*
		Amphinomida	Nereidina *ragworms etc.*
		Eunicida	
		Sternaspida	Glycerina
		Oweniida	
		Flabelligerida	
		Terebellida	
		Sabellida *fan worms etc.*	
		(+ 3 further minor orders)	
	Myzostomaria		
Clitellata	Oligochaeta *earthworms etc.*	Lumbriculida	
		Moniligastrida	
		Haplotaxida	Haplotaxina
			Tubificina
			Lumbricina
	Branchiobdellida		
	Hirudinea *leeches*	Acanthobdellida	
		Pharyngobdellida *worm leeches*	
		Gnathobdellida *jawed leeches*	
		Rhynchobdellida *proboscis leeches*	

- NUMBER OF LIVING SPECIES: about 9,000
- DIAGNOSTIC FEATURES: segmented, coelomate, soft-bodied worms, with body covered by a thin, flexible cuticle which often bears segmentally arranged setae
- LIFESTYLE AND HABITATS: marine, freshwater and terrestrial; mostly free living, some are sedentary or tubicolous; a few commensal and parasitic forms
- SIZE: 5 mm to 6 m but mostly 5 to 20 cm
- FOSSIL RECORD: polychaete tubes or burrows recorded from Precambrian; tiny, toothed fossils (scolecodonts), possibly jaws of polychaetes, known from Cambrian

BELOW: *Lumbricus terrestris* mates on the surface of the soil at night. The sperm from the male genital pores of each worm passes into the spermathecal pores of the other. A few days after copulation, each worm produces several cocoons in which sperm and ova are placed.

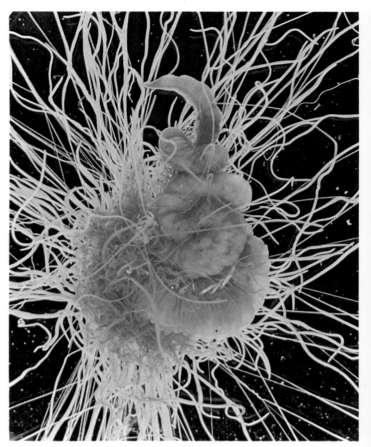

Most terebellids construct a tube to live in, but *Polycirrus calien-drum* lives naked in quiet marine waters among algal roots, pulling itself about by its tentacles. It may be up to 100 mm long.

The Annelida derive their name from the segmented, or annulated, nature of their bodies. The number of segments making up an individual varies considerably: in small forms there may be as few as seven whereas the largest may have as many as five or six hundred segments. Each segment is basically identical and contains excretory organs (nephridia) and portions of the gut, blood system and double ventral nerve cord. Only the reproductive organs are not present in every segment. Many annelids have segmentally arranged setae (sometimes called chaetae)—chitinous bristles used in locomotion. Annelids are present in marine, freshwater and terrestrial habitats. In most species the skin acts as a respiratory surface and since it is covered only by a thin, permeable cuticle it easily loses moisture. Because of this susceptibility to desiccation terrestrial annelids are restricted to habitats that always retain some degree of moisture.

Between the body wall and the gut is a fluid-filled cavity, the coelom. This acts as a hydrostatic skeleton against which the muscle layers of the body wall act. The coelomic fluid is incompressible and contraction of the circular muscles causes a compensating increase in length; contraction of the longitudinal muscles causes a reduction in length. Since the coelom is partitioned by intersegmental septa different parts of the worm's body can elongate and contract independently.

Two different types of life cycle are observed in the Annelida, each characteristic of one of the two classes recognized here (the classification used is a composite one which has been assembled from several sources). Members of the class Aclitellata, all of which are marine, release gametes into the sea where, after fertilization, the ova develop into trochophore larvae. These small ciliated larvae become part of the plankton, where they feed, develop segments and become worm-like. Eventually they sink to the sea bottom and develop into adults. Occasionally the fertilized eggs are retained for a time by the parent and released as well developed larvae.

Most members of the other class, the Clitellata—named because of their saddle or clitellum—occur in freshwater and terrestrial habitats. In these annelids the larval stage has been suppressed and the embryo develops directly into a small adult. All of the members of this class are hermaphrodite. Two individuals exchange sperm, separate and eventually produce cocoons which are secreted by the clitellum. Into each cocoon ova and sperm are deposited. Fertilization occurs and the embryo grows, feeding on the yolk and the albumen which fills the cocoon. Eventually the young individuals emerge from the cocoon. Some of the glossiphoniid leeches (in the Rhynchobdellida) carry the cocoon, and later the young, on the ventral side of their bodies.

The Polychaeta (bristle worms) form the larger of the two subclasses recognized within the Aclitellata. They are all unisexual and have numerous setae on parapodia which are segmentally arranged lobes on the sides of the body. They often have antennae. The polychaetes are the only annelids to have a fossil record. It consists only of worm tubes, impressions of burrows and the occasional individual, and also scolecodonts—tiny toothed or jaw-like fossils thought to be the jaws of predatory polychaetes. From these few traces it has been concluded that many living polychaete families had evolved by the Cambrian period.

Most polychaete families and genera are widely distributed in the world's seas and in some areas they represent a major portion of the bottom fauna. The polychaetes occupy several ecological niches and are often highly adapted for locomotion or food gathering. A survey of the different feeding methods will serve as an introduction to the 17 orders recognized. It should be noted that orders with similar feeding habits are not necessarily closely related: often they have evolved superficially similar feeding structures independently. The polychaetes fall into two loose groups—sedentary forms feeding on detritus and active forms which are browsers, scavengers and predators. (There are also some specialized parasitic forms.) In the sea there is a continuous rain of organic particles onto the sea floor where they are eventually incorporated into the sediment. The sedentary polychaetes exploit this food supply, feeding on the detritus, either in the sediment, on the surface of the sediment or in the water.

Among the simplest polychaetes are the burrowing forms which comprise the orders Capitellida, Cossurida, Orbiniida, Opheliida and Sternaspida. They usually lack antennae and although most have an eversible pharynx it always lacks jaws. They feed on organic matter which they ingest, often non-selectively as they burrow through the sediment. These worms are often very common on intertidal mud flats. Members of the family Arenicolidae (order Capitellida) are known to fishermen as lugworms or sandworms. The U-shaped burrow is visible on the surface as a shallow depression at the head end and a worm cast at the tail end. Detritus in the mud and on the surface is carried down into the burrow where it is ingested by the worm.

Members of the order Spionida and some families of the Flabelligerida and Terebellida feed on detritus on the sea floor. They live under stones, in burrows or tubes and search the surface of the mud or sand with tentacles. Particles are carried to the mouth by ciliated grooves on the tentacles or by muscular contractions. At the mouth particles are sorted and any which are too large are rejected.

The most specialized of the particulate feeders are those which filter suspended material from the water. Members of the order Oweniida, which includes only one family, gather floating particles by means of a frilly membrane around the head. Ciliated grooves pass the food to the mouth. Although the oweniids live in tubes made of sand grains they are capable of moving about the sea floor, carrying their tubes with them.

A lumbricid earthworm dissected to show the reproductive system (left) and the general structure. The blood flows anteriorly in the contractile dorsal vessel and posteriorly in the ventral and subneural vessels. The nerve cord is composed of two types of nerve cells—ordinary fibres and giant fibres; the latter are responsible for fast transmission of impulses from one end of the body to the other. Earthworms do not have any specialized sense organs but do have chemosensory, photoreceptor and tactile cells in the epidermis. Some other annelids have simple eyes and some polychaetes have very specialized eyes. Respiration is through the whole body surface.

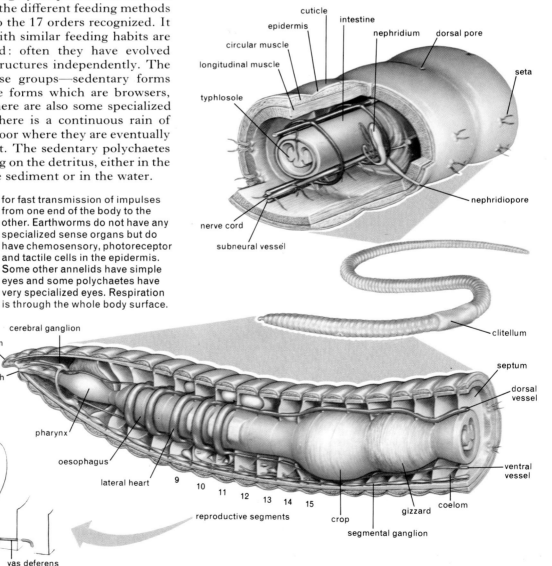

cuticle
epidermis
circular muscle
longitudinal muscle
typhlosole
nerve cord
subneural vessel
intestine
nephridium
dorsal pore
seta
nephridiopore

cerebral ganglion
prostomium
mouth
seminal vesicle
pharynx
oesophagus
lateral heart
9
10
11
12
13
14
15
reproductive segments
crop
segmental ganglion
gizzard
coelom
clitellum
septum
dorsal vessel
ventral vessel

spermatheca
testis
testis sac
ovary
oviduct
vas deferens

Members of the Sabellida, and some families of the Terebellida, filter particles from the sea with a fan of tentacles arranged around their heads. These are the fan-worms, most of which makes tubes of sand but some, the spirorbids and serpulids (both members of the Sabellida), form calcareous tubes. The Sabellariidae (order Terebellida) form extensive reefs with their sandy tubes.

The active polychaetes belong to the orders Amphinomida, Eunicida and Phyllodocida. About half of all known species are included in these orders. Most are active scavengers and predators. A few make tubes but the majority live in temporary burrows in the sand or mud, live under stones or swim in the plankton. Their bodies may be elongate and worm-like or short and flattened. They all have an eversible pharynx and in many of the predators and scavengers this is armed with one or more pairs of jaws.

Members of the order Amphinomida are slow moving carnivores which feed on attached organisms such as sponges, hydroid coelenterates and sea squirts. Although they lack jaws they have a muscular, rasp-like pad in their pharynx with which they pierce their prey to suck out the body fluids. The family Amphinomidae, the fireworms, are included in this order. These oval, flattened, often brightly coloured animals have sharp spines.

ABOVE: the cosmopolitan serpulid *Spirobranchus gigantica* can be up to 12 cm in length. If disturbed, it rapidly withdraws into its tube, which it then seals with a calcareous operculum.

BELOW: the sea mouse *Aphrodita* lives in soft, offshore sediments. The segmentally-arranged parapodia assist in locomotion. Long, iridescent setae overlie the elytrae on the back.

LEFT: the ragworm *Nereis* is a widespread polychaete which lives among rocks and stones. It is an active predator and bears a proboscis which is armed with a pair of sharp jaws.

ABOVE: *Tubifex* (Tubificidae) are small worms which live head downwards in the mud of rivers and streams throughout the world. They feed on detritus and are common in polluted waters.

A variety of lifestyles is present in the order Phyllodocida. The worms of the suborder Nereidina, which includes the ragworms (family Nereidae) are active scavengers and omnivores, living in shallow burrows and under rocks. The Syllidae, which also belong to this suborder, and members of the order Eunicida form sexual individuals by budding. These headless individuals swarm to the surface to release their gametes. The swarming of these worms can be predicted and in the Pacific, where they are considered a delicacy, they are netted in great numbers. The suborders Phyllodocina and Glycerina include both planktonic and bottom dwelling forms. The planktonic members are transparent and often very beautiful. Also included in the Phyllodocina are the paddle worms, the phyllodocids. These worms have enlarged paddle-shaped lobes along the sides of their bodies with which they can swim actively. The scale worms and sea mice are included in the suborder Aphroditina. Both have the dorsal surfaces of their bodies covered partially or completely by scales (elytrae). Some of these worms are slow moving carnivores and detritus feeders whose tracks can be found on undisturbed sediments. Others live in tubes, formed of silky threads, from which they forage for food.

Relatively few polychaetes have adopted parasitic modes of life but some fish and crustacean parasites are found in the Eunicida and the order Spintherida, which includes a single family, is ectoparasitic on sponges.

Until recently another subclass Archiannelida was recognized to accommodate five families of small, simple interstitial worms, each with a ciliated epidermis. It was thought that this group represented the ancestors of the phylum. Zoologists now consider that these worms are not primitive but have evolved from several distinct groups of polychaetes, and the group is not considered separately here. The class Aclitellata also includes the subclass Myzostomaria which contains only one family, all members of which are parasites of crinoids and other echinoderms. They have flattened, circular bodies with reduced parapodia and hook-shaped setae. Internally, segmentation has been lost and, unlike the polychaetes, they are hermaphrodite.

The most primitive of the Clitellata—those annelids which form cocoons to protect their eggs—are the oligochaetes (earthworms and their allies) which resemble polychaetes in their possession of setae although, as their name suggests, they usually have fewer. They can be divided into two ecological groups which are conceptually useful although they are not closely linked to the current classification whose basis lies in the arrangement of the reproductive system. The first group, the microdriles, are small, usually about 10 mm long and rarely exceeding 50 mm, and are usually aquatic; the other group, the megadriles, are larger, often up to 4 m long and are usually terrestrial.

The ecological niches occupied in freshwater by the microdrile oligochaetes are similar to those of burrowing and browsing polychaetes in marine habitats. Some, such as the Tubificidae (suborder Tubificina) are tolerant of pollution and are often present in large numbers in polluted rivers and streams. The earthworms and the large aquatic species make up the megadriles. They are members of the order Moniligastrida (which includes only a single family from India and eastern Asia) and the haplotaxid suborder, Lumbricina. They live in burrows in the soil, under stones, in forest litter and under moss. Earthworms feed either by ingesting organic matter as they burrow through the soil or by browsing on plant material on the surface. The latter occurs at night and browsing individuals usually keep their tails in their burrows so that if they are disturbed by a predator they can rapidly withdraw to safety.

The number of earthworms in the soil is often very high. Estimates as high as 250 individuals per square metre have been made for good pasture land. Earthworms are beneficial to the soil in a variety of ways: their burrows, which often are several metres deep, help to provide drainage and aeration; their casts, which in many species are deposited on the surface, break down to provide a layer of fine soil; and the feeding of the browsing species results in vegetable matter being rapidly broken down and incorporated in the soil. Although earthworms are susceptible to both desiccation and cold weather they are able to survive adverse conditions. They burrow deep into the soil where it is moist and temperatures are more stable. Here they coil up, secrete mucus which hardens to form a protective, moisture conserving layer, and become quiescent.

Most earthworms belong to the superfamilies Lumbricoidea and Megascolecoidea of the suborder Lumbricina. Unlike other annelids the natural ranges of these superfamilies are limited and apparently reflect the movements of continental masses during the Mesozoic and Caenozoic geological eras. Most of the Megascolecoidea occur in the land masses that made up the ancient continent of Gondwanaland, namely South America, Africa, India, Australia and New Zealand. It is thought that they evolved during the Mesozoic before Gondwanaland broke up. From these moving continents they have invaded North America and Asia. The most successful of these invasions, that of the *Pheretima* group of genera, crossed into Asia from Australia by way of Indonesia. The group now dominates the earthworm fauna of eastern Asia; about 750 species have been described—more than 10 per cent of all known oligochaetes. The Lumbricoidea originated at about the same time in Laurasia before this continent broke up to form North America and Eurasia. The more primitive families of this group, such as the Glossoscolecidae, many of which are aquatic, have spread to South America, Africa and India. The truly terrestrial family, Lumbricidae, is restricted to the Holarctic zoogeographical region.

The natural distributions of a small number of earthworms have been augmented greatly by man. Unintentionally, while transporting plants around the world, he has carried these species to many distant countries. The first records of *Amynthas diffringens* and *Metaphire californica*, both members of the *Pheretima* group, were from a greenhouse in North Wales and California respectively although their homeland is in southeast Asia. Man's actions have had catastrophic effects on the natural earthworm fauna of some areas. In New

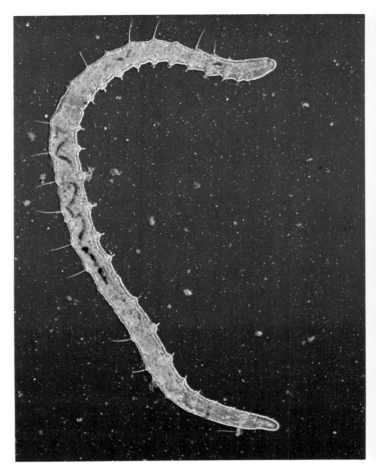

BELOW: *Lumbricus terrestris* is a European lumbricid earthworm. The mouth of its burrow, here shown cut in section, is often marked by the leaves on which the animal browses.

Zealand the native earthworms have been replaced in cultivated areas by European Lumbricidae.

Both of the subclasses Branchiobdellida and Hirudinea have adopted predatory or parasitic modes of life. In both groups locomotion is assisted by the possession of an anterior and a posterior sucker and all species, except those in the Acanthobdellida, have lost their setae. The number of segments is constant—15 in the Branchiobdellida and 33 in the Hirudinea. The Branchiobdellida occur in Europe and North America where they are parasites of freshwater crustacea; the Hirudinea occur throughout the world and have a large range of prey and hosts.

Those leeches in the Hirudinea which are parasitic have pouches in their gut for the storage of large quantities of blood. In these species the saliva contains an anticoagulant to prevent the blood of its host from clotting. Of the four orders of leeches the Acanthobdellida is the most primitive. The leeches of the single genus included in this order are small, worm-like fish parasites from Scandinavia and Lake Baikal in the USSR. They have setae on several of their anterior segments and lack an anterior sucker. The other three orders are cosmopolitan in distribution. The worm leeches, members of the order Pharyngobdellida, devour a wide variety of invertebrates and carrion. The majority live in freshwater and the few terrestrial species return to water to breed.

The jawed leeches (Gnathobdellida) resemble the worm leeches but have toothed jaws. Although some are predators of invertebrates most are active blood suckers. The Medicinal Leech (*Hirudo medicinalis*) belongs to this order as do the land leeches of southeast Asia. Some of the parasitic members of this order live in freshwater, attacking their hosts when they drink. They often enter the nasal cavities where they may cause serious injury. Others are adapted to terrestrial life and in humid areas such as southeast Asia they are present in large numbers wherever the habitat is suitable.

The proboscis leeches, the order Rhynchobdellida, lack

ABOVE LEFT: *Nais* (Tubificina) is a small freshwater oligochaete less than 12 mm long. As well as reproducing sexually, it may also bud asexually to form chains of miniature individuals.

ABOVE: *Piscicola geometra* is a European leech and one of the few piscicolids found in freshwater. When not on a host, it attaches itself to a solid object and sways to and fro, seeking another fish.

RIGHT: the Medicinal Leech lives in freshwater and sucks the blood of mammals, attacking them as they drink. Its saliva contains an anticoagulant to prevent the host's blood from clotting. During the Middle Ages physicians used it for letting blood from their patients, a treatment of dubious value, known as phlebotomy.

jaws but are armed with a muscular proboscis with which they penetrate the body of their prey. This structure is less efficient than the jaws of the gnathobdellids and the proboscis leeches are rarely parasites of mammals. The two families included in this order differ significantly in their distribution and morphology. The Glossiphoniidae, the leaf leeches, live in freshwater and have flattened, leaf-like bodies. Most of them suck the body fluids of snails and other invertebrates, often completing the job by sucking up the soft tissues as well. A few are parasites of fish or even waterfowl. Parasites of the latter enter the nasal cavities and may cause death by choking or loss of blood. The glossiphoniids often exhibit parental care, carrying the cocoon beneath their bodies and even carrying their young for several months. Most of the second family, the Piscicolidae are from marine habitats where they are parasites of fish. They have slim, elongate bodies, often with very large suckers and gills or fins along the sides of their bodies. Little is known, however, of this interesting group of marine parasites.

From this brief description it can be seen that annelids are not only very diverse in structure but also in habitat and lifestyle, and this is particularly the case among polychaetes. The polychaetes were probably the first annelids to evolve, appearing during the Precambrian period. Possibly from this primitive annelid stock the arthropods evolved: a common ancestry is suggested by the fact that both groups have segmented bodies, a double ventral nerve cord and chitin (in the exoskeleton of arthropods and in the setae of annelids); but it is a matter for debate whether these features evolved just the once or several times independently.

SUBKINGDOM Metazoa

PHYLUM Tardigrada

Water bears

ORDER	SUBORDER
Heterotardigrada	Arthrotardigrada
	Echiniscoidea
Mesotardigrada	
Eutardigrada	

- NUMBER OF LIVING SPECIES: about 400

- DIAGNOSTIC FEATURES: coelomate, and superficially segmented, with non-chitinous cuticle, four pairs of stumpy, unjointed legs ending in toes or claws and with muscular bulbous pharynx; head not demarcated

- LIFESTYLE AND HABITAT: mostly free living, usually in water films on leaves of mosses and lichens, but some live on marine and freshwater plants or in bottom sediments; a few commensal and one parasitic species

- SIZE: 0.05 to 1.2 mm, but most less than 0.5 mm

- FOSSIL RECORD: one species from Cretaceous amber

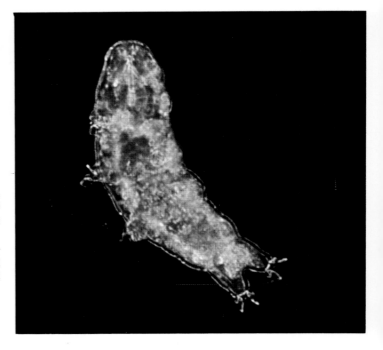

BELOW: some features of tardigrade structure: an external view of a heterotardigrade showing cuticular plates and cephalic cirri; the anatomy of a eutardigrade; and a selection of claw types.

ABOVE: like all the eutardigrades this one lacks cephalic cirri. Its cuticle is smooth, and it has unequal bifurcate claws. The last pair of legs typically trails behind the animal.

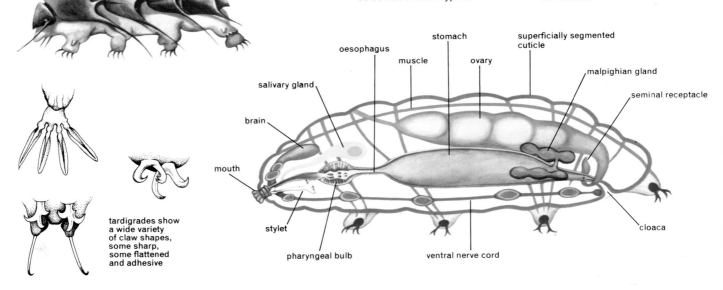

cuticular plate

tardigrades show a wide variety of claw shapes, some sharp, some flattened and adhesive

oesophagus

stomach

superficially segmented cuticle

muscle

ovary

salivary gland

malpighian gland

seminal receptacle

brain

mouth

stylet

pharyngeal bulb

ventral nerve cord

cloaca

The Tardigrada are a group of tiny, specialized animals which have four pairs of stubby clawed legs and move with a bear-like, lumbering gait, a characteristic which gives rise to their common name of water bears. Because of their small size (less than 1 mm) they are frequently overlooked, yet can occur at densities of several millions per square metre in suitable habitats. Most species live in the fluid-filled interstitial spaces of soils and sediments, or in the water films surrounding the leaves of mosses, lichens, liverworts and some flowering plants. Others seem to be commensal on marine invertebrates and one species, *Echiniscus molluscorum*, has adopted a parasitic existence on the giant Puerto Rican land snail, *Bulimulus exilis*.

The affinities of the tardigrades are uncertain. The structure of their special feeding apparatus is similar to both the mastax of rotifers and the pharynx of nematodes. Unlike the aschelminths, however, tardigrades are coelomate. The ladder-like arrangement of the tardigrade nervous system, and the possession of a cuticle which has many similarities to that of insects (although it sometimes retains a mucous coat) place the group somewhere along the annelid–arthropod line.

The diet of tardigrades consists largely of vegetable material, for example bacteria, algae and detritus. Some species are occasionally predatory, and *Milnesium tardigradum* is almost completely carnivorous, feeding on rotifers, nematodes and other tardigrades. Food is taken into the mouth by the combined action of a pair of stylets and a bulbous muscular pharynx (the pharyngeal bulb). The stylets act like hypodermic syringes because they can be projected to penetrate the food material, and the pharyngeal bulb is like a suction pump because it can draw up intercellular fluid and cells into the alimentary canal.

Two distinct types of Tardigrada can be recognized on the basis of external morphology; those with lateral bristles or cirri on their anterior end (Mesotardigrada and Heterotardigrada), and those without such cephalic cirri (Eutardigrada). The order Mesotardigrada contains only one species,

Thermozodium esakii, which differs from heterotardigrades in that it possesses cuticular supports or rods in the pharyngeal bulb. The feeding apparatus is also important in the systematics of the Eutardigrada where the structure of the stylets and their supports, and the shape, number and arrangement of cuticular rods within the pharyngeal bulb differ between species. Many Heterotardigrada have prominent cuticular plates (*Echiniscus*) or lateral cuticular extensions (*Actinarctus*), but Eutardigrada may have a smooth, variously sculpted or punctate cuticle.

The sexes are separate, although the degree of sexual dimorphism is slight. Males have never been recorded for many heterotardigrades and parthenogenesis probably occurs. The single gonad is an elongate sac which lies beneath the dorsal cuticle: it gives rise to two sperm ducts in males and to one oviduct in females. Heterotardigrada show some variation in the size, shape and location of the genital aperture, which is normally situated anterior to the anus. The reproductive ducts in Eutardigrada open directly into the intestine, so that reproductive and digestive systems share a common opening, a cloaca.

Fertilization is probably external, although there have been few observations of mating to confirm this. Many of the semi-terrestrial eutardigrades shed their eggs freely: such eggs are often intricately sculptured and the ornate projections help to trap them in interstitial crevices which provide some protection. Other eutardigrades lay smooth eggs in moulted cuticle, either their own or that of other members of the interstitial fauna. Development of the embryo takes from 6 to 16 days in favourable environmental conditions; however, eggs can remain viable for long periods—a useful adaptation considering the impermanence of many of the habitats colonized by tardigrades.

Postembryonic development is direct—that is, the young closely resemble the adult—in Eutardigrada. In Heterotardigrada juveniles usually lack adult colouration and have a reduced complement of claws or toes and trunk and caudal appendages. Growth is by moulting and most species reach sexual maturity after three or four successive moults. The pharyngeal apparatus is ejected before moulting and for a short time individuals lacking this structure can be found. Some species moult up to twelve times during their life, which may vary in duration from 3 to 30 months, depending on climatic conditions.

Many semi-terrestrial species of Eutardigrada and Heterotardigrada are able to enter a cryptobiotic state, during which they can withstand a wide range of physical and chemical environmental extremes. When the surrounding water film on which the animal depends dries up, the tardigrade contracts into itself to form a flattened, contorted structure known as the tun. On rehydration, which may be after several hours, or even several years, the animal quickly regains its shape and normal life functions. The time taken to achieve complete recovery is dependent on the duration of the cryptobiotic state.

Despite the small size of these animals which makes quantitative sampling difficult there have been several studies of tardigrade populations. Whether found in soil, moss, freshwater or marine sediment, all species so far investigated seem to undergo regular seasonal fluctuations in numbers. There is often both a spring and late summer/early autumn population peak. Climate is probably the major factor limiting population growth: low rainfall can severely reduce population levels by increasing the depredations of fungal attack. Tardigrades are often present in amazingly high numbers and it is interesting to speculate on their role in natural ecosystems: predaceous species may limit the numbers of other components of the interstitial fauna; algal and bacterial browsers could represent the first link in a complex food chain; and in marine environments the detritus feeders such as *Batillipes* could aid decomposition.

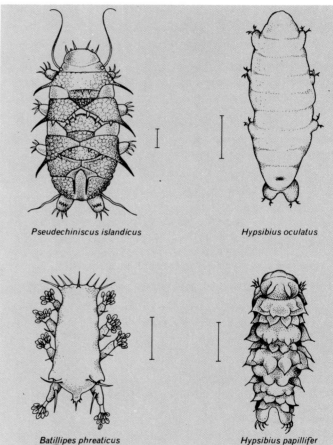

Pseudechiniscus islandicus

Hypsibius oculatus

Batillipes phreaticus

Hypsibius papillifer

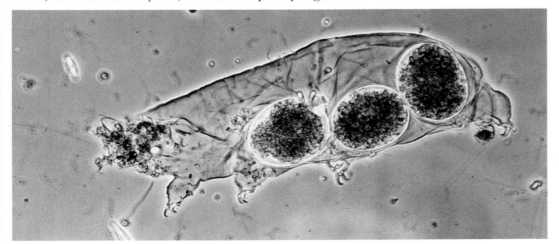

ABOVE: the tardigrades exhibit a wide diversity of form despite the basic similarity in their body plans. Heavily armoured types contrast with those whose cuticles are smooth or papillose. Marine forms like *Batillipes* have many adhesive toes but the freshwater species have grasping claws. The scale bars are 0.05 mm long.

LEFT: a moulted eutardigrade cuticle containing three smooth eggs, each 75 μm long. The cuticle ruptures at the anterior end, and the eggs are laid as the animal crawls out. The moulted cuticle has a full complement of claws and the animal must resynthesize a fresh set after each moult.

SUBKINGDOM Metazoa

PHYLUM Arthropoda

Insects and other arthropods

CLASS
†Trilobita *trilobites*
Merostomata *merostomes*
Arachnida *arachnids*
Pycnogonida *sea spiders*
Crustacea *crustaceans*
Onychophora *velvet worms*
Chilopoda *centipedes*
Diplopoda *millipedes*
Pauropoda *pauropods*
Symphyla *symphylans*
†extinct   Insecta *insects*

- NUMBER OF LIVING SPECIES: more than 1,000,000

- DIAGNOSTIC FEATURES: coelomates with a cuticle usually hardened into plates (sclerites) forming an exoskeleton, and with limbs composed of jointed tubular sections; paired appendages for locomotion, feeding, copulation and sensory purposes, main body cavity a haemocoele

- LIFESTYLE AND HABITAT: marine, freshwater and terrestrial at all depths, altitudes and latitudes; mostly free living but many are plant and animal parasites; the only invertebrates to have developed the power of flight

- SIZE: 0.3 mm to 3 m but mostly between 1 and 10 mm

- FOSSIL RECORD: Cambrian to Recent; particularly abundant in Palaeozoic; probably more than 25,000 species

The spider *Misumena*, a member of the class Arachnida, blends perfectly with its background—an attribute which assists it in ambushing prey. The victim is a bee, one of the representatives of the Insecta. Together with crustaceans, these two classes constitute the majority of the Arthropoda—the largest of all the phyla, containing over 80 per cent of the living species of animals.

The phylum Arthropoda is a successful group of animals by any standards. The main secret of the success of its members lies in the possession of an exoskeleton—the arrangement of hardened plates and tubular sections found in the outer covering or cuticle of the body. These hard plates or sclerites are joined by tough but supple membranes of chitin which allow freedom of movement. The hard parts arise through the formation of chemical linkages between protein molecules to form sclerotin. This hardening process is called tanning and can be seen taking place when an arthropod moults. When it first emerges from the old skin it is uniformly soft and pale (the teneral condition) but then local sclerotization causes darkening and hardening to form the plates or, in the case of limb sections, tubes. The name arthropod, incidentally, is from the Greek 'jointed foot', referring to this sectional structure of the limbs, which the early authors chose to regard as feet. In some arthropods, particularly Crustacea such as crabs and lobsters, skeletal rigidity has been greatly increased by deposits of calcium salts. In others, notably the more advanced insects and arachnids, the outermost layer of the exoskeleton (the epicuticle) is impregnated with wax which provides a waterproof barrier and greatly reduces water loss.

Compared with the cuticle of annelids, the exoskeleton gives more protection, but its chief advantage lies in providing firm anchorages for muscles and leverage points against which they can act. This results in considerable power, precisely directed, in marked contrast to the movement of annelids, which involves contractions of muscle layers, leading to displacements of the whole body which are very wasteful of energy. In arthropods precision of movement is increased by allowing each limb joint to move in one plane only; the whole limb is flexible because it has several joints.

The degree of sclerotization and calcification varies greatly within the phylum. Such insects as beetles are virtually impregnable whereas at the other extreme many larval stages have no hard parts except the jaws and the head capsule to which the jaw muscles are attached. The lack of sclerotization in larvae means that they can have a loose skin and expand considerably in size between moults. In heavily sclerotized insects, growth is only possible in the teneral state, before the cuticle has hardened. The need to moult is the chief disadvantage of the exoskeleton as the process is dangerous—it immobilizes the animal and makes it very vulnerable to enemies and adverse physical conditions. Another disadvantage of the exoskeleton is that it becomes unwieldy as the body increases in size. This is because of the mechanical requirements of limb movement and the need for more massive anchorages. It is notable that the largest arthropods—crabs and lobsters—live on the sea bed where their weight is less of a burden than it would be on land.

Like annelids, arthropods have a ventral nerve cord and a dorsal blood vessel, but the phyla differ in other respects, one being the exoskeleton. Arthropods also have a concentration of sense organs, nervous tissue and feeding apparatus to form a more distinct head region, the substitution of a haemocoele for a coelom as the main body cavity, and the elaboration of the limb series. The limbs of the head region form the antennae and mouthparts; those further back form walking or swimming limbs, the number of which becomes progressively reduced in the more advanced types, leaving (as in arachnids and insects) a rear part of the body without appendages. Often, however, the last few pairs of the limb series are retained in modified form as copulatory organs, the so-called external genitalia. Segmentation is clearly seen in primitive forms but is much less apparent in the more advanced crustaceans, arachnids and insects in which the body is separated into tagmata or divisions with different sets of functions. Insects, for example, have three tagmata—

velvet worm

millipede

primitive crustacean

Evolution of the arthropods has proceeded from long, thin animals with little differentiation of the body, to much more compact creatures, the bodies of which divided into progressively more distinct regions or tagmata. At the same time, the number of legs becomes reduced, allowing greater versatility. The spiders and mites (Arachnida) are the most advanced in this respect.

head, thorax and abdomen—whereas arachnids have two—the anterior prosoma and the opisthosoma.

In all arthropods the sexes are separate and the female normally lays eggs with a good store of yolk. Subsequent development ranges from the most direct, with the young hatching as miniature replicas of the adult, to life histories involving several larval types and resting stages, which may be highly resistant to physical extremes.

Although the basic structure of the arthropod remains notably constant throughout the phylum there are sufficient differences between members to split them into three major classes and eight minor ones. The interrelationships of these groups and the origin of the phylum are matters of debate, because the fossil history is too patchy to provide much help. The traditional view is that arthropods are monophyletic, that is to say all derived from an ancestral stock which already had arthropod features. The simplest classification, used here for clarity of presentation, is to subdivide the phylum directly into classes. However, it has often been split into three subphyla: the Chelicerata (arachnids, merostomes and pycnogonids), Mandibulata (crustaceans, myriapods and insects) and Trilobitomorpha (trilobites and a few related fossils). The Onychophora in this scheme of things are treated as a separate phylum.

In recent years detailed studies of embryology, head structure and locomotion have led to the suggestion that the Arthropoda are polyphyletic, having evolved the exoskeleton several times over quite independently. According to this

advanced crustacean

insect

spider

mite

view, which is likely to gain wide support in the future, living arthropods should be split into three separate phyla: the Chelicerata (comprised as above), the Uniramia (onychophorans, myriapods and insects) and the Crustacea. This leaves the Trilobitomorpha which, for want of evidence of affinity, should probably be treated as a separate phylum.

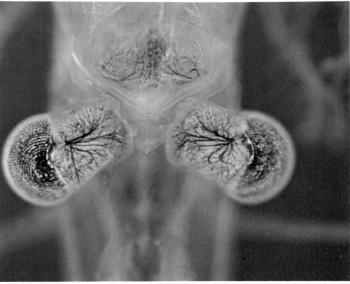

The compound eyes of insects (left) closely resemble those of crustaceans (right) but this does not necessarily indicate a common origin. There is, in fact, evidence that these two groups developed compound eyes independently. If this were the case, it would lend support to the notion that the different arthropods evolved, not from a single ancestral stock but from several; that is they had a polyphyletic, not a monophyletic, origin.

SUBKINGDOM Metazoa
PHYLUM Arthropoda

CLASS †Trilobita

Trilobites

†extinct

Though thousands of trilobites are known from their hard exoskeletons, only a few are known with appendages preserved; partly for this reason the classification at the higher taxonomic ranks has not yet been satisfactorily resolved.

- NUMBER OF LIVING SPECIES: none

- DIAGNOSTIC FEATURES: arthropods with well armoured body divided longitudinally into three regions and with an anterior head, a median trunk and a posterior pygidium; one pair of antennae; all other limbs biramous

- LIFESTYLE AND HABITAT: free living on muddy and sandy bottoms of shallow seas; a few could swim

- SIZE: up to 50 cm but mostly about 5 cm

- FOSSIL RECORD: the dominant fossil group in the Cambrian, gradually diminishing until it disappeared in the Permian; about 4,000 species.

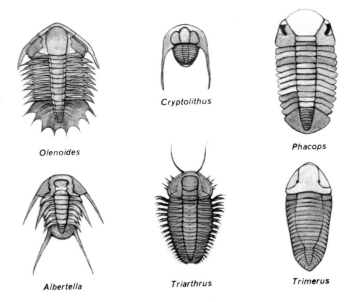

Olenoides

Cryptolithus

Phacops

Albertella

Triarthrus

Trimerus

Early trilobites were small and comparatively simple in design. Later forms were often large (up to 600 mm in length) and protected by elaborate spines. The limbs are only rarely preserved; those shown here on *Triarthrus* have been reconstructed.

The first great surge of animals in the fossil record was caused by the Trilobita, which first appeared in the early Cambrian period 570 million years ago. Rapidly diversifying, trilobites dominated the fossil scene for over 100 million years, but as other groups evolved—particularly the fish which probably fed on them—they declined steadily and became extinct in the Permian period.

The origins of the trilobites—the first undoubted arthropods—are lost in the uninformative strata of the Precambrian. The most primitive forms known, such as *Olenellus thompsoni*, already had most of the typical trilobite features. Flat and oval and rather small (5 cm), these early forms are thought to have cruised around in the mud of the shallow seas in which they lived, eating the worms and other soft animals they came across. Whatever the food supply, there must have been plenty of it, for trilobites were so abundant that sometimes they have been preserved lying on top of each other. They had no jaws, but limbs clustered around the mouth had expanded basal segments which helped to hold and perhaps to crush the food. These limbs, which continued along the trunk, were two-branched or biramous—one branch for walking, the other feathered and probably used as a gill. The head segments were fused to form a shield, with a pair of compound eyes set dorsally and a pair of long antennae in front of the mouth on the underside. At the rear the trunk segments tended to fuse together to form another shield, or pygidium. The name Trilobita refers to the fact that the body is divided lengthwise into three regions, the function of which is obscure.

Between the late Cambrian and the Silurian periods (530 to 400 million years ago), the trilobites showed a considerable adaptive radiation, involving size (*Terataspis* reached 60 cm), number of trunk segments, development of the various spines and so on. The ecological significance of this structural radiation is unknown, but some forms are thought to have given up crawling for swimming, and others could certainly have rolled up into a ball for protection, like pill-bug woodlice. There was a larval stage called a protaspis which consisted mostly of a head, on to which grew first the pygidium and then the intervening body segments, until finally the adult form was reached.

As trilobites were the first arthropods to appear, it is natural to suspect that later arthropods are derived from them. Unfortunately the fossil record is not helpful. The fact that the limbs are biramous suggests a link with crustaceans which also have this arrangement, but the details of the limb structure do not correspond very well. The myriapod/insect line traces back to the soft-bodied onychophorans, which are totally unlike the hard-bodied trilobites. This leaves the chelicerates (see opposite page) as candidates for descent, and there is certainly some similarity between trilobites and the xiphosurans or king crabs. There are fossils, for example *Emeraldella*, which might be intermediate forms. These are sometimes grouped with trilobites proper to make a larger taxon, the Trilobitomorpha.

The class characteristic of the division of the body into three lobes is clearly defined in this well-preserved specimen. The species is *Calymene breviceps* from the Silurian of Indiana.

- NUMBER OF LIVING SPECIES: 5

- DIAGNOSTIC FEATURES: arthropods with body divided into an anterior prosoma bearing feeding and walking limbs, at least some of which are chelate, and a posterior opisthosoma with five or six pairs of gills; compound eyes present

- LIFESTYLE AND HABITAT: sea scorpions swam or burrowed in shallow seas or freshwater; king crabs crawl on the mud or sand bottoms of shallow seas

- SIZE: sea scorpions reached up to 3 m but most were much smaller; living king crabs are up to 40 cm

- FOSSIL RECORD: sea scorpions from late Cambrian to Permian with a peak in Silurian; king crabs Devonian to Recent with an early peak and rapid decline, leaving only relict species at present; 50 to 100 species

The class Merostomata is characterized by the absence of jaws and antennae, the division of the body into a prosoma and opisthosoma or abdomen and the modification of the first pair of limbs as prehensile chelicerae. (The Arachnida have similar features and the two classes are often referred to jointly as the chelicerates.) The prosoma arises through the fusion of the anterior trunk segments with the head, leaving the remaining segments as the opisthosoma, which is without walking limbs.

Merostomes comprise the extinct Eurypterida (sea scorpions) and the Xiphosura (king or horseshoe crabs) which

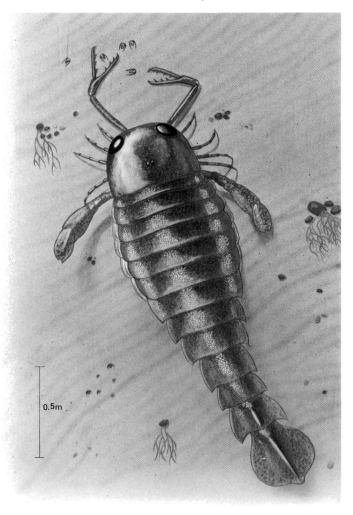

0.5 m

The armoured, horseshoe-shaped carapace and the long posterior telson are typical of king crabs both ancient and modern. Some species are still common in parts of the Atlantic and Pacific.

Pterygotus was one of the largest and most highly developed of the eurypterids, with large chelicerae modified as pincers. It reached a body length of 3 m and lived in Silurian times.

exist today as relict forms. All members of the class were, and are still marine, in contrast to the arachnids which are almost entirely terrestrial.

For once in arthropod phylogeny, the course of evolution seems quite clear, for the modern arachnid scorpions are easily derived from their eurypterid counterparts. The latter were a highly successful group of predators, which caught their prey in well developed chelicerae. The last pair of limbs were modified as paddles and, as befits an active predator, the compound eyes were very large. Remarkably, some eurypterids reached up to 3 m long—making them the largest animals of their age and the largest arthropods of all time. The earlier forms, with tails which possibly had stings, most resembled modern scorpions. Some later eurypterids had the tail developed as a flat telson as an aid to swimming, while others became adapted for burrowing. The group reached its peak in the Silurian period—much later than trilobites—but declined to extinction at the end of the Permian period.

They left behind them an equally old stock, the Xiphosura, which has hung on to the present day in the limited environment of shallow water muds and sands in such places as the Gulf of Mexico. The five surviving species—which all belong to the one genus *Limulus*—are uniform in design and, being specialized for their particular way of life, are well off the main line of chelicerate evolution. The prosoma forms a large, horseshoe-shaped carapace broken only on the dorsal surface by the compound eyes. The opisthosoma is also well shielded and ends in a long spine. Underneath, and gathered around the mouth, are the walking limbs. These are chelate (like the claws of a lobster) and swollen at the base to form gnathobases, which crush the food. The limbs of the abdomen are modified into flattened plates which serve as gills. When these are flapped vigorously, king crabs can just about swim, but normally they crawl in and on the mud, seeking out worms and molluscs which are caught by the chelae and crushed by the gnathobases. The heavy carapace provides effective protection from predators, but if turned on its back by some mischance, the king crab is very vulnerable. It is here that the tail spine comes in useful, as it allows the animal to right itself, providing the bottom is soft.

SUBKINGDOM Metazoa
PHYLUM Arthropoda

CLASS Arachnida

Spiders and other arachnids

ORDER	SUBORDER
Scorpiones *scorpions*	
Pseudoscorpiones (= Chelonethi) *false scorpions*	
Solifugae (= Solpugida) *sun or wind spiders*	
Amblypygi (= Phrynichidia) *tail-less whip scorpions*	
Uropygi *whip scorpions*	
Palpigradi *micro-whip scorpions*	
Ricinulei	
Opiliones (= Phalangida) *harvestmen*	
Araneae *spiders*	Liphistiomorpha
	Mygalomorpha
Acari *mites and ticks*	Araneomorpha

- NUMBER OF LIVING SPECIES: about 50,000

- DIAGNOSTIC FEATURES: arthropods with a body divided into an anterior prosoma and a posterior opisthosoma; prosoma bearing chelicerae, pedipalps and four pairs of legs; compound eyes absent

- LIFESTYLE AND HABITAT: almost entirely terrestrial; mainly free living but many acari are parasites of plants and animals

- SIZE: 1 mm to 20 cm but mostly 2 to 10 mm

- FOSSIL RECORD: Silurian to Recent; wide diversity established by Carboniferous; about 600 species

The Arachnida, the third major class of Arthropods, are basically uniform in structure, almost all predatory by nature, and almost all terrestrial. Like the merostomes, they have the chelicerate arrangement of the body into prosoma and opisthosoma, and have no jaws or antennae; unlike the merostomes, however, their eyes are not compound. The prosoma combines the functions of the head and thorax of an insect. It has six pairs of appendages, the first being the chelicerae, the second the pedipalps and the remaining four the walking limbs. The chelicerae are usually chelate and in some forms the pedipalps are modified in the same way. The walking limbs are normally used as such, but arachnids often seem to be caught short by the absence of antennae and frequently redeploy one or other pairs of the limbs for sensory functions. The abdomen has openings to the genital and respiratory organs and a few highly modified appendages. Primitively it has 13 visible segments, but this number is reduced in higher arachnids such as spiders and mites, and segmentation is lost.

In keeping with their terrestrial way of life arachnids are air breathing, using one of two methods. Lung books are the more primitive. They are gills that have sunk into the body and become closed off from the outside world except for a hole or slit. In this way water loss (the great hazard of terrestrial life) is minimized. The other method of respiration involves air-filled tubes (tracheae) which ramify within the body to the tissues where gas exchange takes place. Different arachnids use one or the other method, or a mixture of the two. A further adaptation to land life is shown by the presence of malphighian tubules. These, like tracheae, are also found in insects, in which they evolved quite independently. The tubules lie in the blood-filled body cavity (the haemocoele) and open into the hind end of the gut. They take up the waste products of metabolism, particularly poisonous nitrogenous

The salticid, or jumping, spiders (order Araneae) use their very well-developed eyesight to judge distances accurately—an asset of particular value when stalking their prey. The dark fangs or chelicerae, are flanked by the pale-coloured pedipalps.

products. These are converted into less soluble and less dangerous products like uric acid which are then voided with the solid wastes from the gut as faeces.

Digestion in arachnids is strongly influenced by their rather limited arrangements at the front end of the gut. They lack jaws and the complicated mouthparts of insects and crustaceans, nor do they have a grinding and straining apparatus so characteristic of the foregut of crustaceans. Instead they have opted for a good deal of external digestion, whereby the prey is held crushed and immobile while enzymes are pumped into it and the resulting soup is subsequently sucked out. Liquid feeding of this kind predisposes to blood sucking and parasitism as practised by the mites and ticks which are members of the order Acari.

Because prey have to be caught, so predators need to be well supplied with sense organs. Most arachnids thus have eyes but with the exception of spiders in some araneomorph

BELOW: the anatomy of a female spider. The front part of the body, or prosoma, contains the brain and many-branched stomach and is linked by a narrow waist to the opisthosoma. This contains the heart, reproductive system, excretory tubules and respiratory organs, in this case both lung book and trachaeae.

RIGHT: scorpions are viviparous, and the newborn young are carried on their mother's back for some weeks. Their pale colour is due to lack of hardening of their cuticle. The tail with the sting is carried erect when the animal is alert, so that poison may be speedily injected into prey that is caught in the claws.

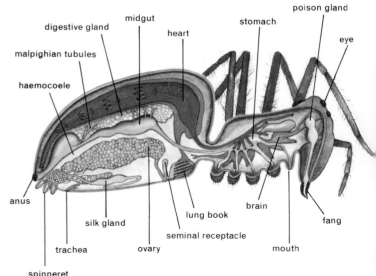

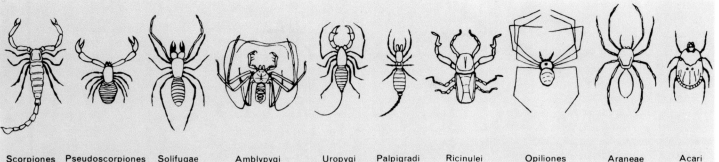

Scorpiones Pseudoscorpiones Solifugae Amblypygi Uropygi Palpigradi Ricinulei Opiliones Araneae Acari

families their vision is poor. Moreover, the absence of antennae severely limits their chemosensory abilities. Arachnids are, however, quite exceptionally sensitive to touch and vibration, which they perceive mainly by means of special sensory hairs called trichobothria.

Reproduction of arachnids is often a process fraught with some risk, as solitary predators like these are hard to approach in safety. Thus the male spider must send out specific signals to the female he is courting in order to convince her that he is a prospective mate rather than a prospective meal. This problem has led to the evolution of rich and bizarre ritual courtships in such groups as the scorpions and spiders. Only in the harvestmen (Opiliones) is insemination internal, that is, a penis is used by the male to insert the sperm directly into the female. In the other arachnid orders either a packet of sperm in the form of a spermatophore is transferred between the sexes or, in the case of spiders, the male uses his pedipalps to transfer sperm. Typically of arthropods, the egg is large and yolky but, atypically, there is often a degree of maternal care and even feeding of the young by the mother. Except in the mites development is direct, with no larval stage.

The fossil history of arachnids stretches back into the Palaeozoic, over 400 million years ago. Being among the first land animals, and probably the first predators, they must have fed on the centipedes and millipedes which evolved around this time, although we may suppose that primitive insects were also available.

Palaeophonus, a fossil scorpion from the Silurian period 420 million years ago, is thought to be fairly directly descended from the eurypterids and its existence shows the

In the members of the ten orders of arachnids, chelate appendages and long legs are common. There are no antennae. Palpigrades, pseudoscorpions, opilionids and many spiders are very small (less than 4mm long), whereas scorpions often exceed 50mm. Acari range from pea-sized ticks to mites measuring under 1mm in length.

extreme age of the order Scorpiones. In spite of its antiquity the group is fairly successful today, with 600 species and a worldwide distribution in the tropics and subtropics. The most notable features of the scorpion are chelate pedipalps or claws and the much feared 'sting in the tail'. Scorpions produce two kinds of poison; a fairly mild one, as is produced by the small black European species *Euscorpius italicus* (the effects of which may be no worse than the sting of a hornet) and a neurotoxic venom, for example of *Androctonus australis* from North Africa, which is dangerous and sometimes lethal to man.

Although scorpions are most often thought of as animals of deserts and dry places, there are some forest species, although they are not often seen. All are active predators, either stalking their prey or waiting in ambush until something comes close enough to be grabbed by the claws. It is then macerated by the chelicerae and consumed (external digestion is less important here than in other arachnid orders). Favourite foods include cockroaches, grasshoppers and crickets, whose presence is detected by the trichobothria on the pedipalps, and perhaps by the rather mysterious comb-like organs (pectines) on the underside of the body. Poison is not always used to subdue prey—in some species its purpose seems to be defensive, and it may be employed against such enemies as birds and baboons.

ABOVE: sun spiders are active at night. They lack poison and kill their prey, in this case a gecko, with massive, biting chelicerae. The very long pedipalps have a sensory function.

BELOW: pseudoscorpions are very small predators which catch their prey with chelate (pincer-like) pedipalps. Most species live in leaf litter, rotting logs and in other sheltered, damp places.

Because scorpions are normally solitary predators, they have a problem of mate recognition, and sperm transfer is preceded by an elaborate courtship, in which the male grasps the female by her pedipalps and 'dances' to and fro with her. In some species the pair may also hold their tails aloft, entwined. The ritual serves both to ensure mate recognition and to find a suitable spot for the spermatophore to be deposited. Once this has been accomplished the male manoeuvres his mate until she is in a position to enclose the packet in one of her genital openings on the underside of her abdomen. The fertilized eggs are not laid but develop inside the mother, where in some species they are supplied with nutrients from her body. The hatched young eventually make their way out through the genital openings and climb onto her back, where they stay, relatively safe from predators, until after their first moult.

Maternal care is also well developed in the Pseudoscorpiones, so named because although superficially similar, they are no more closely related to scorpions than to other arachnids. They lack the tail and sting but have pedipalps modified as claws. They are very small (mostly less than 5 mm) and therefore seldom seen, but can be common in leaf litter. They eat springtails (the insect order Collembola) and other soil organisms of similar size, which they paralyse with poison from their pedipalps. Some species are associated with man, for example, *Cheiridium museorum*, a fact first noted by Aristotle. This species preys on indoor pests such as booklice.

Pseudoscorpions use the silk produced by their chelicerae to construct chambers for moulting and brooding the young. A special feature of their biology is that they often attach themselves to the hairs of larger animals for dispersal; this phenomenon is called phoresy and is found elsewhere in the animal kingdom.

At the other end of the size scale, ranging up to 5 cm, are the Solifugae, known variously as false spiders, sun spiders and wind spiders. At speed, these creatures look like light brown and highly animated balls of fluff, for they are completely covered in long sensitive hairs. They are not poisonous but make up for it by having, relative to their size, the most formidable biting apparatus in the animal kingdom. A bite is extremely painful, takes a long time to heal, and can penetrate even the most tank-like armoured insect. They feed on anything up to lizards in size, and use the pedipalps and first pair of legs to reconnoitre for prey. They occur worldwide in the warmer latitudes and about 600 species have been recognized.

Four small and seldom seen orders—the Amblypygi, Uropygi, Palpigradi and Ricinulei—may now be mentioned. They comprise some 200 species, all of which are tropical and mostly rather uncommon. Uropygids occur naturally in caves and have adapted easily to houses, where they climb quickly over walls looking for cockroaches.

A more familiar order is that containing the harvestman or daddy-longlegs—the Opiliones. These forms are characterized by having the prosoma and opisthosoma fused together and surmounted by a characteristic tubercle from which look out a pair of eyes, one each side. The tubercle is often decorated by spines. Typically all the limbs are extremely long, but there are soil living forms with short legs.

The best known of the arachnids are the spiders which comprise the order Araneae, with about 35,000 species. They form a very well defined set of animals, characterized by the presence of abdominal appendages for silk spinning, the modification of the chelicerae into poison fangs, and of the pedipalps of the male for transferring sperm to the female. They are easily distinguished from other common arachnids by a narrow constriction or waist which marks the division between the prosoma and opisthosoma.

The only outward sign of body segmentation in spiders is seen in the primitive suborder Liphistiomorpha whose members live obscurely in burrows and caves in the Far East. Also rather primitive are their notorious tropical relatives the Mygalomorpha, or 'bird-eating' spiders, also (wrongly) called tarantulas. Although it is true that these spiders do occasionally overpower small birds, the name is a misnomer, as their diet more usually consists of insects and lizards. They are, however, extremely large and very hairy; the leg span can reach about 25 cm, so tales about spiders the size of dinner plates are not entirely without foundation. Some species live in the open, usually in forest, but others construct silk-lined burrows with trap doors, from which they dart out to seize their prey, and then retire at high speed back into the burrow to consume it.

Most spiders belong to the suborder Araneomorpha, which contains all the orb-web and hunting spiders, and many lesser families. The Thomisidae, for example, are squat creatures that ambush their prey by sitting on flowers or elsewhere, often very well concealed by their colour and pattern. They are known as crab spiders because of their appearance and tendency to walk sideways. The hunting or wolf spiders of the family Lycosidae take a more positive approach by actively running down their prey. They are completely nomadic, and the mother carries her young in a brood sac. The wolf spiders are much smaller than Mygalomorpha but some are large enough to have generated a good deal of folklore. Species of the genus *Tarentula* (the true tarantulas) have a bite which is

TOP: harvestmen, like almost all arachnids, are predatory but they will also eat carrion and plant material. The long legs are used for detecting food as well as for walking.

ABOVE: the primitive mygalomorph spiders can be large enough to kill small birds and mammals but insects are the usual diet. The females may live for a very long time—some for up to 20 years.

BELOW: a female lycosid spider, *Pisaura mirabilis*, carrying her brood sac. When the young hatch they live for a short time in a nursery web, spun by the female, before dispersing.

ABOVE: a female *Latrodectus mactans*, one of the black widow spiders. The powerful venom is a nerve toxin and causes acute pain, muscle spasm and paralysis. It can be fatal to children.

ABOVE RIGHT: most mites are very small but this red velvet mite, *Dinothrombium*, reaches a length of 10 mm. Found in the tropics, it is often seen on the ground surface after rain.

supposed to cause the victim to start a wild, uncontrolled dancing, from which the dance known as the Tarentella of southern Italy is derived. Perhaps the blame should really have fallen on one of the black widow spiders of the genus *Latrodectus*, which are very poisonous and about the only spiders capable of causing human fatalities. *L. mactans* occurs commonly in the southern USA. Black widow spiders are a particular hazard because they are rather small and so easily overlooked, and are also not averse to human habitations of a rustic nature, such as privies. Common house spiders in temperate regions, for example *Tegenaria domestica*, mostly

A female garden spider, *Araneus diadematus*, at the centre of her orb web. When waiting for prey she retires to a shelter at one edge of the web; vibrations of a struggling victim entangled in the sticky threads alert the spider which then rushes out to subdue her prey with poison and a wrapping of silk.

belong to the family Agelenidae. Their very long legs, overall hairiness and creepy habits make them generally loathed, but they are quite harmless and in fact beneficial because of all the pestiferous insects they catch in their sheet webs.

Rather more popular are the orb-web spiders, of which *Araneus diadematus* is a common European species. The orb-web is the most advanced of the many types of web evolved by spiders, and stands as a monument to the constructional skills of the arthropods, which although small, are capable of very complex behaviour. Although most orb-web spiders construct fixed traps for catching prey, some make mobile traps. One species, for example, secretes a sticky ball on the end of a short thread. It then suspends itself from the vegetation and twirls ball and thread beneath it like a circling fly paper. When an insect is caught the thread is hauled up and the prey sucked dry. Another species spins a criss-cross lattice of threads between two legs in such a way that when the legs are opened, the net increases greatly in size. The spider then waits for a victim to crawl within range, instantaneously opens the net and traps the creature beneath it. Perhaps most bizarre of all is the behaviour of *Scytodes thoracica*. This spider squirts poisonous gum from a distance in a series of sticky threads on to its unfortunate victim which is then bound to the floor while the poison takes effect.

Unlike the spiders, which form such a well defined group, the mites and ticks of the order Acari are extremely varied, with few characters in common, though all are small in size and have a fused prosoma and opisthosoma. The Acari are by far the most successful arachnids, when judged by ecological diversity, abundance or structural versatility. Chelicerae and pedipalps have combined to form mouthparts adapted for a wide range of diets, and life histories involving two or three larval stages have developed. Minute size has allowed colonization of such miniscule habitats as mammalian hair follicles, in which the mite *Demodex folliculorum* flourishes. Less extreme specializations are found in the many ectoparasites, such as the mite *Sarcoptes scabei*, which causes scabies, and the ticks, which suck blood and transmit a whole suite of diseases. There is a vast range of free living mites, many adapted for living in the soil and litter where they are often the dominant animals. One group has even made a success of life in freshwater, the only arachnids to do so. Many of the free living forms feed on decaying plant material and associated microorganisms, but there is a large number of predators, and some herbivores, like the red spider mites, *Tetranychus*, which are agricultural pests.

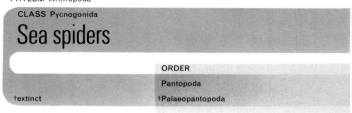

SUBKINGDOM Metazoa
PHYLUM Arthropoda

CLASS Pycnogonida

Sea spiders

	ORDER
	Pantopoda
†extinct	†Palaeopantopoda

- NUMBER OF LIVING SPECIES: about 500

- DIAGNOSTIC FEATURES: arthropods with a body very small in relation to legs, and divided into head, trunk and vestigial abdomen; head with a prominent proboscis and paired chelifores, palps and egg-bearing legs; trunk with four to six pairs of long walking legs

- LIFESTYLE AND HABITAT: exclusively marine; free living and usually found crawling on encrusting hydroids and bryozoans

- SIZE: body length 2 to 10 mm but legs much longer (up to 6 cm) in some species

- FOSSIL RECORD: Devonian to Recent; only 2 species

The Pycnogonida (sea spiders) is a minor class of arthropods chiefly notable for a weird body design and an exclusively marine distribution. The 500 or so species (there are many yet to be described) have a worldwide distribution and occur from the intertidal zone to the ocean depths. Although some species can swim, they are typically sedentary animals of the ocean floor. They crawl slowly over bryozoans and hydroid coelenterates encrusting seaweeds and rocks and it is on these that they feed. Many are cryptically coloured to match their background, whereas others are whitish.

The body of pycnogonids is quite distinct from that of other arthropods, because it consists almost entirely of legs and head. The 'abdomen' is a vestige and the trunk segments are no more than basal sections for the attachment of the walking limbs. Usually there are four pairs of limbs but some species have five or six. The head consists of a large and mobile proboscis with a pair of palps, an eye tubercle and two chelate appendages called chelifores. The legs are usually long, and are well supplied with claws to allow the animal to cling on firmly to avoid being swept away. A unique feature is a pair of egg-bearing legs used by the male to carry the eggs until they hatch. These limbs also have a cleaning function, and when carrying eggs the male becomes encrusted with algae and bryozoans, as he is unable to groom himself. The eggs are fertilized externally while they are being transferred to the male and hatch either into protonymphs with three pairs of appendages or into miniature replicas of the adults. There is no planktonic stage and in view of their limited swimming abilities, pycnogonids must rely on being carried about on weed for dispersal.

Internally, pycnogonids are notable for the absence of excretory and respiratory systems, and for the method of digestion. Cells of the gut wall absorb food and then become detached and float freely about in the gut. Food is absorbed from them by other, still attached, cells, until finally all the food is gone and the floating cells are eliminated through the anus, with the waste products of digestion enclosed within them. The mid-section of the gut is remarkable for the way in which it ramifies far into the limbs (a condition also seen in spiders). Food reaches the gut by way of the proboscis, which is well supplied with muscles and is highly mobile. In many species it is used to probe into the hydroid or bryozoan and suck out the juices, after a certain amount of external digestion. Alternatively the chelae may be used to break off pieces of food which are then stuffed into the proboscis.

The affinities of the Pycnogonida are not clear. They are often classified with the Arachnida and Merostomata in the subphylum Chelicerata because of the number of legs and other features. These similarities, however, may be superficial and the fossil evidence is inconclusive, so probably they are best considered as an isolated arthropod stock of ancient origin, obscure affinities and very limited success.

ABOVE: the sea spider *Nymphon*, one of the genera with very long legs. A notable feature of sea spiders is that the male carries the fertilized eggs — seen here as flesh-coloured masses under the body; the eggs are held in place by a special pair of legs which are also used in grooming.

BELOW: *Pycnogonum littorale* is one of the sea spiders that is very compact in body form. It is often found crawling over dahlia anemones (*Tealia*) on which it preys, chewing away the flesh of its victims with sharp jaws which are situated at the extreme tip of the proboscis.

SUBKINGDOM Metazoa
PHYLUM Arthropoda
CLASS Crustacea

Crabs and other crustaceans

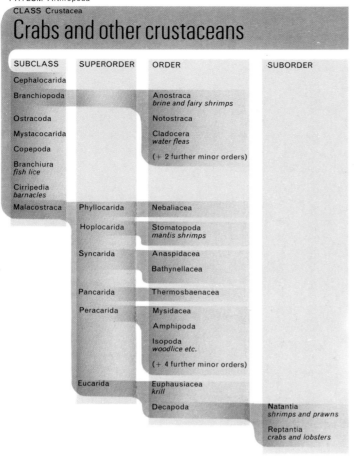

SUBCLASS	SUPERORDER	ORDER	SUBORDER
Cephalocarida			
Branchiopoda		Anostraca *brine and fairy shrimps*	
Ostracoda		Notostraca	
Mystacocarida		Cladocera *water fleas*	
Copepoda			
Branchiura *fish lice*		(+ 2 further minor orders)	
Cirripedia *barnacles*			
Malacostraca	Phyllocarida	Nebaliacea	
	Hoplocarida	Stomatopoda *mantis shrimps*	
	Syncarida	Anaspidacea	
		Bathynellacea	
	Pancarida	Thermosbaenacea	
	Peracarida	Mysidacea	
		Amphipoda	
		Isopoda *woodlice etc.*	
		(+ 4 further minor orders)	
	Eucarida	Euphausiacea *krill*	
		Decapoda	Natantia *shrimps and prawns*
			Reptantia *crabs and lobsters*

- NUMBER OF LIVING SPECIES: about 30,000

- DIAGNOSTIC FEATURES: arthropods with two pairs of antennae, and most with gills and a carapace; three pairs of mouthpart appendages present; limbs typically biramous

- LIFESTYLE AND HABITAT: mostly free living and marine but some freshwater and terrestrial forms; some are parasitic

- SIZE: mostly 1 to 10 mm but decapods are usually larger and some have a leg span reaching 3 m

- FOSSIL RECORD: Cambrian to Recent, decapods becoming frequent from Permian; 5,000 to 10,000 species

Although well represented in freshwater, and even found on land, Crustacea are pre-eminently the arthropods of the sea. The class has exploited the open water and the sea floor with equal success, and has done so for a very long time. By the end of the Carboniferous period, 280 million years ago, primitive members of most of the modern subclasses had already appeared and diversified. They were small swimming forms with the typical biramous crustacean limb. The trunk limbs were of the phyllopodous type and each one functioned as gill, oar and food collector simultaneously. This was possible because the tiny organisms making up the diet were strained out of the water by the same rhythmic actions that aerated the gill elements of the limb and propelled the animal forward. Filter feeding of this kind allowed the single-celled plants and animals of the plankton to be exploited as one of the world's major food resources. Filter feeding remains important in present day crustaceans, and its practitioners in their turn are the major element in the diet of pelagic fish such as herring and mackerel.

Another line in crustacean evolution led to the appearance of the Malacostraca, with stalked eyes, a carapace enclosing the cephalothorax and an abdomen ending in a tail fan. From this stock came the decapods along one evolutionary line, and the amphipods and isopods along another. The tendency here was towards a bottom living existence using the mandibles to chew up large particles of food, and the limbs differentiated for walking, swimming and handling food.

The crustaceans have become highly successful and show great ecological diversity, ranging from free living predators, herbivores and scavengers to internal parasites. In some of these parasites adults are so modified that their identity can only be established by their larvae, which are of the nauplius type, typical of Crustacea. The nauplius is a dispersal stage, adapted for a planktonic life during which it is carried around by the ocean currents. To this end it is very small and well supplied with spiny limbs, with which it can prevent itself from sinking.

The most primitive living Crustacea were not discovered until as recently as 1955 when they were found in muddy sand off Long Island, USA. *Hutchinsoniella macrantha* of the subclass Cephalocarida comes close to our idea of what the earliest crustaceans must have been like: it has phyllopodous and multipurpose limbs, an unspecialized head end and numerous thoracic segments. It is only 3 mm long and specialized for living in the interstices of mud and sand by reduction of the abdominal appendages to simple spines. We may suppose that this species has survived because it is adapted to

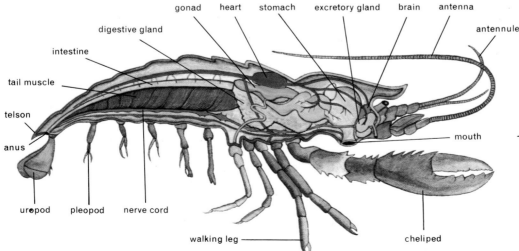

Details of the internal anatomy of a typical decapod crustacean. The limbs are adapted to various tasks: the first thoracic limbs bear chelae for food handling, and the abdominal limbs (pleopods and uropods) are for swimming, walking and reproduction.

(labels on dissection illustration: digestive gland, gonad, heart, stomach, excretory gland, brain, antenna, antennule, intestine, tail muscle, telson, anus, uropod, pleopod, nerve cord, walking leg, mouth, cheliped)

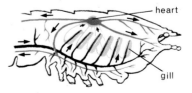

(labels: heart, gill)

the gills, which have been removed in the dissected view, are outgrowths from the bases of the thoracic limbs; they lie under the carapace where a stream of water passes over them; blood takes up oxygen from the water and is then pumped around the body by the heart

a very particular way of life as a member of the interstitial fauna of the sea floor.

The discovery of *Hutchinsoniella* caused great excitement among zoologists, for it is rarely that such an ancient and interesting relict form turns up. It is obviously closely related to the more primitive members of the subclass Branchiopoda, such as the fairy shrimp, *Chirocephalus* (in the Anostraca), which has a long series of phyllopodous limbs hardly differentiated from each other. Other primitive anostracans are

The brine shrimp has primitive swimming limbs called phyllopodia (inset), each of which has a series of lobes and fringes of hairs for rowing. The length of this animal is about 10 mm.

the brine shrimps (*Artemia*), which live in highly saline waters. As an adaptation to such places, which usually occur in arid areas and are prone to dry up, *Artemia* has evolved a resistant egg which can survive in dried-out mud for years. The eggs hatch into typical nauplii larvae, which soon metamorphose into the adult. Another inhabitant of ephemeral waters is *Triops cancriformis* (in the Notostraca), which looks like a cross between a trilobite and a king crab in miniature, because its carapace forms a dorsal shield.

The most specialized Branchiopoda are the Cladocera or water fleas which are mostly found in freshwater. They have a carapace forming a bivalve shell which encloses the whole

Many different copepods live in marine plankton. At the base of the long antennae, which are developed as oars to propel the animal through the water, is a single prominent eye. (× 10)

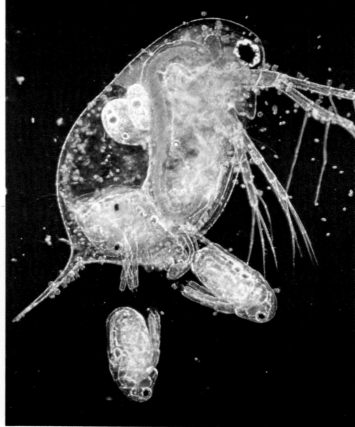

ABOVE: water fleas (*Daphnia* and related genera in the Cladocera) release free swimming young when conditions are favourable, but lay resistant eggs when drought or starvation threatens. (× 20)

body except for the head. The trunk limbs are reduced to four pairs and lie inside the shell, the greatly enlarged antennae being used for swimming. Food is filtered by the limbs as it is pumped through the carapace.

A bivalved carapace is also a feature of the Ostracoda, but in this case it encloses the head as well as the body. Locomotion and feeding are carried out by the large antennae. Ostracods occur both in the sea and freshwater and number only about 200 species. They have the curious distinction of having the largest sperms in the animal kingdom. The male *Pontocypris* is microscopic (0.3 mm) and yet produces sperms 6 mm long! An even smaller group, both in size (0.5 mm) and number of species (three) is the Mystacocarida, discovered in 1943. Its members, like the Cephalocarida, live interstitially between grains of sand.

At the other end of the scale, with 5,000 species, is the subclass Copepoda, which has free living and parasitic members. The free living forms are very common in the plankton and *Calanus finmarchicus* has been claimed as the world's most abundant animal. These free-living copepods have built their success on a conservative plan, featuring a great enlargement of the first antennae, which row the animal along with a characteristically jerky movement, and a filtering apparatus made up of the mouthparts. The single eye of the nauplius larva is retained in the adult (hence the name *Cyclops* for one of the common freshwater forms). Copepods feed on the primary producers of the plankton—flagellated protozoans, diatoms and other single-celled green plants. Parasitic copepods range from species which are no more than surface scavengers to those in which the female is little more than a pair of gonads attached to the host, with the shrunken male attached to her. The chief hosts are fish, but bottom-living annelids and molluscs are also parasitized. In all cases there is a normal nauplius larva.

Compared with the copepods, the parasitism practised by the subclass Branchiura is less extreme, but even so the body is highly specialized for an ectoparasitic existence on marine and freshwater fishes. The fish louse *Argulus americanus* is flattened and oval, with suctorial mouthparts and compound eyes. It frequently leaves its host.

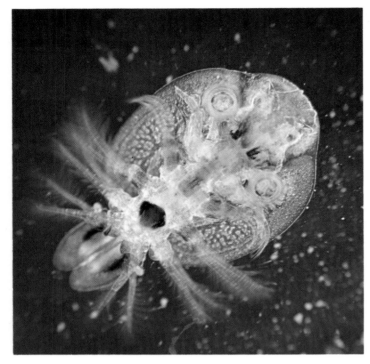

Argulus (subclass Branchiura) is a parasite of freshwater fish. It clings to the surface of its host with suckers that appear as pale brown rings near the head. It is also an active swimmer.

Extreme modification of the crustacean plan is again seen in the barnacles (subclass Cirripedia) that encrust intertidal rocks and foul the bottoms of ships. Two main types occur—stalked and unstalked. The common goose barnacle *Lepas anatifera*, a stalked form found hanging from driftwood and other floating objects in the sea, may reach up to 10 to 15 cm. The stalk represents an elongation of the head, the rest of the body being enclosed in a laterally flattened shell recalling that of a mussel, which can be closed by a powerful muscle. The two halves of the shell open to allow the long feathery thoracic limbs to reach out and filter the water. In the closely related *Balanus*, one of the common intertidal barnacles, the head does not form a stalk but instead the plates of the shell are cemented directly to the substrate. Within the shell the animal effectively lies on its back, catching food by waving its feathery legs in the water. Barnacles are hermaphrodite, and produce a nauplius larva which, after structural alterations, settles and metamorphoses into the adult.

Many cirripedes have become parasitic and rival the copepods in this way of life. The habit is foreshadowed by those barnacles which attach themselves to whales and other living creatures. All the parasites retain the nauplius larva. A typical life cycle is shown by *Sacculina*, one of a large group of cirripedes parasitic on crabs. As a larva it attaches itself to the body of the crab, gains entry, becomes a formless mass and attaches itself to the gut. From here it extends as a series of roots to all parts of the host's body, while the central mass differentiates and comes to project through the cuticle of the crab. Here it forms a brood sac from which the larvae issue. *Sacculina* does not kill but instead castrates its host. Justice of a kind is sometimes at hand, however, as *Sacculina* is itself castrated by a parasitic isopod!

Parasitic isopods belong to the great subclass Malacostraca which contains the remainder and majority of Crustacea, including the familiar forms such as prawns, crabs, lobsters and woodlice. The systematics of the Malacostraca are very complicated, so successful has diversification been within the group. Although there are exceptions, the typical malacostracan design takes the form of a carapace enclosing the thorax which has eight segments and corresponding pairs of limbs. There is an abdomen of six segments with the limbs forming pleopods which are used principally for swimming. The abdomen ends in a telson which, with the uropods that flank it, forms a tail fan. This is operated by powerful muscles in the abdomen (which make very good eating) to provide a backwards tail flick which is the characteristic escape reaction of the group. The eyes are compound, and are usually stalked and mobile.

The most primitive malacostracans, the Nebaliacea, are extremely ancient, with a fossil history back to the Cambrian period. They are filter feeders with primitive phyllopodous limbs and represent a small relic branch from the base of malacostracan evolution. The Malacostraca themselves produced two major lines (the Peracarida and the Eucarida) and several minor developments, one of which led to the orders Anaspidacea and Bathynellacea, another to the Thermosbaenacea. These three orders comprise relict species in relict habitats. *Thermosbaena mirabilis*, for instance, has managed to survive to the present day only in hot springs reaching temperatures of 45°C. In the process it has lost tolerance to any temperature below 35°C and is now trapped in an evolutionary cul-de-sac.

Rather more successful is the order Stomatopoda, the mantis shrimps. These are very advanced but stereotyped predators with raptorial forelimbs like those of the praying mantids. They are large (30 cm) and very powerful. Most have fairly slender forelimbs and catch their prey by grasping it, but in some the limb is massive and the animal strikes with such force that it actually smashes the victim. Such a strike is an effective way of dealing with hard-shelled crabs and molluscs. When attacking a crab, they aim first at its claws, to put them out of action and render the creature defenceless.

LEFT: when mating, the acorn barnacle *Balanus crenatus* uses its long, thread-like penis to reach its neighbour. At the same time, it continues to filter food from the water with its feathery thoracic limbs.

ABOVE: goose barnacles (*Lepas*) feeding. The main part of the body is enclosed in a pair of shells borne on a thick stalk. The stalk can be seen most clearly supporting the smallest animal just below the main group.

Returning to the major lines of malacostracan evolution, the Peracarida include seven orders, the least specialized of which, the Mysidacea, are filter feeders, with a carapace fused to the anterior thoracic segments. The order has 450 species, some of which may be very abundant in shallow seas. Of the remaining six orders, four are obscure but two, the Amphipoda and the Isopoda, are so successful that they have at least 10,000 species between them. Neither has a carapace, the eyes are not stalked and the young, instead of being left to float free in the plankton, develop in a brood pouch attached to the mother. This protects the young from heavy predation but reduces their chances of dispersal. Both groups are adapted for walking rather than swimming and are not filter feeders.

Amphipods are common creatures of the sea and shore. They are flattened from side to side and have very diverse limbs, adapted for feeding, walking, swimming and jumping. On the shore, sand hoppers (*Talitrus*) occur in enormous numbers, living in burrows by day and coming out at night to feed on seaweeds and shore-line debris. The various species of *Gammarus* (water shrimps) are very common. They live in fresh- and seawater, where they swim around on their sides and eat decaying vegetation. Before mating the male is often seen clinging to the smaller female.

All six superorders of the subclass Malocostraca have primitive shrimp-like members, typically adapted for swimming and filter feeding, although the Hoplocarida use raptorial forelimbs to catch their prey. The Malocostracan body plan is extensively modified in the peracarids and eucarids, giving rise to such specialized groups as woodlice (Isopoda), and crabs and lobsters (Decapoda).

PHYLLOCARIDA (*Nebalia*, Nebaliacea)

HOPLOCARIDA (*Squilla*, Stomatopoda)

SYNCARIDA (*Anaspides*, Anaspidacea)

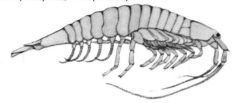

PANCARIDA (*Thermosbaena*, Thermosbaenacea)

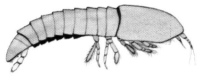

PERACARIDA (*Mysis*, Mysidacea)

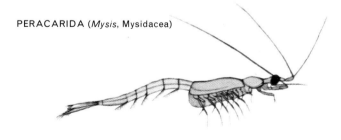

EUCARIDA (*Meganyctiphanes*, Euphausiacea)

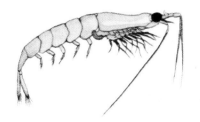

In contrast to the amphipods, the isopods are flattened dorso-ventrally and the limbs are uniform in design. There are numerous free living marine species, such as *Idotea*, which are common among seaweeds on the shore, and a whole range of parasites, some of which rival the copepods and cirripedes in their extreme modification. *Glyptonotus*, a giant free living form from the Antarctic, is almost the size of a football. *Eurydice pulchra* is a carnivore with a voracious appetite which swims in the surf, and *Asellus* is the water louse of freshwater. *Ligia* is the sea slater which is very common on rocky coasts and quaysides and is often found with amphipods feeding on seaweed. It is basically a terrestrial animal. Related species are more fully terrestrial and are known as woodlice or sowbugs. Woodlice are the only wholly terrestrial crustaceans. They score heavily over the various terrestrial crabs by having a brood pouch. In this the young can develop, housed, as it were, in their own private pond, whereas the crabs need to find water to complete their development. Woodlice are very prone to desiccation and so are restricted to the damper land habitats, but even so a number of species are found only in deserts, where they burrow by day and emerge after dark when the dew has fallen. Some of them, the pill bugs, can also roll up to help reduce water loss.

The Eucarida, which make up the other main branch of the Malacostraca, differ from the peracarids in having no brood pouch and in having the carapace fused with all the thoracic segments. The least specialized forms, the Euphausiacea, have retained the primitive features of the nauplius larva and filter feeding. All euphausids live in the plankton of the open seas where they can be very abundant. *Euphausia superba* is fairly large (7 cm) and red in colour. It swims in dense swarms in Antarctic seas at depths of up to 2000 m. It forms the principal food of the large whales, and was called 'krill' by Norwegian whalers.

ABOVE: *Leander serratus* is a common species of prawn found in sandy, shallow European seas. The long antennae and shorter antennules give tactile warning of approaching danger.

BELOW: this lateral view of the marine isopod *Cymodoce truncata* clearly shows the seven pairs of walking limbs characteristic of the group. The well-armoured body is about 10 mm long.

The remaining eucarids are in the order Decapoda which, with 8,500 species, is the most numerous of the Crustacea. Decapods are so named because they have five pairs of thoracic legs. This has come about because the first three of the original eight pairs have become maxillipeds—auxiliary mouthparts. Outgrowths from the limbs and the body wall function as gills, housed under the carapace, and water is pumped through the gill chamber by a branch of the second maxilla, which is modified as a paddle. The nervous system is well developed and life histories are complex, with the nauplius being replaced by other larval stages.

The suborder Natantia contains the prawns and shrimps which have well developed pleopods (abdominal limbs) for

ABOVE: crabs are the most diverse and successful of the decapods. Their eyes are borne on stalks and may be very prominent as in this ghost crab, *Oxypode gaudichaudii*, from the Pacific.

BELOW: terrestrial isopods, or woodlice, moult the rear half of the exoskeleton several days before shedding the front half; this ensures that they are never totally immobilized.

swimming. *Leander*, the prawn, can be separated from *Crangon*, the shrimp, by its narrower, taller appearance and the long rostrum projecting forward between the stalked eyes. *Leander* has been much studied because of its ability to change colour to match its background.

The suborder Reptantia contains all the rest of the Decapoda, which tend to be rather large animals, macrophagous (not filter feeding), often with massive exoskeletons heavily impregnated with calcium salts. The body tends to be flattened dorso-ventrally, and many species have developed the first walking limbs as chelate claws or pincers, as in the lobsters and crabs. The lobster, *Homarus*, can reach a weight of 25 kg, which must make it the most ponderous of all living arthropods. The two claws are different—one has blunt edges and is used for crushing whereas the other has sharp edges for cutting. The squat lobsters, *Galathea*, look like small lobsters with the abdomen reflexed under the body. They can be found under stones on the seashore at very low tides.

The true crabs are notably successful and there are many species in the group. The carapace has become flattened, with the small abdomen reflexed and tucked underneath. The Shore Crab, *Carcinus maenas*, shows the typical features of the group. It is a carnivore and feeds largely on molluscs and annelids. The spider crab *Maia* has a rather lumpish body covered in spines, which soon become encrusted with weed so camouflaging the animal. Spider crabs tend to be big, and *Macrocheira* from Japan has a leg span of 3 m, making it, in one sense, the largest of the living arthropods.

The typical *Carcinus* shape has become much modified in the hermit crabs, so called because of their habit of living solitary lives in discarded mollusc shells which provide good protection. The only snag is that the animal eventually grows too large and has to find a bigger home, a process not without its dangers, for in adapting to this way of life the abdomen has become soft and asymmetrical to fit into the coiled shell. Hermit crabs are abundant marine creatures, and quite a number have invaded the land to some extent. *Birgus latro*, the Robber or Coconut Crab, is a related form that has grown too large for any shell. The abdomen is still asymmetrical but is secondarily hardened to provide adequate protection. This crab is largely terrestrial and lives in burrows on sandy beaches in the Pacific region. It is the subject of many unlikely travellers' tales, the most popular of which is that it climbs coconut trees, snips off the fruit and then climbs down to feast off the contents. Recent reports suggest that it is actually a rather poor climber with no interest in coconuts, but it is certainly an accomplished thief and scavenger.

SUBKINGDOM Metazoa
PHYLUM Arthropoda

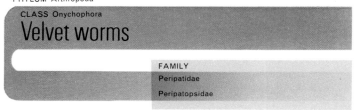

CLASS Onychophora
Velvet worms

FAMILY
Peripatidae
Peripatopsidae

- NUMBER OF LIVING SPECIES: about 120

- DIAGNOSTIC FEATURES: atypical, soft-bodied arthropods with an elongate worm-like body bearing 14 to 43 pairs of stubby legs, each with a pair of claws; head with a pair of fleshy annulated antennae and jaws

- LIFESTYLE AND HABITAT: free living in dark and moist terrestrial habitats in warm latitudes

- SIZE: 1 to 10 cm

- FOSSIL RECORD: sparsely represented, a marine form from the Cambrian being the most notable; possibly 2 species

ABOVE: although they superficially resemble annelids, velvet worms are classified with the arthropods on the basis of their internal anatomy. Their gait is rather like that of millipedes.

BELOW: many onychophorans bear hatched young rather than eggs. *Macroperipatus insularis* gives birth to surprisingly large off-spring, one of which is seen here emerging head first.

Few animals have proved such a puzzle to zoologists as the class Onychophora or velvet worms. As they show a mixture of annelid and arthropod characters, they were originally hailed as the missing link between these two phyla. Ideas today are rather different. Recent research has shown a close link between these 'walking worms' and the many-legged myriapods, which themselves have a clear affinity with the insects. It is thus that the onychophoran *Peripatus* and its allies find themselves in the arthropods despite their lack of hard exoskeleton and jointed limbs. The onychophoran–myriapod–insect assemblage makes up the proposed phylum Uniramia (see page 135) with a common ancestry from some early proto-annelid stock.

Peripatus is a long thin animal with a soft velvety cuticle, many pairs of stubby unsegmented legs and a pair of antennae. The main body cavity is a haemocoele, respiration is by means of tracheae, and the dorsal blood vessel has slits in each segment and functions as a heart. These are all arthropod features. On the other hand, it has layers of circular and longitudinal muscle in the body wall, very simple head structure and a flexible cuticle—all annelid characters. It is specialized in limb structure, mouthparts, antennae and in the ability to produce large quantities of gum.

Detailed analysis of the way onychophorans walk has shown that they have three gaits—top, middle and bottom gear—which differ in the number and order of the legs on the ground at any one time and the speed of each propulsive stroke. Bottom gear, used for starting and accelerating, has most legs on the ground and a short stroke. Top gear has most legs off the ground and a long stroke.

Onychophorans are described as cryptozoic, because they live in damp dark places and shun light. They prey on insects and worms which they subdue with gum, spat from openings on either side of the mouth. Their soft cuticle allows them to squeeze through very narrow spaces and to move about easily in leaf litter and rotting wood. The sexes are separate. Fertilization is usually internal, but in *Peripatopsis capensis* a rather remarkable (and seemingly haphazard) strategy is adopted by the male, who deposits spermatophores here and there over the female's body. The sperm have to bore through the body wall and find their own way to the gonads. In many species the eggs develop inside the mother.

The class has a worldwide but very discontinuous distribution, occurring in parts of the tropics and south temperate regions, including Australia. Such a distribution is held to indicate a very ancient origin, a notion supported by the fossil record, which includes the onychophoran-like *Aysheaia*, from marine deposits in the Cambrian period.

SUBKINGDOM Metazoa
PHYLUM Arthropoda

CLASSES Chilopoda, Diplopoda, Pauropoda and Symphyla

Centipedes, millipedes and their allies

Centipedes CLASS Chilopoda

ORDER

Lithobiomorpha

Scolopendromorpha

Scutigeromorpha

Craterostigmomorpha

Geophilomorpha

Millipedes CLASS Diplopoda

ORDER

Onsicomorpha *pill millipedes*

Proterospermomorpha *flat-back millipedes*

Opisthospermomorpha *iulids*

Penicillata *polyxenids*

(+ 3 further minor orders)

Pauropods CLASS Pauropoda

FAMILY
Pauropodidae
Eurypauropodidae

Symphylans CLASS Symphyla

FAMILY
Scutigerellidae
Scolopendrellidae

- NUMBER OF LIVING SPECIES: more than 10,000

- DIAGNOSTIC FEATURES: elongate arthropods with body divided into head and trunk, the latter bearing ten or more pairs of legs

- LIFESTYLE AND HABITAT: free living in moist terrestrial habitats

- SIZE: pauropods and symphylans usually 1 to 8 mm; millipedes and centipedes up to 30 cm

- FOSSIL RECORD: pauropods, none; symphylans, possibly one Oligocene specimen; millipedes from Devonian onwards and centipedes from Carboniferous; 100 to 150 species overall.

These four classes—Chilopoda, Diplopoda, Pauropoda and Symphyla—superficially resemble each other because all have many legs and a trunk which is not clearly differentiated into thorax and abdomen. Embryological studies suggest that they are quite closely related and in some classifications are grouped together as the subphylum Myriapoda. The first fossils are those of millipedes from the Devonian period, but most fossils occur in amber of Tertiary age, between 20 and 50 million years ago. Analysis of the way myriapods walk points up close affinities with the onychophorans, and their embryology supports such a link. The same sources of evidence also link them closely with the insects, which are thought to have evolved from stock rather similar to living myriapods, by reduction of limb number and tagmatization (division of the body into parts with different functions).

Features shared by all myriapods, apart from the multiplicity of legs, include the presence of antennae, tracheae and malphighian tubules. The eyes, if present, are seldom compound and usually consist of groups of ocelli (simple eyes) without multiple facets. The large yolky eggs may be laid at random in the soil but are often cared for and guarded by the mother, who in some cases builds special egg shelters. Development after hatching varies, but typically the young start with fewer than the adult number of segments and during development more are added on at the rear.

Myriapods are terrestrial and cryptozoic, being found particularly in soil, litter and rotting wood. They have a worldwide distribution and are common in both temperate and tropical regions. Their design, though differing considerably from class to class, is conservative in each case, and they show little ecological diversity. Of the 10,200 species, 8,000 are millipedes, 2,000 centipedes, 120 symphylans and the rest pauropods.

ABOVE: *Scutigera* is a common centipede in the tropics, where it is often found in huts and dwellings. It can run very fast on its long legs and preys on quite large insects such as cockroaches.

BELOW: *Lithobius* is a centipede found in temperate regions. Its flattened sinuous body allows it to move rapidly through crevices in litter and dead wood, where it searches for prey.

This Malayan pill-millipede, an oniscomorphan, is in the process of unrolling following a threat from a would-be predator. When fully rolled, its head and legs are hidden in its carapace.

The Chilopoda (centipedes) differ from the other myriapods in having the genital openings at the rear end of the body. There is one pair of legs on each segment behind the head. The first pair is modified as poison claws that lie in a curve, almost enclosing the mouthparts which consist of mandibles and two pairs of maxillae. The body is flattened and the legs are long, those at the rear more so than those further forward. This arrangement allows the animal to develop a long stride without tripping itself up.

Chilopods are split into five orders. The most familiar in temperate regions, the Lithobiomorpha, are relatively short—they commonly reach a length of 5 cm—and have only 15 pairs of legs. A typical species, *Lithobius forficatus*, feeds on a wide range of cryptozoic animals, including woodlice and insect larvae. The Scutigeromorpha, an entirely insectivorous group, have extremely long legs and compound eyes—both adaptations for an active hunting life in the open. Some species of *Scutigera* live in houses and eat household pests such as cockroaches.

The Scolopendromorpha are the large tropical centipedes, for example, *Scolopendra morsitans*, which has 21 pairs of legs and is handsomely coloured with alternating bands of green and yellow. The largest species may reach 30 cm long and have a dangerous bite, although human fatalities are very rare. Their prey ranges from insects to gecko lizards, mice and even scorpions.

Very different in general appearance from the other centipedes are the Geophilomorpha, which are fairly small and very long, with as many as 170 pairs of very short legs. They lead very retiring lives in the soil, in caves and even on the seashore. Their thin, supple bodies and short legs are adaptations to living in soil crevices, which they can enlarge after the manner of the earthworm, using the legs as anchor points and shortening the body which swells and forces the soil particles aside.

The Diplopoda (millipedes) are in many ways opposites to centipedes. Their genital openings are at the front of the body rather than at the rear. The body is stiff and inflexible, the animal slow and vegetarian. Body stiffening is achieved by calcification of the cuticle, the fusion of pairs of segments into diplosegments and, in the Opisthospermomorpha, the adoption of a cylindrical cross-section. The mouthparts are distinctive, the single pair of maxillae having fused with each other and the appendages of the first trunk segments to form a single plate, the gnathochilarium, which functions as a lower lip to the mouth and manipulates the food. The first trunk segment is thus bereft of its appendages, and the next three are also atypical because they are not fused as diplosegments. Millipedes do not quite deserve their name, as they never have many more than 200 pairs of legs, and may have as few as 13. In most cases the young hatch as larvae with six legs, adding more segments and legs with each moult. This hexapod larva is one of the few exceptions to the rule that the only six-legged arthropods are insects (some mites also have only six functioning legs).

The slow movements of millipedes make them an easy target for predators, but they are not defenceless. Many forms possess repugnatorial glands from which unpleasant and caustic fluids are exuded when danger threatens. In some cases this appears as a fine spray, which carries some distance. As hydrocyanic acid and chlorine help to make up the brew, the effects are predictably damaging and uncomfortable to would-be predators. Another strategy is seen in the order Oniscomorpha, which can roll up into a ball when attacked, just as do the pill-bug woodlice. This adaptation, which has involved extensive modification of the body structure, also serves to limit desiccation by closing off the relatively moist ventral surface. The Penicillata, for example *Polyxenus lagurus*, which look rather like minute bottle brushes because they are covered in tufts of short hairs, use the hairs as irritant barbs. Rather than having stink glands, some millipedes are simply distasteful, and advertise the fact by striking colour patterns, which predators learn to associate with unpleasant taste. Distinctive odours are produced for the same reason.

The oniscomorph millipede, *Glomeris*, is an inhabitant of leaf litter and rotting wood, where it plays an important part in the decomposition of dead vegetation, as do many other millipedes. Some, however, attack living plants and a few are agricultural pests. *Blaniulus guttulatus*, one of the Opisthospermomorpha, is probably the worst offender. The big iulid millipedes, recognizable by their cylindrical segments, are common in the tropics, where they can often be seen walking around in daytime.

The remaining two classes of myriapods are the Pauropoda and Symphyla. Pauropods are tiny soil-living species with diplosegments and anterior genital openings and thus closely resemble diplopods. They differ in having three-pronged antennae, and mandibles with one segment instead of two. The symphylans are also small soil dwellers. They are built along centipede lines, but are whitish in colour with only 12 pairs of legs (centipedes have more than 12 pairs). They resemble diplopods in having anterior genital openings, but do not have diplosegments or a gnathochilarium. They may be very abundant and pestiferous in greenhouses.

One feature of particular interest in the myriapods is their locomotion, for they show very clearly how different ways of life, body forms and limb movement are interlinked. Moreover, a study of the gaits of millipedes and centipedes has suggested that in both they are derived from those of onychophorans. Centipedes, leading the lives of active predators,

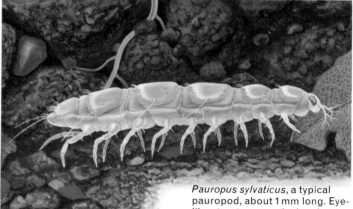

Pauropus sylvaticus, a typical pauropod, about 1 mm long. Eye-like structures on the head are sensory but not image-forming.

are built for speed. Their long legs allow big strides to be taken, while the gait used resembles the top gear of onychophorans, with most of the legs off the ground at any one time, moving forward for the short, rapid backstrokes. This combination gives high speed, but long strides mean low power and, with most of the legs off the ground, there is some loss of stability. Millipedes, forcing their way through the soil and litter, are the bulldozers of the animal world, built to provide maximum thrust. To this end the legs have a very short travel, while the gait used is the bottom gear of onychophorans, with most of the legs on the ground at any one time. This combination gives the necessary power, but at the cost of speed. There is also a risk that the forces involved will cause the body to telescope, so it must be built like a ramrod. The calcification of the exoskeleton and the fusion to form diplosegments can thus be seen as adaptations to increase the rigidity of the body, allowing it to push through the soil.

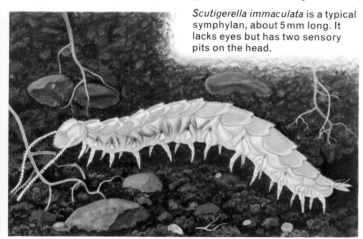

Scutigerella immaculata is a typical symphylan, about 5 mm long. It lacks eyes but has two sensory pits on the head.

BELOW: a large iulid millipede from Trinidad. The powerful legs lie largely beneath the stiffened body, enabling the animal to force its way easily through loose soil and leaf litter.

SUBKINGDOM Metazoa
PHYLUM Arthropoda

CLASS Insecta (= Hexapoda)

Insects

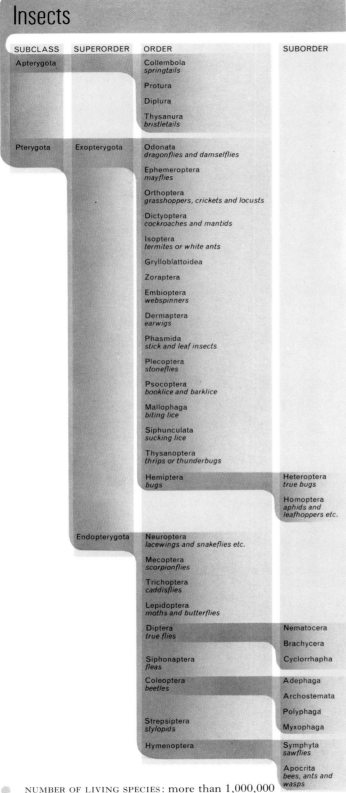

SUBCLASS	SUPERORDER	ORDER	SUBORDER
Apterygota		Collembola *springtails*	
		Protura	
		Diplura	
		Thysanura *bristletails*	
Pterygota	Exopterygota	Odonata *dragonflies and damselflies*	
		Ephemeroptera *mayflies*	
		Orthoptera *grasshoppers, crickets and locusts*	
		Dictyoptera *cockroaches and mantids*	
		Isoptera *termites or white ants*	
		Grylloblattoidea	
		Zoraptera	
		Embioptera *webspinners*	
		Dermaptera *earwigs*	
		Phasmida *stick and leaf insects*	
		Plecoptera *stoneflies*	
		Psocoptera *booklice and barklice*	
		Mallophaga *biting lice*	
		Siphunculata *sucking lice*	
		Thysanoptera *thrips or thunderbugs*	
		Hemiptera *bugs*	Heteroptera *true bugs*
			Homoptera *aphids and leafhoppers etc.*
	Endopterygota	Neuroptera *lacewings and snakeflies etc.*	
		Mecoptera *scorpionflies*	
		Trichoptera *caddisflies*	
		Lepidoptera *moths and butterflies*	
		Diptera *true flies*	Nematocera
			Brachycera
		Siphonaptera *fleas*	Cyclorrhapha
		Coleoptera *beetles*	Adephaga
			Archostemata
			Polyphaga
		Strepsiptera *stylopids*	Myxophaga
		Hymenoptera	Symphyta *sawflies*
			Apocrita *bees, ants and wasps*

- NUMBER OF LIVING SPECIES: more than 1,000,000

- DIAGNOSTIC FEATURES: arthropods with body divided into head, thorax and abdomen; head with paired antennae and three pairs of mouthpart appendages; thorax with three pairs of legs and often two pairs of wings

- LIFESTYLE AND HABITAT: predominantly free living—many can fly—in terrestrial and aquatic habitats (none is truly marine)

- SIZE: 1 mm to 30 cm but mostly between 1 and 10 mm in length

- FOSSIL RECORD: Devonian to Recent, with major expansion in late Palaeozoic and early Mesozoic; about 12,000 species

These bristletails, *Petrobius maritimus*, can be found crawling on rocks in the splash zone on sea shores. Other species occur inland, under stones, among dead leaves and in buildings. They are nocturnal insects and feed on vegetable matter. (× 0.7)

The class Insecta is the only arthropod group with three pairs of legs and with the body divided into head, thorax and abdomen. In the embryo the head has six segments, but these become fused as development proceeds. The appendages of the first and third segments have been lost. Those of the second form the antennae, and those of the fourth, fifth and sixth segments form the mandibles, maxillae and labium respectively. Laterally there is typically a pair of compound eyes and dorsally a set of three ocelli. Thus the primary functions of the head are sensory perception and the manipulation and uptake of food.

The thorax has three segments, the pro-, meso- and metathorax, each bearing a pair of legs. In the Pterygota the meso- and metathorax each bear a pair of wings, unless these have been secondarily lost, as is quite often the case. The thorax is the powerhouse of the insect because locomotion is its main function. Embryologically the abdomen has eleven segments (except in Collembola), and some primitive insects retain this number in the adult stage, but there is a reduction in more advanced forms. Abdominal appendages exist only as rudiments, except for those of the terminal segments, which are modified as external genitalia for sperm transfer. The abdomen functions chiefly in reproduction, digestion and excretion. The respiratory system consists of tracheae, which open to the exterior by way of a series of slits arranged, one per segment, on each side of the thorax and abdomen. The nervous system has, by arthropod standards, a well developed brain, but insects have only a very limited capacity for adaptive behaviour, so much of their activity is stereotyped. This is not to say, however, that it is necessarily simple, as the courtship rituals of flies and 'dancing' of bees testify.

Body organization, physiology and behaviour have combined to make the insects very successful. So great is the ecological diversity of the group that there are nearly one million known species and an estimated one to four million yet to be described. At least 85 per cent of all known animals are insects and some species occur in astronomical numbers.

The origins and explanation of this success lie back in the Silurian period when the first plants invaded the land from the sea. The terrestrial environment was far less uniform than that of the sea and soon generated a highly diverse flora which provided a rich series of ecological opportunities for herbivores and decomposers able to exploit them. The requirements for this included biting/chewing mouthparts, resistance to water loss and extreme temperature fluctuations, high mobility, good sensory perception and small size. The arthropod body design, as refined along the myriapod–insect line, provided the best available answer at the time, and proved highly adaptable. Insect mouthparts, for instance, radiated from the basic biting/chewing type to those used for seizing, rasping, lapping, piercing and sucking, with many variations on each theme.

To the herbivores and decomposers were very rapidly added predators and parasites, and as the plants diversified so did the insects. A boost to insect diversity came about through co-evolution with flowering plants, resulting in a host of mechanisms for ensuring cross pollination by insects. An even larger (and much earlier) evolutionary development was the invasion of freshwater, in which insects have been much more successful than their marine relatives, the Crustacea. The sea is the one environment the insects have failed to penetrate—because crustaceans were there first. With the same basic body design but much longer to evolve the appropriate detailed adaptations, crustaceans have left no room for insect invaders, which are restricted to skimming the surface of the oceans and living along the shore line.

The classification of insects is, at present, far from stable, because of conflicting opinions among systematists over evolutionary relationships within the group. The system adopted here divides the insects into the wingless Apterygota and the usually winged Pterygota. Some authorities, however, regard the four apterygote orders and the pterygotes as having reached the insect condition independently from different starting points; that is, they regard the group as representing a level of organization rather than a set descended from one insect ancestor. Accordingly they describe the Collembola, Protura, Diplura, Thysanura and the Pterygota as classes and not orders. Much of the doubt surrounding this issue stems from the patchy nature of the insect fossil record. The first known insect is the collembolan *Rhyniella praecursor*, from the Devonian period. Other apterygotes appear in the Carboniferous period, but none of the fossils indicates the group's origins. Nor is anything known of the origin of winged insects, which appear and become plentiful in the coal measures of the Carboniferous. The more advanced winged insects, such as the scorpionflies, lacewings and beetles, originated as long ago as the Permian period but only became abundant at the end of the Cretaceous.

The Apterygota are the primitively wingless insects. The members of the orders Collembola, Protura and Diplura are cryptozoic and lead obscure lives in soil and litter. The Collembola or springtails are the most important, as they occur in immense numbers and have some economic impact, for example, *Sminthurus viridis* which attacks lucerne (alfalfa). Collembolans differ from the other primitive insects in having only six segments to the abdomen. On the fourth segment is the forked springing organ, the operation of which gives these insects their common name. The fourth group of apterygotes, the Thysanura or bristletails, have the primitive quota of eleven abdominal segments, each with a pair of vestigial appendages. They have always been of interest as possible ancestors of the Pterygota, with which they share several features. They are nocturnal animals and live in rocky places or domestic situations. *Petrobius maritimus* is very common on sea cliffs; in contrast the Silverfish, *Lepisma saccharina* is a household pest.

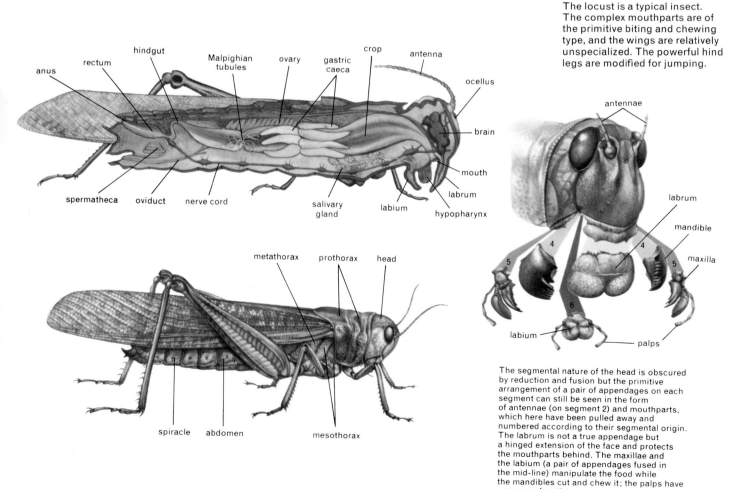

The locust is a typical insect. The complex mouthparts are of the primitive biting and chewing type, and the wings are relatively unspecialized. The powerful hind legs are modified for jumping.

The segmental nature of the head is obscured by reduction and fusion but the primitive arrangement of a pair of appendages on each segment can still be seen in the form of antennae (on segment 2) and mouthparts, which here have been pulled away and numbered according to their segmental origin. The labrum is not a true appendage but a hinged extension of the face and protects the mouthparts behind. The maxillae and the labium (a pair of appendages fused in the mid-line) manipulate the food while the mandibles cut and chew it; the palps have a sensory function

All the remaining insects are either winged or clearly derived from types with wings. They are split into two groups, the Exopterygota and the Endopterygota. The difference lies in the structure of the young stages. Exopterygotes have young in which the wings appear as external buds and which increase in length with each moult until full adult size is reached. Such young are called nymphs and usually resemble the adult in general form, but change slightly with each moult or ecdysis. Each stage is called an instar. By contrast endopterygotes have young in which the wing buds develop internally, out of sight. The young are known as larvae, and are often totally different in structure from the adults. So extreme is the reorganization required during the transition from larva to adult that a special rebuilding stage—the pupa—is included in the life cycle. All the more advanced insects are endopterygotes.

The development of wings and flight was one of the great turning points in the evolution of insects, since it gave them much greater scope for success in the terrestrial environment. Although there are no fossils to show how wings evolved, there are plenty which show what primitive winged insects looked like. The two large pairs of wings were of similar size and each was criss-crossed by a complicated network of veins. The wings could not be flexed backwards nor held flat along the body—the palaeopteran condition. Most of this stock became extinct, but the dragonflies (Odonata) and mayflies (Ephemeroptera) survived. As in other exopterygotes, the young are nymphs, but in both cases are considerably modified for an aquatic existence. Dragonfly nymphs are predators and seize their prey in a modified, extendable labium. Dragonfly adults, in spite of the primitive nature of the wings and the associated musculature, are very good fliers, being adapted for catching other flying insects on the wing. Mayflies are chiefly known for the brevity of their adult lives and the unique fact they have a pre-adult stage (sub-imago) which can fly.

The remaining exopterygotes and all the endopterygotes are able to flex and to fold the wings back over the abdomen—the neopteran condition. Thus the wings' utility is greatly increased. They can, for example, be hardened and used for protection, or patterned and sculpted to provide camouflage, as in moths. Hardening of the wings is seen in the Orthoptera, Hemiptera and to an extreme degree in Coleoptera, where the forewings form elytra and play little part in flight. This

ABOVE: insect wings are modified in many ways, either to increase the efficiency of flight or so that one pair can serve some other purpose. Although they are superb fliers, Odonata, like the Green Damselfly (*Lestes sponsa*), seen far left, show the primitive condition where two pairs of equal sized, non-overlapping wings move up and down in opposite directions. Lepidoptera, like many insects, move both pairs in unison. To aid this, either the wings overlap or, as in the Spurge Hawk Moth (*Celerio euphor-* *biae*), a bristle (the frenulum) projects from the hind wing and locks with the forewing. Diptera, such as this dungfly *Scatophaga stercoraria*, use only the front pair of wings for flight; the hind pair are modified into tiny balance organs, called halteres, that move up and down, like the wings, but in opposite phase. In Coleoptera the front wings are hardened into elytra which protect the hind wings when at rest. The Cockchafer (*Melolontha melolontha*) shows how the elytra are held raised during flight.

reduction of two pairs of functioning wings to one pair is a recurring theme in the more advanced insects. It seems that to bring out the potential for speed and precision available in the insect wing system, a functional one-wing arrangement has to be adopted. This means that either the two pairs become linked to operate as one, or one pair is dispensed with for flight and becomes available for other uses, such as the elytra of beetles or the balance organs (halteres) of flies. Most of the neopteran exopterygotes have mouthparts of the basic insect design, with mandibles for biting and chewing, maxillae acting as ancillary jaws, and a labium which manipulates and contains the food while it is chewed. The most important orders are the Orthoptera, the Dictyoptera and the Isoptera.

The Orthoptera are a large group of 17,000 species which include the grasshoppers, crickets and mole crickets. All have forewings which are somewhat hardened to form tegmina, and well developed hind legs for jumping. Many of the males have loud and distinctive songs with which they attract

LEFT: a mayfly dun, the sub-imago stage of *Ephemera danica* (order Ephemeroptera), holds its wings in a typical 'butterfly' position over its back. Earlier nymphal stages are aquatic.

BELOW: *Mantis religiosa* (order Dictyoptera) is the most common praying mantis in Europe. This one is feeding in its usual inverted position on a cricket. Over 1,800 species are known.

mates. Some orthopterans are important pests, none more so than the locusts, the largest members of the group.

Pests of a different kind are to be found in the Dictyoptera (the cockroaches and mantids). Cockroaches feed on a wide range of detritus and are one of man's principal commensals. They are also one of our oldest living insects, having existed almost unchanged since the Carboniferous period. Praying mantids are predators adapted for seizing their prey in raptorial forelimbs which they hold at rest as if in an attitude of supplication.

The Isoptera, or termites, have been described as 'socialized' cockroaches. They are remarkable animals, with peculiar digestive arrangements and an extremely elaborate social organization. They live in large colonies organized around a male and female (the king and queen). This pair is attended by several castes of workers, who forage for food, build the nest and tend the young. Primitive termites are able to digest wood with the help of protozoans which live in the gut, an ability which they have inherited from the cockroaches. It has allowed the termites to become severe pests of wooden structures in the tropics and subtropics. But it is not just man-made structures that suffer—all vegetable material is acceptable to termites. Their role in the ecology of the tropics can hardly be overstated. Reliance on gut symbionts gives way in the more advanced termites to the cultivation of fungus gardens, in which indigestible vegetable matter is 'farmed' to produce termite food (the fungus). These fungi tend to be specific to particular species of termite and probably cannot exist on their own for long. Termite nests can be massive in size and very elaborate in design. Those of *Macrotermes bellicosus* may be 7 m high and last for decades.

The Dermaptera, or earwigs, is a distinctive order whose members have pincers at the rear, with which they defend themselves vigorously. The Phasmida are also readily identified, because neither stick nor leaf insects could easily be mistaken for anything else, except some of the more cryptic mantids; but, unlike mantids, phasmids are strictly vegetarian. The final order showing affinities with the Orthoptera—the Plecoptera—is distinct from the others in having an aquatic larvae. Characteristically squat and flattened, these insects, the stoneflies, tend to live in cold, clear running water. The adults live unspectacular lives among riverside vegetation and fly rather reluctantly. Their colouring is as subdued as their behaviour.

The remaining exopterygotes have little in common. The Psocoptera, the barklice and booklice, are small creatures with specialized mouthparts for scraping up their food. Barklice are rather dumpy creatures with wide heads and long filiform antennae. They are very common on tree bark if there is a good growth of the lichens on which they feed. Booklice, as their name implies, live in houses where they delight in eating the paste of book bindings.

The Mallophaga and the Siphunculata are respectively the biting lice and sucking lice. Both are small ectoparasites which have lost their wings. The biting lice live mainly among bird feathers and find nourishment in skin exudations and fragments, whereas the sucking lice prefer mammalian hosts and have piercing and sucking mouthparts for feeding on blood. The Human Louse *Pediculus humanus* is an infamous siphunculate which transmits epidemic typhus and relapsing fever. It occurs all over the body, but a related species *Pthirius pubis* occurs only in the pubic region. The eggs of many species are cemented to the hosts' hairs and are known as nits.

BELOW: *Pediculus humanus* clinging by its claws to the human hairs among which it lives. Lice have specialized mouthparts for sucking blood and can transmit a number of unpleasant diseases.

RIGHT: many species of aphids are devastating plant pests. They feed on sap by piercing vessels in the stem with their mouthparts. Voracious ladybirds are among their chief predators.

Sucking lice have no special means of dispersal, but biting lice often indulge in phoresy, clinging on with their mandibles to parasitic Diptera which can fly from host to host.

The Thysanoptera, or thrips, are so small that few people are aware of their existence, although 5,000 species have been described. They are very abundant, and can cause considerable economic loss to crops, which they damage by piercing the tissues and sucking the sap. Many live inside flowers, and can there be seen as slender-bodied creatures of minute size. Their wings are very distinctive, being reduced to narrow straps fringed with long fine hairs.

Sucking mouthparts are also the hallmark of the last exopterygote group, the Hemiptera, which contains many interesting insects and, for its size (55,000 species), probably has more pests than any other insect order. The aphid *Myzus persicae*, apart from doing a great deal of direct damage by sucking plant sap, transmits more than 50 different virus diseases. Although most Hemiptera are plant feeders, quite a number have become predatory and some

BELOW: the heteropteran *Rhodnius prolixus* is one of the assassin bugs. It is known to transmit Chaga's disease—an often fatal illness in man.

RIGHT: cicadas are the largest homopterans; some have a wing spread of 15cm or more. Only the males sing but both sexes can hear well.

even parasitic in that they suck the blood of birds and mammals. The mouthparts have been modified for piercing and sucking and the projecting labium or rostrum acts as a grooved sheath in which the stylet-shaped mandibles and maxillae can move up and down. The rostrum hinges at its base and can be reflexed under the body in most species, but in the assassin bugs (family Reduviidae) it is fixed and stands out in a rather menacing curve.

Two suborders of Hemiptera—the Homoptera and the Heteroptera—are distinguished by wing structure. The forewings are always more leathery than the hindwings, evenly so in the Homoptera, but only in the basal half in the Heteroptera. A further difference between the two groups is that the Homoptera hold their wings tent-like over the body, whereas the Heteroptera fold them flat with the membranous outer areas overlapping each other.

The Heteroptera or true bugs vary a great deal in size and shape, and show considerable ecological diversity. Among the plant feeders are many pests, including the cotton stainers (*Dysdercus*), notable for the damage they do to cotton. Whole families of bugs have taken to the water, including the water boatmen (Notonectidae) and the giant water bugs (Belostomatidae), which can exceed 11 cm in length and are among the largest of all insects. They feed on fishes and frogs which they subdue with a poisonous bite. They fly readily on dark nights and are disconcertingly common in parts of the tropics. The domestic environment is the home of the bed bug *Cimex*, which is a human parasite, hiding in crevices by day and sucking blood by night.

Many Heteroptera are notable for their bright colours, which serve to warn potential predators of their distastefulness. Stink glands increase the disincentive. By contrast the Homoptera are odourless and subdued in colour. For protection they rely on small size, cryptic colouration and glands which produce waxy secretions in the form of a dense powder or twisted filaments; these secretions form a protective coat on the surface of the animals. The Homoptera include the froghoppers, the aphids and the cicadas. The nymphs of the froghoppers protect themselves by producing a froth of bubbles known as cuckoospit, which protects them both from enemies and from desiccation. The adults, as their name implies, are good jumpers. The aphids, otherwise known as greenfly or plantlice, are among the most successful of all insects in temperate regions.

Very different to the other Homoptera are the cicadas, which are large, brightly coloured and very noisy. Their songs serve to attract mates and surpass those of all other insects in power and lack of musical merit. The nymphs live underground and suck sap from the roots of trees. In North America some species take as much as 17 years to complete their development, which is synchronized, all the nymphs in one locality being the same age. As a result there is a mass emergence of adults every 17 years.

The remaining nine orders of insects form the Endopterygota, with development by means of larva and pupa rather than nymph. The institution of the pupa makes it possible to have a developmental stage very different in structure from the adult, an opportunity which the more advanced insects have exploited with success. The basic strategy that has evolved is the separation of functions between larva and adult. The larva becomes a machine of insatiable appetite and rapid growth, the adult a highly tuned apparatus for reproduction and dispersal. The contrast between the two stages is well seen in the Lepidoptera and becomes extreme in the more advanced Diptera and Hymenoptera where the larva is a helpless grub living among its food.

The endopterygote orders make up the vast majority of insects. The Coleoptera (beetles) take pride of place with 300,000 described species; the Lepidoptera (butterflies and moths) come next with 120,000, followed by the Hymenoptera (ants, bees and wasps) with 100,000 and the Diptera (flies) with 58,000. All of these figures are gross underestimates and certainly the majority of Diptera and Hymenoptera have yet to be described and named.

The fossil history of the endopterygotes begins in the Permian period 250 million years ago, in which remains of Neuroptera and Mecoptera are found. The Neuroptera (lacewings, antlions and snakeflies) have survived as a rather isolated offshoot, but studies of wing venation suggest that the Mecoptera (scorpionflies) have played a more central part in endopterygote evolution. Loss of some veins and modification of others has led to the Trichoptera (caddisflies) and Lepidoptera in one direction and to the Diptera and Siphonaptera (fleas) in another. Wing structure throws no definite light on the origin of the Hymenoptera or the Coleoptera, but it is likely that they are derived from an endopterygote ancestor not unlike a neuropteran. The first beetles appeared in the Permian period and there are traces of Hymenoptera from the Jurassic, but the explosion of diversity in the main orders did not begin until the end of the Cretaceous period, 70 million years ago, when flowering plants began to evolve.

The most primitive of the Neuroptera, to judge from both the fossil record and wing structure, are the alderflies and snakeflies. Alderflies, *Sialis*, are common inhabitants, as larvae, of the muddy bottoms of ponds and lakes, where they live as active predators. The heavily built and sombre coloured adult has wings which show their primitive nature by having many cross veins. The closely related snakeflies, *Raphidia*, are specialized in having the prothorax much extended to form a long neck which, to some eyes at least, gives them a snake-like appearance.

The lacewings are beautiful insects with a delicate tracery of veins in the wings. Some, such as *Chrysopa carnea*, are vivid green in colour, but many are brown. Both larvae and adults feed voraciously on aphids. Neuroptera in the family Mantispidae have raptorial forelimbs which, in form and function, recall the limbs of praying mantids; this provides a classic example of convergent evolution, made the more

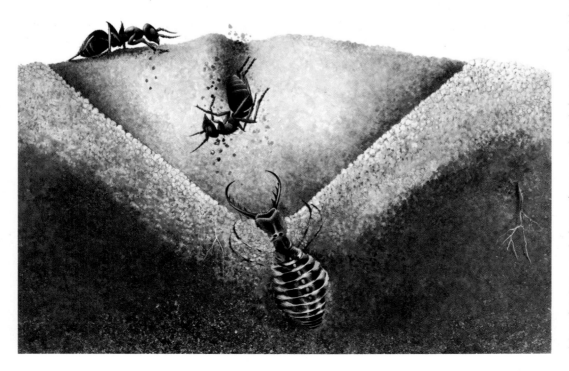

TOP: a male scorpionfly feeding avidly on the chrysalis of a Small Tortoiseshell butterfly (*Aglais urticae*). Its biting mouthparts are carried at the tip of its elongated, beak-like head.

ABOVE: the wings of this lacewing (*Chrysopa flava*) show clearly the fine cross venation that indicates the antiquity of the Neuroptera. Their wings are very long giving a slow, rather clumsy flight and a fluttering appearance.

LEFT: the antlion larva hides by burying itself up to the head at the bottom of its pit-fall trap. Like the other Neuroptera it has biting jaws, and the length of these ensures that it need not emerge far from the sand to catch its startled prey.

RIGHT: the spectacular proboscis of the Convolvulus Hawk Moth enables it to feed on nectar from flowers with deep corolla tubes. Large eyes are typical of moths and butterflies, and suitable flowers are located by sight.

striking by the existence of a similar trend in mantis shrimps (Crustacea). Mantispids are parasites of other insects. The active young larva of *Mantispa*, for example, searches for an egg-cocoon of wolf spiders (*Lycosa*). On entering the cocoon, which is being carried by the parent spider, the larva feeds off the young spider brood until, fully fed, it pupates inside the egg cocoon, from which it eventually hatches.

The antlions form a third group of Neuroptera. Superficially the adults look rather like dragonflies, from which they can be separated by the prominent antennae and the fact that they usually fly at night. The larvae typically live at the bottom of conical pits which they dig in sandy places. Unwary insects, particularly ants, may slip into one of these pits and arrive at the bottom in a miniature landslide, there to be greeted by a pair of formidable jaws which make certain there is no return.

The Mecoptera are a distinctive group because of the scorpion-like tail which is present only in the male and serves to clasp the female during mating. The adults are predatory on flies and other insects, whereas the caterpillar larvae live a retiring life in moss and soil, though they too are said to be carnivorous.

From the Mecoptera one line of evolution leads to the Lepidoptera and the Trichoptera (caddisflies). The differences structurally between the two orders are slight, particularly in wing venation, but adult caddisflies are as drab and reluctant to fly as butterflies and moths are active and brilliant. Caddisfly larvae are aquatic and mostly live in cases built from debris or sand grains. The adults have long wings covered in hairs and mouthparts for lapping up liquids. There are about 5,000 species and they are an important element of the freshwater fauna.

The Lepidoptera by contrast have mouthparts for sucking up liquids, flat scales covering the wings and larvae that are almost always terrestrial. The mouthparts are highly modified from the primitive cockroach condition, with vestigial mandibles, the labium present only as palps, and the maxillae developed into a long double-channelled tube or proboscis. This is normally coiled up, hidden between the labial palps, but when the animal is stimulated to feed it can be straightened and in some species extended to very long distances. In the Convolvulus Hawk Moth (*Agrius convolvuli*) it measures 14 cm, and is longer than the body. The proboscis is an adaptation for feeding on the nectar of flowers, and its evolution has had much to do with the success of the Lepidoptera, as it allows them to make the most of the rich food resources made available by flowering plants in the cause of cross pollination.

The name Lepidoptera means 'scaly-winged' (as Trichoptera means 'hairy-winged') and refers to the detachable scales that cover the wings and provide the brilliant colours. They are flattened hairs and in fact grade into more typical hairs on the body of a moth or butterfly. The larvae of butterflies and moths are very constant in structure, with powerful biting jaws and usually eight pairs of legs. The first three pairs are the true thoracic legs, the rest being prolegs peculiar to the larval stage, the purpose of which is to grip on to surfaces tightly. Many caterpillars are covered with

spines, sometimes poisonous, to discourage predators, or are distasteful and warningly coloured. All can spin silk and use this not only to spin cocoons to house the pupal stage but also to draw leaves together as shelters or to construct cases out of debris. Almost all caterpillars are vegetarian and some do great damage to crops. The best known Lepidoptera are the butterflies. They can be regarded as a rather special family of moths, with large wings, naked pupae (known as chrysalids), slender bodies and clubbed antennae. However, the differences from moths are not clear cut in the skipper butterflies, which are unusual in that they have heavy bodies, cocoons and unclubbed antennae.

From the Mecoptera another line of evolution has led to the Diptera, in many ways the most specialized of the insects. Flies can be distinguished from all other insects by having the hindwings modified as halteres, which operate as miniature gyroscopes to allow the fly to monitor its flight performance. The wings are very narrow at the base and their venation is much reduced. Dipteran powers of flight are exceptional. The mouthparts of the more primitive adults are suctorial, but exhibit seemingly endless modifications for lapping, piercing and sucking. Chewing mandibles are, however, absent. Blood sucking is the major occupation for many female Diptera in the adult stage, and in the process some of the most serious human diseases, particularly malaria and yellow fever, are transmitted from host to host. Health problems are also caused by flies which carry germs from unsavoury places and contaminate human food. The common House Fly, *Musca domestica*, is a major culprit. Larvae of certain other flies are serious agricultural pests. Fruit, although edible after attack by Mediterranean Fruit Fly (*Ceratitis capitata*), cannot be sold and there are worldwide quarantine laws to prevent infected fruit being distributed. Altogether the flies have an enormous impact on human welfare, bearing in mind that malaria alone affects more than 200 million of the world's population of 4 billion. Fly larvae

characteristic of these primitive flies. The most important Nematocera are the mosquitoes (family Culicidae). Only the females suck blood, which they require in order to mature the eggs. The males, in common with very many other flies, take nectar from flowers. Mosquito larvae are aquatic, like most Nematocera. Other important families are the Ceratopogonidae (biting midges), Simulidae (black flies) and Tipulidae (crane flies).

The Brachycera ('short-horns') contains numerous families of medium to large flies, with much shorter antennae than Nematocera. The bee flies (Bombyliidae) are very like bumblebees with long tongues for taking nectar. The Asilidae, or robber flies, are predators on other insects, which they seize in their strong legs and suck dry. An important brachyceran family is the Tabanidae—the horse flies and clegs—which are large and accomplished blood suckers. They are particularly troublesome to horses and cattle, but feed also on a wide range of hosts from crocodiles to monkeys.

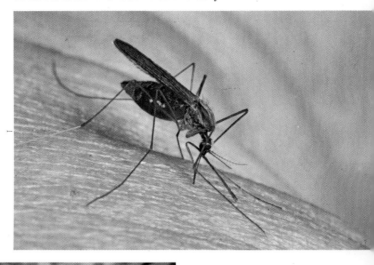

ABOVE: the mosquito has a very elastic abdomen to enable it to take in a large quantity of blood at a single feed. Mosquitoes locate their hosts by sensing exhaled carbon dioxide and then following it to its source.

LEFT: like bees, adult syrphids feed on nectar from flowers and are important pollinators. The larvae have a variety of habits, some feeding on aphids and others on decaying organic matter. This species is *Scaeva pyrastri*.

ABOVE RIGHT: *Archaeopsylla erinacei* is the specific flea of the European Hedgehog. Typical siphonapteran features include long legs and a flat body. (×30)

RIGHT: the stag beetle *Lucanus cervus* (suborder Polyphaga) is unusually large for a temperate zone species and may exceed 5cm in length. Only the male has greatly enlarged mandibles. The delicate wings are covered by the armoured elytra that typify the Coleoptera.

are legless, but are often quite active because they can wriggle. Many of them are aquatic or parasitic, but most live in decaying vegetation, dung or corpses.

The Diptera are divided into three suborders, the Nematocera, Brachycera and Cyclorrhapha. Nematocera means 'thread horns' and refers to the long thread-like antennae

The third group of Diptera, the Cyclorrhapha, is the largest and many families display a bewildering variety of life styles. The larvae or maggots have specialized mouthparts and taper towards the head end. The Syrphidae (hover flies), Drosophilidae (fruit flies), Calliphoridae (bluebottles) and Muscidae (house flies) are among the more important families.

The Tachinidae are parasites, laying their eggs inside other insects, which are eventually killed by the developing larvae. The Hippoboscidae, which are highly modified as ecto-parasites, include some wingless forms, for example, *Melophagus ovinus*, the Sheep Ked.

Although the adults of Siphonaptera and Diptera differ markedly from each other, the larvae of the former resemble those of primitive flies. They live on organic debris in the nests and dens of their hosts, which are birds and mammals. The adults are wingless, and are adapted as ectoparasites, feeding on blood. They are laterally flattened, well supplied with claws and can jump with exceptional power. Fleas are vectors of various diseases, notably bubonic plague, and have thus played a notorious part in history.

Although the Diptera may be the most highly specialized insects, the beetles or Coleoptera are certainly the most numerous in terms of species. They are as old as any of the endopterygotes and thus have had a long time to diversify. Their success is not based on specialization but rather on a generalized design—they have the same kind of biting/chewing mouthparts seen in the cockroach. Their environment is the ground rather than the air and they fly relatively little, although they are able to disperse very effectively on the wing in many cases. They are heavily built, with the abdomen and hind wings well protected by the elytra (hardened forewings) and probably more than any other arthropods they rely on exoskeletal armour to keep attackers at bay. The diet of beetles runs the full gamut from plant feeding through predation to carrion feeding, but parasitism has few exponents, and blood sucking none at all.

The success of beetles is not confined to land as they have invaded the freshwater habitat in large numbers and acquired several adaptations for swimming and respiration. They have also been outstandingly successful in exploiting dead wood as a food resource, either feeding on it directly, or introducing a fungus on which they themselves can feed. Beetle larvae

have ordinary mandibulate mouthparts and a well developed head. The body form varies considerably, from that of active predatory types with well developed legs to the cockchafer variety which still has legs but is very fat and incapable of much movement. Even more extreme is one legless type found in the weevils.

Of the two main suborders of beetle, the Adephaga and the Polyphaga, the Adephaga are thought to be the more primitive. Both larvae and adults are active predators. The largest family is the Carabidae or ground beetles, to which the tiger beetles (Cicindelidae) are closely related. Tiger beetle larvae live in burrows and pop up to grab at passing prey. Many adephagans are aquatic, for instance the whirligigs (Gyrinidae) and the Dytiscidae, which contains the big predatory diving beetles.

The Polyphaga is a much larger and diverse group and contains a number of large families, such as the rove beetles (Staphylinidae). These are mostly small, with short elytra which do not entirely cover the abdomen. The Scarabeidae mostly feed on dung and carrion although some, like the cockchafer (*Melolontha*), are vegetable feeders. This family includes the largest of all beetles. The ladybirds (Coccinellidae) are well known because of their bright colours (they are distasteful to predators). They feed on aphids and other Homoptera and have been used to control these pests. The leaf beetles (Chrysomelidae) feed on plants. Several of them are pests, none more so than is *Leptinotarsus decemlineata*, the Colorado Beetle, which devastates potato crops.

The largest beetle family of all, the weevils (Curculionidae), has more than 40,000 species, many of which feed on or in seeds and various parts of plants. Several are serious pests, for example, the Grain Weevil, *Sitophilus granarius*. Weevils are very distinctive because the mouthparts are borne on a beak or rostrum, from which the elbowed antennae project.

A very enigmatic group thought to be closely related to the Coleoptera are the Strepsiptera (stylopids). These are tiny insects which parasitize other insects and have the forewing reduced to form a club.

The final endopterygote order is the Hymenoptera, made up of the sawflies, ants, bees and wasps. An interesting feature of the group is that it ranges from quite primitive to very advanced forms, with a transition from biting to sucking mouthparts and great reduction in wing venation. The order is highly successful, with 100,000 described species and considerable ecological diversity.

The sawflies, forming the suborder Symphyta, can be distinguished from the other suborder, the Apocrita, by their more complete wing venation, the absence of a waist between thorax and abdomen, and by their larvae which are plant feeding caterpillars, very lepidopteran-like but with more than five pairs of prolegs. They have the habit of curling up into a flat spiral when disturbed.

The Apocrita itself is split into two groups: parasites, such as the ichneumons and chalcids which parasitize other insects, and stinging forms such as the ants, bees and wasps. This second group is one of the most remarkable assemblages among the invertebrates, for their social life is rivalled only by termite and vertebrate societies.

By no means all the stinging Apocrita are social. The primitive stock is thought to have been similar to the solitary digging wasps of today, and to have developed along three lines: one led to the social wasps, one to the solitary bees and thence to the social bees, and a third to the ants. Social life originated either through maternal care for the young or through sister care (cooperative rearing of broods), and was facilitated by the habit of provisioning cells with a food store for the larvae. Initially the food was collected, an egg laid and the larvae left to their own devices, but continuous provisioning also developed, so that the young were attended as they grew. The next and critical stage was the suppression of fertility among the offspring and their retention to form a worker caste, with the mother as the single egg laying queen. This is the level seen in bumble bees and social wasps, where a colony produces fertile males and females only in the autumn and then dies out, leaving young fertilized queens to found new colonies in the spring. The final stage was the development of a permanent colony, as seen in the Honey Bee, *Apis mellifera*, in which mechanisms have evolved whereby a failing queen can be replaced or the colony can split if it becomes too large. Hand in hand with these developments have gone changes in diet from active predation to nectar collecting. Nectar is a high energy food which can be stored, and is thus very valuable. To ensure the best possible supplies an elaborate communication system has developed in the form of 'dances' in the honey bee, so that 'scouts' can inform other workers of the location, in terms of direction and distance, of the best nectar sources.

The ant *Lasius fuliginosus* builds an elaborate nest at the base of a decaying tree. As spring advances, the column moves upwards to take advantage of the higher temperatures in the summer nest, and then moves down again to avoid the rigours of the winter.

LEFT: the female *Urocerus gigas*, the Giant Woodwasp, implants its eggs into timber through a long ovipositor. The larvae are sometimes common enough to be pests of cut timber.

BELOW: the Hairy-legged Mining Bee (*Dasypoda hirtipes*) does not form social colonies but lives solitarily. It digs burrows in which to lay its eggs and stocks them with food for the larvae.

RIGHT: several ant species are symbiotic with other insects. Here wood ants (*Formica*) are collecting sugary secretions from aphids which, in return, receive protection from predators.

Pentastomids

Pentastomids are small parasitic animals of doubtful affinity which live as adults in the respiratory passages of birds, mammals and reptiles, and as larvae in intermediate hosts. They are elongate and worm-like with little external structure except for a simple mouth flanked by two pairs of rudimentary claws. Their affinities are obscured by their adaptations as endoparasites, but the form of the larval stage and moulting of the cuticle are held to indicate arthropod affinities, and for want of better evidence they are often classified as Arthropoda. Recent opinion, however, considers them to represent a separate phylum but they are not sufficiently significant to warrant individual treatment here. Only about 70 species are known.

SUBKINGDOM Metazoa

PHYLUM Pogonophora

Beard worms

ORDER	FAMILY	GENUS
Athecanephria	Oligobrachiidae	Oligobrachia (+ 3 further genera)
	Siboglinidae	Siboglinum
		Siboglinoides
Thecanephria	Polybrachiidae	Zenkevichiana (+ 6 further genera)
	Sclerolinidae	Sclerolinum
	Lamellisabellidae	Lamellisabella
		Siphonobrachia
	Lamellibrachiidae	Lamellibrachia
	Spirobrachiidae	Spirobrachia

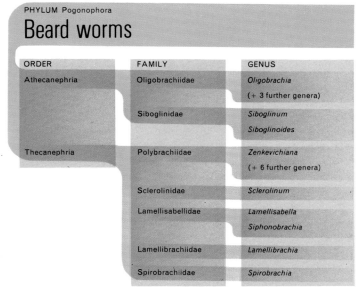

- NUMBER OF LIVING SPECIES: about 160

- DIAGNOSTIC FEATURES: coelomate worms that live in self-secreted chitinous tubes; body divided into a cephalic lobe bearing tentacles, a bridle, a trunk and a segmented opisthosoma; setae present; gut absent; closed blood vascular system present

- LIFESTYLE AND HABITAT: all marine; sedentary bottom dwellers in soft substrates often at great depths

- SIZE: 5 cm to 1.5 m long and 0.1 to 2.5 mm in diameter; mostly 8 to 15 cm long

- FOSSIL RECORD: none substantiated

The Pogonophora are long, very thin tube-dwelling invertebrates which live mainly in the deep sea. They are relatively new to science, being first discovered at the beginning of this century in material collected by the Siboga Expedition to the East Indies. The first genus was accordingly named *Siboglinum*. They are superficially similar to polychaete annelids and were once grouped with them but they were later recognized as a separate phylum which was named Pogonophora after the beard-like tuft of tentacles present in many species.

Since their original discovery pogonophores have been recorded from all five oceans and from some coastal waters. They are generally deep water animals—80 per cent of all known species occur between 200 and 4,000 m—but they are also found in shallower continental waters where the temperature remains low all the year. Although the distribution of some genera is extensive—*Siboglinum*, for example, is cosmopolitan—each species tends to have a small range. Nearly all pogonophores live with their tubes buried, or partially buried in soft sediments, such as oozes and muds. Two species are exceptions however: *Sclerolinum brattstromi* has been found inside rotten wood, paper and even leather in some Norwegian fjords and *Lamellibrachia barhami* builds a stiff tube which stands proud of the muddy substrate.

Siboglinum eknani isolated from its tube. The coiled tentacle extends beyond the burrow to absorb oxygen. In this male, brown spermatophores are visible inside the trunk. (×20)

All pogonophores are remarkably similar anatomically. At the front end a cephalic lobe bears one (in *Siboglinum*) to more than a hundred long tentacles. Where tentacles are very numerous, as in *Spirobrachia*, they are held in a corkscrew curve. Behind the cephalic lobe is the bridle, a ridge of thickened cuticle which is important in tube building. Most of the length of the body consists of the trunk, the surface of which is covered with papillae. These occur in two metameric rows anteriorly but gradually become less regular further back. Papillae are thought to assist in movements of the animal within its tube and the plaques—hardened plates associated with the papillae—are probably also important in this respect. The girdle region of the trunk is characterized by epidermal ridges in which rows of short setae are embedded, the number of which varies according to the species. The setae are very similar to those of polychaetes and they function in a like manner—helping the animal to maintain its position in the tube.

Until 1963 the trunk was thought to be the posterior portion of a pogonophore. But they are now known to possess a distinct rear end, the opisthosoma, which is easily broken off. The opisthosoma consists of 5 to 23 segments each bearing a number of setae, which are larger than those of the girdle. There has been some debate as to the function of the opisthosoma but it seems likely that it protrudes from the bottom of the tube and acts as a burrowing organ. The tube itself is a chitinous structure which ranges in length from a few centimetres to 1.5 m (in *Zenkevichiana*) but is only 0.1 mm to 3 mm wide. It is usually stiff and may have a banded appearance with light and dark rings.

The most unusual feature of pogonophores is the complete absence of a gut. This has led to many theories concerning methods of feeding. It is possible that the animals gain all their nourishment by directly absorbing dissolved organic matter from the surrounding water. It has been shown experimentally that there is a net uptake of these materials and the extreme attenuation of the body means that the surface to volume ratio is very high so that maximum absorption can be achieved.

The sexes are separate in pogonophores and can be distinguished externally by the position of the genital apertures. In most species the females produce a few large yolky eggs which are held in the tentacles after spawning.

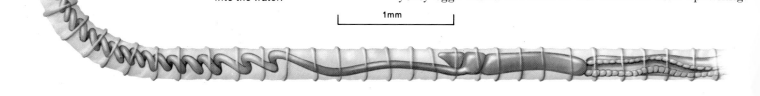

Siboglinum in its tube. The tube is flexible and much longer than the animal. It is embedded in the bottom deposits, just protruding into the water.

1 mm

tentacles

bridle

cephalic lobe

trunk

metameric papillae

A complete beard worm of the genus *Lamellisabella*. The greatly elongated trunk bears metameric papillae at the front end which are used to hold the animal in its tube. The dilated opisthosoma, often missing from collected specimens, is probably used for burrowing and as an anchor.

opisthosoma

trunk

papillae

The male releases sperm packaged in spermatophores which float around until they become entangled in the tentacles of another pogonophore. In time the spermatophore dissolves and fertilization takes place. The fertilized eggs remain within the maternal tube and develop into larvae which all lie with their heads pointing down towards the parent. When they have grown large enough they leave the parental tube and settle nearby. Some species are known to produce a large number of small eggs and in this case there is no evidence of brood protection; it is not known if these species have a free-living larval stage.

Ever since their discovery there has been considerable debate on the systematic position of the Pogonophora. The situation is greatly complicated by the absence of a gut which makes it impossible to distinguish the dorsal and ventral surfaces of the animal. Originally the pogonophores were divided up among various polychaete families, but they are much too similar to each other to be split up in this way and this scheme, together with the idea that they were linked with annelids (and other protostomes), was soon dropped. The apparent division of the body into three parts, each with a coelomic cavity, suggested instead that pogonophores were allied to the chordates and there was some indication that the coelom developed in a typically chordate (that is deuterostome) way. However, the discovery of the opisthosoma meant that the coelom is not after all divided into three and reinvestigations showed that it might not develop as originally thought either. So a relationship to the annelids could be possible and the segmentation of the opisthosoma makes this likely. Recent comparisons between polychaete and pogonophore setae show remarkable similarities which are unlikely to reflect convergent evolution.

Scanning electron micrographs of setae from a polychaete annelid (left) and a beard worm (right).

The marked resemblance suggests that the two groups are probably closely related. (× 2,000)

If pogonophores are on the annelid line then they must have evolved from some 'proto-annelid' stock during the early Cambrian period. Fossil pogonophores cannot be identified with certainty but a well known early Cambrian fossil, *Hyolithellus micans*, bears a marked resemblance to extant pogonophores. So it seems that, although they are new to man, the Pogonophora are in fact a very old, well-established part of our fauna.

girdle bearing fine setae

SUBKINGDOM Metazoa

PHYLUM Chaetognatha

Arrow worms

GENUS
Eukrohnia
Bathyspadella
Heterokrohnia
Krohnitta
Pterosagitta
Sagitta
Spadella

● NUMBER OF LIVING SPECIES: about 75

● DIAGNOSTIC FEATURES: coelomate, transparent and worm-like, with one or two pairs of lateral fins and a tail; head armed with chitinous hooks; straight gut; ventral nerve glanglion usually conspicuous; blood vascular system absent

● LIFESTYLE AND HABITAT: all marine and free swimming at all depths; planktonic except for one benthic genus

● SIZE: range from 4mm to 10cm, but most less than 2cm

● FOSSIL RECORD: one reported fossil doubtful

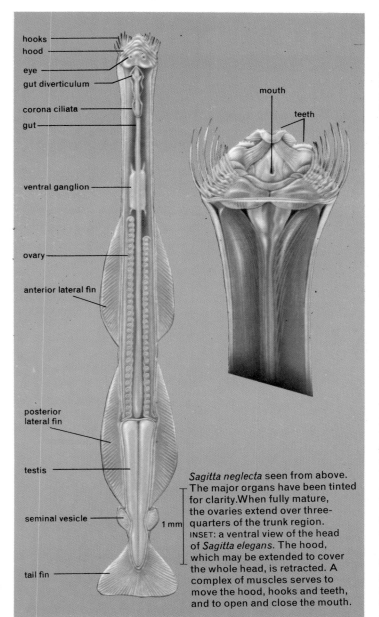

Sagitta neglecta seen from above. The major organs have been tinted for clarity. When fully mature, the ovaries extend over three-quarters of the trunk region. INSET: a ventral view of the head of *Sagitta elegans*. The hood, which may be extended to cover the whole head, is retracted. A complex of muscles serves to move the hood, hooks and teeth, and to open and close the mouth.

Labels on main figure: hooks, hood, eye, gut diverticulum, corona ciliata, gut, ventral ganglion, ovary, anterior lateral fin, posterior lateral fin, testis, seminal vesicle, tail fin, 1 mm

Labels on inset: mouth, teeth

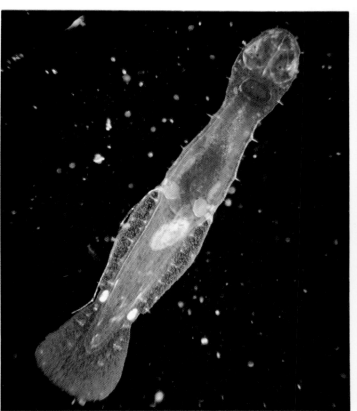

Spadella cephaloptera is a well known neritic and mainly benthic species. The pink colouration of the gut and around the head is due to staining, which has made the corona ciliata on the neck conspicuous. The eyes and sensory tufts are also visible.

The Chaetognatha are small, transparent, bilaterally symmetrical and torpedo- or arrow-shaped animals, hence their common name arrow worms. They are entirely marine and, with the exception of one benthic genus, are planktonic. They are almost ubiquitous in the seas and oceans—nearly every plankton net tow will be found to contain specimens often in large numbers. Although they occur so commonly, chaetognaths are inconspicuous animals; only a handful of species are coloured and most species are, when alive, completely transparent and measure generally less than 20 mm in length.

All species of chaetognaths are closely similar in appearance: they have one or two pairs of lateral fins, a rather fish-like tail and consist of three anatomical regions—the head, the trunk and the tail segment. The head is armed with curved chitinous hooks which are used to seize prey and to push it by alternating movements of the hooks into the mouth. There are one or two paired rows of teeth which serve to cut the cuticle of prey during the process of swallowing. There is a pair of small, simple eyes, incapable of image formation, which are usually pigmented though in a few deep sea species pigment is lacking. Chaetognaths are hermaphrodite: the trunk region contains the simple straight intestine and the two ovaries; the tail, separated from the trunk by a septum, contains the testes and when the animal is fully mature bears two characteristic structures, the seminal vesicles, which are diagnostic for each species. Scattered over the body are small sensory tufts which have been shown to be sensitive selectively to vibration and enable the animal to detect and capture prey.

All members of the phylum are so similar that few attempts have been made to establish categories higher than genera. The most recent attempt to do so has not met with general acceptance and will not be discussed here. There are seven genera of living chaetognaths. One supposed fossil genus, *Amiskwia*, a fossil from the middle Cambrian, has been described as a chaetognath but on anatomical grounds seems more likely to be a pelagic nemertean. Of the other genera one, *Spadella*, is benthic usually in inshore waters; the remaining six genera are all planktonic, but at different levels in the sea. Two genera—*Bathyspadella* and *Heterokrohnia*—are exclusively deep living; *Eukrohnia*, which is mainly deep living, has one species (*E. hamata*) which is also epipelagic—that is, lives near the surface—in cold waters; *Sagitta*, *Pterosagitta* and *Krohnitta* are predominantly epipelagic though *Sagitta macrocephala* is found only at depths between 500 m and the sea floor. The largest species are *S. gazellae* (105 mm long) and *S. maxima* (98 mm long).

The distribution of the pelagic species is worldwide: the numbers of species are highest in tropical surface waters and there are more species in the Indo-Pacific region than in the Atlantic region. The shallow-living species may be conveniently classified according to the water conditions in which they live: there is a small number of neritic inshore forms, usually only one species in any area; a more numerous group of nearshore species adapted more or less to mixed water conditions; and a large group of oceanic species adapted to stable water conditions of the open ocean. There are, in addition, a few mesopelagic and bathypelagic species.

Chaetognaths are active carnivores; their main food is small crustaceans, but they also prey on other chaetognaths and larval fish in the plankton; indeed almost the only impact the group has on man is that they are a major predator of larval herring. In their turn chaetognaths are preyed on by fish and larger crustaceans.

Chaetognaths have great powers of regeneration; it is claimed that a head will regenerate from the trunk if cut off, the eyes appearing first and then follow the mouth and hooks. For many years it was almost impossible to keep arrow worms alive in a laboratory but most of the problems have been overcome in recent years, at least for relatively hardy inshore species which have now been kept successfully throughout a whole life cycle.

In the coldest waters chaetognaths complete a life cycle in one, sometimes two, years; in temperate waters the life cycle is about six weeks and in tropical waters it is shorter. The number of broods per year increases from the poles to the equator. Fertilization is internal, and in most species the eggs are simply released into the sea where they develop into free swimming larvae which resemble miniature adults. In *Pterosagitta*, however, the eggs float together in gelatinous masses. Some species of *Eukrohnia* have brood pouches which contain the eggs and the resulting larvae.

Although chaetognaths are widespread common animals they were not observed until 1768 when Slabber, a Dutch naturalist, found one on the Dutch coast and described it in the following year. The invention of the tow net in 1829 made possible the study of planktonic organisms and from that time on chaetognaths were caught in quantity and studied by scientists. In keeping with the zoological ideas of the nineteenth century, much work was done to try to establish the relationships of the chaetognaths to other animal groups. Charles Darwin who collected chaetognaths from South American waters during the voyage of the Beagle wrote in 1844 that the chaetognaths are 'remarkable for obscurity of affinities' and one may well agree with him today. During the succeeding years they have been supposed to be related to at least eight invertebrate groups as well as to the chordates. Consideration of embryology and anatomy indicate such conflicting relationships that current opinion tends to regard the group as widely separated from all other groups and without any certain affinities.

Eukrohnia bathyantarctica is an uncommon deep-sea species. The ovaries are short. The seminal vesicles are transferred intact during copulation in this species, and the remains of a vesicle can be seen protruding from the female opening of one specimen

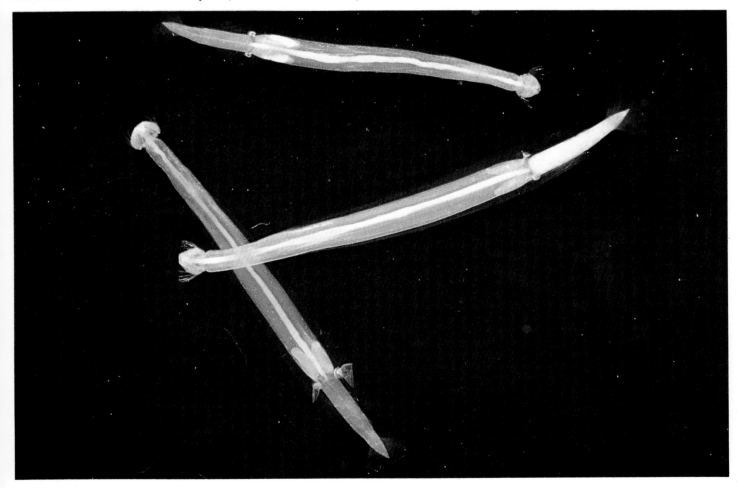

SUBKINGDOM Metazoa

PHYLUM Hemichordata

Acorn worms

CLASS	ORDER	FAMILY	GENUS
Enteropneusta *acorn worms*		Harrimaniidae	*Saccoglossus*
			(+ 4 further genera)
		Spengelidae	
		Ptychoderidae	*Balanoglossus*
			Ptychodera
			Glossobalanus
Pterobranchia	Rhabdopleurida		*Rhabdopleura*
			(+ 2 fossil genera)
	Cephalodiscida		*Cephalodiscus*
			Atubaria
			(+ 2 fossil genera)
†Graptolithina *graptolites*	†extinct		

- NUMBER OF LIVING SPECIES: about 90

- DIAGNOSTIC FEATURES: coelomate, solitary and worm-like (enteropneusts) or colonial and tubicolous (pterobranchs); body divided into proboscis, collar and trunk

- LIFESTYLE AND HABITAT: all marine; enteropneusts burrow in soft substrates in shallow waters; pterobranchs usually tube dwellers in deep waters

- SIZE: enteropneusts 2 cm to 2.5 m (mostly 10 to 50 cm); ptero branch colonies up to 10 m in diameter with zooids up to 14 mm but usually less than 5 mm

- FOSSIL RECORD: enteropneusts unknown as fossils; pterobranchs possibly rare from Ordovician to Recent, though possibly related graptolites well represented from mid-Cambrian to Carboniferous; over 1,000 species

ABOVE: *Glossobalanus* about to burrow into shell gravel. The characteristic division of the enteropneust body into three distinct regions—the head, collar and trunk—shows clearly.

BELOW: *Saccoglossus* lives in a coiled burrow in muddy sand. Food is extracted from the substrate which it eats, the residue being expelled to form a cast on the surface at the end of the burrow.

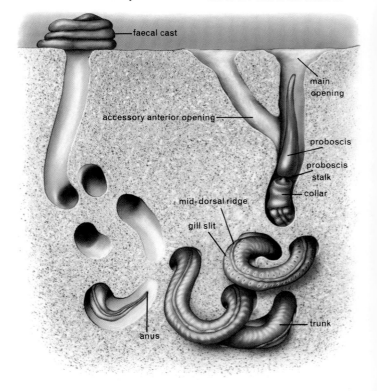

Hemichordates form a small group of marine invertebrates whose living members are divided into the two classes Enteropneusta and Pterobranchia. The enteropneusts are worm-like and grow from 9 cm to 1.5 m long. Their body is divided into three parts—the proboscis, the collar and the trunk—and the positioning of the proboscis onto the collar, reminiscent of an acorn in its cup, gives rise to their common name, acorn worms. Enteropneusts are inhabitants of shallow water and the characteristic casts of burrowing species such as the giant *Balanoglossus gigans*, or *Saccoglossus* can often be seen at low tide. Other species do not form a permanent burrow but lived buried in sand or mud, among seaweed or under stones. Pterobranchs are sessile, deep water animals which occur mainly in the southern hemisphere, although *Rhabdopleura* has been dredged up off European coasts. Except for *Atubaria*, all pterobranchs live in secreted tubes. In the colonial *Rhabdopleura* these tubes interconnect and the individuals are joined to each other with stolons but in *Cephalodiscus* there is a looser aggregation.

Although hemichordates may be described as a minor phylum in terms of their abundance and number of species, they are interesting because they display certain features indicative of the chordates and were once thought to be on the direct evolutionary line of the vertebrates. Even the name Hemichordata suggests this half-way state.

The proboscis of an acorn worm is a short cone attached by means of a narrow stalk to the collar which is short and cylindrical and extends forwards to overlap the proboscis stalk. The mouth is situated ventrally in the collar region. Most of the body is made up of the trunk. Anteriorly there is a longitudinal row of gill slits arranged on each side of the

mid-dorsal line. Outside of these rows the trunk contains the gonads, and in some species, such as *Ptychodera flava*, the body is drawn out into genital wings in this region. The coelom is also divided into three regions—one cavity in the proboscis, a pair in the collar and a pair in the trunk. This tripartite division of the coelom in the enteropneusts is typical of the deuterostome invertebrates.

Methods of feeding depend on the habitat. Many burrowing species are deposit feeders—they gain nourishment from organic matter in the sand and mud. Others are suspension feeders—the proboscis is covered with sticky mucus which traps plankton and detritus floating in the surrounding water. Ciliary currents are then used to carry the food to the mouth. Large particles can be rejected by bringing the collar over the mouth thus interrupting the current. The gut consists of a buccal tube in the collar which passes into the pharynx with its gill slits, then into the oesophagus and finally the intestine

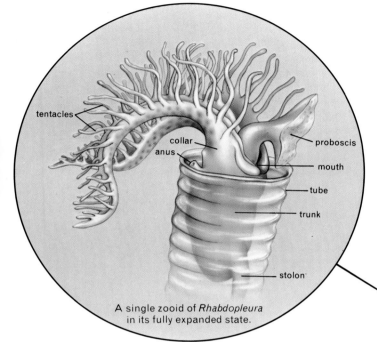

A single zooid of *Rhabdopleura* in its fully expanded state.

proper. A long thin diverticulum is associated with the buccal tube and passes forwards from it into the proboscis. This structure was once thought to be a notochord but histological investigations show it to be part of the gut.

The gill slits probably evolved as a feeding mechanism but in the hemichordates there is a secondary function of gaseous exchange. The gill slits themselves are ciliated and induce a stream of water in through the mouth and out through the gills. The nervous system consists mainly of epidermal

ABOVE: the pterobranch *Rhabdopleura* lives in small erect tubes. Individuals in the colony are connected by a continuous stolon.

BELOW: the graptolite *Didymograptus*. The thecae were interconnected and resemble the tubes of colonial pterobranchs.

nerve cords and is relatively primitive. The collar nerve cord, however, becomes internal (separated from the epidermis above) during development and, in some cases, is hollow.

The sexes are separate in acorn worms. The females produce masses of mucus-embedded eggs which are fertilized externally. Currents soon break up the clusters and the eggs float around singly. Early development is remarkably similar to that of echinoderms and, in those species with indirect development, results in a planktonic tornaria larva which is very similar to the bipinnaria larva of starfish. Other species develop directly without the larval stage and one species, *Saccoglossus kowalevskii*, hatches as a young worm.

Pterobranchs appear very different from enteropneusts but the body is divided into the same three regions. The proboscis is shield-shaped and rather insignificant, being overshadowed by the greatly expanded lophophore-like collar. In *Rhabdopleura* there are two recurved arms, with heavily ciliated tentacles, and *Cephalodiscus* may have five to nine arms. The tentacles are believed to function in feeding. Gill slits are less prominent in pterobranchs—*Cephalodiscus* has one pair only and *Rhabdopleura* none. The anus is not terminal, as in acorn worms, but, because of the U-shaped gut, opens on the dorsal side of the collar. Such an arrangement is a reflection of the sessile habitat of pterobranchs.

Much research has been carried out concerning the phylogenetic relationships of hemichordates. Typical chordate features are the gill slits and the dorsal nerve cord, which is hollow in some species. The absence of a notochord,

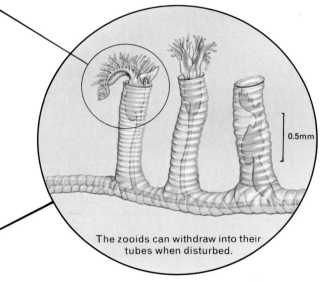

The zooids can withdraw into their tubes when disturbed.

however, excludes them from the phylum Chordata. Pterobranchs are more primitive than enteropneusts and accordingly give less indication of chordate affinities. In the extreme case, *Rhabdopleura* has no gill slits and no hollow nerve cord.

A close relationship to the echinoderms is indicated by embryology and the tornaria larva; pterobranchs also show marked similarities to the lophophorates. This would place hemichordates less near the origin of vertebrates and closer to the first deuterostomes themselves. One scheme proposes a pterobranch-like ancestor giving rise independently to the echinoderms, the hemichordates and the chordates. Pterobranchs are rare as fossils and the soft-bodied enteropneusts have no fossil record. However, the graptolites—a group of fossils encountered from the mid-Cambrian to Carboniferous—are thought by some to be related to the pterobranchs by virtue of the similar mode of growth. The chordate and echinoderm affinities of hemichordates cannot be doubted but their evolutionary position apparently brings us no closer to understanding the origin of the vertebrates.

SUBKINGDOM Metazoa

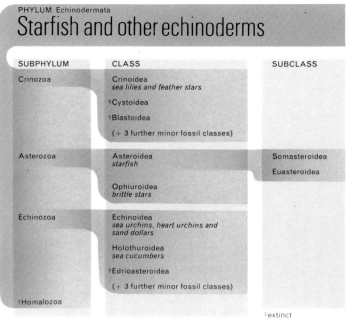

PHYLUM Echinodermata
Starfish and other echinoderms

SUBPHYLUM	CLASS	SUBCLASS
Crinozoa	Crinoidea *sea lilies and feather stars*	
	†Cystoidea	
	†Blastoidea	
	(+ 3 further minor fossil classes)	
Asterozoa	Asteroidea *starfish*	Somasteroidea
		Euasteroidea
	Ophiuroidea *brittle stars*	
Echinozoa	Echinoidea *sea urchins, heart urchins and sand dollars*	
	Holothuroidea *sea cucumbers*	
	†Edrioasteroidea	
	(+ 3 further minor fossil classes)	
†Homalozoa		

†extinct

- NUMBER OF LIVING SPECIES: about 6,000

- DIAGNOSTIC FEATURES: mainly pentaradiate as adults; lacking a head but with a calcareous endoskeleton and often bearing spines; tentaculate tube feet associated with a water vascular system present

- LIFESTYLE AND HABITAT: attached or errant, free living marine animals almost exclusively benthic as adults

- SIZE: 5 mm to 1 m in diameter, average 8 cm

- FOSSIL RECORD: Precambrian to Recent, with greatest abundance in middle to late Palaeozoic; about 13,000 species

The phylum Echinodermata—the name literally means 'spiny-skinned'—is a wholly marine group whose members include the familiar starfish and sea urchins often found on the shore. It is not a large group by comparison with some in the animal kingdom but it has a long fossil history which began in Precambrian times nearly 600 million years ago. Echinoderms have exploited almost all types of sea bed from fine silt to rocks and coral reefs and they are represented in all the world's seas and oceans at all depths. Virtually all species are bottom dwellers as adults, living either on or in the substratum. Some are actually attached to the sea bed, as were many more now extinct, and only a few species of sea cucumber have evolved to live a pelagic life. By contrast with the vast majority of adults, the developing larvae are principally planktonic.

The precise evolutionary position of the phylum is not certain. Echinoderm embryology and manner of development, however, suggest affinities with the chordates but few

biologists today seriously believe that forms similar to any modern species were chordate ancestors. So far as echinoderm origins are concerned, it seems possible that they may have arisen from worm-like ancestors somewhat resembling present-day Sipuncula.

The echinoderms are unique in the animal world because they display certain features which have no parallel elsewhere. These features serve as a most useful means of making an accurate diagnosis of the group. Most striking is their five-sided adult symmetry which is termed pentamerism. Its evolutionary significance is not well understood but some scientists have suggested that the five-sided or five-rayed body plan provides extra strength. The endoskeleton is peculiar and consists of many small crystals of calcium carbonate, each one of which is perforated many times by living tissue. These skeletal elements are embedded in the body wall to form what is technically termed the test. When the animals die the spaces within each ossicle may become invaded with mineral and thereby fossilize very well.

Inside the test is a large body cavity which is a true coelom. This houses the gut and reproductive organs. During its development part of the coelom becomes separated off to form the water vascular system which is one of the most distinguishing characteristics of the echinoderms. The water vascular system performs several important functions including respiration, transport of materials within the body, food collection and locomotion. The tube feet, which are the outward manifestations of the water vascular system, are involved in most of these functions. These organs are often suckered to assist with adhesion and they are normally arranged on the body surface in five double rows, each double row being known as an ambulacrum. In all species the ambulacra converge on the mouth so that in a number of cases they can function as food grooves, transporting collected food material for ingestion. In the crinoids the mouth and ambulacral surfaces point away from the substratum but in all other living groups of echinoderm, apart from the holothuroids, the mouth and ambulacra are placed next to the sea bed. The arrangement in the holothuroids may be likened to a sea urchin which has grown very tall and then fallen over on one side so that the animal has come to lie on three of the ambulacra, with two more well off the substratum and the mouth pointing forwards with the anus behind.

Crinoids and some holothuroids are filter feeders. In the former group the many fine tube feet are used very effectively to strain small particles from the seawater. In the holothuroids which filter feed, the special tube feet around the mouth have become divided many times to act as straining organs. In other sea cucumbers these organs sweep up detritus settled on the sea bed and pass it to the mouth. The starfish are efficient predators which use their tube feet to handle the prey and open shells. The ophiuroids are scavengers and carrion feeders which have found many ways of life, often very successful ones if the sizes of their populations are

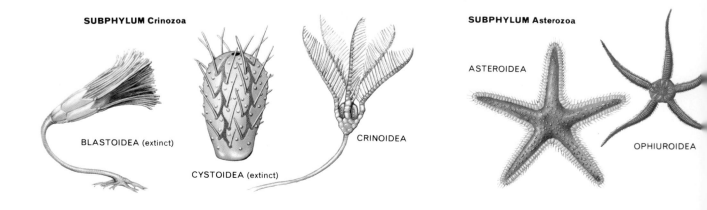

SUBPHYLUM Crinozoa

BLASTOIDEA (extinct)

CYSTOIDEA (extinct)

CRINOIDEA

SUBPHYLUM Asterozoa

ASTEROIDEA

OPHIUROIDEA

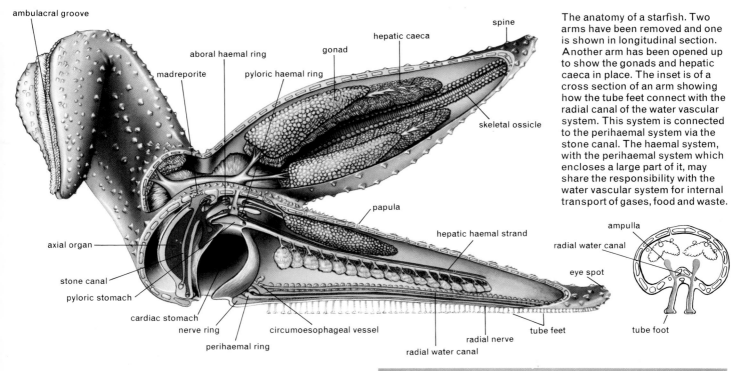

ambulacral groove
aboral haemal ring
madreporite
pyloric haemal ring
gonad
hepatic caeca
spine
skeletal ossicle
axial organ
papula
hepatic haemal strand
ampulla
radial water canal
stone canal
eye spot
pyloric stomach
cardiac stomach
nerve ring
circumoesophageal vessel
tube feet
tube foot
perihaemal ring
radial nerve
radial water canal

The anatomy of a starfish. Two arms have been removed and one is shown in longitudinal section. Another arm has been opened up to show the gonads and hepatic caeca in place. The inset is of a cross section of an arm showing how the tube feet connect with the radial canal of the water vascular system. This system is connected to the perihaemal system via the stone canal. The haemal system, with the perihaemal system which encloses a large part of it, may share the responsibility with the water vascular system for internal transport of gases, food and waste.

taken as a guide. The echinoids have developed several ways of feeding: the regular or round sea urchins scrape and browse on rocks for encrusting animals and plants, using a special chewing apparatus known as the Aristotle's lantern; some attack larger plants such as seaweeds; the sand dollars burrow but retain their chewing apparatus for dealing with the food; but the heart urchins simply ingest the sand in which they burrow and digest from it any suitable organic matter.

Because of the unusual symmetry, echinoderms have no head and in consequence the nervous system is not aggregated in one place but is spread more or less diffusely within the outer layer of the body. However, a radial nerve cord is associated with each ambulacrum and its function is to co-ordinate the activities of the tube feet. Each radial nerve cord connects with a ring nerve which surrounds the oeso-phagus and allows coordination between ambulacra and body rays.

Like so many sedentary bottom-dwelling marine inverte-brates, the echinoderms generally reproduce by means of planktonic larvae. The sexes are normally separate and ferti-lization of the egg takes place in seawater. The larva develops as a bilaterally symmetrical organism which floats free and

Diadema setosum is a tropical echinoid that is often found on coral reefs. Its long spines are barbed and penetrate the skin of predators. They are also sensitive to light.

feeds on microscopic marine plants. This allows dispersal. The larval phase terminates by settlement on the substra-tum, followed by metamorphosis into a juvenile which resembles the adult in most ways apart from sexual maturity. Echinoderms generally live from one to seven years.

Within the Echinodermata there is a wide range of fossil and living forms. The five-sided adult body plan shown by most members of the phylum is adapted to suit various ways of life.

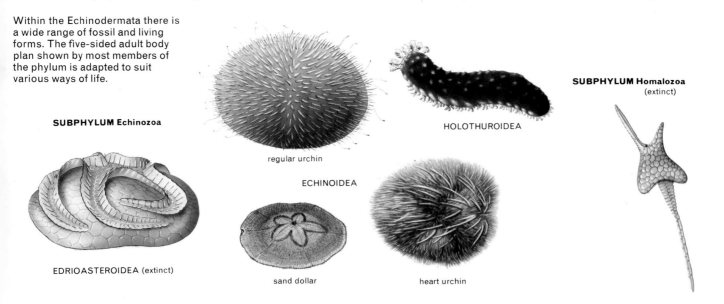

SUBPHYLUM Echinozoa

SUBPHYLUM Homalozoa
(extinct)

HOLOTHUROIDEA

regular urchin

ECHINOIDEA

EDRIOASTEROIDEA (extinct)

sand dollar

heart urchin

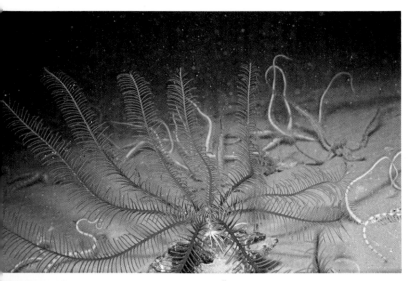

LEFT: *Antedon mediterranea* is a free living crinoid. It feeds by raising its arms so that the tube feet can strain suspended food particles from the water. In the background some ophiuroids can be seen feeding in a similar way.

RIGHT: a fossil stalked crinoid, *Dialutocrinus milleri*. The cup-like body is supported on the stalk composed of many ossicles, and is surmounted by a circlet of branching arms.

BELOW: *Comanthina schlegeli* is another free living crinoid. It occurs in coastal waters of the Indo-west Pacific seas. The many branching arms bear rows of pin-nules which give the animal the feathery appearance from which it gets its common name.

The classification of the phylum has been revised in recent years and most experts now agree that the Echinodermata are best divided into four subphyla of which the first three have living representatives but the last is composed entirely of fossil species. Each of these subphyla contains several classes and has fossil representatives. The significance of the echinoderm fossils in the history of the Earth's seas will be appreciated further when it is pointed out that of the 19,000 or so known species only 6,000 are alive today.

Within the subphylum Crinozoa only the class Crinoidea (sea lilies and feather stars) has representatives living today. The word crinoid is derived from the Greek word for lily. The crinoids represent the most primitive existing echinoderms. In the case of the stalked sea lilies the adults are permanently attached to the sea bed by means of a jointed stem which has a special holdfast at its lower end. The free living feather stars, or comatulids as they are sometimes known, lack a stalk as adults although a rudimentary one is present in the final larval stage. Comatulids grip the substratum by means of special adhesive appendages called cirri which are many jointed and moveable, terminating in a claw-like segment. From the cup-shaped body or theca the arms, arranged in multiples of five, radiate from the opening of the mouth. The anal opening lies to one side of the theca outside the ring of arm bases. The arms, like the cirri and thecal walls, are supported by skeletal ossicles and the arms and the cirri can be freely moved by local muscle action. The arms bear a row of small branches termed pinnules on either side of their upper surface, and both the arms and the pinnules bear the tube feet which themselves are branched. Consequently when the animal spreads its arms a large filtering area of fine tube feet is made available to strain particles from the passing seawater. Each ambulacrum therefore sends branches into each pinnule and functions as a food groove along which captured food particles are sent to the mouth, propelled by the action of the many cilia which cover the surface.

The crinoid water vascular system has not evolved the individual ampullae which are associated with the tube feet of starfish and sea urchins and which provide the necessary fluid pressure to extend each tube foot. Instead the pressure is provided by a complex system of local muscle contraction in the radial water canal. The radial nerve cord lies close to each radial water canal and serves to integrate and coordinate the various movements of the tube feet and pinnules. The reproductive organs are carried on the arms too. Pelagic larvae generally develop after fertilization followed by an attached larval stage known as the pentacrinule larva. From this either the stalked adults grow or, in the case of the comatulids they break free.

The class Cystoidea is known only from fossils which are widely distributed in early Palaeozoic rocks. A spherical or bladder-shaped theca was apparently borne on a stem and displayed radial symmetry in most cases. The mouth was borne at the apex of the body and from it radiated five food grooves bordered by paired food-collecting appendages or brachioles. It is believed that the way of life was similar to that of the stalked crinoids, but there are many anatomical differences which separate the two classes. The class Blastoidea is also exclusively fossil. The blastoids were stalked and their bodies were compact and globular, hence the name which means bud-like. Again the mouth was surrounded by five grooved structures bearing food collecting brachioles. A peculiar internal structure known as the hydrospire lay beneath each groove. Together with the brachioles the hydrospire is thought to have been responsible for respiration.

The subphylum Asterozoa includes two classes, the Asteroidea and the Ophiuroidea. The Asteroidea comprises the starfish and is divided further into two subclasses—the Somasteroidea which, except for one genus *Platasterias*, seems to be exclusively fossil, and the Euasteroidea which contains all the extant starfish together with some fossil genera. The somasteroids are thought by many zoologists to be the most primitive asteroids because of the simple level of organization shown by their water vascular system and skeletal arrangement. Their bodies are simple star shapes often with rather leaf-shaped rays. The large oral opening lies on the underside and the upper surface is supported by smaller skeletal plates than are present below. It is suggested that most of the fossil somasteroids lived as burrowers in sand and that in life only the ray tips protruded above the substratum. They probably fed on other burrowing organisms.

The more modern Euasteroidea have evolved the water vascular system in such a way that it acts both as an efficient locomotor mechanism and assists with respiration and transport of materials. The underside of the body is clearly marked by the ambulacral furrows with their rows of tube feet, which, in the more advanced euasteroids, are suckered. Each tube foot is associated with a well developed ampulla from which water can be driven by muscle power to protract the associated tube foot. The ampullae, which are lacking in the crinozoans, are associated with the radial water canal which in turn is linked with all the other radial canals by means of the circumoesophageal vessel. From this vessel ducts ascend to the upper side of the animal and connect with the madreporite, a modified ossicle somewhat resembling the rose of a watering can. The precise function of the madreporite and its associated vessels is poorly understood.

ABOVE: the euasteroid *Pycnopodia helianthoides* is a sun-star from the west coast of North America. Each of the many arms carries tube feet on the underside. The white spot on the upper surface is the madreporite.

BELOW: *Linkia laevigata* is a starfish which inhabits shallow waters of the tropical Indo-west Pacific. The blue pigment makes this species conspicuous amid the coral reefs and may serve as protection from intense sunlight.

The euasteroid starfish represent some of the most ecologically significant invertebrate predators. To this end the pentameric body plan has some advantages in that stimuli can be received equally well from any direction since tactile and chemoreceptors cover the whole outer surface of the body. They are particularly well concentrated in the ray tips where special non-locomotory but sensory tube feet occur. On receiving chemical stimuli emanating from suitable food material many starfish are able to home on to the prey although there are some species which need to blunder into prey before they can perceive it. Once in contact with the prey it can be dealt with in various ways.

Some starfish such as *Astropecten*, which in any case lack suckers on their tube feet, swallow the prey whole, and strange lumps can often be made out inside live animals. Small animals such as gastropod molluscs and worms may be taken in this way. Other species, such as the notorious crown-of-thorns starfish (*Acanthaster*) climb on top of their prey and spread out their stomach membranes through the open mouth and wrap them closely against the prey which is digested outside the body. After the meal is consumed the stomach can be retracted. In other starfish like *Asterias* the tube feet are used for prising open the shells of bivalves such as mussels. For some time it was thought that a chemical was released which paralysed the bivalve, but it now seems that the starfish achieves its purpose by recruiting groups of tube feet to pull on the shell for limited periods after which another group can take over. In this way the starfish can outlast the bivalve which in the end relaxes just sufficiently enough to allow the shells to gape a fraction of a centimetre so that the stomach membranes can be everted and insinuated into the bivalve shell where digestion takes place. After the meal the stomach is withdrawn.

The various orders and families of euasteroids have evolved a limited variety of body shapes and life styles. Some are flattened and live on or under sand, whereas others have more rounded rays and live among rocks or corals. Most species have only five rays, but some have twenty or more. In some species the body is soft and flexible whereas in others it is hard and rigid. The flexibility of the test depends on how tightly its component ossicles are bound together by connective tissue. Although all echinoderms have an endoskeleton, some parts of it, notably the spines, protrude through the skin to show on the outside of the body. The large body cavity of the starfish lies within the test and houses the gut and reproductive organs. A blind ending pouch or digestive caecum enters each ray from the gut. Similarly each ray houses a paired gonad which opens to the exterior through a short duct. Apart from the gonopores and the madreporite the upper surface of the starfish also bears the anus and numerous little coelomic out-pouchings called papulae.

Although closely related to the starfish the brittle stars (Ophiuroidea) show some notable differences. Possibly the most outstanding are those relating to the sharp division between the arms and the disc. The gut and other viscera are entirely confined within the disc itself. There is no anus, and the madreporite is reduced to a single opening or madrepore on the under side. The arrangement of the tube feet is also different. They lack both suckers and ampullae. The brittle stars move by flexing their arms and to achieve this there is a special system of arm ossicles which resemble vertebrae. They are connected by muscle blocks which serve to bend and flex the joints. Associated with each arm is a radial nerve cord and a radial water canal. In most cases there are only five arms, but a few species have more. The arms are generally well covered with spines which are not readily moveable with respect to the ossicles which support them. The brittle stars have successfully invaded a variety of habitats. Although most are scavengers or carrion feeders some live on the outside of other animals, for example sea fans, and come nearer to being parasites than do any other members of the phylum.

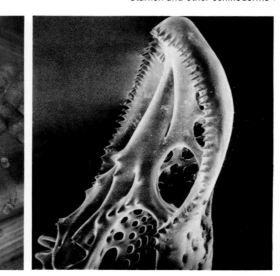

The distribution of the types of pedicellariae on the tests of sea urchins varies between species. The tridentate type is easily recognized by its long, slender jaw blades which are normally held shut unless the animal is disturbed mechanically. The ophiocephalous type is smaller and rounded, and under very high magnification can be seen to be armed with many pointed, gripping teeth. The trifoliate pedicellariae are the smallest and have minute, petal-shaped jaws. Their stems are often bent or twisted so that the jaws can touch the skin.

The subphylum Echinozoa contains the class Echinoidea to which the sea urchins belong. This is a diverse group showing three markedly different life styles. Their most conspicuous and unifying features are the moveable spines carried on small 'ball and socket' joints on the outside of the rigid test. The spines may be adapted for a great variety of functions including locomotion and defence. As well as spines, most regular sea urchins carry on their tests four different types of tiny appendages called pedicellariae. These function like minute tongs which can seize upon animals and inert objects which stray on to the test.

The tridentate and ophiocephalous types of pedicellariae appear modified to grip and hold on to the legs or other parts of animals which might cause damage to the urchin or settle upon it and clog it up. They respond to tactile stimuli by opening their jaws and converging on that part of the test which is irritated. The globiferous pedicellariae have venom glands on their jaws and generally respond to chemical stimulation of the test such as is provided by the chemicals on the surface of predators of the urchin such as starfish. They open their jaws in response to these chemicals and if a specialized chemoreceptor on the inside of the jaws strikes more surfaces bearing the chemicals the jaws snap shut and inject venom into the tissues of the intruder usually through a hollow venom tooth. As the intruder flees from the urchin

LEFT: *Asterias rubens* is a northeast Atlantic starfish. Here it is using its suckered tube feet to force apart the valves of a mussel so that it can feed on the flesh inside.

BELOW: the European brittle star *Ophiothrix fragilis* is a typical member of the Ophiuroidea. In this individual some of the tips of the arms are regenerating after being damaged.

RIGHT: basket stars, such as *Astrophyton*, are another type of ophiuroid with branching arms. They feed rather like crinoids, using their numerous tube feet to strain food from the water.

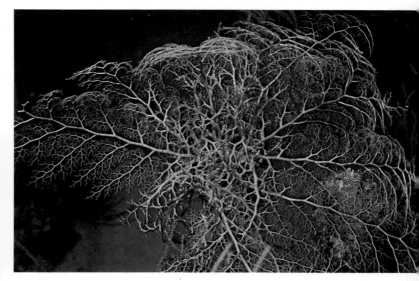

this sort of pedicellaria breaks off from the test and accompanies the victim continuing to apply the venom. The trifoliate pedicellariae are the smallest and differ from the other types in that they open and close their jaws spontaneously and spend most of the time mouthing over the test epithelial cells which they probably keep free from fine sediment and bacteria. Interestingly they appear to carry small protozoans on their jaws which probably consume the bacteria they have collected. Thus it is that the coordinated activities of various types of pedicellariae appear to keep the test free from obstruction by other organisms and inert material, and help to defend the animal against predators who have managed to penetrate the spiny armament.

The test, which is often wrongly called the 'shell', is not a shell but a true endoskeleton as in all echinoderms. It is typically globular and rigid in the regular sea urchins although in one group which is composed largely of deep sea animals, the test is flexible. The rigidity characteristic of the Echinoidea as a whole is achieved by the precise interlocking of the test ossicles which will remain fitted together even after the connective tissue has rotted away after death. The test in the sand dollars is flattened and has numerous reinforcing struts which run vertically between the upper and lower surfaces. In the heart urchins it is heart-shaped. In both these groups there is a departure from the pentameric symmetry and their adults show a functional front end with clear left and right sides, though they retain the characteristic five ambulacra radiating from a point on the upper surface. The test houses within it all the viscera including the gut and reproductive organs. Only the tube feet of the water vascular system protrude from it. In the sea urchins the tube feet assist with locomotion and posture and in the regular (spherical) sea urchins they are sometimes used like the guy ropes of a tent to hold the animal in one position. They may also serve to hold small pieces of weed and stones on the test so as to camouflage it. All except the heart urchins have the Aristotle's lantern for scraping food off the substrate.

Sand dollars and heart urchins burrow in various types of sand and gravel. The sand dollars do not cover themselves completely and can move through the substrate in the quest for food which they consume with the aid of their teeth. The various slots and indentations in the periphery of some sand dollar tests have probably been evolved to allow sand and gravel particles to pass through them so they can move through the substrate more easily. All the spines seem to be similar in this group and may assist with burrowing and manoeuvring food over the test. Although the heart urchins are not as flattened as the sand dollars they are more extensively modified for burrowing. The chewing organ is lost completely and the animals just take in sand or gravel

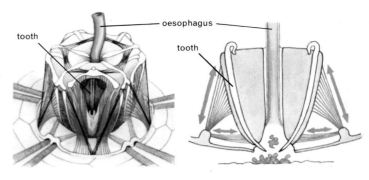

Aristotle's lantern is a complex structure composed of 40 skeletal elements bound together with muscle and connective tissue. This supports five teeth which grow continuously to compensate for wear. Muscles move the teeth up and down so that encrusting organisms may be scraped from the substrate into the mouth.

through the scoop-like mouth. They then digest off any organic material that may be contained. The tube feet around the mouth are considerably modified to become organs specially equipped for maintaining ventilation shafts to the burrow and for pumping water up and down these shafts.

The sea cucumbers (Holothuroidea) show some similarities with the sea urchins, but the two groups differ in several important ways. The holothuroid skeleton is flexible because the ossicles of the test are embedded loosely in a thick body wall. In most types the tube feet which lie on the underside are important for locomotion whereas those on the upper side of the body have degenerated and may perform secondary functions such as respiration. In all the sea cucumbers tube feet specially adapted for feeding are retained around the mouth. In some species all tube feet, except those round the mouth, have been lost and in such cases the test ossicles become important for movement and are of an interesting and special shape. They may be hooked or anchor-shaped, and as waves of peristaltic contraction pass along the body of the animal the hooks become engaged in the substrate and provide a point of purchase allowing movement forward.

As well as living on the surface of sand or mud and burrowing in it some sea cucumbers also live in rock crevices. A few, such as *Pelagothuria* and *Planktothuria*, have become sufficiently modified to live as free floating animals, a rarity in echinoderms. Many sea cucumbers have certain special modifications such as the respiratory trees which are additional areas for gas exchange located in the posterior region of the body cavity and have a common opening with the anus. In some types there are other structures known as Cuvierian organs which seem to have become adapted for defence. When stimulated by a predator these organs may discharge sticky secretions in the form of white threads. These may serve to entangle and deter other animals.

The remaining classes of the Echinozoa are all exclusively fossil. The class Edrioasteriodea, for example, appeared in the earliest part of the Cambrian period. Some of these mysterious animals were carried on a stem while others were directly attached to rocks or shells. The mouth was carried on the upper surface along with five ambulacral and five interambulacral areas. The ambulacra appear to have been roofed over with plates which were possibly liftable to allow for the extension of the tube feet to trap food particles. As well as a mouth there also seems to have been an anus and a gonopore on the upper surface. The underside was concerned with attachment.

The remaining subphylum—the Homalozoa—originated in the middle Cambrian period and includes some of the oldest known echinoderms. Though its members are all extinct, their fossils show that they were asymmetrical, stalked or sedentary, and often had a disc-like body composed of rigid or flexible plates. Around the edge of the disc ran a food-collecting groove leading to the mouth and some species also carried a flexible feeding arm. The Homalozoa are classified here as a subphylum of the echinoderms but some authorities maintain that they show a number of chordate characteristics; when viewed in that context they are referred to as calcichordates.

LEFT: *Psammechinus miliaris* is a regular echinoid living on the shore and in the shallow waters around northwest Europe. Like many other sea urchins the species frequently covers the test with small pieces of stones, shells and bits of weed.

TOP: fossil sand dollars from Iran show clearly the radial pentameric symmetry basic to echinoderms.

ABOVE: *Plagiobrissus grandis*, a heart urchin, burrowing into sand. The adaptations of heart urchins for life under the sand include long and short, backward-pointing spines and a scoop-like mouth for taking in the substrate.

BELOW: this specimen of the sea cucumber *Holothuria argus* is from the Great Barrier Reef of Australia. The defensive white threads have been discharged by the Cuvierian organs.

RIGHT: *Pawsonia saxicola*, a sea cucumber, showing three types of tube feet: active, for walking, on the underside; degenerate on the dorsal side; and branching oral tube feet used in feeding.

SUBKINGDOM Metazoa

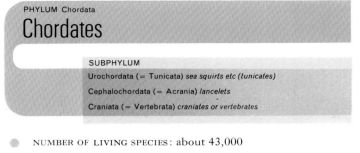

PHYLUM Chordata

Chordates

SUBPHYLUM
Urochordata (= Tunicata) *sea squirts etc (tunicates)*
Cephalochordata (= Acrania) *lancelets*
Craniata (= Vertebrata) *craniates or vertebrates*

- NUMBER OF LIVING SPECIES: about 43,000

- DIAGNOSTIC FEATURES: post-anal tail, containing a notochord, present at some stage of the life history; tail contains muscles, usually in blocks, and a hollow, dorsal nerve cord expanded anteriorly to a hollow brain; pharynx with visceral slits and an associated endostyle or thyroid gland

- LIFESTYLE AND HABITAT: primitively marine but many have invaded freshwater and land; almost all are free living; sessile species found only in the tunicates

- SIZE: 5 mm to 31 m but mostly between 10 cm and 1 m

- FOSSIL RECORD: Cambrian to Recent; vertebrates are extensive from Ordovician; calcichordates (about 60 species) occur from mid-Cambrian to mid-Devonian; probably more than 17,000 species

Though quite unlike more typical chordates, these filter-feeding adults of the tunicate *Clavelina* have developed from tadpole-like larvae, whose chordate affinities are shown by the presence of a notochord and dorsal nerve cord in the tail.

Although people have long recognized the difference between invertebrates and vertebrates, the evolutionary relationship of vertebrates to other animals has only begun to emerge more recently. In the 1840s, the brilliant German anatomist Johannes Müller discovered that the animals known as lancelets (for example, amphioxus) could be compared in detail with primitive jawless fish. Later, in the burst of zoological activity that followed the publication of Darwin's *Origin of Species* in 1859, two outstanding discoveries were made by the Russian embryologist Kowalevsky. In 1866 he showed that the acorn worms (*Balanoglossus* and its relatives) had gill slits, and in 1867 he showed that tunicates in their early development were like vertebrates, although they had previously been regarded as molluscs. Another Russian, Metschnikoff, discovered in 1870 that in their embryological development acorn worms (now placed in the Hemichordata) were like echinoderms. As a result of these discoveries zoologists came to recognize the phylum Chordata, comprising the vertebrates, tunicates and lancelets (that is the subphyla Craniata, Urochordata and Cephalochordata); and beyond the Chordata a wider grouping could be dimly perceived. The animals within this larger grouping were referred to as deuterostomes, because in all of them the mouth developed embryologically as a secondary perforation. In addition to the chordates this grouping included the hemichordates and echinoderms—both with the rank of phylum. Thus such diverse animals as man, starfish and acorn worms were all seen as related.

The unifying features of the chordates are found in the tail and the front part of the gut (pharynx). In many non-chordates, such as earthworms and crustaceans, the anus is at the extreme rear end of the animal. In chordates, however, the anus is some way forward. Behind it there is a tail, which is thus referred to technically as post-anal. Such a tail exists, at least in the first stages of life-history, in all chordates except a few specialized tunicates; even in humans it is well developed during the foetal stage. The tail contains along its central axis, at least during early life, a stiffish rod of large cells called the notochord (whence the name 'chordate'). In vertebrates this notochord is always present in the

The Grey Reef Shark (*Corcharinus menisorrah*) shows many primitive chordate features. Gill slits are present in the pharynx, and the tail is post-anal. In common with other advanced chordates they have jaws and teeth so that large prey may be eaten.

Until the reptiles appeared, all the chordates had been aquatic or dependent on water for breeding successfully. The evolution of a watertight skin and eggs enabled animals like this Bearded Dragon Lizard (*Amphibolurus vorbatus*) to live wholly on land.

embryo but in most adults it is replaced by the backbone. Above the notochord is the dorsal nerve cord (spinal cord) which is fundamentally a hollow tube, having been formed by a downfolding of the surface layer of the embryo. At its front end the dorsal nerve cord is expanded into a brain. On each side of the notochord, or the backbone that replaces it, chordates have muscles that move the tail from side to side. Primitively these muscles are divided into muscle blocks, which, to anyone who has eaten a fish, are familiar as repeated zig-zag segments of flesh.

The other unifying features of chordates are found in the front part of the gut. In the embryo this originates in two parts. The actual mouth cavity (buccal cavity) arises as an inpouching of the surface layer of the embryo. Immediately behind the buccal cavity is the gut region known as the pharynx, represented in man by the throat. In all chordates the pharynx has so-called visceral slits, at least in early embryos. In primitive aquatic chordates these are holes, or

Lions are advanced chordates, the adults lacking some of the primitive characteristics. There is still a post-anal tail but the notochord of the embryo persists in the adult only as discs of cartilage between the vertebrae. Only the embryo has gill slits.

gill slits, in the walls of the pharynx, and water can flow through them. In advanced land vertebrates, on the other hand, most of the visceral slits exist only in the embryo and they are never open holes. Each one consists of an internal pit in the wall of the embryonic pharynx corresponding to an external pit on the surface. The earholes of mammals arise from the most anterior pair of visceral slits.

Another feature of the pharynx in all chordates is a gland called the endostyle, or its equivalent the thyroid gland. The endostyle as such is found in tunicates, lancelets and the larvae of lampreys. It is a furrow along the ventral mid-line inside the pharynx and is lined with glandular and ciliated cells. From its front end a pair of ciliated bands (peripharyngeal bands) pass upwards round the wall of the pharynx to meet in the dorsal mid-line. In adult lampreys the larval endostyle has turned into the thyroid gland which is also found in all other vertebrates. In mammals the thyroid gland covers the front surface of the larynx.

These peculiarities of the chordate pharynx are connected with the original habit of feeding by filtration. In tunicates, for example, the endostyle secretes a sheet of mucus on each side. The front end of each sheet is held by one of the peripharyngeal bands whose cilia pull the sheet continuously upwards. Where the two sheets meet in the dorsal mid-line they are rolled up together into a mucous rope which is pulled backwards by cilia into the oesophagus. Thus, the pharynx is completely lined with a bag of mucus. Water entering the mouth passes into this bag and can only escape through the gill slits by first diffusing under pressure through the mucus. Tiny food particles are thus caught in the mucus and eventually pulled back in the mucous rope into the oesophagus. The same method of catching food is used by lancelets and larval lampreys. In tunicates and lancelets the current of water is pumped by cilia on the gill slits and this is the most primitive method. In larval lampreys, however, the current is pumped by flaps of muscle at the front end of the pharynx. All adult living vertebrates have abandoned this primitive method of feeding—adult agnathan fish are blood suckers while all other vertebrates have jaws.

Asymmetry is a curious feature of living primitive chordates. In a larval amphioxus the first gill slits to appear are those of the left side, which are formed in series from front to back. The second group of gill slits appears on the right side, and they are formed simultaneously opposite the left gill slits already there. A third group of gill slits is added later, on both sides of the animal behind those already present. Similarly there are many detailed asymmetries in the tunicate pharynx.

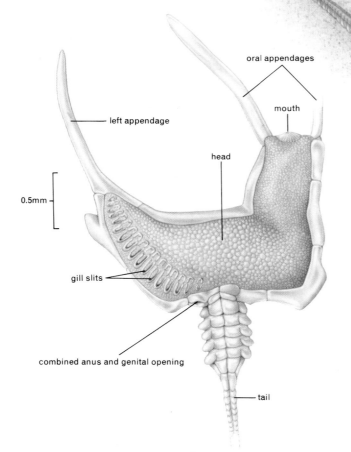

oral appendages

left appendage

mouth

head

0.5mm

gill slits

combined anus and genital opening

tail

ABOVE: dorsal view of a fossil of *Cothurnocystis elizae*, a primitive calcichordate of about 450 million years ago, and a reconstruction of the animal as it might have appeared in life.

BELOW: *Mitrocystites mitra*, an advanced calcichordate, lived in a shallow sea in Czechoslovakia about 470 million years ago. On the left is an imprint of the animal, preserved in a siliceous nodule, and to the right, a rubber cast of the fossil showing the dorsal surface of the head and the tail in positive relief.

Which group of animals did the chordates evolve from? It is widely believed that the hemichordates, because of their gills slits, are the closest living relatives of the chordates and that the echinoderms because of their embryology are also related, but more distantly. However, there is almost no group of multicellular animals which has not been seen by somebody as the source of the chordates. Candidates, defended even in recent years, have been the nemertean worms and the arthropod/annelid/mollusc group. On the other hand there is now strong fossil evidence that the echinoderms are the closest living relatives of the chordates, and that the hemichordates are more distantly related. Some resemblances of chordates to other groups (such as the segmentation shown by both fish and arthropods) have almost certainly been acquired independently.

The evidence of close echinoderm relationship comes from a strange group of ancient marine fossils which can be called calcichordates. These have a calcareous skeleton, as the name suggests, and this is fundamentally of echinoderm type. Indeed until recently most palaeontologists have regarded them as echinoderms and they are usually placed in the subphlylum Homalozoa of that phylum. However, they do possess a large number of striking chordate features.

An example of a rather primitive calcichordate is *Cothurnocystis* from Ordovician rocks of marine origins in Scotland. It consists of a head and a tail. The head is shaped rather like a mediaeval ankle boot in outline (*cothurnos* is Greek for boot). There is a frame of large plates round the head and attached to this frame were upper and lower plated skins. The head probably lay on the sea floor, and there are spikes round the lower edge whose function was probably to lift the head slightly above the sea bed, like the legs on a low table.

There are a number of openings in the head. The most striking is a series of slits in the left side of the upper surface. The detailed structure of these slits suggests that they functioned mechanically as outlet valves and that they were therefore gill slits, as originally suggested by the Swedish zoologist Gislen in 1930. The fact that they exist only on the left side suggests a comparison with the first gill slits of a larval amphioxus. Another opening lay just left of the tail, behind the gill slit farthest to the right. This was the combined anus and reproductive opening, situated where sperm, eggs and faeces would be washed away by the current from the gill slits. The gonads and non-pharyngeal gut probably lay near the right-hand posterior angle of the head. The mouth was the largest opening of all, at the anterior end of the head. The brain was just in front of the tail, in a special basin in the skeleton.

The tail of *Cothurnocystis* was complicated. The part nearest the head had a relatively loose skeleton enclosing a large space that could have contained muscles and a notochord in life. The notochord would have continued backwards into the rest of the tail, where it was probably lodged in an internal groove in the skeleton. The tip of the tail always ends abruptly as if part had broken off. The part of the tail that appears hindmost in fossils seems to be adapted for bending downwards into the sea floor. *Cothurnocystis* probably moved rearwards across the sea floor, pulling itself by side-to-side motion of its tail, the end of which would have been stuck into the mud for grip. The shape of the various spikes on the head would have prevented the animal sliding forwards.

More advanced calcichordates were more like modern chordates for they were almost bilaterally symmetrical, with internal evidence of right gill slits as well as left ones. The right gill slits would correspond to the second group of gill slits in a larval amphioxus and to the right gill slits of a vertebrate or tunicate. These advanced calcichordates had an exceedingly complicated fish-like nervous system. Within the advanced calcichordates, the evolutionary divergence of vertebrates, tunicates and lancelets can already be seen. The calcite skeleton has probably been lost independently in the three living chordate subphyla.

The calcichordates therefore suggest that echinoderms are the closest living relatives of chordates and this implies that the hemichordates are less closely related. This would mean that echinoderms are descended from ancestors with gill slits, but have lost the slits subsequently.

Within the chordates it is usually thought that the lancelets are more closely related to vertebrates than are the tunicates. This idea, however, is based on a vague overall resemblance and may be mistaken, for lancelets are now known to have a curious echinoderm-like way of innervating the main muscles: a strip of muscle tissue extends from each muscle to the nearest nerve. Tunicates and vertebrates, on the other hand, send nerves to the muscle. This suggests that tunicates and vertebrates are perhaps more closely related to each other than to the lancelets.

The picture of chordate relationships given here is largely new. It is based on fossils and agrees well, though not completely, with classical theory which was based on study of the living animals.

The evolutionary relationships of the chordate subphyla may be inferred from study of the fossil calcichordates. From those like *Cothurnocystis* more advanced symmetrical forms evolved. Three groups can be recognized among these, distinguished by features which separate the three living chordate subphyla.

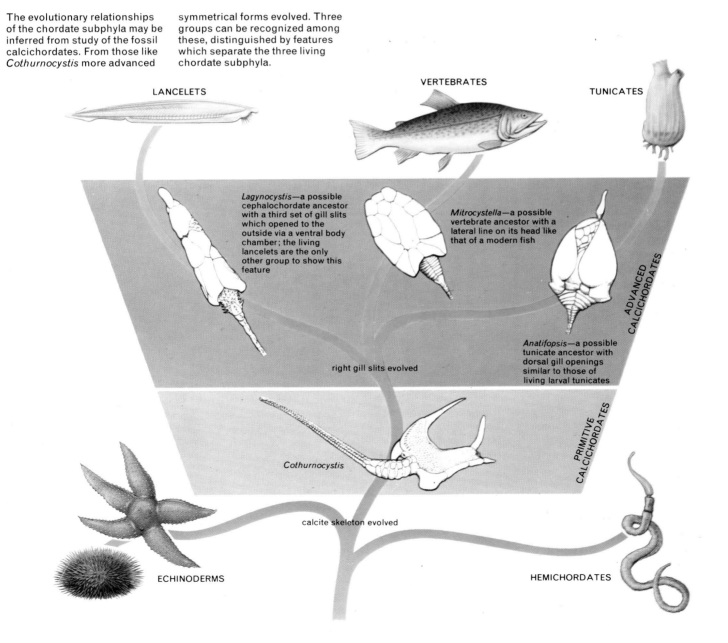

LANCELETS

VERTEBRATES

TUNICATES

Lagynocystis—a possible cephalochordate ancestor with a third set of gill slits which opened to the outside via a ventral body chamber; the living lancelets are the only other group to show this feature

Mitrocystella—a possible vertebrate ancestor with a lateral line on its head like that of a modern fish

ADVANCED CALCICHORDATES

Anatifopsis—a possible tunicate ancestor with dorsal gill openings similar to those of living larval tunicates

right gill slits evolved

PRIMITIVE CALCICHORDATES

Cothurnocystis

calcite skeleton evolved

ECHINODERMS

HEMICHORDATES

SUBKINGDOM Metazoa
PHYLUM Chordata

SUBPHYLUM Urochordata (= Tunicata)

Sea squirts and other tunicates

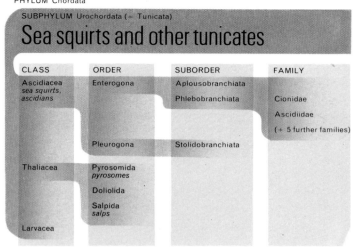

CLASS	ORDER	SUBORDER	FAMILY
Ascidiacea *sea squirts, ascidians*	Enterogona	Aplousobranchiata	
		Phlebobranchiata	Cionidae
			Ascidiidae
			(+ 5 further families)
	Pleurogona	Stolidobranchiata	
Thaliacea	Pyrosomida *pyrosomes*		
	Doliolida		
	Salpida *salps*		
Larvacea			

- NUMBER OF LIVING SPECIES: more than 2,000

- DIAGNOSTIC FEATURES: chordates with no segmentation or bony skeleton but with a notochord, at least in the larval stage; usually with a dorsal atrial cavity (adults only), and a test containing tunicin (related to cellulose)

- LIFESTYLE AND HABITAT: all marine, at all depths; solitary or colonial; sessile or pelagic

- SIZE: solitary forms from a few millimetres to 20 cm; colonial forms usually a few centimetres in diameter, but one species several metres long

- FOSSIL RECORD: fewer than 10 species; Cambrian to Quaternary

The Urochordata (or Tunicata) are marine animals of great theoretical interest to scientists, and often of considerable delicacy and beauty when seen alive. Their alternative names refer to two interesting characteristics of the group: first, the existence of a notochord—a supporting rod of specialized cells—in the tail and, second, the tunic or test—a cellulose-like coating of the surface of the body. The notochord exists only in the early or larval stage of the life history of tunicates and although it does not persist in the adults of most groups it is, nevertheless, one of the clues indicating a relationship to the vertebrates, all of which also possess a notochord at some stage of their lives. The tunic is a unique feature, however, and the existence in animals of a substance chemically so similar to the cellulose of plants remains something of a biological puzzle.

The affinity between the tunicates and the other chordates, including the vertebrates, has been recognized for over a hundred years, following the discovery of the ascidian tadpole larva, but the precise relationships are still a matter of dispute among scientists. Unfortunately tunicates have left little fossil record, since they have no hard skeleton.

Two very different lifestyles have evolved among the tunicates, for although the majority have adults which are firmly attached to objects in the sea and are quite unable to move about, some swim or drift passively throughout their entire life. Underlying the differences there remain the fundamental similarities in the structure of the animals, and many of these structural features are related to the method of feeding. The minute plant cells which abound in the sea form their food and have to be filtered out from a current of water which the animal passes through its pharynx.

Among the Ascidiacea (the familiar sea squirts), which is the largest class of tunicates, the body has at one end a tubular inhalant siphon into which water is drawn, and at some other point an exhalant opening to allow escape of the used water. The inhalant siphon leads into the voluminous branchial sac—a modified pharynx—which has its walls

ABOVE: a solitary ascidian, *Ciona intestinalis*. Younger specimens are more delicate and transparent. When disturbed, or when removed from the water, the animal will contract and force water from both siphons—hence the common name of sea squirt.

BELOW: a cutaway drawing of *Ciona* showing the branchial sac (pharynx) through which the food-bearing water current passes, the large atrial cavity that leads to the exhalant siphon, and the viscera occupying the lower third of the animal's body.

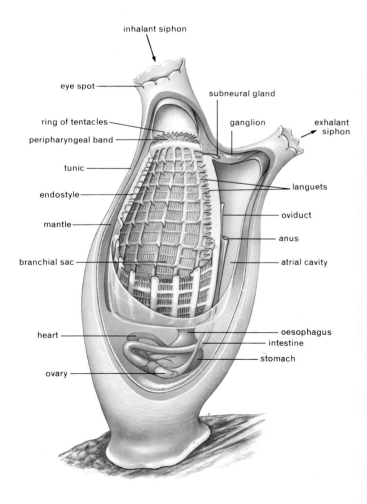

perforated by many small openings (called stigmata) through which the water passes. It is while the water is flowing from the sac outwards through the stigmata that the food particles are strained out, being trapped on sheets of sticky mucus secreted by cells lining a specialized groove (the endostyle) running along the length of the branchial sac. The compacted mucus and food are subsequently drawn back into the oesophagus to be passed to the stomach and intestine for digestion. Both the water and the mucus with its entrapped food are propelled, not by muscular action, but by the beating of innumerable minute hair-like cilia which line the branchial sac. This very delicate food-gathering mechanism is surrounded and protected by another cavity, the atrial cavity, lying between it and the muscular body wall. Once the water has escaped through the stigmata it is collected in the atrial cavity before being finally discharged from the exhalant opening. This continuous current of water, in addition to

A larva of an ascidian sea squirt looks very much like the tadpole larva of a frog. When it finds a suitable site it settles head down on the substrate and develops into its adult form.

carrying the ascidian's food, brings the essential oxygen in solution, and it too is extracted in the branchial sac whose walls are richly supplied with blood vessels.

The filter-feeding habit and the rather inactive life of these animals do not require them to have highly developed eyes, ears or other elaborate sense organs. The animal must be able to detect the presence of enemies, however, and so it is sensitive to touch and vibration, reacting by closing the siphons, stopping the feeding current and contracting the body. It may also sense variations in light intensity and the presence of certain chemical substances in the seawater.

Ascidians are hermaphrodite, with male and female gonads in the same individual, situated either near the stomach or on each side of the body. Some species shed their eggs directly into the sea, to be fertilized by sperm discharged from other individuals. Some retain the eggs within the atrial cavity where sperm drawn in with the feeding current may reach and fertilize them; this internal fertilization is followed by incubation of the embryos, which are thus protected from the dangers of a free life in the sea, at least for a time. In either case the embryos develop, within a

period varying from hours to a few days, into tadpole larvae. These are small independent animals consisting of a rounded body and a long flat muscular tail. Indeed they are superficially very like the tadpoles of frogs, but much smaller. The larva is in some respects more highly organized than its parents, for the body often includes a complex light-sensitive eye (ocellus) and a gravity-sensitive otolith. It requires these sensory organs during its free-swimming life, even if that may last for only a few hours, since the larva of sessile tunicates is responsible not only for distributing the species to new places, but also for choosing a suitable site where it can settle and metamorphose into the adult form.

The other classes of tunicates—the Thaliacea and Larvacea—share the same basic urochordate characters with the Ascidiacea just described, yet have evolved along very different lines. Members of the thaliacean order Pyrosomida live mainly in the warmer waters of the world, and are remarkable for the phosphorescent light which they emit. They exist only as delicate transparent colonies in the shape of a hollow cylinder closed at the front and open at the rear. The zooids, as the small individuals of the colonies are called, are arranged round the walls of the cylinder, with their inhalant siphons opening on the outer surface and their exhalant siphons discharging water, which is moved by ciliary activity, as in the ascidians, into the cavity of the cylinder. The water is pumped in by all the zooids and

BELOW AND BELOW RIGHT: a colony of the ascidian *Botryllus* is characteristically shaped like a star. The exhalant siphons of the individual members of a colony open into a common cloaca.

RIGHT: a colourful compound sea squirt from the Australian coast. The zooids are grouped in double rows and discharge the filtered water through large openings at the top of the colony.

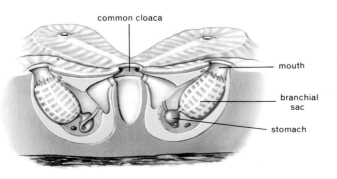

common cloaca

mouth

branchial sac

stomach

emerges as a single jet from the open rear end, propelling the colony forward through the sea. In the Doliolida and Salpida it is the individual and not the colony which is designed for swimming by jet propulsion, and the jet is generated by contraction of hoop-like muscles encircling the body.

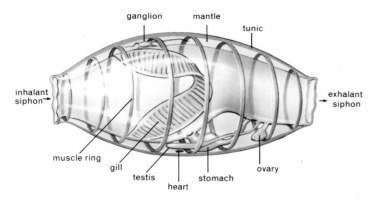

The sexual stage of *Doliolum* is a solitary animal only a few millimetres in length. The eggs develop into complex colonies of asexual zooids, some of which break free as sexual individuals.

The members of the class Larvacea are perhaps the strangest of all these groups of pelagic tunicates. These small animals are superficially like the tadpole larvae of ascidians, divided into a main body and a muscular tail in which the larval notochord persists. They are much more complicated in structure, however, and live freely within a hollow shell-like house secreted from the surface of the body. Rhythmic movement of the tail draws a stream of water in through openings in the wall of the house, then the combined action of cilia and mucus extracts food particles as the water flows through the pharynx. From time to time the larvacean abandons its house and secretes a new one.

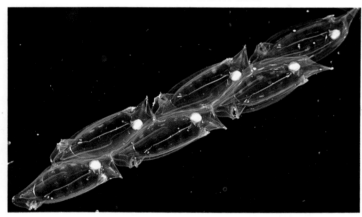

Within the tunicates, as we have seen, individuals in some species live as independent animals and in others are zooids living communally as members of a colony. A good example of the solitary type is *Ciona intestinalis*, a species widely distributed in European and American waters which has also been accidentally introduced into new areas. It has a tubular body up to 10 cm in length and shows the typical ascidian structure in what is probably its least modified form. Much of the body is taken up by the very large pharynx containing many rows of minute stigmata and constituting a filter so efficient that it can trap food particles as small as bacteria. The two siphons, inhalant and exhalant, are close together at the upper end of the body and in the limited space below the pharynx are crowded the stomach, intestine, gonad and heart. The heart of this species, incidentally, has been carefully studied as it is large and easily seen through the transparent tissues especially in young specimens, and shows another peculiar feature of the tunicates, for it regularly

reverses the direction of its beat, and pumps blood around the body first in one direction and then, after a momentary pause, in the opposite direction. The value to the animal of this periodical reversal of blood flow has never been satisfactorily explained.

In several families of the Ascidiacea the simple ground plan as exemplified by *Ciona* has been modified, with the introduction of the process of budding which leads to the development of a colony. *Sidnyum turbinatum* is a colonial species of this type, which is not uncommon on the lowest part of the foreshore. One consequence of colony formation is a reduction in size of the zooids which measure only about 5 mm in this species. They differ from *Ciona*, too, in having

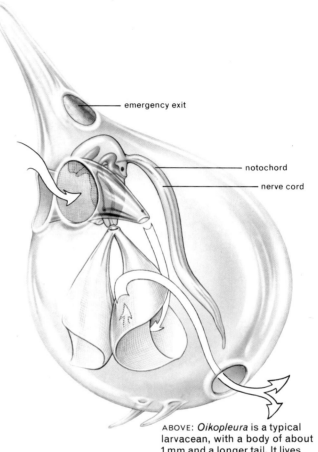

ABOVE: *Oikopleura* is a typical larvacean, with a body of about 1 mm and a longer tail. It lives in a transparent house that has openings through which water is drawn by tail movements. Food is extracted from the water by filters that lead to the mouth.

LEFT: salps exist in two forms: the solitary and (shown here) the aggregate form, which represents the sexual phase.

their body divided into sections; a thorax containing the pharynx, an abdomen in which the stomach and intestine are situated and a post-abdomen which accommodates the gonads and heart. When one of the zooids is about to generate new zooids by budding, the upper parts, that is the thorax and abdomen, break down and the remaining lower part, the post-abdomen, divides into a series of small rounded buds containing no recognizable organs but only a mass of unorganized cells. It is from each of these buds that a new zooid will develop, when the cells become specialized and rearrange themselves to create a new thorax, abdomen and post-abdomen. Thus each of the old zooids can create several new ones, and so the colony grows in size.

In other species, like *Diplosoma listerianum*, a species common in harbours, the zooid has only two divisions, thorax and abdomen, and is even smaller, about 1 mm long. The pharynx is reduced to only four rows of stigmata and the abdomen accommodates gut, heart and gonads. The life

ABOVE: the body of the Australian ascidian *Pyura pachydermatina* is borne on a long stiff stalk. The inhalant siphon points downwards and the exhalent one is directed upwards, so that the animal appears to stand on its head.

LEFT: species of *Clavelina* are among the most beautiful and delicate of compound ascidians.

history, too, is modified, for the egg breaks out of the zooid and completes its development to the tadpole larva while embedded in the jelly-like substance of the colony.

The tunicates are sometimes considered a rather insignificant group of animals, perhaps because they are usually small and are of little obvious use to mankind, although a few kinds are collected for food in the Mediterranean and some Asiatic countries. They are, nonetheless, of some importance in the general economy of the sea. Indeed on certain parts of the coastline of Australia and South Africa they are the dominant life form, so abundant that they are commonly used as bait by fishermen, and elsewhere they may outnumber all other kinds of animals of their own size, and constitute an important part of the regular food of commercial fish. For instance the population of a large species which covers rocks on the Chilean coast has been found to number over 300 animals per square metre, and these animals average 350 g each. In offshore waters, too, ascidians can be very abundant,

numbering up to 5,000 individuals per square metre. But it is as fouling organisms on ships, buoys and marine structures that ascidians have the most immediate economic impact. Ascidians, like other forms of fouling, reduce the efficiency of vessels by increasing their drag with consequently increased fuel consumption. *Ciona*, which was cited as an example of a simple ascidian, is so successful in colonizing fresh surfaces in the sea, that 10,000 specimens, weighing a total of 140 kg, have been estimated to grow on each square metre of a plate hung in the sea. On floating stuctures they interfere with maintenance and have to be laboriously removed before painting is possible.

It is, above all, the efficiency of the tiny tadpole larva in finding places to settle, and the fast rate of growth of the adult, facilitated by its efficient food-gathering mechanism, that are responsible for these ascidians being so successful in the face of competition from the many other marine animals which occupy the same niche in the life of the sea.

SUBKINGDOM Metazoa
PHYLUM Chordata

SUBPHYLUM Cephalochordata (= Acrania)

Lancelets

FAMILY	GENUS
Branchiostomidae	*Branchiostoma amphioxus*
Asymmetronidae	*Asymmetron*

- NUMBER OF LIVING SPECIES: 14

- DIAGNOSTIC FEATURES: fish-like chordates with well developed notochord and dorsal hollow nerve cord in adults but vertebral column absent; muscles segmented into myotomes

- LIFESTYLE AND HABITAT: all marine, bottom dwellers in shallow waters; normally burrowers but can swim

- SIZE: up to 5 cm long

- FOSSIL RECORD: none

The subphylum Cephalochordata contains just two genera of small marine invertebrates which are commonly known as lancelets because of their body shape. The name Cephalochordata refers to the extension of the notochord into the head; in craniates it stops at the base of the cranium and in tunicates it is confined to the tail. Eight species belong to the genus *Branchiostoma* which is often referred to as amphioxus and there are six species in the genus *Asymmetron*. The adult amphioxus is fish-like in appearance; it lacks pigmentation and is therefore pale in colour. Individuals are usually quite small, never exceeding a length of 5 cm.

Cephalochordates are found in shallow waters in the tropics and subtropics and some species extend into temperate regions. Essentially burrowing animals, they can also

Two *Branchiostoma* buried in shell gravel with their anterior ends protruding. They spend most of their lives like this, filtering food from the water. Much of the internal anatomy is readily seen through the transparent body wall. The muscle myotomes, fin ray boxes and pale, square gonads are all clearly visible.

locomotion and feeding. The body is elongated and flattened from side to side. Because the skin is transparent the underlying muscle blocks, or myotomes, can be seen through the body wall. These effect to throw the body into folds which

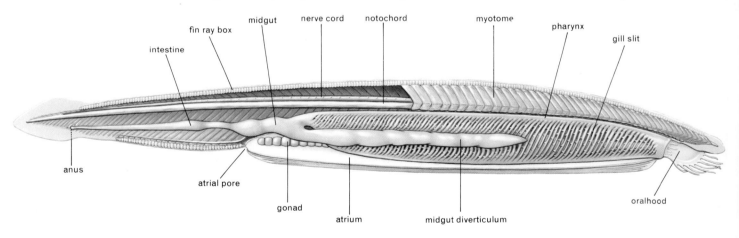

Branchiostoma dissected to show the internal anatomy. The body wall and gonads of the right side have been removed and the muscle myotomes have been cut or removed to show the pharynx, gut, nerve cord and notochord intact. Water that is taken into the pharynx passes through the gill slits and leaves through the atrial pore. The anus opens on the left side of the body.

swim freely when necessary but their distribution seems to be determined by the texture of the deposit which may be sand, gravel or shell. The distribution is very patchy, however, even in areas with apparently suitable substrates. In favourable locations the population may reach enormous numbers and definite 'amphioxus grounds' can be distinguished.

Because amphioxus illustrates some basic chordate characteristics it has been much studied. Details of its functional anatomy can be largely explained in terms of

pass back down the body as the animal moves forward in the water. Amphioxus does not often swim, however, and it is not capable of fast movements. It lacks the well developed fins of fishes and cannot maintain stability. It is true that the lateral wall of the atrium projects downwards into so-called metapleural folds but these probably function to protect the delicate tissues underneath when the animal pushes its way into the substrate.

The anterior end is distinguished only by the presence of a series of buccal cirri around the mouth. These form a sieve which rejects large particles from the food currents and they also have a sensory function. Amphioxus is a filter feeder—it extracts small particles from a stream of water drawn in through the mouth by the action of cilia which are situated on the gill bars of the pharynx. On the floor of the pharynx is the endostyle. This produces mucus which is then swept up across the gill bars where it traps food particles. Strings of this sticky material are then drawn into the midgut by further ciliary currents.

One of the puzzles in cephalochordate anatomy is the excretory system. This consists of sac-like nephridia that lie above the primary gill bars, and each nephridium is studded with solenocytes. Amphioxus frequently displays primitive features whose evolution can be traced through the vertebrates, but the vertebrate kidney cannot have its origins in the cephalochordate solenocytes. These are elongated flame cells which are comparable to those found in flatworms, annelids and molluscs—all quite remote in evolutionary terms—and their origin remains obscure.

The sexes are separate in amphioxus. The gonads lie along the base of the myotomes, on both sides in *Branchiostoma* and on the right side only in *Asymmetron*. When the gametes are ripe the gonads rupture and the genital products are released into the water through the atriopore. Spawning usually occurs around sunset on warm days, often following storms. There is no obvious reproductive behaviour and it is not known whether spawning occurs during the periodic swimming excursions or while the animals are buried.

After fertilization the egg develops in a few hours into a free swimming larva which remains partially planktonic for

Amphioxus plays an important part in our understanding of vertebrate organization. Its simple structure makes it relatively easy to recognize its basic chordate characteristics and it points the way to many features found in man. Most important of these is the segmentation of the muscle blocks. It is thought that this evolved in order to provide a set of muscle blocks which could contract serially during locomotion. Amphioxus differs from the vertebrates in not having a vertebral column. Instead it retains the notochord in the adult stage. This is rod-like and forms a skeletal support equivalent to the backbone.

Amphioxus is often described as a primitive organism. But although it is of a simple plan it is highly specialized for its way of life. The notochord, for example, differs in fine structure from that of other chordates. Changes in hydrostatic pressure within its sheath effect the stiffness of the

A diagram of the anterior end of *Branchiostoma*, sectioned to show the major organs. At the front the body wall and part of the pharynx have been cut away to show the feeding apparatus. The membrane around the nephridia has also been removed. Tiny cilia on the gill bars draw in water bearing food particles. Larger particles are excluded by the buccal cirri which fold over the mouth. Stray food is gathered by the cilia of the wheel organ. The food is trapped on mucus from the endostyle and moved up and around the pharynx by the gill bar cilia; it then collects in the epipharyngeal groove and passes backwards to the intestine.

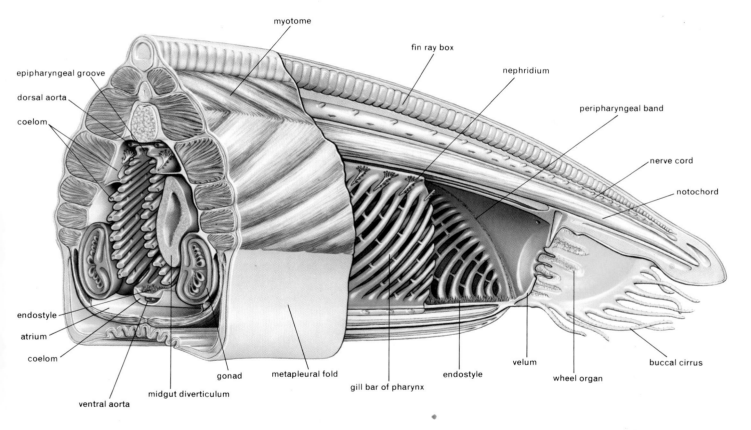

several months during which time it feeds and grows. During the daytime it rests on the bottom and feeds but at sunset it begins swimming actively and enters the plankton. Dispersal of the species is achieved by this stage in the life cycle. Metamorphosis occurs when the larva is about 5 mm long and involves adaptations to enable it to adopt a more or less sedentary lifestyle.

A supposed third genus of cephalochordate, *Amphioxides*, has now been shown to be a larval form. It is greatly enlarged and may contain rudimentary gonads. It is believed that these 'giant larvae' develop because they remain pelagic and, lacking contact with the bottom, cannot metamorphose.

organ and are believed to be related to locomotion. The notochord is thus not a primitive structure but instead a highly specialized hydrostatic skeleton.

The actual relationship of cephalochordates to the vertebrates is unclear. Unfortunately there is no fossil record to assist the investigation but it seems certain that cephalochordates are not on the direct evolutionary line of the vertebrates. Nevertheless, amphioxus does give us a clear indication of what the early chordates must have been like. Perhaps the next step was an extended larval stage with maturation occurring in the plankton. The existence of the amphioxides larva suggests that this might be possible.

SUBKINGDOM Metazoa
PHYLUM Chordata

SUBPHYLUM Craniata (= Vertebrata)

Birds, mammals and other vertebrates

CLASS
Myxini *hags*
†Heterostraci *heterostracans*
Cephalaspidomorphi *lampreys etc.*
Elasmobranchiomorphi *sharks etc.*
†Acanthodii *spiny sharks*
Osteichthyes *bony fish*
Amphibia *newts, frogs etc.*
Reptilia *lizards, snakes etc.*
Aves *birds*
Mammalia *mammals*

†extinct

- NUMBER OF LIVING SPECIES: more than 40,000

- DIAGNOSTIC FEATURES: chordates with an enlarged brain enclosed in a protective case or cranium; body typically formed of distinct head, trunk and tail

- LIFESTYLE AND HABITAT: almost exclusively free living in all habitats; primitively marine but many have invaded freshwater and land, and some have mastered active flight

- SIZE: 1 cm to 31 m but mostly between 10 and 150 cm

- FOSSIL RECORD: early Ordovician to Recent; uncommon before Devonian, becoming more abundant from Triassic onwards; at least 16,000 species

It is difficult to comprehend that familiar animals such as mammals, birds, reptiles, amphibians and fish are all grouped together as a single subphylum—the Craniata—and are given the same classificatory status as the subphylum Urochordata, for example. This quirk of classification belies their obvious importance and success in the modern world fauna. Most members of the Craniata occupy high positions in the food chain and they are found in almost every terrestrial and aquatic environment.

The unifying characters of the group suggest some of the reasons for its success. Importantly, there is an elaborate brain—an expansion at the anterior end of the dorsal hollow nerve cord—and this is primitively divided into three portions, each associated with a complex sense organ. The forebrain is intimately connected with the nasal apparatus which is extraordinarily adept at detecting air- and water-borne chemical signals. The midbrain receives nervous impulses from paired eyes which, unlike the eyes of many invertebrates, can detect shapes as well as movement and direction of light. The hindbrain is associated with the ear. Primitively the ear is concerned with balance and with the detection of acceleration and turning movements, but in those members of the group which live on land the reception of air-borne sound waves is an additional faculty superimposed on these basic functions.

The delicate sense organs and the brain are surrounded by a protective skeleton of bone or cartilage which forms the braincase or cranium—the feature which is reflected in the taxonomic name for the group. In all classes of the subphylum except one—the hags (Myxini)—the spinal cord also is protected by a bony or cartilaginous vertebral column which additionally has an important stiffening function; this attribute is the basis for the alternative, but strictly less accurate, name for the group—Vertebrata. Despite its shortcomings, the term 'vertebrates' is widely used as the common name for the Craniata, as it contrasts usefully with that used for all other animals—the invertebrates. It should

be remembered, however, that craniate and vertebrate are not strictly synonymous terms.

Vertebrates also have a complicated musculature system which, coordinated by the brain, allows elaborate movement. In the more primitive vertebrates locomotion is effected by waves of contraction passing down the sides of the body to cause lashing of the tail which acts as the propulsive organ. Movement in all but the hags and the lampreys is stabilized by paired fins which themselves have a complex skeleton. In land-dwelling vertebrates—the tetrapods—the paired fins are modified to form limbs with distinct hands and feet. The limbs have taken over the function of providing the propulsive force.

Vertebrates are large animals compared with primitive chordates such as lancelets and sea squirts, and this has two immediate consequences: first, the pharynx is no longer used as a filter feeding device but is primarily equipped with gills and used as a site of gaseous exchange (in tetrapods, of course, it is the lungs—ventral outgrowths of the oesophagus—which are used in respiration); second, vertebrates have a large muscular heart to pump blood, containing oxygen and food, to all parts of the body.

One final distinctive feature of vertebrates is the thick skin which is always composed of many layers of cells and is commonly associated with further coverings such as scales, scutes, feathers or fur. This covering of the vertebrate body forms an effective barrier between the animal and the environment it lives in.

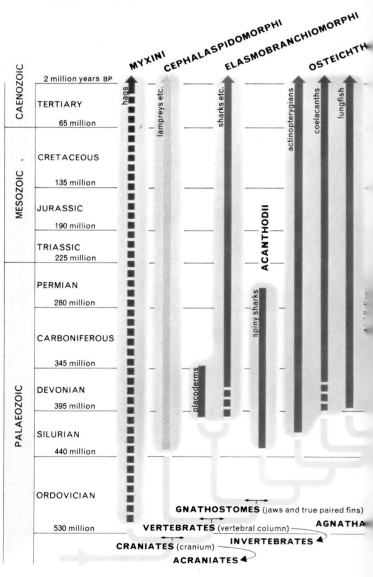

A great deal of attention has been devoted to attempts to understand the evolution of the Craniata and the inter-relationships of the various subgroups. From these points of view our understanding of vertebrates is far greater than that of any invertebrate group. We recognize ten classes, six of which are aquatic, breathe with gills and bear fins instead of limbs. These are loosely referred to as fish to separate them as a group from tetrapods (amphibians, reptiles, birds and mammals).

Throughout the course of vertebrate history, which covers some 500 million years (from the lower Ordovician), there have been many major modifications of vertebrate structure. Four of these are particularly noteworthy. The origin of jaws from modifications of gill arches was of paramount importance. Members of three vertebrate classes (Myxini, Cephalaspidomorphi and the extinct Heterostraci) have no jaws and are collectively termed agnathans. Undoubtedly the vertebrates with jaws—the gnathostomes—gained a distinct advantage, being able to take larger, more nutritious prey. Accompanying the development of jaws there was an improvement in the locomotory ability, with the origin of paired fins which served to increase both stability and manoeuvrability.

The shift from the water to the land, as tetrapods evolved from fish, is a second major modification, facilitated by several structural adaptations of which the limb structure is the most obvious. The fish fin became modified to form a strong yet flexible limb. Together with modification of the musculature and the girdles, particularly the pelvis, the limbs could be used to raise the body off the ground and propel the animal forwards.

A third significant innovation was the development of the cleidoic egg (literally, 'closed box' egg) seen in reptiles, birds and mammals. This egg is surrounded by protective amniotic membranes which enable the embryo to develop within its own independent aqueous environment. No longer was it necessary for the animals to live in or near water; they could in fact become fully terrestrial throughout their life cycle.

A fourth important development apparently occurred twice within the course of vertebrate history. Two groups, birds and mammals, acquired the facility of endothermy (loosely, but misleadingly referred to as warm-bloodedness). Endothermy—the physiological maintenance of a constant body temperature—provided a very stable internal environment and allowed the extraordinary development of the brain, unparalleled in the animal kingdom.

The interrelationships between the various vertebrate groups are understood in basic outline. These are shown in the accompanying chart where the known geologic ranges are displayed together with the hypothesized ranges. One class of vertebrates, the extinct heterostracan fish, are very difficult to place in this scheme and have been omitted. They occurred in lower Ordovician times and were thus the earliest evidence of craniate animals; they are very poorly known, however, and comparisons between them and other members of the subphylum are therefore difficult.

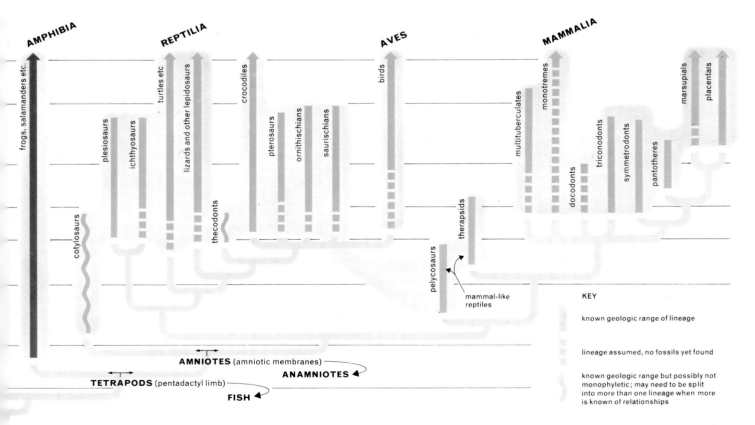

The interrelationships between craniate groups, shown as named vertical lines, are displayed on this chart. It is not exactly a true family tree as no ancestors are designated, but it attempts to present hypotheses of nearest relatives, such relationships being defined by the pale blue connecting lines. For example, it implies that marsupials are the nearest relatives of the placental mammals—these two can be looked upon as sisters; and the nearest relatives of these sisters—the cousins, if you like—are the extinct pantotheres. Some of the groups shown, for instance the extinct rhipidistians, cotylosaurs and thecodonts, are undoubtedly mixed-bag assemblies; that is to say, some thecodonts are probably more closely related to members of other reptile groups than they are to other thecodonts presently classified with them. Grey areas delineate the individual classes of the subphylum into which the various lower groups are usually assembled. While mammal-like reptiles are included with the other reptiles, it is generally considered that they are phylo-genetically more closely related to mammals. Some key events in craniate evolution are indicated, together with the all-embracing terms that are applied to the animals existing before and after the event. For instance, all the groups from Amphibia onwards primitively possess four legs instead of fins and are referred to as tetrapods.

SUBKINGDOM Metazoa
PHYLUM Chordata
SUBPHYLUM Craniata

CLASSES Myxini and Heterostraci

Hags and heterostracans

Hags CLASS Myxini

ORDER	FAMILY	GENUS
Myxiniformes	Myxinidae	*Myxine* hagfish
		Eptatretus slime hags
		(+ 3 further minor genera)

Heterostracans CLASS †Heterostraci

ORDER
†Cyathaspidiformes
†Pteraspidiformes
†Psammosteiformes
(+ 6 further minor fossil orders)

†extinct

● NUMBER OF LIVING SPECIES: Myxini: 32. Heterostraci: none

● DIAGNOSTIC FEATURES: craniates without jaws or true paired fins. Myxini: eel-shaped body lacking scales but with numerous mucous glands; single nostril at tip of snout. Heterostraci: body with scales, and head encased in bony shield; single gill opening

● LIFESTYLE AND HABITAT: Myxini: all marine, worldwide except tropics; bottom dwelling, usually in burrows. Heterostraci: found in marine, brackish and freshwater deposits

● SIZE: Myxini: up to 80 cm, average 50 cm. Heterostraci: up to 1.4 m, average 20 cm

● FOSSIL RECORD: Myxini: none. Heterostraci: early Ordovician to late Devonian but most common in late Silurian and early Devonian; about 200 species

LEFT: the underside of *Myxine glutinosa*, a hag from the Atlantic Ocean, shows the simple, jawless mouth with the touch-sensitive tentacles that surround it. A row of mucous glands can be seen along the side of the body.

BELOW: hags extricate themselves from tight corners in an unusual way: the tail is contorted into a knot, and this is used as a stop through which to draw the head. The process is facilitated by the production of copious mucus.

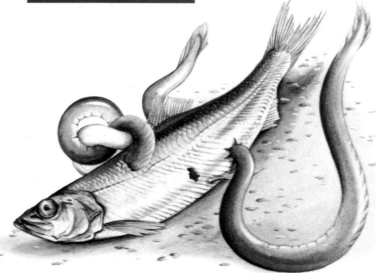

Pteraspis was a common heterostracan which lived in estuaries and streams in the early Devonian. The lack of paired fins suggests rather poor swimming ability. It was about 20 cm long.

The Myxini (hags) are the most primitive of living craniates. They have no paired fins, no jaws, and there is no evidence of a lateral line (a series of sense organs in the skin responsible for detecting vibrations in the water). The body is eel-shaped and the skin is scaleless.

Hags have gained some notoriety because of their unpleasant methods of feeding and defence. A hag will often feed by literally eating its way through a dead or dying fish, leaving just skin and bones. For this purpose the hag has pointed horny teeth on the tongue which bite against a single tooth in the roof of the mouth. In some Californian and Japanese fishing grounds hags cause a great deal of harm by attacking fish caught in bottom-set nets. The commonly used term 'slime hag' refers to the defence mechanism whereby the hag will encase itself in a thick envelope of gelatinous slime in response to mechanical or chemical irritation.

The hagfish (*Myxine*) spends most of its time buried in soft mud with just the snout protruding. Water is drawn in through the single nostril at the tip of the snout, passes a single nasal organ, which senses the water, and enters the roof of the pharynx. From here the water passes laterally through a series of gill pouches where gaseous exchange takes place.

Excepting smell, the senses of hags are poorly developed. The eye is non-functional. There is only one semicircular canal in the ear and this may explain the fish's rather erratic swimming behaviour. Little is known about the reproductive biology. Individuals have a gonad which is bisexual but at maturity each fish acts only as male or female. The eggs are large and yolky and are contained within a horny envelope which is attached to rocks or seaweed by many thin tendrils or filaments. The young hatch as miniature adults.

One group of ancient jawless fish, the Heterostraci, has been suggested as relatives. There is, however, no good reason for this belief although it is true that both groups are of a primitive structure. Heterostracans are now considered to be a distinct group but little is known about them and they contribute insignificantly to our knowledge of the modern fish fauna. They do not therefore warrant separate treatment and they are considered here only for convenience and not because of any biological affinities. The heterostracans were a diverse assemblage of small littoral and fluviatile fish in which the head was covered with an armour of acellular bone and the body was invested with large scales. They were first known from the early Ordovician period (about 500 million years ago) of central Australia; they lasted until the late Devonian (about 340 million years ago) and represent the earliest record of craniate animals. It is probable that they were mud-grubbers, scooping up mud and digesting the organic debris which had settled on the bottom.

SUBKINGDOM Metazoa
PHYLUM Chordata
 SUBPHYLUM Craniata

CLASS Cephalaspidomorphi

Lampreys and other cephalaspidomorphs

ORDER	FAMILY	GENUS
Petromyzontiformes *lampreys*	Petromyzonidae	*Petromyzon* (+ 8 further genera)
†Cephalaspidiformes *cephalaspids etc.*		
†Anaspidiformes *anaspids etc.*		
	†extinct	

- NUMBER OF LIVING SPECIES: 31

- DIAGNOSTIC FEATURES: craniates without jaws or paired fins; circular mouth, surrounded by sucking disc in modern forms; single nostril on dorsal surface of head; body scaleless in modern forms

- LIFESTYLE AND HABITAT: mostly freshwater; adults usually ectoparasitic on bony fish and marine mammals

- SIZE: 15 to 90 cm but mostly between 30 and 50 cm

- FOSSIL RECORD: Silurian to Recent but most common in early Devonian; about 160 species

The lampreys, order Petromyzontiformes, and their extinct relatives, the cephalaspids and anaspids, are primitive vertebrates which have no jaws or true paired fins (although fleshy flaps were present in some cephalaspids). These three orders, which are grouped together as the Cephalaspidomorphi, are unique among craniates because they have a single nostril which opens with a hypophysial duct onto the surface of the head.

The eel-shaped lamprey is highly specialized. Adults of many lamprey species are ectoparasitic and attach themselves to fish or marine mammals by means of a sucker developed around the mouth. Once attached, the horny teeth on the tongue rasp away the flesh of the host. The pharynx and associated gill pouches of the lamprey are separate from the gut and this specialization allows feeding and breathing to take place at the same time.

Lampreys spawn in freshwater where the eggs hatch into a larval stage known as the ammocoete. The ammocoete has been of theoretical interest to scientists because it is superficially similar to the cephalochordate amphioxus and, by inference, to the hypothetical primitive vertebrate. It lies buried in mud and feeds by filtering microscopic food particles from the water. At metamorphosis there is a reorganization of the pharynx and the mouth with the change from microphagy to ectoparasitic feeding. The parasitic habit is of considerable antiquity since fossil lampreys are known from the late Carboniferous (about 280 million years ago). In non-parasitic species only the ammocoete stage feeds.

The head of a lamprey is equipped with a fleshy sucking disc around the mouth. The arrangement of the horny teeth is used to identify the different species. The two shown are *Petromyzon marinus*—the Sea Lamprey (left) and *Lampetra fluviatilis*—the River Lamprey.

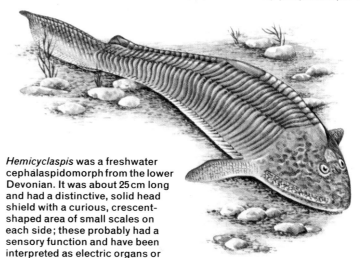

Hemicyclaspis was a freshwater cephalaspidomorph from the lower Devonian. It was about 25 cm long and had a distinctive, solid head shield with a curious, crescent-shaped area of small scales on each side; these probably had a sensory function and have been interpreted as electric organs or as localized lateral-line areas.

The parasitic phase of the Sea Lamprey (*Petromyzon marinus*) has been of special importance to North American fisheries. Since the opening of the Welland Ship Canal in 1829, Niagara Falls no longer forms a natural barrier and lampreys have invaded all of the Great Lakes, reaching Lake Superior by the late 1940s. As the lampreys moved further into the American continent so their habits changed, with the result that the adults now attack fish within the Great Lakes and run up tributary streams to spawn. Coincident with the invasion there was a severe depletion in commerical fish stocks. For instance, between 1952 and 1961 the catch of Lake Trout in Lake Superior fell from 2,022 metric tons to 167 metric tons per year causing the virtual collapse of the fisheries. The lampreys also attacked other fish species resulting in a considerable imbalance in the food web.

Effective control of this predator has largely centred on attacking the adults as they run up the streams to spawn. Mechanical barriers and electrified fences have been placed across the mouths of the known spawning streams. In the streams themselves selective larvicides, designed to eradicate the ammocoete, have been used with some success. It is evident, however, that lampreys can never be eliminated from the Great Lakes and lamprey control will continue to be a major aspect of fish management.

A River Lamprey (*Lampetra fluviatilis*) of northwest Europe uses its sucker to attach to the stony river bottom. The water current passes through the seven large 'gill-ports' in an ebb and flow pattern; this is a characteristic peculiar to lampreys.

SUBKINGDOM Metazoa
PHYLUM Chordata
SUBPHYLUM Craniata

CLASS Elasmobranchiomorphi

Sharks and other elasmobranchiomorphs

SUBCLASS	INFRACLASS	ORDER
†Placodermi		†Arthrodira
		†Antiarchi
		(+ 4 further minor fossil orders)
Chondrichthyes *cartilaginous fish*	Elasmobranchii *elasmobranchs*	†Hybodontiformes
		Heterodontiformes *Port Jackson sharks*
†extinct		Lamniformes *true sharks*
		Squaliformes *dogfish sharks*
		Rajiformes *rays and skates*
		(+ 6 further minor orders)
	Holocephali *holocephalans*	Chimaeriformes *ratfish or chimaeras*
		(+ 2 further minor fossil orders)

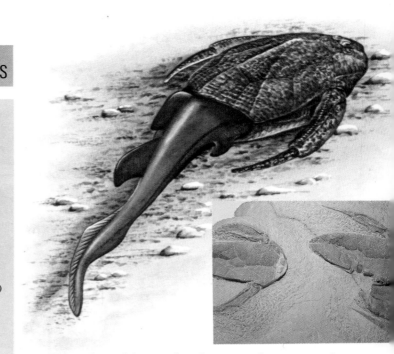

- NUMBER OF LIVING SPECIES: about 630

- DIAGNOSTIC FEATURES: craniates with jaws, paired fins and carti-laginous endoskeleton; large nasal organs present; mouth under-slung and body covered with tiny placoid scales in modern forms; internal fertilization in all species

- LIFESTYLE AND HABITAT: mostly marine; some swim in midwater but most live on or near the bottom

- SIZE: 20 cm to 15 m but mostly between 1 and 2 m

- FOSSIL RECORD: Devonian to Recent; about 600 species

The class Elasmobranchiomorphi is best known by one of the contained groups, the sharks, but it also includes the rays, the ratfish and the extinct placoderms. Members of this class make up only 3 per cent of the total fish fauna, yet they command a disproportionate attention. The sheer elegance of shark locomotion coupled with unpredictable behaviour and the lurid, often grossly exaggerated accounts of shark attack continue to hold a fascination for Man.

A Sand Shark (*Odontaspis taurus*) shows typical shark dentition of many pointed teeth that gradually move forwards into a functional position and are soon shed. The protruding snout, with openings to the large nasal organs, is typical of all elasmobranchiomorphs.

Elasmobranchiomorphs show two important character-istics. First, the nasal organs are large, resulting in a keen sense of smell; they are housed in a protruding snout which overhangs the mouth. Second, members of this group practice internal fertilization and the males have modifications of the pelvic fins, the claspers, for use during copulation. Elasmo-branchiomorphs are gnathostomes, that is, they have true jaws—these are thought to have developed by modification of the anterior gill arches (skeletal supports alternating with the gill slits).

The Placodermi comprised a varied group of Devonian fish. Although their history was brief the large number of fossils found suggests that they were very successful in both the sea and freshwater. Placoderm literally means 'plate-skinned'. In most species the head was encased in a shield composed of many bony plates and this articulated by means of a neck joint with a shield which covered much of the trunk. The upper jaw was firmly braced against the head shield. During jaw opening the head shield was raised and the lower jaw depressed, resulting in a very wide gape.

The cartilaginous fish (subclass Chondrichthyes) are so-called because the internal skeleton is composed entirely of cartilage. In a few of the larger sharks, however, the cartilage may become calcified and take on the physical properties of

Representatives of the two main orders in the subclass Placodermi. The pectoral fins of *Bothriolepis* (Antiarchi) (left) took the form of bony 'arms', used possibly as props; fossils are common in late Devonian deposits but usually only head and trunk shields are preserved. *Coccosteus* (Arthrodira) (right) of middle Devonian lakes was about 4.5 cm long; the bony head shield, armed with teeth of bone, not enamel, gave protection at the expense of flexibility.

bone. The skin is covered with many tiny placoid scales—tooth-like structures with a flattened base embedded in the skin—and is often referred to as shagreen. Each placoid scale has a core of dentine and an enamel coat, and in structure there is no real difference between this particular type of scale and the shark tooth.

Sharks and rays are together called elasmobranchs—literally 'plate-gill', which is a reference to the fact that the gills open separately to the outside. Of the two elasmobranch types the sharks have received the greater attention. For the scientist the shark is of interest because of its almost perfect streamlined form and the efficiency of its swimming capabilities. The upturned and asymmetrical (heterocercal) tail provides a forward and slightly downward thrust and the pectoral (front) fins act as variable hydrofoils to regulate pitching movements. There are no unnecessary protrusions from the streamlined form and even the tiny placoid scales are each individually sculptured to reduce turbulence. Sharks lack a swimbladder (a gas-filled buoyancy apparatus found in bony fish), but are often neutrally buoyant because of large amounts of oil stored in the liver, a fact which used to be of commercial importance.

Sharks are predators and their dentition is ideally suited to capturing and holding struggling prey. The teeth are set in rows and each row lies transversely across the jaw. New teeth are constantly formed, used and are then shed. Tooth replacement may be very rapid: in the Lemon Shark, for instance, a tooth may be replaced every eight days. The teeth are bound to the jaw by tough connective tissue but it is not uncommon for whole rows of teeth to be ripped out when the shark is struggling with a particularly formidable adversary.

The Palaeozoic fossil record of sharks is very fragmentary and consists mostly of isolated teeth and spines. Mesozoic sharks are slightly better known and *Hybodus* (in Hybodontiformes) was a common inhabitant of Jurassic and Cretaceous seas. *Hybodus* had two dorsal fins, each preceded by a prominent spine and the teeth modifications suggest a particularly catholic diet. The teeth at the front of the jaw were sharp and presumably used for seizing and cutting, but towards the back they become progressively lower-crowned and more suited for grinding. The modern Port Jackson Shark (*Heterodontus philippi*: Heterodontiformes) of the Pacific has retained the fin spines and the differentiated dentition of the hybodont antecedents. The diet of *Heterodontus* includes crustaceans, molluscs and sea urchins.

Most sharks belong to the order Lamniformes. They have no fin spines and the teeth are all alike in any one species. Some of the most powerful swimmers are members of this group. The Mako (*Isurus paucus*) is a fine example of the streamlined form, ideally suited to sustained swimming in the open waters of the Indo-Pacific. *Carcharodon carcharias*, the notorious Man-eater Shark, is a close relative of the Mako and is the largest of the predatory sharks. The authenticated record is 6.3 m but it is probable that fossil *Carcharodon* reached 13 m. But the true giants are the Basking Shark (*Cetorhinus maximus*) and the world's largest fish, the Whale

Shark (*Rhinocodon typus*) at 15.2 m. Both of these oceanic sharks are plankton feeders.

Many species of shark are found around coral reefs and in coastal waters where prey is generally more plentiful. Common examples include the Lemon Shark (*Negaprion brevirostris*), Grey Nurse (*Odontaspis arenarius*), cat sharks (for example, *Scyliorhinus*) the Sand Shark (*Odontaspis taurus*) and hammerheads (*Sphyrna*). The last two mentioned are particular villains of bathing beaches.

ABOVE: the harmless White-tipped Reef Shark (*Triaenodon apicalis*) possesses the typical streamlined elegance of true lamniform sharks. The notch at the tip of the tail, found in many species, is thought to reduce water turbulence. The large claspers behind the pelvic fins indicate that this individual is a male.

BELOW: the Port Jackson Shark is a primitive species found in the seas around Australia. It has retained several features of its Mesozoic ancestors, such as the two spines, one in front of each dorsal fin, and two different types of teeth—pointed ones in the front of the jaws and pad-like ones at the back.

ABOVE: the Wobbegong (*Orectolobus maculatus*) is a squaliform of Australian waters. Bordering the mouth is a frill of skin flaps which serves to break up the outline of the head.

BELOW: some of the rajiforms, such as the Giant Devil Ray (*Manta birostris*), live in surface waters. They 'fly' through the sea, straining microscopic food through large, modified gills.

The smallest shark (*Squaliolus*, 20 cm long), the deepest living shark (*Centroscymus coelolepis*, at 2,700 m), a luminous shark (*Isistius*) and the familiar Spiny Dogfish (*Squalus acanthias*) are all examples of squaloid sharks (Squaliformes). This group is morphologically intermediate between the true sharks and the rays. Shark-like characters are the streamlined body, the teeth and the side gills. But like the rays there are certain features associated with a bottom-dwelling existence. There is no anal fin, the head is flattened and the spiracle (a modified gill slit) is large and is used for water intake.

True rays (Rajiformes) show further adaptations to life on the bottom. The gill openings are on the ventral surface and the eyes and the spiracle have moved to a position on top of the head. The most obvious feature of all, however, is the pair of enormously expanded pectoral fins which have become attached to the sides of the head. Approximately one-third of all ray species are members of the family Rajiidae, the skates. The skate spends most of its time lying partially buried in bottom sediments relying on cryptic camouflage for protection. When it moves it does so by gentle undulation of the pectoral fins, the reduced tail being ineffectual in propulsion. The upper and lower jaws carry many low-crowned teeth, which are closely packed together and used to crush hard-shelled invertebrates.

Among the rays, sawfish (*Pristis*) are the most shark-like. Here the tail still provides the thrust and the pectorals are only slightly expanded. Distinctively the snout is elongated and is armed with enlarged placoid scales which form the 'saw'. This is flailed amidst schools of fish, the injured being eaten. Electric or torpedo rays (for example, *Torpedo*) also use the tail for locomotion but are better known because of their ability to discharge electric shocks—up to 200 volts in some species. The charge is produced by special muscle bands on either side of the head and seems to be used for defence. However, an unwary crab or prawn may become prey if, by chance, it should trigger the release of a shock.

Sting rays (for example, *Trygon*) and devil rays (for example, *Manta*) are the most advanced of the rays. Members of this group use the large pectoral fins as wings and literally fly through the water. The sting of the sting ray is a poisonous spine in the position of a dorsal fin. Many sting rays live in the freshwaters of Florida, South America, India and Australia. Among the rays, devil rays are giants with a 'wing span' of some 7 m and a weight of up to 1,360 kg. Like the largest sharks, devil rays are oceanic fish and strain plankton with special horny outgrowths of the gill arches.

Ratfish or chimaeras (Chimaeriformes) are inhabitants of deep water and so are rarely seen except in those few places (such as New Zealand) where they have been marketed for food. A ratfish has a fleshy snout which widens to a deep body and then narrows to a whip-like tail. The sluggish movement is accomplished by gentle undulation of the dorsal fin and balance is maintained by constant movement of the pectoral fins. The technical name of the embracing taxon (Holocephali) means solid-head and refers to the fact that the upper jaw is firmly fused with the braincase. This specialization is correlated with the feeding method whereby hardshelled invertebrates are crushed between large grinding plates in the upper and lower jaws.

Unlike the sharks and the rays the gill openings of the ratfish are covered with a flap of skin, analogous to the operculum of bony fish. The male ratfish is equipped with a peculiar median clasper on top of the head. This structure, which resembles a mediaeval spiked mace, is erected before copulation and may be used to grasp the female. Immediate relatives of the ratfish are recorded from early Jurassic rocks but fragmentary remains of more distantly related Holocephali have been found in late Devonian deposits. Thus this highly specialized lineage is of considerable antiquity.

Dasyatis is one of the sting rays, a group whose members inhabit coastal waters of tropical and subtropical seas. Like most rajiforms, it spends much of its life buried in sand. The venomous spine on the tail is used chiefly for defence.

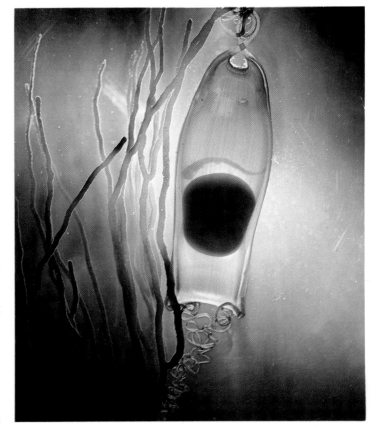

Many species of cartilaginous fish lay eggs surrounded by a horny egg-case. This is one from a squaliform—the Nurse Hound (*Scyliorhinus stellaris*). The embryo develops within the protective coat, feeding from its attached yolk sac.

A distinctive feature of all cartilaginous fish is the universal practice of internal fertilization, the male using pelvic claspers to grasp the female and to guide sperm into the oviduct. But once the egg has been fertilized further development may differ considerably from species to species. Ratfish, some sharks and dogfish (for example, *Heterodontus* and *Scyliorhinus*) and skates are oviparous: the egg is encased within a horny capsule produced by the shell gland in the oviduct wall and the egg case is then laid. Usually this becomes attached to seaweed by filaments at the ends (the familiar 'mermaids' purses' cast up on the shore).

Most sharks and rays, however, are ovoviviparous: the egg is retained and develops within an enlarged portion of the oviduct, the uterus. Within this safe confine the foetus develops and feeds on its own yolk sac (receiving no nourishment from the mother) before being born. In a few sharks and in many rays true viviparity (as seen in placental mammals) has arisen whereby the young are retained in the uterus and are nourished by the mother. This is done in several ways. The embryo may be bathed in 'uterine milk', a nourishing fluid secreted by the uterine wall. In the Smooth Butterfly Ray (*Gymnura micrura*) the uterus develops special finger-like outgrowths which penetrate the spiracle of the foetus. The uterine milk is thus passed directly to the foetal gut by way of this 'placenta'. Another type of 'placenta' is developed in smooth dogfish (*Mustelus*) in which the wall of the yolk sac becomes attached to the uterine wall and food passes directly from the mother's blood stream into the circulatory system of the foetus.

There is no doubt therefore that the foetal care shown by some sharks and rays is comparable with that shown by mammals. This type of behaviour is hardly the hallmark of a 'primitive vertebrate', a title attributed to the sharks in many of the textbooks.

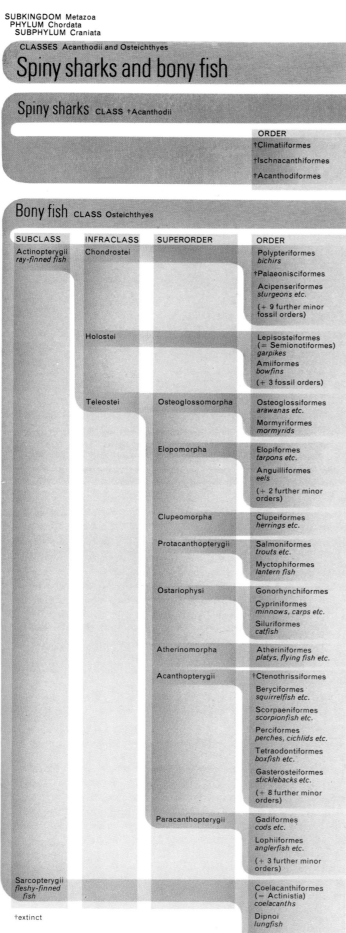

SUBKINGDOM Metazoa
PHYLUM Chordata
SUBPHYLUM Craniata

CLASSES Acanthodii and Osteichthyes

Spiny sharks and bony fish

Spiny sharks CLASS †Acanthodii

ORDER

†Climatiiformes

†Ischnacanthiformes

†Acanthodiformes

Bony fish CLASS Osteichthyes

SUBCLASS	INFRACLASS	SUPERORDER	ORDER
Actinopterygii *ray-finned fish*	Chondrostei		Polypteriformes *bichirs*
			†Palaeonisciformes
			Acipenseriformes *sturgeons etc.*
			(+ 9 further minor fossil orders)
	Holostei		Lepisosteiformes (= Semionotiformes) *garpikes*
			Amiiformes *bowfins*
			(+ 3 fossil orders)
	Teleostei	Osteoglossomorpha	Osteoglossiformes *arawanas etc.*
			Mormyriformes *mormyrids*
		Elopomorpha	Elopiformes *tarpons etc.*
			Anguilliformes *eels*
			(+ 2 further minor orders)
		Clupeomorpha	Clupeiformes *herrings etc.*
		Protacanthopterygii	Salmoniformes *trouts etc.*
			Myctophiformes *lantern fish*
		Ostariophysi	Gonorhynchiformes
			Cypriniformes *minnows, carps etc.*
			Siluriformes *catfish*
		Atherinomorpha	Atheriniformes *platys, flying fish etc.*
		Acanthopterygii	†Ctenothrissiformes
			Beryciformes *squirrelfish etc.*
			Scorpaeniformes *scorpionfish etc.*
			Perciformes *perches, cichlids etc.*
			Tetraodontiformes *boxfish etc.*
			Gasterosteiformes *sticklebacks etc.*
			(+ 8 further minor orders)
		Paracanthopterygii	Gadiformes *cods etc.*
			Lophiiformes *anglerfish etc.*
			(+ 3 further minor orders)
Sarcopterygii *fleshy-finned fish*			Coelacanthiformes (= Actinistia) *coelacanths*
			Dipnoi *lungfish*
			†Rhipidistia *osteolepids etc.*

†extinct

At a conservative estimate there are 18,050 species of bony fish (class Osteichthyes) which constitute about 98 per cent of the ichthyofauna. As a group, they are found in almost every aquatic environment from the torrents of mountain streams to the still, black depths of ocean trenches. The Icefish (*Chaenocephalus aceratus*: Perciformes) lives in the waters beneath the Antarctic ice where temperatures may tip to minus 2°C, and at the other extreme the Desert Pupfish (*Cyprinodon milleri*: Atheriniformes) can tolerate water temperatures of 42°C in Death Valley, Nevada.

Fossil osteichthyans are found as far back as the late Silurian (about 410 million years ago). To the scientist interested in the origin of vertebrate life on land one group of bony fish, the subclass Sarcopterygii (which includes lungfish and coelacanths), is of theoretical interest as members of this category are more closely related to tetrapods than to other bony fish.

Primitively the head of a bony fish is invested with dermal bones (bones that develop in the skin) which both interlock and articulate with one another. This provides the head with a protective yet flexible covering. Some of these bones are particularly characteristic of bony fish: an operculum covers the gill openings; ventrally the throat membrane is strengthened by bony rods known as the branchiostegal rays; there is a series of tooth-bearing jaw bones—the premaxilla and maxilla of the upper jaw and the dentary of the mandible; and the fin webs are supported by bony rays or lepidotrichia, thought to have been derived from scales.

Before dealing with the osteichthyans proper, brief mention must be made of the Acanthodii or 'spiny sharks', a small assemblage of Palaeozoic fish. Little is known about acanthodians and consideration of them under a separate heading is not appropriate. However, although their true affinities are unclear, like bony fish they do have branchiostegal rays and a structure resembling an operculum. They were small (usually less than 30 cm) and superficially like sharks. Unlike sharks, however, their eyes were large and, characteristically, their fins were formed by a leading spine which supported a web of skin. It is probable that acanthodians fed in surface waters (both marine and freshwater) by straining plankton.

The Osteichthyes contains two subclasses, Sarcopterygii and Actinopterygii, which are distinguished from one another by the structure of the paired fins. Sarcopterygii means 'fleshy fin'; in these fish the supporting skeleton and associated musculature of the fin project outside the body, thus forming a fleshy base. This contrasts with the Actinopterygii (meaning 'ray fin') where only the dermal fin rays may be seen externally, the endoskeleton and muscles

---

• NUMBER OF LIVING SPECIES: Acanthodii: none. Osteichthyes: more than 18,000

• DIAGNOSTIC FEATURES: craniates with jaws and paired fins. Acanthodii: streamlined body covered with small scales and with prominent spines. Osteichthyes: bony endo- and dermal skeleton, including an operculum; primitively with lungs (modified as a swimbladder in most forms)

• LIFESTYLE AND HABITAT: Acanthodii: mostly freshwater, active swimmers. Osteichthyes: mostly marine but some groups important elements of freshwater faunas; mostly active swimmers

• SIZE: Acanthodii: 5 cm to 1.5 m, average about 15 cm. Osteichthyes: 1 cm to 4.5 m, but mostly 2.5 to 30 cm

• FOSSIL RECORD: Acanthodii: Silurian to Permian; about 60 species. Osteichthyes: Silurian to Recent, with majority in Tertiary; at least 6,000 species

remaining within the body. The bony fish primitively had paired lungs, a condition still seen in some, but in most actinopterygians the lungs are replaced by a swimbladder which is located immediately above the stomach and helps regulate the buoyancy of the fish.

Complete remains of actinopterygians are first found in Devonian deposits and are represented by members of the order Palaeonisciformes. An early palaeonisciform, such as *Rhadinichthys*, was streamlined with a heterocercal tail produced as a long dorsal axis. The snout was blunt, the eyes large and the long jaws carried many pointed teeth. Clearly, these early actinopterygians were predators. The body was covered with rhomboid-like scales which interlocked with one another and the surface of each was covered with ganoine, a shiny, enamel-like layer.

BELOW: *Climatius* was a Silurian member of the class Acanthodii. A spine formed the leading edge of each fin, except that of the tail, and the body was covered with very small scales.

RIGHT: the bichirs are African freshwater fish which show many primitive characteristics such as heavy skull bones and thick, rhomboid scales. The long dorsal fin is made up of 15–20 finlets.

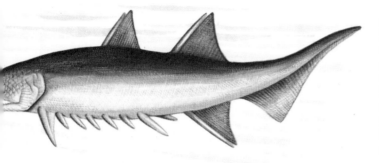

By the Cretaceous period the Palaeonisciformes had died out, but two groups of living fish, the bichirs (Polypteriformes) and the sturgeons (Acipenseriformes), seem to be near relatives of this ancient stock. The eel-shaped bichirs (for example, *Polypterus*) are African freshwater fish which have the ganoid type of scale. They also retain lungs in their primitive form and are air breathing at times of oxygen depletion in their environment. Sturgeons (for example,

Acipenser) seem to be late derivatives of the palaeonosciform stock. Their endoskeleton is almost entirely cartilaginous. The scales have been reduced so that only a few remain on the heterocercal tail and there are a few enlarged scales (scutes) on the body. Sturgeons have given rise to the taxon name Chondrostei ('cartilaginous bony fish') but in a sense this term is a misnomer since most members of the group are, in fact, bony.

During the Mesozoic era actinopterygians were represented by a varied array of fish, most of which have usually been grouped as the infraclass Holostei. The Holostei represents several independent lineages, each of which showed improvements in locomotion and feeding methods: the dorsal lobe of the tail shortened, resulting in a hemiheterocercal tail which produced a nearly horizontal thrust; the upper jaw became more flexible and in the lower jaw there was a change in the orientation of the adductor mandibulae muscle (the muscle which closes the jaw) which resulted in a more powerful bite.

Three of the holostean lineages are of interest to the student of living fishes. The garpike (*Lepisosteus*: Lepisosteiformes) is the surviving member of a lineage which can be traced back to the Triassic period. This fish, an inhabitant of North American freshwaters, is elongate and covered with thick ganoid scales. Its jaws are very long and armed with many pointed teeth used to grasp prey, which is usually fish or crustaceans although there are reports of the Alligator Gar (*Lepisosteus spatula*) taking ducks.

The Atlantic Sturgeon, *Acipenser sturio*, like all living members of the order, is a bottom dweller with a small, ventrally situated mouth. This young individual may grow up to 3 m long.

The Alligator Gar spends most of its life hovering among reeds. It darts suddenly from concealment to catch any passing prey. A large individual may reach a length of 3 m.

LEFT: *Osteoglossum bicirrhosum* is one of the arawanas which occur in South American freshwaters. Very large, circular scales and long, upturned jaws are typical features of many osteoglossiforms.

BELOW: moray eels, like many of the anguilliforms, spend most of their lives in crevices in the rock. The elongate body enables them to move in both directions with equal ease. They are rapacious predators.

RIGHT: *Argyropelecus lychnus* is one of the deep sea hatchetfish, found down to depths of 4,000 m. Like many abyssal fish its flesh is translucent and contains bioluminescent organs (photophores) in rows on the underside and flanks.

A second holostean lineage (Amiiformes), which may also be traced back to the Triassic, is represented today by the Bowfin (*Amia calva*). The Bowfin is confined to the freshwaters of eastern North America. Its swimbladder, like that of *Lepisosteus*, is highly vascular and can be used as a lung.

The third of the main holostean lineages includes a number of small herring-like fishes, known only as fossils, in which certain structural improvements leading through to the teleost fish took place; there was an increase in jaw mobility; the scales lost their ganoine covering and became circular, so forming the typically thin, cycloid scales of teleosts; and the tail became perfectly symmetrical (homocercal) so providing a horizontal thrust.

By the end of the Jurassic the first members of the Teleostei had appeared and today the group contains an estimated 18,000 species. The Osteoglossomorpha, the most primitive of the eight teleost superorders, are predominantly freshwater fish and include in the Osteoglossiformes the Goldeye (*Hiodon alosoides*) of North America and the more specialized arapaimas and arawanas of South America, Africa and the Malayan–Australasian region. These specialized osteoglossiforms are somewhat elongate and have large circular scales. *Arapaima gigas*, found in the Amazon drainage system, grows up to 4 m long and is the world's largest freshwater fish. The tongue of osteoglossomorphs is particularly well equipped with teeth (hence the group's name meaning 'bony tongue') and this bites against a toothed area on the base of the braincase.

The famous game fish *Tarpon atlanticus* (order Elopiformes) of the east coast of North America and the true eels (order Anguilliformes) are examples of a second primitive superorder, the Elopomorpha. The group is characterized by a special type of larval stage, the leptocephalus, which during its growth actually shrinks before metamorphosing into the adult form. The leptocephalus is a planktonic stage, thought to have a distributive function; at least this seems to be its relevance in the life history of the European Eel (*Anguilla anguilla*). Mature eels leave European rivers and are then assumed to swim across the Atlantic at great depth to the spawning grounds in the Sargasso Sea. After spawning the adults presumably die. The larvae then drift in the Gulf Stream back to Europe, taking three years to complete the journey, before metamorphosing into elvers to ascend the rivers and eventually breed themselves.

Salmon (*Salmo*, Atlantic salmon; *Oncorhynchus*, Pacific salmon), smelts (for example, *Osmerus*) and the pikes (*Esox*) are important food and game fishes of the northern hemisphere. They represent primitive members of the super-

order Protacanthopterygii. Their southern hemisphere counterparts are the galaxiads and the southern smelts (for example, *Aplochiton*). The group has both marine and freshwater representatives with some, such as salmon, spending their adult lives in the sea before migrating up rivers to spawn. A commonly found feature of these fishes is the adipose (fatty) fin, a small fleshy lobe which projects from the dorsal surface of the body immediately in front of the tail. Among advanced salmoniforms there are deep sea examples such as the hatchetfish (for example, *Sternoptyx diaphana*) which, in their dim environment, produce light (bioluminesce) and display a ghostly translucence.

Many of the teleosts dealt with so far were, in older classifications, united in the Isospondyli, a reference to the fact that each vertebra along the length of the backbone is similar to its neighbour. In members of the superorder Ostariophysi, however, the anterior vertebrae are modified to form a bony connection between the swimbladder and the ear. This chain of small bones (the Weberian ossicles) is thought to improve hearing ability. About 73 per cent (5,000 species) of all freshwater fish are ostariophysans which are

classified into two main orders, Cypriniformes (consisting of the Characoidei and Cyprinoidei) and Siluriformes.

The characoids constitute an exclusively freshwater group, confined today to South America and Africa. They are often very colourful (for example tetras, *Hyphessobrycon*) and are favourites with aquarists. The group also includes some South American villains such as the piranhas (for example, *Serrasalmus*) and the Electric Eel (*Electrophorus electricus*). The latter can discharge shocks up to about 500 volts and this ability is used to navigate and kill prey as well as for defence.

Cyprinoids are small freshwater fish of North America, the Old World and southeast Asia. Common examples are the minnows (for example, *Notropis*, *Barbus*), carps (for example, *Cyprinus*) and loaches (for example, *Noemacheilus*). Teeth are absent from the jaws, but instead, there are often enlarged teeth on the posterior gill arches. Cyprinoids have developed a suction feeding mechanism, analogous to that seen in acanthopterygians.

The Northern Pike (*Esox lucius*) is a freshwater salmoniform species with a circumpolar distribution. The median fins are at the back of the body and, with the tail, provide a strong thrust for rushing out to strike at passing prey.

The catfish (Siluriformes) are ostariophysans with the head slightly flattened and a ventral mouth usually surrounded by touch-sensitive spines or barbels. The dorsal and pectoral fins are preceded by stout spines and in a few species (such as *Heteropneustes fossilis* of tropical Asia) the pectoral spine may be extremely poisonous. Many catfish have no scales but in some (for example, *Corydoras*, a common aquarium fish) the scales are enlarged and form an armour. The Walking Catfish (*Clarius batrachus*) is one of several species with modifications of the gill chamber enabling them to breathe air during their brief excursions onto land. Catfish occur in freshwater all over the world.

The catfish, such as this South American freshwater genus *Pimelodella*, are usually recogniz-able by the presence of long, touch-sensitive barbels around the mouth.

The flying fish *Exocoetus volitans* (Atheriniformes) is one of several species which literally swim out of the water and glide into surface level air currents, using their enlarged pectoral fins spread out as aerofoils.

The superorder Atherinomorpha is a small group (about 825 species which include the well known Guppy, *Poecilia reticulata*) of predominantly freshwater fish which are structurally intermediate between the protacanthopterygians and the acanthopterygians. The male Guppy, like the males of many other species of atherinomorphs, has an anal fin which is modified to form an intromittent organ called the gonopodium. In most species, the young develop internally and receive maternal nourishment. (Internal fertilization and viviparity are rare phenomena in the bony fish world.) Most atherinomorphs have a dorsally positioned mouth and a flattened skull which are adaptations for taking food which floats on the surface of the water.

Squirrelfish, of which there are many species, are members of the order Beryciformes. They are brightly-coloured inhabitants of tropical waters. Because part of the dorsal fin is composed of spines instead of soft rays, they are regarded as primitive.

Of all bony fish, members of the Acanthopterygii rank as the most sophisticated. Acanthopterygians can remain at any level in the water, or rise and sink seemingly without effort; they can brake and turn in their own body length and have a remarkably efficient jaw mechanism. These attributes can be related to certain morphological features. The swimbladder is closed off from the gut but buoyancy can be regulated by either secreting or absorbing gas. The pectoral fins are held high on the flank and function as brakes and the pelvic fins are usually found directly beneath them. But perhaps the most significant feature is the protrusible upper jaw; the toothless maxilla acts as a lever to push out the enlarged premaxilla. This is done very quickly and has the effect of creating a momentary but powerful suction.

Acanthopterygian means 'spiny fin' and this is a reference to the spines which are developed in the pelvic, anal and dorsal fins (the dorsal fin is often divided into anterior spiny and posterior soft-rayed portions). Spines are often found on many of the head bones and even the scales have a fine comb-like (ctenoid) margin.

Primitive acanthopterygians, such as *Ctenothrissa*, have been found in marine deposits of Cretaceous age and the marine environment has remained the stronghold of the spiny-finned teleosts. By the Eocene epoch representatives of most of the modern orders had appeared. The 10,000 species of living acanthopterygians are divided among 13 orders and about 230 families, but there is as yet little understanding of their interrelationships and hence, of how to classify them. It is generally recognized that the order Beryciformes (for example squirrelfish, *Holocentrus*) represents a primitive assemblage and that members of the Tetraodontiformes (for example pufferfish, *Tetraodon*), are advanced. Between these two extremes, however, the astonishing variety of morphologies has clouded the task of deciphering their evolution.

Most acanthopterygians are streamlined with large, well-toothed mouths (for example, rockfish: Scorpaeniformes; basses, wrasses and barracudas: Perciformes) and are usually inhabitants of coastal waters. Other predaceous acanthopterygians (mackerels and tunas: Perciformes) are more typical of the high seas and show particular adaptations, such as a narrow tail, for fast, sustained swimming. Some tunas (for example, the Atlantic Tunny, *Thunnus thynnus*) attain speeds of 60 to 90 km per hour over short distances. At the opposite end of the scale defence, rather than speed, seems to be the most obvious strategy. The boxfish (*Ostracion*: Tetraodontiformes), for instance, is encased in an inflexible armour formed by modified scales whereas the sticklebacks (Gasterosteiformes) combine enlarged scales with prominent pelvic spines.

Acanthopterygians are common members of the coral reef environment with representatives of three perciform families being prevalent—the jacks, the damselfish and the butterfly-fish. The last two mentioned have bands and spots of brilliant colours (blue, yellow, red and black predominate) and are the fashion models of the fish world. This conspicuous colouration may serve to break up the outlines of the fish in its naturally colourful environment. The intertidal fish fauna is dominated by sculpins (Scorpaeniformes) and blennies (Perciformes), many species of which remain in pools between high tides.

Certain families of acanthopterygians are common members of freshwater faunas. The true perches (family Percidae: Perciformes) are northern hemisphere forms which often occupy top positions in the food webs of lakes and rivers. Their southern hemisphere counterparts are the cichlids (family Cichlidae: Perciformes) which have many representatives in Africa (for example, *Haplochromis*) and South America (for example angelfish, *Pterophyllum*).

The superorder Paracanthopterygii is a relatively small group (about 1,000 species) whose members, as the name suggests, have paralleled the acanthopterygians in many ways. There is, for instance, a protrusible upper jaw mechanism but this is constructed differently from that seen in acanthopterygians. Spines are generally present, although inconspicuous, and the pectoral fins are set high on the flank with the pelvics beneath or even in front of them.

The paracanthopterygian lineage can be traced back to late Cretaceous marine forms such as *Sphenocephalus*. Most modern members are also inhabitants of the sea. Many of them such as the Atlantic Cod (*Gadus morhua*), hake (*Merluccius*) and the Haddock (*Melanogrammus aeglefinus*) are commercially valuable species. Members of this particular group, the Gadiformes, can usually be recognized because they have three dorsal fins, chin barbels and inconspicuous scales. In advanced paracanthopterygians, such as the anglerfish (Lophiformes) scales are usually absent altogether. The general trend of paracanthopterygian evolution is towards exploiting the benthic environment, and even the streamlined cods live just above the bottom.

ABOVE: the Copperband Butterfly (*Chelmon rostratus*) is a deep-bodied perciform fish which lives in coral reefs. The bright, banded colour pattern is, in fact, a form of camouflage in that it serves to break up the outline of the body.

BELOW: many of the species of anglerfish are inhabitants of deep water. In several of them certain specialized fin rays are highly modified to form fishing lures which are often luminescent; these attract prey into the very large mouth.

The second major division of bony fish, the Sarcopterygii, includes the coelacanths, lungfish and the rhipidistians. All of these types made their appearance in the Devonian period and were common members of the Palaeozoic fish fauna, particularly in freshwater environments. Today there are only seven surviving species (one coelacanth and six lungfish), but the importance of the group lies in their structural and physiological adaptations which were probably present in the ancestor of amphibians and thus all tetrapods.

Until 1938 the order Coelacanthiformes was thought to have been an ancient lineage which had died out in Cretaceous seas. But in that year a commercial fisherman, trawling off the coast of East London, South Africa, noticed a strange looking fish among his catch. Many features of this steel-blue fish suggested that here was a living coelacanth. This coelacanth (*Latimeria chalummae*) is a large (about 1.5 m), plump-bodied fish with heavy circular scales. The head is covered with thick bones which give it an ancient appearance. The pectoral, pelvic, anal and posterior dorsal fins each have fleshy bases. The anterior dorsal fin is sail-like and the tail has equally developed dorsal and ventral lobes with a small central lobe; these features are characteristic of coelacanths.

Since the first discovery of *Latimeria*, another 85 or so coelacanths have been caught, but all of these have been taken from deep water off the Comores Islands. It is probable therefore that the East London specimen was a stray. Only a few coelacanths have been brought to the surface alive and none has survived more than a few hours. It has been noticed, however, that the pectoral fin can be rotated through 180°, an attribute expected in a relative of the group from which the tetrapods probably evolved.

Latimeria has a swimbladder instead of lungs, and this is filled with fat. In many fossil coelacanths the swimbladder was very large and had bony (ossified) walls but the reason for this obvious specialization has yet to be explained. Quite recently it has been discovered that the modern coelacanth is ovoviviparous. The body cavity of a large female was opened up to reveal five advanced embryos in the right oviduct, each with its own attached yolk sac for nourishment.

The coelacanth *Latimeria chalumnae* lives in the deep waters of the Mozambique channel between Africa and Madagascar. Most of the specimens have been collected from depths between 150 and 400 metres from areas with rocky substrates. The lobed fins are supported by muscle and bones and are highly manoeuvrable.

Lungfish (Dipnoi) are highly specialized members of the Sarcopterygii which, like the coelacanths, seem to have evolved very slowly throughout their long history. Devonian types, such as *Dipterus*, were very similar to the living lungfish except that their skeleton was more heavily ossified and there were two distinct dorsal fins. Recent lungfish are eel-shaped with a single long dorsal fin which is confluent with the caudal and anal fins.

There are three living genera, all found in freshwater: *Neoceratodus forsteri* (southeast Queensland, Australia); *Protopterus*, with four species (eastern and central Africa) and *Lepidosiren paradoxa* (central South America). The best known feature of lungfish is the presence of lungs—a pair in *Lepidosiren* and *Protopterus* but one only in *Neoceratodus*. All three use the lung(s) as a supplement to the gills for respiration and during certain periods the South American and African lungfish use the lungs exclusively. Correlated with the functional lungs are special pulmonary vessels

Modern lungfish are found in widely separated areas. The African and South American forms are most closely related and this reflects the idea that the continuity between these two continents persisted longer than their connection with Australia.

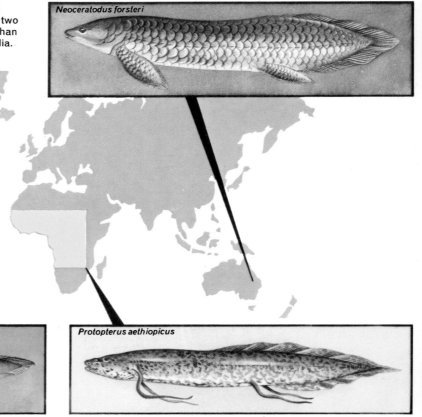

Neoceratodus forsteri

Lepidosiren paradoxa

Protopterus aethiopicus

Eusthenopteron foordi, a late Devonian rhipidistian, had many features in common with the early tetrapods. In particular, the bones of the pectoral limbs were strikingly similar to those of an amphibian, indicating that it was probably able to crawl over dry land, searching for pools in times of drought.

supplying them, and a heart in which the atrium is divided to left and right chambers: these features represent an early stage in the development of the double circulation found in more advanced vertebrates and result in partial separation of oxygenated from deoxygenated blood.

Protopterus and *Lepidosiren* inhabit stagnant waters and swamp areas which are depleted of oxygen and, more importantly, subject to periodic drying up. In both genera the gills are poorly developed and great reliance is placed on the lungs for respiration.

In contrast to coelacanths and lungfish, the Rhipidistia were a short-lived group, lasting only about 130 million years (Devonian to Permian). Rhipidistians were predominantly freshwater types which in many respects were similar to early amphibians. In all probability the two groups shared a common ancestor. *Eusthenopteron*, the best known of the rhipidistians, was a plump-bodied fish of about 60 cm in length. The two dorsal fins and the anal fin were placed

posteriorly on the body, close to the tail, and these fins must have acted together to provide a very powerful thrust. The mouth was large, with many teeth, some of which were enlarged to form tusks. The enamel coat of a tusk was infolded into the underlying dentine and this forms an important resemblance to the teeth of labyrinthodont amphibians.

The structure of the paired fins and the vertebral column suggest that *Eusthenopteron* and its relatives could make short excursions onto land. The paired fins had an endoskeleton very like that of a primitive amphibian. For instance, in the pectoral fin there were bones corresponding precisely to the humerus, radius and ulna of the front limbs of tetrapods. Nevertheless, typical fish fin rays were also present. The vertebral column was composed of strong interlocking vertebrae, very similar in structure to the vertebrae of primitive amphibians. Rhipidistians had true internal nostrils and it is assumed that they also had lungs, which they used like the lungfish. They are often found in deposits which geologists interpret as having been laid down in arid conditions, perhaps with seasonal drying up of pools and streams. An ability to move on to land, even if temporarily, would have been of distinct selective advantage to these forerunners of land vertebrates.

Lepidosiren, the South American lungfish, breathes almost entirely through its lung; its gills are very poorly developed, and it may drown if held under water. During times of drought it buries itself in the mud, breathing air and living on stored body fat.

SUBKINGDOM Metazoa
PHYLUM Chordata
SUBPHYLUM Craniata

CLASS Amphibia
Amphibians

ORDER	SUBORDER	FAMILY
†Labyrinthodontia		
†Phyllospondyli		
†Lepospondyli		
Caecilia (= Apoda)		Caecilidae
Urodela (= Caudata) *newts and salamanders*	Cryptobranchoidea	Hynobiidae
		Cryptobranchidae *giant salamanders*
	Sirenoidea	Sirenidae *sirens*
	Salamandroidea	Salamandridae *true salamanders and newts*
		Proteidae *mudpuppies etc.*
		Amphiumidae *Congo eels*
	Ambystomatoidea	Ambystomatidae *mole salamanders*
		Plethodontidae
Anura (= Salientia) *frogs and toads*	Amphicoela	Leiopelmidae
	Opisthocoela	Pipidae
		Discoglossidae
	Anomolcoela	Pelobatidae *burrowing toads*
	Procoela	Bufonidae *true toads*
		Dendrobatidae
		Rhinodermatidae
		Hylidae *tree frogs*
		(+ 7 further minor families)
	Diplasiocoela	Ranidae *true frogs*
		Rhacophoridae *flying frogs*
		(+ 2 further minor families)

†extinct

The Green Frog (*Rana clamitans melanota*—Ranidae) from eastern North America is a common species in shallow water. The big eyes and prominent eardrum are evidence of the well-developed senses of sight and hearing characteristic of anuran amphibians.

- NUMBER OF LIVING SPECIES: about 2,900

- DIAGNOSTIC FEATURES: craniates with a permeable skin across which respiration can take place; life cycle includes shell-less egg which gives rise to a distinct larval form

- LIFESTYLE AND HABITAT: free living in fresh water and on land (some are arboreal) but all are dependent on water or damp conditions

- SIZE: 2 to 125 cm but mostly between 5 and 15 cm

- FOSSIL RECORD: Devonian to Recent but virtually unrepresented from Triassic to early Cretaceous; about 280 genera

The name Amphibia implies an ability to live with equal ease both on land and in water. In fact, very few species in the class Amphibia are able to do this. Many are totally aquatic and are unable to survive on land while others are entirely terrestrial in their habits and live in water only during the first weeks of life. What does typify virtually all amphibians is their dependence on water, if not in the form of a pond or stream, at least as generally damp and humid surroundings.

In many respects the amphibians represent an intermediate stage between the fishes and the reptiles in the conquest of the land by the vertebrate animals. The paddle-like fins of fish have been replaced by complex jointed arms and legs that hold the body clear of the ground and propel the animal by walking, running, jumping or climbing. Amphibians differ from reptiles in having a soft, moist skin, rather than a dry, scaly one, and in producing eggs that are not protected by a hard shell. It is because both their skin and the membrane covering their egg are permeable to water that amphibians are restricted to habitats where water is plentiful, whereas reptiles have successfully colonized most terrestrial environments including the driest deserts.

Amphibians also differ from reptiles in having a life cycle that incorporates a larval stage. The amphibian egg does not develop into a miniature version of the adult but into a creature—for example the tadpole of frogs—which undergoes a radical change in form, or metamorphosis, to reach the adult state. Although adult amphibians may be terrestrial or aquatic in their habits, their larvae are invariably confined to water. In most species the larval environment is a pond or stream, but in a few species the larva develops in the water enclosed within the egg membrane, eventually emerging as a fully-formed but very small adult.

The skin of amphibians, unlike that of other vertebrates, plays an important part in respiration, providing a surface at which oxygen is taken in and carbon dioxide is expelled from the body. To be effective in this respect their skin must be kept moist and permeable to water. The skin is not the only route by which respiratory gases enter and leave the body: like their fish ancestors, amphibians typically possess lungs, though these have been discarded in the evolution of some groups. Many species also breathe through the inner surfaces of their mouths, rhythmically sucking in and expelling water or air through the mouth or nostrils. Larval amphibians possess external gills that project from just behind the head and through which dissolved gases are exchanged between the blood and the water.

Caecilians spend most of their life burrowing in soft earth, emerging only after rain. They are voracious carnivores and rush from their burrows to seize such prey as worms, insects or mice. This typical caecilian is *Geotrypetes grandisonae* from Ethiopia.

The female of the Marbled Newt (*Triturus marmoratus*), a salamandrid from France and Spain. The newts, like most other urodeles, have small eyes and poor ears, relying mostly on smell to locate food. They breed in water but spend most of the year on land.

Amphibians keep their skin moist by means of mucus glands. In some species additional skin glands secrete toxic substances that afford protection against predators. The males of some species possess hedonic glands whose secretions have an aphrodisiac effect on the female during mating.

The class Amphibia is divided into three orders that differ markedly from each other in body form and habits. The Urodela comprise the newts and salamanders; they have elongated bodies and relatively small limbs and are typified by the retention into adult life of the long tail which is found in all larval amphibians. The Anura comprise the frogs and toads; the adults have no tail, their bodies are short and they have very large hind limbs which in many species are used for jumping. Members of the third order, the Caecilia, are rarely-seen, burrowing or swimming animals found only in the tropics; they are worm-like in shape with greatly elongated bodies and they are entirely limbless.

Caecilians look and behave like large earthworms. Their skin is covered in tiny scales that betray their fish ancestry and in some species the body is marked by a series of circular grooves, giving them a segmented appearance. Caecilians have very small eyes that are covered over with skin and which are capable of only rudimentary vision. Characteristic of this order is a pair of small sensory tentacles on the head. Caecilians differ from the other two orders of Amphibia in that the male possesses a penis-like organ by which sperm is transferred directly into the female's body.

The living amphibians—a much smaller group than the reptiles, birds or mammals—are the relics of a group whose first representatives lived during the late Carboniferous period about 300 million years ago. Although living amphibians are generally small, delicately-built creatures, early labyrinthodont amphibians like *Eryops* and *Ichthyostega* were stout, heavy animals that retained the tough, scaly skin of their fish ancestors. The amphibian fossil record is unfortunately incomplete; after animals like *Eryops*, that lived during the Permian period, about 250 million years ago, there is a huge gap before we find very modern-looking frogs in Cretaceous deposits about 100 million years old.

It is widely believed that the Amphibia are derived from a group of fish called the rhipidistians that possessed lungs. These fish lived in the Devonian period which was followed by the Carboniferous, a time of swampy conditions that favoured the ability to leave the water and crawl around on the land. It is generally thought that the living amphibians are the descendents of a single line that divided early in its history to give rise to the three living orders. However, at least one authority has suggested that the Urodela may have evolved quite independently from the other two orders from a different group of lunged fishes. Unless more fossils are discovered the early evolutionary history of the amphibians must remain a subject for much speculation .

Eryops was a primitive amphibian from the Permian period. This and similar creatures probably inhabited swamps and the edges of rivers and lakes. They were heavily built and must have been very slow and clumsy in their movements on land.

With the exception of some species living in the northern part of South America, the order Urodela is confined to the northern hemisphere. In retaining a long tail in the adult stage, members of this group are much more similar in form to the typical amphibian larva than are the frogs and toads. A number of species have taken the retention of larval features into the adult stage further by being permanently aquatic and, in some species, by possessing external gills. These latter species are typically found in stagnant water where oxygen is scarce. This reversal of the general trend of amphibian evolution, which has been towards increased emancipation from water, has occured in at least four urodele families and is an example of convergent or parallel evolution. Most of those species that are fully terrestrial in the adult stage must return to water each year to breed.

The Hynobiidae is an Asiatic family whose members live in fast-flowing streams where oxygen is abundant. They are able to get most of their oxygen through their skin and possess only small lungs. The Siberian Newt (*Ranodon sibericus*) is able to withstand the intense cold of its high altitude habitat.

The cryptobranchids or giant salamanders are permanently aquatic animals. Living in rivers and streams where a good flow of water provides a good oxygen supply, they do not have external gills. However, the American Hellbender (*Cryptobranchus alleghaniensis*) has a number of folds in its skin that aid oxygen uptake by increasing the surface area of the body. Some species in the genus *Megalobatrachus* from China and Japan may reach 1.5 m in length.

The Sirenidae is a family of permanently aquatic creatures that live among the mud and matted vegetation at the bottom of lakes and pools. In such habitats oxygen is scarce and they have external gills. Their bodies are long and eel-like and they have very small limbs. The Siren (*Siren lacertina*) reaches a length of 1 m.

The Salamandridae is a family of many and varied species. With a few exceptions, they are terrestrial as adults. The European Salamander (*Salamandra salamandra*) produces poisonous skin secretions and has a dramatic warning colouration of black with yellow spots or stripes. The North American Red-spotted Newt (*Notophthalmus viridescens*) leaves the water at metamorphosis as a vivid red creature called a Red Eft. Its skin produces a highly poisonous substance that can prove fatal to its enemies. After three or four years of terrestrial life the Red Eft becomes a greenish adult with small red spots that spends most of the rest of its life in water. The closely-related European newts (*Triturus*) spend most of the year on land, returning to water for two or three months of each year to breed. The Iberian Ribbed Newt (*Pleurodeles waltl*) possesses a remarkable anti-predator device. Along each side of its body is a row of wart-like protuberances. When the animal is grasped the sharp ends of its ribs are protruded through these warts. Newts of the genus *Taricha* that live in rivers and streams in western North America show remarkable powers of homing. Displaced several miles from their home stream, they will cross mountain ridges and other streams, taking as long as three years to do so, to get back to their home.

The Proteidae is another family of salamanders that have reverted to the larval form. Living in mud or stagnant water, they possess external gills. The American Mudpuppy (*Necturus maculosus*) lives in mud. The Olm (*Proteus anguinus*) a very rare creature from Yugoslavia, lives only in subterranean caves. Never exposed to light, it has greatly reduced eyes and is virtually blind. It is entirely white in colour and has a flattened snout which it uses to burrow its way through sand and mud which form its habitat.

The Amphiumidae is represented by a single genus. The American Congo Eel (*Amphiuma means*) has a long eel-like body and very small limbs. It possesses lungs but not external gills. The Ambystomatidae is a family of salamanders in which adults may be terrestrial or aquatic. Many of them are very beautifully patterned and coloured. *Dicamptodon ensatus*, from California, is the largest land-living salamander, reaching 27 cm in length.

The plethodontids form the largest group of salamanders, found mostly in North America. They breathe only through their skin and mouth, lacking external gills and having lost the lungs of ancestral amphibians. Some live in fast-flowing

The Cave Salamander from North America is a very agile climber, often using its long, prehensile tail to cling to rocky ledges in its cave habitat. It has recently been shown to be able to detect the direction of the earth's magnetic field.

upland streams; others are totally terrestrial. The Cave Salamander (*Eurycea lucifuga*) hides in caves by day but may emerge into the open at night. Members of the genus *Batrachoseps* are called worm salamanders; they burrow in the ground and have long, thin bodies and small limbs. The American Olm (*Typhlomolge rathbuni*) found in caves in Texas, is a pinkish-white creature with virtually useless eyes. Another American cave species, the Grotto Salamander (*Typhlotriton spelaeus*) has fully functional eyes when young; however, if it remains within a dark cave as it grows older, its

eyes gradually degenerate. *Pseudotriton ruber* is a bright red salamander that is not at all distasteful or poisonous, but it is thought to be protected against predators because it resembles the poisonous Red Eft.

A species in which the adult has retained the larval form but which is capable of reproduction is said to show neoteny. The most well-known neotenous salamander is the Mexican Axolotl (*Ambystoma mexicanum*). In the wild this is found both as an aquatic form with external gills and as a form which lacks gills. So different are the two forms that for a long time they were regarded as distinct species. However, if neotenous Axolotls are injected with an extract of thyroid or pituitary glands they lose their larval features and metamorphose into the terrestrial form.

LEFT: the European or Fire Salamander lives wholly on land, apart from a short return to the water by the female to bear her young. Salamanders are generally active at night but may sometimes appear in the daytime, especially after rain.

BELOW: the Red-spotted Newt is found over almost all of eastern USA. The poisonous immature stage—the Red Eft—unlike the adult (which lives in ponds and ditches) can often be discovered wandering about in full view of potential enemies.

RIGHT: the aquatic form of the Mexican Axolotl sports enormous external gills. This neotenous form may be due to a shortage of iodine in the environment. Iodine is necessary for the synthesis of the thyroid secretion that would stimulate metamorphosis.

In most urodeles mating and development of the egg and larva take place in water. Thus most species that are terrestrial as adults must return to a pond or stream each breeding season. In some species, such as the North American salamander *Ambystoma tigrinum*, this involves a spectacular migration as thousands of individuals converge by night on a breeding pond. Breeding is usually highly seasonal, and mating and egg laying are often completed within a few days or weeks. Breeding usually occurs in spring; at this time ponds and streams are full and there is an abundance of small animals for the larvae to eat. Amphibian eggs and larvae provide a tasty and easily-caught meal for many predators and in most species very large numbers of eggs are laid to offset such losses.

Some of the more terrestrial salamanders among the plethodontids and ambystomatids lay their eggs in damp places under rocks and stones. *Ambystoma opacum* lays its eggs in such places in the autumn and spends the winter coiled around them. Most urodeles, however, show little parental care towards their eggs or larvae.

A few species, such as the European Salamander, are ovoviviparous. The eggs develop in the genital tract of the female. As they hatch she goes to a pond and the tiny gilled larvae are 'born' into the water. This is the only time that an adult of this species enters water.

The courtship of many urodeles is bizarre and complex and can be seen as an adaptation to the fact that male urodeles lack a penis by which sperm could be transferred into the female's body. In the more primitive urodeles the eggs are fertilized outside the body, eggs and sperms simply being shed into the water. This is an inherently inefficient mechanism, and many eggs must remain unfertilized. In the hynobiids and cryptobranchids, both primitive families, the female lays her eggs in sacs and the male then merely sheds sperm onto them. An evolutionary advance on this procedure is shown by one hynobiid, *Ranodon sibericus*, in which the male deposits his sperm in a capsule called a spermatophore onto which the female places her egg sacs. In all other urodeles efficiency of fertilization is increased by the spermatophore being taken up into the female before the eggs are laid.

The Smooth or Common Newt (*Triturus vulgaris*) typically lives in water only during the spring breeding season. Besides his nuptial crest, the male also sports conspicuous dark spots all over his body. The Smooth Newt is the most numerous (but not the most widespread) urodele in Western Europe.

Males of the Pyrenean Mountain Salamander, *Euproctus asper*, ambush females and then wrap their tails around them in such a way that their genital openings meet. The spermatophore is then extruded directly into the female. The male European Salamander grasps the female from below and stimulates her by rubbing her with hedonic glands around his head. After a time he deposits a spermatophore and then flicks his body to one side so that the female simply falls onto the spermatophore. In this species the female is held in amplexus, as it is called, throughout the mating process but in many urodeles amplexus is maintained only while the female is being aroused.

Amplexus may involve the male grasping the female from above or from below, with his fore or hind limbs or both, and in some species, also with his tail. During amplexus he stimulates her with glandular secretions, either rubbing his hedonic glands over her nostrils or wafting odour towards her with his tail. In the American plethodontid *Eurycea bislineata* the male possesses two prominent teeth at the front of the mouth; with these he scratches the female's skin before rubbing his aphrodisiac secretions into the wounds. Amplexus ends when the female signals her sexual receptivity to the male. He releases her, walks in front of her and then deposits a spermatophore; she then follows him and picks up the spermatophore in her genital opening. In some plethodontids such as *Plethodon jordani* the female follows the male in a characteristic way, keeping her nose above the base of his tail in a movement called the 'tail-straddling walk'. Plethodontids mate on land, whereas most salamandrids and all the ambystomatids mate in the water.

The European newts (*Triturus*) and a few closely related genera differ from other urodeles in that their courtship behaviour does not involve amplexus at all. When a male encounters a female he attempts to block her path and displays to her with a variety of movements. When she responds by approaching him he moves away and a spermatophore is transferred in a way similar to that seen in other urodeles. *Triturus* species are unique among urodeles in showing marked sexual dimorphism, the breeding male differing from the female in having a large, brightly coloured crest along his back and tail.

Many urodeles mate in large groups, individuals gathering in ponds from a wide area. This creates the conditions in which a remarkable form of behaviour, called sexual interference, can occur. In many respects mating is a competitive activity, individuals increasing their reproductive success at the expense of others. In *Ambystoma maculatum* and *A. tigrinum* males will intrude on the courtships of other males and will deposit their spermatophores on top of those of their rivals, making the rival spermatophores inaccessible to females. In *A. tigrinum* and in a number of plethodontid species males 'pretend' to be females, causing rival males to deposit spermatophores that are wasted because no female is present to pick them up.

Sexual interference leads to the evolution of counter-measures or sexual defence: in plethodontids males aggressively drive rival males away before courting females and in some species the male tries to lure the female away from the rest of the mating group before initiating spermatophore transfer behaviour. These curious behaviour patterns are an evolutionary consequence of the spermatophore method of sperm transfer. Spermatophores are used by a variety of animals, such as insects and arachnids, but, among amphibians, they are unique to the urodeles.

ABOVE: during amplexus the male Iberian Ribbed Newt (*Pleurodeles waltl*) grasps the female from below. Transfer of sperm is achieved in this posture.

BELOW: Palmate Newts (*Triturus helveticus*) often lay their eggs in the fold of a leaf. The male differs from the female in having a distinctive pattern of spots.

LEFT: the South African Clawed Toad is permanently aquatic and well adapted for fast swimming. In the dry season it burrows into the mud and remains dormant until its pond refills. It has the ability to alter the density of its body colour to match that of the background.

BELOW: this small Forest Tree Frog (*Leptopelis natalensis*) is another African species. It has disc-like suckers on its toes for gripping and climbing, and large protruding eyes for detecting its insect prey.

RIGHT: the Edible Frog (*Rana esculenta*) is widespread throughout Europe, Asia and northern Africa. It is always found in or near water and will quickly jump back to it if disturbed on dry land. Its name is derived from the fact that its legs have long been prized as a culinary delicacy.

BELOW RIGHT : the webbed feet of the Costa Rican Flying Frog (*Agalychnis spurrelli*—Hylidae) enable it to glide between trees. This is an adaptation similar to that of a number of rhacophorid frogs. Such species are largely independent of water, returning to it only to lay eggs.

The order Anura, which comprises the frogs and toads, is by far the largest group of amphibians and contains about 90 per cent of all living species. Its members are characterized by the complete absence of a tail, though this is present in anuran larvae—usually called tadpoles. Adults have very long hind-legs consisting of four, rather than three articulating sections. In many species the hind legs are very muscular and powerful, giving the animal the ability to leap prodigious distances. Most frogs and toads have large protruding eyes that enable them to see all around without having to move the head, and in some the tongue can be flicked out, like a chameleon's, to strike prey some distance from the body. There are anurans which are permanently aquatic as adults, though none has gone as far as some urodeles in retaining the external gills of the larval condition.

The family Pipidae includes the South African Clawed Toad or Platanna (*Xenopus laevis*) and the Surinam Toad (*Pipa pipa*). These have a flattened body and webbed hind-feet that enable them to swim very effectively. They lack a tongue and feed largely on debris, shovelling decaying matter into their wide mouths with their hands. The fingers of the Surinam Toad have highly sensitive star-like processes at the tip with which the animal feels for its food in the mud. In the dry season, when their ponds dry up, Clawed Toads burrow into the mud and go into aestivation—a state of suspended animation similar to hibernation. Species of the discoglossid genus *Barbourula* from the Phillipines are aquatic toads that have webbed fingers as well as toes. The Pelobatidae or burrowing toads are adapted for digging into mud or sand, where they spend the day, coming out to feed at night. Species of the American genus *Scaphiopus* and the European genus *Pelobates* have a sharp, horny growth on the inside of the foot with which they scrape away the soil as they dig.

Many species of frogs are expert climbers, notably the Hylidae or tree frogs. Many climbing species have disc-shaped pads at the tips of the fingers and toes which act as

adhesive suckers, enabling them to climb smooth vertical surfaces. Some of the tree frogs are very small, with a body length of only 2 to 3 cm. The most remarkable locomotory adaptation is shown by members of the family Rhacophoridae, the flying frogs, which are able to glide across wide gaps between trees. They have very long fingers and toes with webs of skin between them. As they leap into space they spread these wide so that the webs provide uplift, enabling the frog to control the speed and direction of its fall. One such species, *Rhacophorus schlegelii*, is found in Japan.

Many anurans are beautifully patterned and brightly coloured. One European discoglossid, the Fire-bellied Toad (*Bombina bombina*), is a dull grey colour on its back and is

usually well camouflaged against its background. If disturbed, however, it rears up on its legs exposing a vivid red and black pattern on its underside and this acts as a warning to predators. The skin of this species also produces poisonous secretions. The arrow-poison frogs are dendrobatids with vividly patterned skins that produce extremely toxic secretions which have long been used by South American Indians to treat the tips of their arrows.

The Ranidae (true frogs) is a large family with a worldwide distribution. Its members typically have a slender build, large eyes and are quick and agile in their movements. The European Common Frog (*Rana temporaria*) is a typical example. For most of the year it lives on land and is rarely

seen, but in the spring it often congregates in huge numbers at ponds to breed. The members of another large family, the Bufonidae or true toads, are of heavier build and are slower and more clumsy in their movements. They have warty skins that produce poisonous or irritant slime.

The larvae or tadpoles of frogs and toads are generally rather different in appearance from the larvae of newts and salamanders. The body is spherical in shape, the legs appear only late in development and the external gills are often covered by a flap of skin. The shape of the anuran tadpole seems to be an evolutionary consequence of the adult body form. Frogs and toads have short bodies, having lost many vertebrae in the course of their descent from ancestral amphibians. For much of their life tadpoles eat plant food, for which they require a long gut and as they have a short backbone, this long gut cannot be accommodated in a long body but must be contained within a spherical one. As a result the body is not streamlined, so that anuran tadpoles tend to swim in a characteristically wriggling fashion, propulsion being provided by the tail.

The most primitive anurans, the Leiopelmidae, retain a vestige of the ancestral tailed condition. Members of the genus *Leiopelma* from New Zealand have no tail but do have vestiges of the muscles that wag the tail in the tadpole. The male Tailed Frog (*Ascaphus truei*) has what appears to be a short tail, but which is in fact a specially evolved structure with which he pushes his sperm into the genital opening of the female during mating.

With their dependence on ponds and other bodies of water for breeding, many frogs and toads are very vulnerable to the activities of man, who drains areas of wetland and pollutes others. As a result many species, such as the British Natterjack Toad (*Bufo calamita*), have become very rare. In spring when they migrate in large numbers to their breeding sites, many frogs and toads are killed on roads. In some parts of the United States enlightened authorities close stretches of road for a few days each spring to protect migrating frogs.

In their reproduction the frogs and toads face essentially similar environmental problems to the urodeles but they have solved these problems in very different ways. While all but the more primitive urodeles practice internal fertilization, almost all anurans fertilize the eggs outside the female's body, the only exception being the Tailed Frog. Furthermore, while much of the evolutionary elaboration of mating behaviour in urodeles has centred around the mechanism of sperm transfer from male to female, in anurans the main evolutionary development has been the sophistication of techniques for locating and attracting mates. A feature of the mating behaviour of many frogs and toads is the use of mating calls by which males attract females.

The mating calls of anurans are often very loud. In some species many males call together, creating a chorus which can be deafening when heard nearby. The call is produced by the movement of air back and forth between the lungs and mouth, over the vocal cords, and it may be a series of deep croaks, a prolonged trill, or a more complex sound rather like the call of a bird. Many species possess vocal sacs, bag-like extensions of the mouth cavity which are inflated with air during calling and which act to amplify the sound. Some species have a single vocal sac under the floor of the mouth, others have two, one on each side of the mouth. Small species

generally produce high-pitched calls, effective over a relatively short range, while large species produce deep, booming notes which can be heard from a considerable distance. Associated with their use of sound in communication, frogs and toads generally have well-developed and highly sensitive ears. The sound is picked up by a large tympanum, or ear-drum, which can often be seen at the surface of the skin just behind the eye. In many species the mating call serves the dual function of not only attracting females but also keeping other males away. Many male frogs and toads are territorial in the breeding season, attacking rivals who ignore their calls and come very close.

Once a male and a female have come together they join in amplexus, the male grasping the female from behind. If she accepts him as a suitable mate she starts to lay her eggs and he extrudes his sperm into them as they emerge into the water. Many frogs, like *Rana temporaria*, release their eggs suddenly in one large amorphous mass. Most toads, like *Bufo bufo*, are more controlled, slowly releasing their eggs in strings which they wrap around water weeds. Males may

The enormous inflatable vocal sac of the male Red Banded Frog (*Phrynomerus bifasciatus*) greatly increases the range at which the females can hear his mating call. This species is found in southern Africa. Its vivid colours serve to deter predators.

compete for females, trying to dislodge rivals who are already in amplexus. In a number of species, including the two European species mentioned above, the result of this inter-male rivalry is that, when they actually spawn, females tend to be paired with males who are comparable to them in size. This size-matching makes for more efficient fertilization of the eggs, since the eggs and sperm are less likely to become dispersed before they meet.

In some anurans calling and mating may be completed on just one night of the year. In other species it takes longer and males tend to defend quite large territories. Males with the biggest territories, or those in which conditions are best for egg development, tend to attract the most females over the course of a breeding season.

In many frogs and toads one of the parents cares for the eggs or larvae. In the hylid *Flectonotus goeldii* the female places the developing eggs in a hollow on her back and carries them around with her. Females of the hylid Marsupial Frog (*Gastrotheca marsupialis*) have a pouch of skin on their backs into which the eggs are placed. Most remarkable of all is the Surinam Toad, females of which develop spongy skin on their backs during the breeding season. The fertilized eggs are pressed into this skin where they develop in little enclosed pockets. They pass through the tadpole stage in these pockets, eventually emerging through their mother's skin as tiny adults.

In some species it is the male who cares for the offspring. The male Midwife Toad (*Alytes obstetricans*, family Disco-glossidae), wraps the eggs, which are laid in strings, around his hind-legs and carries them around with him. In some of the dendrobatids, for example *Phyllobates bicolor*, males carry tadpoles around on their backs, regularly entering water to keep them moist. The male Mouth-breeding Frog (*Rhinoderma darwinii*, family Rhinodermatidae) carries the young throughout the entire course of their development in a greatly enlarged vocal sac.

Like the urodeles, anurans require water in which their eggs and larvae can develop. Most species use ponds or streams but some have evolved elaborate and ingenious alternatives. The male Smith Frog (*Hyla faber*) builds a mud basin which fills with rain water, providing a private pool in which his progeny can grow. Some species, like the hylid *Phyllomedusa rhodei*, deposit their eggs in pools of water that form in the junctions between the stems and leaves of

ABOVE LEFT: a pair of European Common Frogs (*Rana temporaria*) in amplexus by a clump of spawn laid by another pair. The mating may be completed in a single day.

ABOVE: the male Midwife Toad cares for his eggs, carrying them on his back and seeing that they are kept moist. When the eggs

are ready to hatch, the male takes them to the water, and the tadpoles emerge.

BELOW: the foam nest of the Grey Tree Frog (*Chiromantis xeram-pelona*) of southern Africa keeps the eggs moist, so enabling this species to live far away from standing water.

certain plants. Members of the African rhacophorid genus *Chiromantis* produce large quantities of foam around their eggs to keep them moist. The foam nest is produced by the female with the assistance of as many as three males.

The eggs and tadpoles of many anurans provide easy prey for a variety of predators. Many young may die when the temporary pools in which they have been laid dry out before they reach maturity. However, most frogs and toads are long-lived creatures and have many opportunities to breed during their lifetime, so compensating for such losses.

SUBKINGDOM Metazoa
PHYLUM Chordata
SUBPHYLUM Craniata

CLASS Reptilia

Reptiles

SUBCLASS	ORDER	SUBORDER
Anapsida	†Cotylosauria *stem reptiles*	
	Chelonia	Pleurodira *side-necked turtles*
†Synapsida *mammal-like reptiles*	†Pelycosauria	Cryptodira *tortoises, turtles and terrapins*
	†Therapsida	
Lepidosauria	†Eosuchia	
	Rhynchocephalia *Tuatara*	
	Squamata	Lacertilia *lizards, geckos and iguanas*
Archosauria	†Thecodontia	Ophidia *snakes*
	Crocodylia *crocodiles, gavials and alligators*	
	†Pterosauria *pterodactyls*	
	†Saurischia *dinosaurs*	†Theropoda
		†Sauropoda *brontosaurs*
†Euryapsida *plesiosaurs etc.*	†Ornithischia *dinosaurs*	†Ornithopoda
		†Stegosauria
†Ichthyopterygia *ichthyosaurs*		†Ankylosauria
	†extinct	†Ceratopsia

- NUMBER OF LIVING SPECIES: about 6,300

- DIAGNOSTIC FEATURES: cold-blooded tetrapod craniates which breathe using lungs, have a dry, scaly skin and lay large, shelled (cleidoic) eggs

- LIFESTYLE AND HABITAT: live in all types of terrestrial habitats, the majority in tropical regions; many secondarily freshwater and marine species

- SIZE: 6 cm to more than 9 m (some dinosaurs reached 27 m) but mostly between 20 and 100 cm

- FOSSIL RECORD: middle Carboniferous to Recent with a peak in Jurassic and Cretaceous at the end of which they declined abruptly; perhaps 2,000 species

The thin Grass-green Whip Snake (*Ahaetulla prasinus*), a colubrid from the Far East, has the large gape and mobile, sensory tongue that typify the group. It is a mildly venomous tree dweller.

The Reptilia ('creepers') consist of four living groups of mainly terrestrial tetrapod vertebrates—the turtles and tortoises, the lizards and snakes, the single species of Tuatara in New Zealand, and the crocodiles, alligators and gavials. These, however, are only the remnants of a once vast radiation of an enormous variety of reptiles that dominated the Earth, the skies and the seas during the Mesozoic era, a radiation since supplanted by their two great descendant groups, the birds and the mammals.

The essential biological features of reptiles are a series of adaptations which protect them sufficiently from the hazards of life on land. Reptiles, unlike amphibians, can lead a fully independent terrestrial existence without the need for free-standing water even for breeding, and can survive the great temperature fluctuations that occur on land. Perhaps the most remarkable of these adaptations is the reptilian egg which is a type known as cleidoic or amniotic. This has a leathery or calcareous shell resistant to water loss, and a series of folds or membranes completely surrounding the developing embryo. Unlike the simple amphibian and fish eggs, the cleidoic type provides the embryo with sufficient food, oxygen intake for respiration and storage of excretory products that it can develop to a large enough size to hatch directly as a small, but fully formed version of the adult. It is then immediately able to adopt a terrestrial way of life.

Several structures of adult reptiles are important in restricting water loss. The skin consists of multiple layers of cells produced by division of the epidermal cells. As the

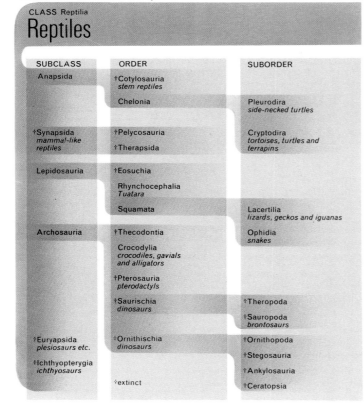

The ability of reptiles to breed independently of water is a key evolutionary advance. This has been achieved by the development of the cleidoic egg which, in some cases, hatches within the female's body.

LEFT: a cluster of crocodile eggs hatching out. Crocodile eggs are protected by a hard, white shell and are contained in a crude nest. When ready to hatch, the young break out by using the egg-tooth on the tip of the snout.

ABOVE RIGHT: the Green Turtle (*Chelonia mydas*) laying its eggs in a hollow excavated in a sandy beach. The eggs hatch in four to eight weeks, and the young then make their way to the sea.

RIGHT: a Blue Tongue Skink (*Tiliqua scincoides*) giving birth to her young. The Blue Tongues are ovoviviparous, the female retaining the eggs in her body until they hatch.

cells age the waterproofing protein keratin is laid down in them, they flatten and then die, but are retained as a thick outer layer which is impervious to water. As a result, the rate of water loss through the skin of a typical reptile is a small fraction of the loss through amphibian skin. Water loss during excretion of waste material is restricted by the use of the virtually insoluble substance called uric acid which is passed out of the reptile in solid form, and by the presence of a special chamber called the cloaca. This is a pocket-like invagination of the ventral body wall, into which the excretory ducts, rectum and reproductive ducts all open, and its walls can reabsorb most of the water from the urine and faeces. Reproduction itself involves internal fertilization with part of the wall of the male cloaca developed as a protrusible penis that is inserted directly into the female's cloaca and thus prevents desiccation of the seminal fluid. Most reptiles are oviparous—they lay eggs in the normal way—but some lizards and snakes retain the eggs in the female's body so that the young develop internally within the egg membranes and are born as young active individuals—a condition which is known as ovoviviparity.

The second great physiological problem of life on land is how to survive the wide temperature range compared with the relatively stable temperature of water. Only birds and mammals have overcome this problem completely, but reptiles have a surprisingly effective range of methods to retain their own body temperatures within fairly narrow limits—high enough for active life, but not too high to cause damage to their metabolism. Reptiles are ectothermic, which means they must use the sun's heat directly to raise their temperature rather than do this by their own metabolic processes. Typically reptiles follow a daily regime of basking in the sun broadside on with outstretched ribs during the early and late parts of the day. In the middle of the day they lie head on to the sun to reduce the surface area exposed to it, and may seek shade, burrow, or in the case of crocodiles, lie in water.

The terrestrial habitat is far more varied than the aquatic environment and so the variety of different kinds of animal that could evolve, once independence from water was achieved, proved enormous. Nothing illustrates this better than the range of methods of feeding and locomotion adopted in the different groups of reptile. The ancestral reptile was a small, insectivorous animal with numerous sharp pointed teeth and sprawling limbs for running over the ground. From this basic type evolved such diverse reptiles as limbless snakes with huge poison fangs; grazing, galloping dinosaurs; marine turtles with bird-like beaks and a swimming ability equal to the best of the fishes; and even the flying, fish-catching pterodactyls.

The variety of habitats open to reptiles includes fairly normal dry land areas such as forest, savannah and uplands which are today occupied largely by lizards and snakes. A number of forms, however, can live under desert conditions. Even more interesting, the impermeability to water of the reptilian skin has allowed some members of all the major groups to become secondarily aquatic, both freshwater and marine, and to compete successfully with fish. Apart from the extinct pterosaurs very few reptiles have become successful fliers, although the gliding lizards (*Draco*) which have elongated ribs supporting a gliding membrane are adept gliders, and certain snakes are said to be able to glide from tree to tree.

The major limitation to reptile distribution is due to their inability to generate very much heat to keep warm. Most species are therefore tropical and the number of both species and individuals decreases rapidly through the temperate zones. Nevertheless one lizard, *Lacerta vivipara*, is found within the Arctic Circle in Scandinavia.

Reptile skulls may be classified into four types depending on the number and position of spaces, or fenestrae, between some of the cranial bones. Three fenestrated forms arose from the ancestral anapsid type; diagrams of each type are shown, together with an actual example. The evolution of fenestrae allowed the development of more powerful jaws and gave a wider choice of potential foods. When muscles contract they swell laterally; in an enclosed skull muscle swelling is restricted and the jaws remain weak, but fenestrae allow the jaw muscles to bulge out of the skull, thus producing a powerful bite.

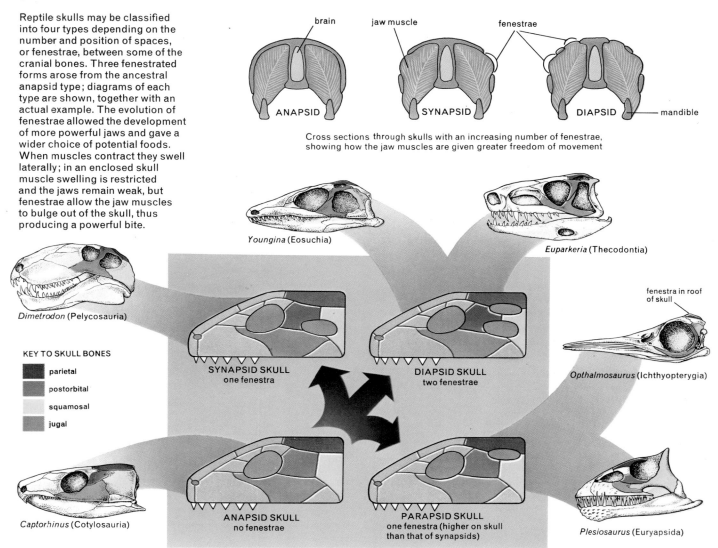

Cross sections through skulls with an increasing number of fenestrae, showing how the jaw muscles are given greater freedom of movement

Youngina (Eosuchia)

Euparkeria (Thecodontia)

Dimetrodon (Pelycosauria)

fenestra in roof of skull

Opthalmosaurus (Ichthyopterygia)

KEY TO SKULL BONES
- parietal
- postorbital
- squamosal
- jugal

SYNAPSID SKULL
one fenestra

DIAPSID SKULL
two fenestrae

ANAPSID SKULL
no fenestrae

PARAPSID SKULL
one fenestra (higher on skull than that of synapsids)

Captorhinus (Cotylosauria)

Plesiosaurus (Euryapsida)

The main character used to classify reptiles concerns the bones of the skull which in the earliest, most primitive types, form a complete roofing behind the eyes. These early reptiles, the Anapsida, first appeared in the late Carboniferous period as small, lizard-like forms, the Cotylosauria, with long jaws bearing numerous sharp-pointed teeth, and well developed limbs and tail. The origin of these ancestral or 'stem' reptiles was probably from a group of equally small, rather terrestrially-adapted labyrinthodont amphibians called the Seymouriamorpha sometime during the early or middle Carboniferous. What does seem certain is that all the later reptile groups evolved from this one central stock.

The tortoises and turtles in the order Chelonia are also anapsids, and although in many respects they are the most archaic of living reptiles, in other ways they are the most highly specialized. The shell which protects the body is a development of both the horny external scales of typical reptiles and the underlying bony scales found in a variety of extinct reptiles. Both layers form a series of large interlocking plates. The dorsal covering or carapace, which is fused to the reduced number of vertebrae and ribs, joins the ventral abdominal shield or plastron, leaving spaces for the head, the limbs and the much reduced tail to protrude. The jaws are short and powerful and all traces of teeth have gone, to be replaced by a horny beak with a sharp cutting edge.

The side-necked or pleurodire turtles of the southern continents can withdraw their heads into the confines of the shell by bending their necks horizontally. This suborder includes several rather bizarre forms such as the South American Matamata (*Chelys fimbriata*) which has a snorkel-like extension carrying its nostrils high above its head. All the remaining chelonians belong to the Cryptodira which withdraw their heads into the shell by a vertical movement of their necks. There are three main types of cryptodire chelonian: fully terrestrial forms often called tortoises; freshwater turtles and terrapins; and the highly adapted marine turtles.

The familiar tortoises (family Testudinidae) have high-domed shells and well developed limbs for terrestrial locomotion, and are among the few reptile groups that are herbivorous. The giant tortoises (*Testudo elephantopus* and *T. gigantia*) of the Galapagos and Indian Ocean Islands can grow to 150 cm in length and weigh 250 kg. The freshwater

turtles belong to several families and most are in fact amphibious because they spend part of their time on the fringes of rivers and lakes, where their eggs are laid. They are carnivorous and typical food includes insects, worms, small fish and amphibians. One of the most completely aquatic families is that which comprises the soft-shelled turtles (Trionychidae) which lack the horny plates covering the bone of the shell. They emerge from the water only during the breeding season to bury their eggs in sand.

Most marine turtles, including the familiar, edible Green Turtle (*Chelonia mydas*), belong to the family Cheloniidae. Their shells are flattened and streamlined and the bones of both the fore and the hind limbs are also flattened and covered by a continuous web of skin obscuring the individual digits. Swimming in these forms is analogous to the flight of birds, the paddles moving up and down rather than to and fro. They are fast and highly manoeuvrable. The Green Turtle itself is strictly herbivorous as it feeds on seaweeds, but the related Loggerhead Turtle (*Caretta caretta*) catches fish and other marine animals. Marine turtles lay their eggs on sandy beaches and to do this they often undertake prodigious migrations. The Leathery Turtle (*Dermochelys coriacae*) is unique among all chelonians because the bony plates of its carapace do not fuse to the vertebrae, but are embedded in the thick leathery skin. It is the largest of all turtles, some individuals reaching a length of 180 cm and about 500 kg in weight.

In the late Carboniferous period there appear in the fossil record some small reptiles which are very like the stem reptiles, but which, unlike Anapsida, have a large space, or fenestra, in the roofing or temporal bones of the skull, low down behind each eye. This condition allowed greater room for the jaw musculature and hence an improved jaw movement. These forms are the first members of the subclass Synapsida, the mammal-like reptiles, and are of great interest for two reasons: they constituted the first large evolutionary radiation of vertebrates on land, and it is from them that the mammals eventually arose.

Early, primitive members of the group are called Pelycosauria and although many of them grew to a large size (as much as 240 cm long) they still had the weak jaws and clumsy, sprawling gait of their lizard-like ancestors. The more primitive pelycosaurs were rather crocodilian in habit with long jaws and numerous sharp teeth for fish-eating. Others, however, evolved a much more terrestrial mode of life, and one group, the edaphosaurs, were the very first vertebrate group to adapt to a completely land-based herbivorous diet, a particular habit that was to be repeated continually by different groups of animal right through to the grazing mammals of today. Similarly, another group that includes the finback (*Dimetrodon*) evolved a wholly new habit for the first time—that of large terrestrial carnivore. *Dimetrodon* had enlarged fangs or canine teeth towards the front of the jaws for killing prey, and the other teeth were sharp-edged for cutting up flesh. The dorsal processes of the *Dimetrodon* and also of the edaphosaurs were enormously long and supported a 'sail' along their backs to assist the animals in their ectothermic temperature regulation.

BELOW LEFT: this particularly active pleurodire is *Chelodina longicollis*, an Australian snake-necked turtle. When it is fully extended, the neck may be longer than the shell.

RIGHT: suitably adapted for a marine life, the small Hawksbill Turtle (*Eretmochelys imbricata*—Cheloniidae) occurs throughout the tropical seas. Its mottled, horny plates are the source of the so-called tortoise-shell.

BELOW: the Starred Tortoise (*Geochelone elegans*) is one of the testudinids which occur in particularly hot regions in Asia. It is active only in the early morning and late evening.

BELOW RIGHT: several features of the Trionychidae family separate its members from other turtles. This Florida Softshell (*Trionyx ferox*) has the typical, reduced shell, and the head bears fleshy lips and ends in a proboscis.

therapsids were finally replaced by the archosaur reptiles—particularly dinosaurs—but some of the small, highly evolved therapsid insectivores survived as the very first of the true mammals.

The next important group of reptiles to evolve from the stem reptiles was originally classified as the Diapsida because its representatives have two temporal fenestrae on each side of the skull, behind the eye (contrast the condition in Synapsida with one and Anapsida without a fenestra). The taxon is no longer recognized in most classifications but the term diapsid has been retained as usefully descriptive of the skull type. Diapsid reptiles first appeared in the upper part of the Permian in South Africa in the form of a small lizard-like animal called *Youngina*. In this fossil reptile—a member of the Eosuchia—the fenestrae were already fully developed, indicating that the size and strength of the jaw muscles had

Dimetrodon, the finback pelycosaur, is one of the earliest mammal-like reptiles (Synapsida). Although it was ultimately an ancestor of the mammals, it had a primitive reptilian structure with scales, sprawling limbs and poorly differentiated teeth.

The pelycosaurs died out in the middle of the Permian period, but not before giving rise to the Therapsida—more advanced mammal-like reptiles with stronger jaw muscles, a dentition containing different types of teeth in different parts of the jaw bones, and far more agile limbs. Therapsid reptiles dominated the late Permian and first half of the Triassic periods, and produced a great variety of types. There are herbivores in large, roaming herds, big carnivores feeding on them, and a series of small insectivorous forms which occupied much the same position as do the lizards of today. As they evolved, the therapsids gradually became more and more like mammals, developing such features as a false or secondary palate (separating the air and food passages), complex molar teeth, a larger brain, upright stance, and narrow feet. Towards the end of the Triassic the

increased for a more active, more voracious, insect-eating life. During the late Permian and the first half of the Triassic period, while the mammal-like reptiles were at their peak as the dominant terrestrial vertebrates, the diapsid reptiles remained small and rare. However, during the second half of the Triassic they began to increase in numbers and diversity and gradually they replaced the mammal-like reptiles to become dominant themselves, a position they were to continue to hold for the next 120 million years or so. There were two quite distinct groups of diapsids which diverged from one another soon after the *Youngina* stage of evolution.

The Tuatara is the sole surviving rhynchocephalian. It is a 'living fossil', having remained more or less unchanged for about 200 million years. Individuals may live for over 50 years. They are nocturnal, feeding mainly on insects and worms.

One line, the Lepidosauria remained relatively conservative and led to the rhynchocephalians, the lizards and the snakes. The other diapsid line, the Archosauria, was initially the most successful and advanced group for it included the dinosaurs and their relatives.

Among the Lepidosauria, a lizard-like form about 30 cm long called *Sphenodon punctatum*, the Tuatara, is unique today. It is restricted to damp forests on certain offshore islands in New Zealand and because of its rarity is heavily protected by the New Zealand government. The only living representative of the Rhynchocephalia, the Tuatara is a relic of a brief but considerable radiation of reptiles during the Triassic period. In many ways its structure is close to that of its distant ancestor *Youngina*, for the skull still has the basic unmodified diapsid structure, unlike that of the lizards and snakes.

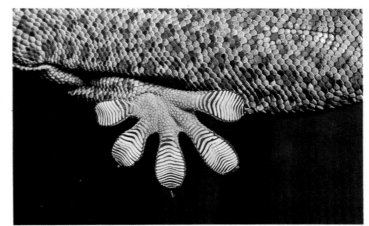

ABOVE: the phenomenal climbing ability of the gecko is due to the structure of the underside of the foot: there are rows of fine scales or lamellae which are covered by a mass of microscopic subdivided bristles.

LEFT: Gould's Monitor (*Varanus gouldi*) occurs in sandy parts of Australia and is probably the fastest lizard. It can adopt a bipedal stance, particularly when indulging in fighting behaviour with other individuals

BELOW: the 'wings' of a flying dragon are supported by ribs. They can be folded alongside the body, making the animal's outline less conspicuous, or extended to allow it to glide between the trees of the tropical rain forest in which it lives.

All the other living lepidosaurs are members of the Squamata—lizards and snakes—which are distinguished from other reptiles by several anatomical features including modification of the basic diapsid condition of the skull to allow various kinds of movement (kinetism) between different sections of the skull itself. A sensory organ in the roof of the mouth called Jacobson's organ is highly specialized for smelling, the tongue is slender and mobile and highly sensitive, and there are paired copulatory organs in the male. The lizards form a group called the Lacertilia, most of which have four well developed limbs. In contrast to the snakes (Ophidia) they have moveable eyelids, an eardrum on the surface of the head behind the lower jaw, and numerous small scales protecting the belly. The snakes have no sign of limbs externally although in a few there are traces of the pelvis deep within the body. Their eyes are protected by a permanent, transparent covering or spectacle, they lack an eardrum, and the belly is covered by a single row of transversely expanded scales.

The earliest known lizards occurred in the late Triassic although the group must have had a fairly long history before this period, since some of these early forms are highly specialized for gliding flight, with enormously elongated ribs supporting a gliding membrane. By the Cretaceous most of the modern lacertilian families had already differentiated. As a result of the changes to the skull the lizard snout can move up and down independently of the rest of the skull bones, giving a faster and more powerful bite. Most lizards feed on insects and other small invertebrates, although there are a few specialist herbivores, and some of the larger members feed on fish, birds and small mammals. Locomotion, by means of the four well developed but sprawling limbs, is

often very rapid, but there are some specializations of the locomotory system in various forms. When moving really fast some lizards adopt a bipedal gait using only the enlarged hind legs, and the Basilisk (*Basiliscus basiliscus*), an iguanid from Central America, can actually run some distance over the surface of water in this manner. The geckos have special foot pads which allow them to run up smooth vertical surfaces. Several gliding types are known, including particularly the flying dragons (*Draco*) which have folds of skin along the flanks that can be extended like narrow wings. On the whole lizards tend not to be aquatic, although the Nile Monitor (*Varanus niloticus*) has a crocodile-like tail for swimming in rivers.

There are about twenty families of lizards, several of which contain relatively few members of rather limited distribution. Perhaps the most typical lizards are the Lacertidae, small and agile forms with a worldwide distribution, except America and Australia, and typified by the British Common Lizard (*Lacerta vivipara*) and the elegant European Green Lizard (*L. viridis*). However, the commonest lizards of all are the skinks (Scincidae) which occur in all the continents. They are characterized by a thick, fleshy tongue and heavily built bodies. All stages of limb reduction are found among the members of this group and some have a superficial appearance of snakes. Their bodies are covered by large, overlapping scales, and the desert living forms are invariably good burrowers. The smallest of the lizards belong to the family Gekkonidae, the geckos, which are nocturnal and are often associated with human habitation. The European Gecko (*Phyllodactylus europaeus*) is only 7 cm long and is common in parts of Italy and France. By contrast, the Tackay (*Gekko gekko*) of the Far East is over 30 cm long; it eats fledgling birds, other lizards and mice as well as the more usual insects. The characteristic noise of the geckos is produced by a clicking of the tongue.

Two families of lizards with typically spiny scales are the Agamidae of the Old World and Australia and the New World Iguanidae. Among the agamids are the desert dwelling *Uromastyx* or mastigures of the Sahara, with powerful claws for burrowing and a marked ability to change colour. They are exclusively herbivorous and particularly like dates. The Spiny Lizard (*Moloch horridus*) of the Australian deserts feeds on ants and also from Australia comes the bizarre Frilled Lizard (*Chlamydosaurus kingi*) which is 60 cm long and bears a bright red and blue frill some 20 cm in diameter around its neck. The iguanids have a dorsal crest along the body and tail. The true iguanas (*Iguana iguana* and *I. delicatissima*) of tropical America are up to 180 cm long and live in burrows by the river banks. The Marine Iguana (*Amblyrhynchus cristatus*) occurs on the Galapagos Islands, living by the shore and browsing on seaweeds. *Phyrosoma cornutum* is the Texan Horned Toad, so-called because of its short, squat body. It has the unusual ability to fire squirts of blood out of its eyes at potential attackers. The largest lizards, the monitors, are members of the Varanidae whose diet is generally vertebrate prey. The Komodo Dragon (*Varanus komodoensis*) reaches over 3 m long and is now found only on the Sundra Archipelago in

BELOW: the Marine Iguana is the only sea-going lizard and occurs solely on the Galapagos Islands. It is large, reaching a length of 175 cm, and swims well by using its tail. There are specialized salt-secreting glands in the nose to remove excess salt.

RIGHT: the Cylindrical Skink (*Chalcides chalcides*) inhabits western Mediterranean regions. The limbs are reduced, and only three toes remain on each. When travelling rapidly, this lizard moves in a snake-like fashion with the legs off the ground.

ABOVE: the Australian Frilled Lizard, one of the agamids, lives in trees and feeds on insects and small mammals. The frill is used for courtship and to intimidate predators. Frill-raising is often accompanied by opening the mouth and hissing.

ABOVE RIGHT: Wiegmann's Worm Lizard (*Blanus cinereus*) lives in Spain and northwestern Africa. The totally limbless body has an extraordinary resemblance to an earthworm due to the annulation of the loose skin and the lack of other external features.

Indonesia. The Nile Monitor (*V. niloticus*) is over 180 cm long and has a crocodile-like existence, eating fish. The only poisonous lizards are the Gila Monster (*Heloderma suspectum*) of the deserts of the southern United States, and the Beaded Lizard (*H. horridum*) of western Mexico. These are the only two species in the family Helodermatidae. Like so many poisonous animals they are brightly coloured, with reddish tubercules all over the body.

The chamaeleons (Chamaeleonidae) are perhaps the most specialized of all lizards. Their limbs are vertical and the feet form grasping organs. The tail is long and prehensile for grasping branches. The chamaeleon's tongue lies coiled in

A chamaeleon feeding. The limbs and the feet are specialized, having two toes on one side and three on the other for grasping

branches. The highly extensible tongue permits the animal, in spite of its very slow movements, to capture insects.

the floor of the mouth and can be ejected by suddenly pumping blood into it, to capture flies on its sticky, club-shaped end. Retraction of the tongue is by contraction of muscles within it. The eyes can act independently of one another giving the animal a large range of vision while remaining absolutely stationary. The chamaeleon's ability to alter its colour to an extraordinary extent for camouflage is legendary: the skin has cells containing black, yellow, red and white pigments, and by contracting different pigment granules to different extents a wide range of overall colour is possible.

A final family of lizards is the Amphisbaenidae or worm lizards (for example *Amphisbaena*) of Africa and South America. Hind limbs are absent and the forelimbs are either greatly reduced or missing. The body is cylindrical in shape, the eyes and ears are hidden by folds of skin, and the skull is massively built. All these features are associated with its burrowing habit, and several authors have suggested that the differences between amphisbaenids and other lizards are so great that they ought not to be classified together.

The snakes are a much less diverse group morphologically than the lizards, never having, for example, the various spines and crests possessed by many lizards. Snake vertebrae have been found in early Cretaceous rocks but the detailed history of the group is not well known, perhaps because its members tend to live in habitats unsuitable for fossilization. It is generally believed that snakes evolved from a lizard-like group which took to a burrowing existence; this would account for the total loss of limbs and the compact bony protection of the braincase. The usual method of locomotion in snakes is effected by the muscles between the numerous ribs, which cause alternate waves of contraction to pass down the sides of the body. These waves push against objects or irregularities of the ground and drive the animal forwards. On smooth ground alternative means of propulsion are possible, particularly sidewinding where parts of the body are lifted off the ground, bent sideways and replaced. This often occurs in desert forms and results in a rapid movement of the snake in a forwards and sideways direction. Other snakes can move in a straight line, the large ventral body scales being raised and lowered slightly in a progressive sequence to work the animal forwards.

In connection with a highly specialized mode of feeding, the skull of snakes is very much modified from the basic diapsid type of its ancestors. Several of the bones of the skull are free to move independently, particularly the quadrate on which the mandible or lower jaw hinges. This bone can swing from side to side as well as to and fro about its connection to the rest of the skull. The front ends of the two mandibles are connected only by extremely elastic connective tissue. The result is that snakes can consume very large animals whole, often prey that is greater in size than the predator's head itself. The oesophagus is distensible and the ribs are unattached ventrally so such bulky food can pass easily into the stomach.

A snake's skull is, however, too weak to be able to cope with active, struggling prey and so hand in hand with the skull changes have evolved special means of killing the prey before ingestion. Boas and pythons kill by constricting the prey with coils of their bodies. Many other snakes have evolved the more efficient method of injecting toxic venom into the prey tissues, using enlarged, sharp fangs with grooves along the crowns. Back-fanged snakes such as the African Boomslang (*Dispholidus typus*) have enlarged fangs at the back of the tooth-bearing upper jaw bone or maxilla and are generally less dangerous because of the difficulty of inserting the fangs into the prey. On the other hand, the front-fanged snakes such as the cobras, mambas and vipers have fangs at the front of the maxilla which can be rotated when in use to point directly forwards and deliver a deep bite with ease. The poison glands themselves are modified salivary glands which produce toxic proteins and are connected by a duct to the base of the grooved or hollow fangs. The snake's tongue is very slender, delicate and mobile and when not in use is protected by a sheath in the floor of the mouth. It is highly sensitive to touch, taste and smell. The pit vipers or rattlesnakes have sensitive heat receptors set in small pits at the front of the head, and these are used to detect and locate warm-blooded prey.

BELOW: in the mountain forests of Central and South America lives the Boa Constrictor (*Constrictor constrictor*). It feeds on small vertebrates and, being harmless to man, it is often domesticated on South American farms to keep down rats and mice.

RIGHT: the British Grass Snake (*Natrix natrix helvetica*) is semi-aquatic. It is a subspecies of the Ringed Snake, which is found commonly throughout Europe, North Africa and Asia Minor and is one of the most northerly of all the snakes.

The African Egg-eating Snake devouring a chicken's egg. The mouth and neck are sufficiently extensible to allow an egg which is much larger than the head to be swallowed and passed entire to the stomach, where the contents are digested.

The most primitive living snakes are the boas (Boidae) and pythons (Pythonidae), which still have vestigial pelvic and hind limbs within the body. These families include the largest snakes, and both the Reticulated Python (*Python reticulatus*) of India and Malaysia and the South American Anaconda (*Eunectes murinus*) reach 9m in length. Both families consist of non-venomous forms and kill large prey by constriction. The boas are viviparous—the young develop internally and are born as active individuals. The pythons lay eggs—that is they are oviparous—but are remarkable for the way in which the mother incubates the eggs by coiling around them and keeping them warm. The most widespread snake family is the Colubridae which includes non-venomous forms, like the Grass Snake of Europe (*Natrix natrix*), which rely on biting to kill and capture their prey. The Egg-eating Snake (*Dasypeltis scabra*) of Southern Africa specializes in eating birds' eggs and has special processes on its front few vertebrae to break the shells. Also in this family are

the mildly venomous back-fanged snakes, many of which are tree dwellers, including the remarkable Flying Snake (*Chrysopelea ornata*) which can glide at an angle of about 45° to the ground by spreading its body.

The highly venomous snakes belong to four families. The vipers (Viperidae) occur in the Old World, and have very long fangs, often almost as long as the head, that can be erected to point forwards. The related rattlesnakes (Crotalidae) are found in the Americas. The rattle itself consists of a series of dry epidermal rings near the tail, which rattle together as the snake moves and presumably act as a warning

LEFT: Western Diamond-backed Rattlesnakes (*Crotalus atrox*) are the most notorious of the rattle-snakes, being highly excitable and aggressive. They feed on small mammals which they catch by making a characteristic forward lunge of the body.

ABOVE: the Egyptian Cobra (*Naja naja naja*) in a threat posture. The hood, raised and supported by the ribs of the neck, spreads out as a threat signal. From the raised position of the body, the strike is forwards or downwards, never upwards.

A second evolutionary line, quite distinct from the lepidosaurs, also led from early *Youngina*-like diapsids to appear in the fossil record in the early Triassic as the archosaur reptiles. These, like the lepidosaurs, were rather insignificant at first but by the late Triassic were well on the way to supplanting the mammal-like reptiles as the dominant group of terrestrial vertebrates. All the archosaurs retained the diapsid condition of the skull unmodified and their teeth were set into sockets in the jaw. Other than these there are few features which universally distinguish the Archosauria because of the enormous diversity of the group. One very marked evolutionary trend found throughout the archosaurs is a tendency for enlarged limbs, particularly the hind limbs, which improved their locomotory ability. The earliest and most primitive members of the archosaur radiation are called the Thecodontia, most of which were between 60 and 90 cm long, agile in build and already had relatively powerful hind limbs. Several were adapted to a crocodile-like mode of life with long bodies, powerful tails and long jaws adapted for fish-eating. By the late Triassic four distinct lines of archosaurs had evolved from the thecodonts—the Crocodylia, the flying Pterosauria and the two superficially similar but unrelated dinosaur groups, which constituted the orders Saurischia and Ornithischia.

device to would-be predators. The Elapidae are the cobras, mambas and kraits. The King Cobra (*Naja hannah*) of India is the largest of the poisonous snakes—it reaches as much as 380 cm—and is also the most aggressive and dangerous. The kraits (*Bungarus*) are found only in southeast Asia and the mambas (*Dendroaspis*) are exclusively tree snakes of Africa. Finally the sea or water snakes (Hydrophiidae) of the Pacific regions have laterally flattened tails to assist their swimming and several species occur far from the nearest land. Their venom is the most poisonous of all and they use it to kill their prey which consists chiefly of fish.

The Cuban Crocodile (*Crocodylus rhombifer*) from the West Indies can be recognized as a member of the family Crocodylidae by the open notch in the upper jaw for reception of the fourth tooth of the lower jaw, which is visible even when the jaws are closed.

The American Alligator (*Alligator mississipiensis*) lives in south-eastern USA and it is especially well known in the Everglades of Florida. In the members of the family Alligatoridae, the fourth lower tooth fits into an enclosed pit in the upper jaw.

The Crocodylia are the only archosaurs to have survived to the present day and there are about 25 species, all really rather similar. All are tropical freshwater inhabitants although a few live in brackish water and the Estuarine Crocodile (*Crocodylus porosus*) frequently makes long sea journeys around its southern Asian and northern Australian range. The principle food of crocodiles is fish which are caught by means of the numerous sharp teeth along the elongated jaws. Large individuals of, for example, the Nile Crocodile (*C. niloticus*) will take other reptiles and even large mammals,

A reconstruction of the Jurassic pterosaur *Rhamphorhynchus* in flight. These reptiles were highly modified for flying, with light, hollow bones, a bony sternum for attachment of the flight muscles and the greatly elongated fourth finger, supporting the front edge of the membranous wing. The tail acted as a stabiliser.

which they characteristically knock into the water with the large tail and hold underwater to drown. The carcass is then torn up by a group of crocodiles. Crocodiles have evolved an extensive false or secondary palate in the roof of the mouth so that the respiratory tract is not affected by water entering the mouth during feeding. Locomotion in the water is effected by the long tail which is flattened from side to side and is also sufficiently powerful to be used as a formidable weapon. The digits of the feet are partially webbed.

The body temperature of crocodiles, as in most reptiles, is maintained fairly constant by suitable behavioural means. In the early and late parts of the day crocodiles tend to bask in the sun on the river bank, whereas in the middle part of the day they enter the water to remain cool, lying inactively with little more than the tip of the snout showing. The main sense organs, the eyes, nostrils and ear, are closed by muscular valves. Complete submergence for up to five hours is possible. All the crocodiles are oviparous, and often a crude nest is constructed. In many species some rudimentary parental care is exhibited after the eggs hatch. A Nile Crocodile may live for more than 100 years.

Modern crocodilians are divided into three families, although these differ by little more than the shape of the snout, pattern of scaling and details of the dentition. The true crocodiles, Crocodylidae, have broad but relatively long snouts. The gavials, Gavialidae, are restricted to India and have very narrow snouts which widen at the tip to accommodate the nostrils. Finally the alligators and caimans with broad but rather short snouts form the family Alligatoridae; most of the species are American although the Chinese Alligator (*Alligator sinensis*) inhabits the Yangtse-Kiang river system.

The Pterosauria or pterodactyls evolved directly from the primitive thecodont archosaurs independently and appear in the fossil record in the early Jurassic. Perhaps the most remarkable of all reptiles, they were completely adapted for true flight. The fourth finger of the hand was enormously long and supported a membranous wing for gliding, and probably weak flapping too. Their legs were weakly developed and as they occurred mostly among shorelines, they probably rested on the edges of cliffs from which they could launch themselves directly into the air. The skull is light and

The Indian Gavial (*Gavialis gangeticus*) is the sole member of its family. Its most striking feature is its long, narrow snout armed with numerous sharp teeth, with which it catches fish. The Gavial is more strictly aquatic than any other crocodilian.

as the teeth are long, slender and few in number the main diet was probably sea fish which they picked from the water while still in flight. Like the dinosaurs, the pterosaurs became extinct at the end of the Cretaceous period.

The two distinct dinosaur groups both appear in the Triassic as independent derivatives of the primitive thecodont stock. Despite their superficial resemblance to one another they are well distinguished by several features, particularly the structure of the pelvic girdle: in the Saurischia the ventral pubis bone extends forwards (it is reptile-

Reconstructions of *Allosaurus* (left), a theropod saurischian about 10m long, and *Stegosaurus* (right), the type genus of one of the four ornithischian suborders and reaching a length of at least 6m. The two dinosaur groups are distinguished from each other by a major difference in the pelvic girdle structure.

like) whereas in the Ornithischia it runs backwards and downwards alongside the posterior ischium (bird-like). All the dinosaurs were terrestrial and had a worldwide distribution, but by no means were they all large, some being no bigger than a chicken. One feature common to all was a great improvement in their locomotory ability. The hind legs were very large and the gait was upright rather like that of a mammal. Two groups adopted a permanently bipedal gait, so freeing the forelimbs for such functions as food capture above ground level.

Much recent debate has centred on the question of whether dinosaurs were warm-blooded like mammals and birds and there is much evidence—for example the existence of dinosaurs in areas well outside the tropics—to support this idea. As a whole, the dinosaurs adaptively radiated throughout the Jurassic and Cretaceous periods and came to occupy most of the niches now occupied by the modern mammals. There was a great variety of herbivores and carnivores of differing types and sizes and presumably food preferences.

The Saurischia include the Sauropoda (popularly called brontosaurs) which were gigantic quadrupedal herbivores. They were probably swamp-dwellers and could reach an estimated weight of 30 metric tons. Related to them were the Theropoda which in contrast were all bipedal carnivores. Some like *Coelurus* (about 30 cm long) and the very ostrich-like *Ornithomimus*, were of distinctly modest size. Other theropods were the most dramatic carnivores of all time—such as *Tyrannosaurus*, some 5.4 m tall, with enormous dagger-like teeth but very reduced forelimbs.

All the Ornithischia were herbivorous and had a variety of specialized, often leaf-like teeth for cutting up vegetation. The Ornithopoda were generally the most primitive, even though they were the only bipedal members. *Heterodontosaurus* from the middle Triassic of Africa is the earliest known dinosaur and was a relatively small, exceedingly agile creature. Related to it are *Hypsilophodon*, a tree dwelling type from the Jurassic, and *Iguanodon*, which was almost as large as *Tyrannosaurus* and was probably a browser on high plants—something like the giraffe of today. The hadrosaurs or duck-billed dinosaurs had a bird-like beak and were amphibious in habit. Bizarre crests surmounted the head, perhaps as a means of species recognition.

The origin of the ichthyosaurs is not known, for even at their very earliest appearance in the Triassic, they already possessed the highly modified features of fully marine animals. As can be seen in this reconstruction, the likeness to dolphins is obvious, and no doubt they led a similar kind of fish-eating existence.

All the remaining ornithischians were quadrupedal and occupied the same niche as the herd-dwelling ungulates among modern mammals. They lived together in large numbers and browsed or grazed over huge areas of plain. Because of their vulnerability to predation, they evolved a variety of protective devices. The Stegosauria or plated dinosaurs had large vertical plates of bone set in a row along the middle of the back and the Ankylosauria were built like tanks with thick interlocking scales covering the body. The last of these groups to evolve, the Ceratopsia, or horned dinosaurs, had up to three rhinoceros-like horns on their snouts, and also a massive bony frill extending from the back of the skull over the vulnerable neck region. Despite their great success and diversity, the dinosaurs became extinct at the close of the Cretaceous period with what, in geological terms, was dramatic abruptness.

Two other groups of reptile, the Euryapsida and the Ichthyopterygia, both with the parapsid type of skull, evolved from the primitive stem reptiles, independently of one another and of all the other types. Both were fully adapted to a permanent marine existence although in quite different ways. The Euryapsida included particularly the plesiosaurs, which had both the fore and hind limbs modified

as flat paddles. Locomotion was rather turtle-like, with the paddles moving up and down. The tail, with its vertical fin-like end, was no more than a steering organ. The neck of many plesiosaurs achieved great length in association with the animal's fish-catching habit, and the teeth were long and sharp and set in elongated jaws.

The Ichthyopterygia or ichthyosaurs ('fish lizards'), in contrast, were the most completely water-adapted of all reptiles and had a close superficial resemblance to fish. The body was streamlined and fusiform (cigar-shaped) and locomotion was by means of a tail fin at the hind end which moved from side to side exactly as in typical fish. The limbs were reduced to no more than small steering organs. The jaws became extremely long and slender, and well endowed with sharp teeth suitable for preying upon fish. A remarkable feature of some specimens of fossil ichthyosaurs is the apparent presence of young individuals within the body cavity, indicating that the group was viviparous. Like the dinosaurs, the plesiosaurs and ichthyosaurs became extinct at the close of the Mesozoic era, although from time to time

The plesiosaurs were abundant in the shallow seas of the Jurassic and Cretaceous periods. They were fully aquatic but may have dragged themselves ashore to lay their eggs, as modern turtles do. While they were possibly not the most adept swimmers, their long necks and sharp teeth helped them snap at and grasp fish.

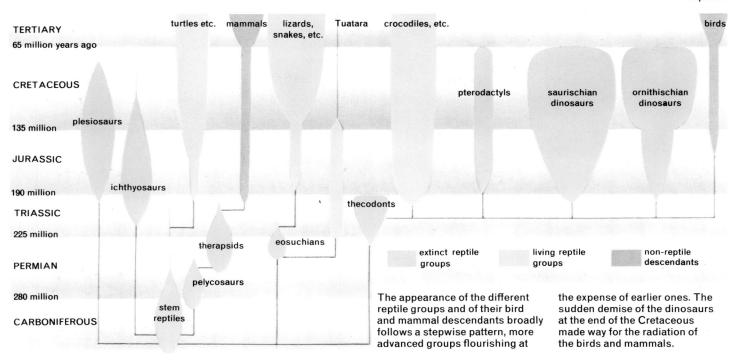

TERTIARY
65 million years ago

turtles etc. mammals lizards, Tuatara crocodiles, etc. birds
snakes, etc.

CRETACEOUS

pterodactyls saurischian ornithischian
dinosaurs dinosaurs

135 million plesiosaurs

JURASSIC

190 million ichthyosaurs

TRIASSIC thecodonts

225 million

therapsids eosuchians

PERMIAN

extinct reptile living reptile non-reptile
groups groups descendants

pelycosaurs

280 million

CARBONIFEROUS stem
reptiles

The appearance of the different reptile groups and of their bird and mammal descendants broadly follows a stepwise pattern, more advanced groups flourishing at the expense of earlier ones. The sudden demise of the dinosaurs at the end of the Cretaceous made way for the radiation of the birds and mammals.

the suggestion is made that the famous 'Loch Ness Monster' of Scotland, if it exists at all, may be a surviving representative of the plesiosaurs.

The fossil record of the reptiles is remarkably complete right from the first appearance to the present day, and so the general pattern of their evolution is clear, although a great many details remain to be described. This pattern consists essentially of a sequence of adaptive radiations, that is, evolutionary profusion of a basic type of animal into a number of slightly different types adapted to different ecological roles. The members of each new adaptive radiation are more biologically advanced than the members of the earlier radiation and replace them. This new radiation in its turn expands, flourishes, and then finally at least most of its members become extinct as yet another group takes over.

Thus, in the case of the reptiles, the stem reptiles of the late Carboniferous underwent a very modest radiation, but by the early Permian they had very largely become extinct and the dominant group were the primitive mammal-like reptiles. These, the pelycosaurs, produced a variety of types including herbivores, large carnivores, insectivores and fish-eaters. However, the pelycosaurs themselves became extinct during the middle of the Permian period and by late Permian times the therapsids or advanced mammal-like reptiles were flourishing. The late Triassic saw the final demise of therapsids and the beginning of the great archosaur adaptive radiation which replaced them and which flourished the longest, right through the Jurassic and Cretaceous periods. However, the archosaurs too finally became extinct, leaving the way clear for the next radiation—that of birds and mammals of the Tertiary era—a radiation which is still going on today. Of course, this is a rather simplified view of reptile evolution and ignores many details such as the complexity of the individual radiations themselves, and the ability of some animals, like the crocodiles, to survive long after the extinction of all their relatives.

Palaeontologists are still far from providing adequate explanations of this pattern. In general it seems that the end of a phase of adaptive radiation is related to some profound alteration to the habitat such as a change in climate. Most of the members of the radiation have evolved over the millenia to become highly specialized, each for one particular ecological niche, and requiring one particular set of environmental characteristics. They are incapable of surviving under the new conditions. However, one, or perhaps a few members of that radiation had remained more generalized and most adaptable of all enabling them to survive the change.

The most dramatic of all the faunal changes during the course of reptile evolution was undoubtedly the extinction of the dinosaurs at the end of the Cretaceous period. Geologically speaking this was an abrupt event and many explanations for it have been proposed. The most likely cause is related to the fact that the supercontinents of the Mesozoic had recently broken up and the land masses recognizable today were beginning to approach their present positions. Undoubtedly this major event must have had enormous effects on the climate of the late Cretaceous, leading probably to cooler, more unstable weather patterns over the whole world. The dinosaurs and other highly specialized reptile groups would not have been capable of withstanding such changes and so the prevailing ecological harmony would have rapidly broken down. Only the relatively small, unspecialized animals capable of protecting themselves from harsh conditions by hiding or burrowing, and relying for food on a variety of small invertebrates survived, to become in their turn the rulers of the Earth.

Excavating a fossil plesiosaur skeleton. The vertebral column, limb girdles and limbs of this 150 million-year-old reptile have been exposed by careful removal of the overlying loose rock. They will now be mapped, numbered and carefully removed.

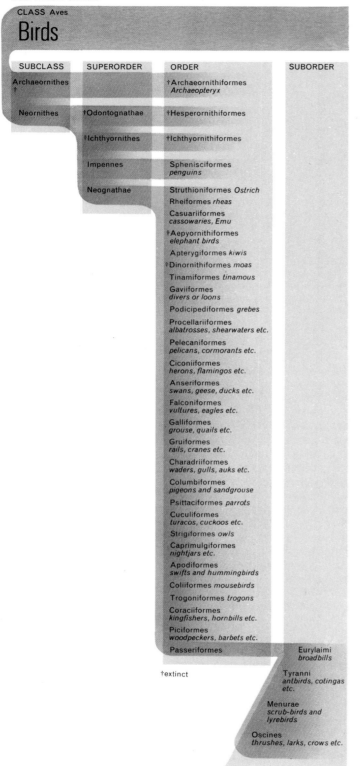

SUBKINGDOM Metazoa
PHYLUM Chordata
SUBPHYLUM Craniata

CLASS Aves
Birds

SUBCLASS	SUPERORDER	ORDER	SUBORDER
Archaeornithes †		†Archaeornithiformes *Archaeopteryx*	
Neornithes	†Odontognathae	†Hesperornithiformes	
	†Ichthyornithes	†Ichthyornithiformes	
	Impennes	Sphenisciformes *penguins*	
	Neognathae	Struthioniformes *Ostrich*	
		Rheiformes *rheas*	
		Casuariiformes *cassowaries, Emu*	
		†Aepyornithiformes *elephant birds*	
		Apterygiformes *kiwis*	
		†Dinornithiformes *moas*	
		Tinamiformes *tinamous*	
		Gaviiformes *divers or loons*	
		Podicipediformes *grebes*	
		Procellariiformes *albatrosses, shearwaters etc.*	
		Pelecaniformes *pelicans, cormorants etc.*	
		Ciconiiformes *herons, flamingos etc.*	
		Anseriformes *swans, geese, ducks etc.*	
		Falconiformes *vultures, eagles etc.*	
		Galliformes *grouse, quails etc.*	
		Gruiformes *rails, cranes etc.*	
		Charadriiformes *waders, gulls, auks etc.*	
		Columbiformes *pigeons and sandgrouse*	
		Psittaciformes *parrots*	
		Cuculiformes *turacos, cuckoos etc.*	
		Strigiformes *owls*	
		Caprimulgiformes *nightjars etc.*	
		Apodiformes *swifts and hummingbirds*	
		Coliiformes *mousebirds*	
		Trogoniformes *trogons*	
		Coraciiformes *kingfishers, hornbills etc.*	
		Piciformes *woodpeckers, barbets etc.*	
		Passeriformes	Eurylaimi *broadbills*
			Tyranni *antbirds, cotingas etc.*
			Menurae *scrub-birds and lyrebirds*
			Oscines *thrushes, larks, crows etc.*

†extinct

● NUMBER OF LIVING SPECIES: more than 8,600

● DIAGNOSTIC FEATURES: warm-blooded craniates with a body covering of feathers; typically with forelimbs modified as wings for active flight

● LIFESTYLE AND HABITAT: free living (most can fly) in all terrestrial and surface water habitats; some are capable of swimming underwater

● SIZE: 6 cm to 2.5 m but mostly between 15 and 50 cm long

● FOSSIL RECORD: Jurassic to Recent but scarce until the Tertiary; more than 500 species

Birds—class Aves—evolved from small, lizard-like, reptilian ancestors belonging to the subclass Archosauria. The precise line of descent is still a matter of dispute but the ancestral stock was probably contained within either of the orders Thecodontia or Saurischia; these were small, rather unspecialized carnivorous animals and possessed many teeth, The fossil record between these reptiles and true birds is, with one important exception, extremely scanty. The exception is an animal called *Archaeopteryx lithographica*, a creature about 60 cm long, first discovered in Bavaria in the 19th century. Almost certainly *Archaeopteryx* would have been classified as a reptile, but for the fact that this fossil 'lizard' had feathers. It is now realized that *Archaeopteryx* is an extremely important 'missing link'. It lived in the Jurassic period—about 150 million years ago. The next fossil birds known are from the Cretaceous, 30 million years later.

Archaeopteryx retained many features which are considered characteristic of reptiles rather than birds—for example, a long tail with free, unfused vertebrae, a reptilian type of articulation between the vertebrae, a short and poorly fused sacrum, teeth, claws on the fingers of the hands, unfused hand bones and a brain which was largely reptilian in character. Against this, *Archaeopteryx* also had many bird-like features that are not normally found in reptiles—for example, fused clavicles (the wishbone), fusion of the upper foot bones into an extra section of the limb, an opposable hind claw, an enlarged eye and associated parts of the brain, and, of course, feathers.

There has been much controversy about how *Archaeopteryx* lived. The dispute centres around whether it was a running or an arboreal animal, using its long clawed forelimbs to clamber up trees. The most recent view is that it could run. The precise function of the wings is not clear, but it seems certain that *Archaeopteryx* was not capable of continuous flapping flight; at best it may have been able to glide weakly. The elongation of the forelimbs is, on this theory, explained in terms of their use in grasping prey (probably insects). The feathers may have developed along the trailing edges of the forelimbs as a barrier—which would have helped prevent the prey escaping. It is anyway likely that the smaller body feathers evolved, not as requirements for flight, but as insulation. The ancestors of *Archaeopteryx* are now thought to have been warm-blooded and therefore would have

The hard, scaly legs of these young Mascarene Reef Herons (*Egretta dimorpha*) are a legacy from their reptile ancestors; the modification of body scales into an insulating layer of feathers, however, represents a very significant advance.

Archaeopteryx as it might have appeared as an arboreal predator; the feathered forelimbs may have been used to trap insects as well as for gliding. Its actual colour in life is, of course, unknown, and here it is shown in a cryptic plumage. Like birds today, either sex may have had bright feathers for attracting mates. Inset is a fossil imprint of the animal clearly showing feather details

required such an insulating layer; scales resembling simple feathers are found on some of the early Jurassic reptiles.

The wing feathers of *Archaeopteryx* are remarkable in that they do not only appear identical to, but are also arranged in an exactly similar pattern to those of modern birds; thus both have slightly different feathers (the primaries) attached to the hand from those (the secondaries) attached to the forearm. For this and other reasons palaeontologists believe that *Archaeopteryx* was either a direct ancestor of modern birds or at least very close to it, rather than on some side alley in evolution.

A further feature suggests that *Archaeopteryx* played a major part in the evolution of birds. The remains of five specimens of *Archaeopteryx* have been found in levels of different age—but all belonging to the Jurassic period—in slate quarries in Germany (a feather discovered in similar deposits in Spain may or may not have belonged to an *Archaeopteryx*). Hence these animals lived over the course of several million years. Therefore we believe that *Archaeopteryx* was successful over a long period, during which no other ancestors of birds have come to light.

Later fossils throw little light on the evolution of birds because they were obviously bird-like in almost all their features. The two earliest species are *Ichthyornis* (Ichthyornithiformes) and *Hesperornis* (Hesperornithiformes) from the early Cretaceous period. *Ichthyornis* was a small, tern-like seabird; it had a well-developed sternum for the attachment of flight muscles and so presumably was an able, flying bird. *Hesperornis* is of some interest in this respect; it was much larger, about 180 cm long; it had very weak shoulder girdles and no sternum, and probably lacked most of the distal part of the forelimbs. Hence it certainly could not fly. Nevertheless (as with many other flightless birds such as the Ostrich) it had some anatomical features characteristic of flying birds. One can therefore be reasonably certain that *Hesperornis* was the descendant of flying forms which had reverted to the flightless state. The legs and feet of *Hesperornis* were well adapted for swimming and so it was probably rather poor at moving on land as is the case with many of the most highly adapted aquatic birds today.

Thus between *Archaeopteryx* and the first Cretaceous birds a considerable amount of modification to the *Archaeopteryx* 'design' had taken place. By the late Cretaceous (about

70 million years ago) a number of other birds were to be found. Most of these seem to have been associated with water but this evidence may be misleading; creatures falling into water tend to fossilize better than those that die on the land. Examination of fossils from the beginning of the next geological period, the Tertiary, shows that by then a great radiation of birds had occurred. Fossils of birds that are the presumed forerunners of the present herons, pelicans, flamingos, geese, gamebirds, rails and vultures have all been found. Most of these fossils have been discovered in Europe and North America. However, with their powers of flight, birds presumably spread around the world at an early stage in their evolution. Indeed recent finds of fossil feathers near Melbourne in Australia are of early Cretaceous age and hence belonged to some of the very earliest known birds.

RIGHT: *Ichthyornis* is less well known from fossil specimens than *Hesperonis*; much of its skeletal structure can only be inferred. Toothed jaws found with the first specimens were identified later as belonging to aquatic reptiles that occurred in the same habitat, and whether or not *Ichthyornis* had teeth is still unknown. The colours shown here are speculative and are based on those typical of modern seabirds.

BELOW: *Hesperonis* was adapted for fast, submerged swimming. The body was streamlined and the large, webbed feet were situated at the hind end to provide maximum thrust. The wing was reduced to merely a thin humerus which may not have been visible externally.

Flight shapes the lives of most birds, allowing them to exploit scattered food sources, to avoid harsh winters by migrating and to escape from predators. However, flight is demanding, and both the structure and habits of birds must be modified accordingly.

We have seen that birds evolved from lizard-like ancestors. It is easiest to view the structure of a bird as a series of evolutionary adaptations to that lizard-like form, each of them contributing to the development of an efficient flying animal. Indeed the demanding nature of flight could be said to have had an overriding influence on almost all features of a bird. Flight requires that birds be compact and strong yet, above all, light in weight.

In comparison with the reptilian skeleton, that of birds shows two basic types of modification. First, the structure has been lightened in almost every way and, second, the weight that remains has been positioned as close as possible to the centre of gravity. The bones of birds are extremely light and this has been achieved by two means. The bones are mostly hollow (and the larger the bird, the more hollow bones there are) but in places of particular stress they may have small cross-struts inside. Further, many bones have been fused together and although this does not necessarily result directly in reduction of weight, fused bones do not need muscles and ligaments to hold them in position and the elimination of these leads to a reduction in weight.

The bones of the skull are greatly reduced, the jaws are lightened and the teeth lost. The tail is greatly reduced in length—all that is left is a small plate of bone, the pygostyle, to which the tail feathers are attached. Over the abdomen the number of vertebrae have been greatly reduced and many of the remaining ones, especially those around the pelvic girdle, have been fused. The limbs also have been lightened by the hollowing of bones and by reduction and fusion. Man has only one method of locomotion and has to support himself on his legs which are therefore positioned under the centre of gravity; however, birds have two entirely separate means of locomotion—walking (or swimming) and flying. In order to be able to perform both these functions efficiently, a bird needs to have both the wings and legs positioned close to the centre of gravity.

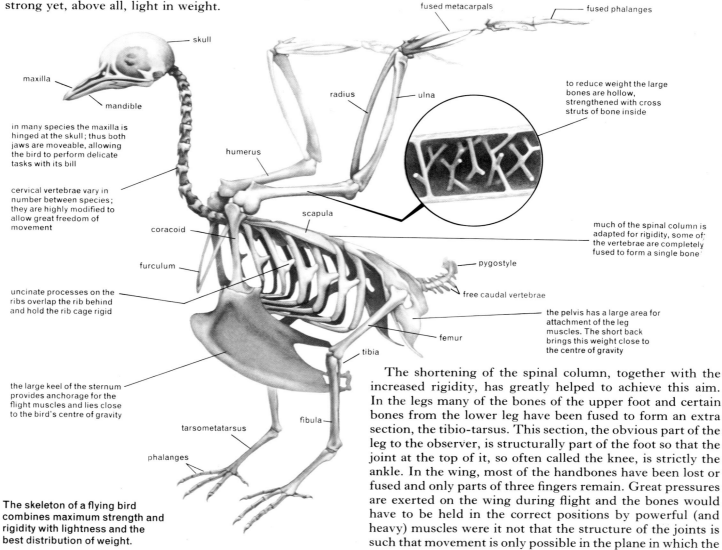

fused metacarpals

fused phalanges

skull

maxilla

mandible

radius

ulna

to reduce weight the large bones are hollow, strengthened with cross struts of bone inside

in many species the maxilla is hinged at the skull; thus both jaws are moveable, allowing the bird to perform delicate tasks with its bill

humerus

cervical vertebrae vary in number between species; they are highly modified to allow great freedom of movement

scapula

coracoid

furculum

much of the spinal column is adapted for rigidity, some of the vertebrae are completely fused to form a single bone

pygostyle

free caudal vertebrae

uncinate processes on the ribs overlap the rib behind and hold the rib cage rigid

femur

the pelvis has a large area for attachment of the leg muscles. The short back brings this weight close to the centre of gravity

tibia

the large keel of the sternum provides anchorage for the flight muscles and lies close to the bird's centre of gravity

fibula

tarsometatarsus

phalanges

The skeleton of a flying bird combines maximum strength and rigidity with lightness and the best distribution of weight.

The shortening of the spinal column, together with the increased rigidity, has greatly helped to achieve this aim. In the legs many of the bones of the upper foot and certain bones from the lower leg have been fused to form an extra section, the tibio-tarsus. This section, the obvious part of the leg to the observer, is structurally part of the foot so that the joint at the top of it, so often called the knee, is strictly the ankle. In the wing, most of the handbones have been lost or fused and only parts of three fingers remain. Great pressures are exerted on the wing during flight and the bones would have to be held in the correct positions by powerful (and heavy) muscles were it not that the structure of the joints is such that movement is only possible in the plane in which the wing is folded; the joints at the elbow and wrist do not allow the wing to bend in the plane of the wingbeat. Likewise, there is little room for many movements in the leg joints.

A further feature of the limbs of birds is that the sections farthest from the body have very little musculature. The major movements of the limbs are brought about by large muscles located at the bases of the limbs and attached by tendons to the bones. The largest muscles are, of course, those associated with flight. These muscles are, size for size, the largest found in the animal kingdom, sometimes reaching as much as 15 per cent of the whole weight of the bird. Once again they are positioned close to the bird's centre of gravity. Large muscles require large areas of bone for their attachment. The large keel to the sternum provides the main site for these muscles and its absence in *Archaeopteryx* is a fairly certain sign that the animal was not capable of powerful flight. Large coracoid bones hold the sternum and the shoulder girdle (to which the wing articulates) apart. Without these the body of the bird would collapse under the strength of the contraction of the flight muscles. There are two pairs of flight muscles, the larger, the pectoralis major, is responsible for the powerful downbeat of the wing and the smaller, the pectoralis minor or supracoracoideus, raises the wing in preparation for the next downbeat.

The respiratory system of birds is designed to produce a plentiful supply of oxygen during the great activity of flight. The lungs themselves are rather small compared with those of mammals, but they are connected to air sacs within the

main sense. Their eyes, conforming with this, are unusually large, compared with those of other animals. Unable to reduce the size of the eye, birds have almost eliminated the muscles which, in mammals, move the eye in its socket, and depend instead on a very wide field of vision and movements of the whole head.

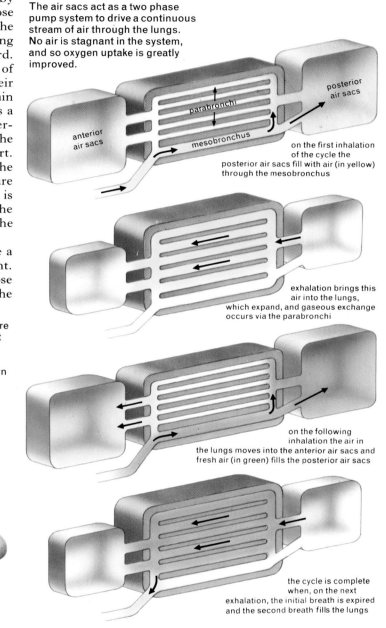

The air sacs act as a two phase pump system to drive a continuous stream of air through the lungs. No air is stagnant in the system, and so oxygen uptake is greatly improved.

anterior air sacs
parabronchi
mesobronchus
posterior air sacs

on the first inhalation of the cycle the posterior air sacs fill with air (in yellow) through the mesobronchus

exhalation brings this air into the lungs, which expand, and gaseous exchange occurs via the parabronchi

on the following inhalation the air in the lungs moves into the anterior air sacs and fresh air (in green) fills the posterior air sacs

the cycle is complete when, on the next exhalation, the initial breath is expired and the second breath fills the lungs

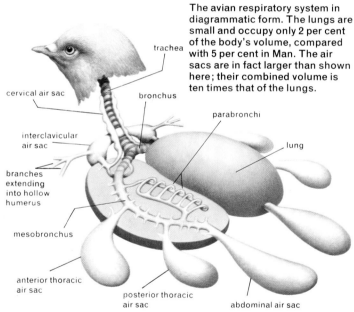

The avian respiratory system in diagrammatic form. The lungs are small and occupy only 2 per cent of the body's volume, compared with 5 per cent in Man. The air sacs are in fact larger than shown here; their combined volume is ten times that of the lungs.

trachea
cervical air sac
bronchus
interclavicular air sac
parabronchi
lung
branches extending into hollow humerus
mesobronchus
anterior thoracic air sac
posterior thoracic air sac
abdominal air sac

rest of the abdomen (and these have small extensions entering the hollow bones). Although the air sacs do not function as lungs they enable the inhaled air to pass straight through the lungs with the result that, in contrast to a mammal where the lungs are cul-de-sacs and only a small proportion of air is changed at each breath, in birds all the air in the lungs is changed and thus a very rapid uptake of oxygen is possible.

The lack of teeth means that food must be ground up elsewhere; this is done in the muscular gizzard situated near to the centre of gravity. Even the breeding biology of birds can be related to the demands of flight. Birds cannot carry large numbers of young with them in flight, rather they must form one egg at a time and then deposit it safely in a nest before forming the next one.

The eyes of birds deserve special mention. A fast moving animal requires exceptionally good information about its surroundings and birds depend heavily upon vision as their

The eyes of an owl are enormous, filling a great proportion of the head, and they enable the bird to see detail even in conditions of very low light. Acute vision is particularly important to fast moving animals, and even non-predatory birds have large eyes.

more rapidly over the upper surface than the lower surface of the wing, producing reduced pressures on the upper surface and hence lift. The amount of lift provided by a wing is related to (1) the speed at which the wing is travelling through the air, (2) the surface area of the wing and (3) the angle of attack (the angle at which a line from the front to the back of the open wing makes to the horizontal). An increase

1.
the aerofoil section of a wing produces lift by constricting the airflow over its upper surface; this air moves faster than the air below causing a decrease in air pressure above the wing; it is this pressure differential that generates the lift

2.
at steeper angles of attack, air passing over the wing's upper surface must travel slightly further in the same time and so moves faster still; the pressure differential increases and more lift is generated

3.
at angles of attack over about 15 the airstream begins to break away from the wing's upper surface; the region of rapid airflow becomes very restricted and very little lift is generated; in addition turbulent eddies form, producing extra drag, and the wing stalls

Cross sections of a bird's wing showing how increasing angles of attack affect the amount of lift that is generated.

4.
at steep angles of attack a smooth upper airstream can be restored, and stalling averted, by raising the alula – the three feathers on the bird's 'thumb'; air passing through the resulting slot speeds up and follows the upper surface once more; spreading the primary feathers produces a series of slots which have a similar but more complex effect

A White-backed vulture (*Gyps africanus*) coming in to land. It has lowered its air speed by excessively increasing the angle of attack of its wings and now, to prevent stalling, has separated its primary feathers and raised its feathered alula. Its legs are extended and bent ready to absorb the shock of landing.

Flight is one of the more striking features of birds and undoubtedly has led to their success as a group; their ability to make use of parts of the world which are only habitable for short periods of the year has enabled them to exploit large areas only sparsely occupied by mammals.

The most striking modification to the lizard-like ancestral forms is, of course, in the forelimbs and their associated muscles. The wings provide both the forward propulsion and the lift which enables the birds to stay airborne. The tail, and sometimes the feet, are at times important accessory features, and may be useful in manoeuvring and landing—though witness the occasional sight of a tail-less swallow hunting insects with little visible reduction in expertise!

The special shape of a wing section (an aerofoil) is such that, when it is moved forwards through air, the air moves

in any of these three features leads to an increase in the amount of lift provided. However, if the angle of attack of the wing becomes too steep (about 15°) the airflow can no longer remain smooth, but becomes turbulent and lift is lost—we say that the wing stalls. The bird's body and wings provide a resistance to the air which is referred to as drag. In order to move forwards a bird must produce sufficient thrust to overcome this drag.

It is perhaps easiest to understand the basic principles of flight by looking first at a bird gliding in still air. If an airborne bird were to stop flapping there are two ways in which it could avoid dropping vertically to the ground. One way would be to glide steadily downwards; this could be done without loss of speed. In many ways this method is comparable with a toboggan sliding down a slope; energy provided by the pull of gravity is used to overcome drag. Alternatively the bird could try to continue to glide without loss of height but then its drag would begin to slow it down. Since lift is partly a function of air speed, the bird would lose lift. It could compensate for this by progressively increasing the angle of attack which provides more lift. Such a technique would be short-lived because soon the stalling angle would be reached and the bird could glide no more. Both these types of gliding are important in a bird's life. The first is used by many soaring birds which glide in upcurrents of air. Provided that the rate at which the air is rising is at least as fast as the rate at which the bird would sink downwards in still air, the bird can remain aloft. The comparison might be with a toboggan sliding down a hillside at the same rate as the hill was rising upwards; the toboggan would not lose height. The second type of gliding is used by most birds when they come into land; by slowly losing speed they can land with the least possible shock to their 'undercarriage'.

In powered, flapping flight, a certain amount of the lift is obtained by the fact that the inner part of the wing remains in the shape of an aerofoil, giving lift nearly all of the time. Extra lift and forward propulsion are achieved by the powered

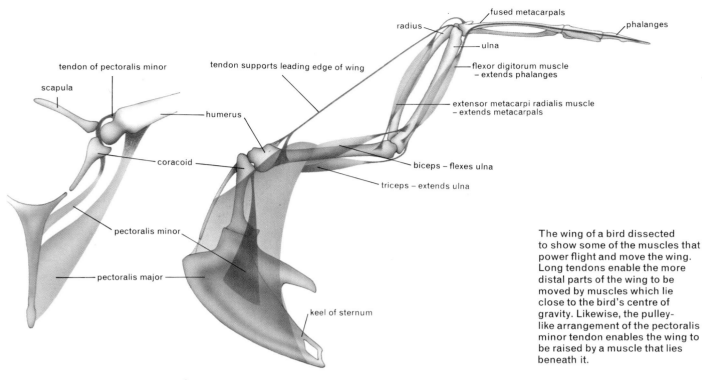

The wing of a bird dissected to show some of the muscles that power flight and move the wing. Long tendons enable the more distal parts of the wing to be moved by muscles which lie close to the bird's centre of gravity. Likewise, the pulley-like arrangement of the pectoralis minor tendon enables the wing to be raised by a muscle that lies beneath it.

downbeat of the wing, mainly the wing tip. It is for this great power that the enormous pectoralis major muscles are used. The recovery of the wing into the raised position ready for the next downwards beat is a relatively easy movement and hence the pectoralis minor muscles used for this are usually much smaller than their counterparts.

In actual flight, the bird moves forwards as it beats its wings so that by the end of the downstroke the wing tip is well ahead of the position where it started the stroke. This angled path of the wing through the air would produce aerodynamic problems for the bird if the wing was a rigid structure. However, the bones which support the outstretched wing are much closer to the leading edge than to the trailing edge and because of this, when the wing is flapped downwards, the unequal forces on the two sides of the bones cause the structure to twist so raising the trailing edge above the leading edge. In a similar way, the quill of the large flight feathers is near to the leading edge and they, too, twist on the downbeat so as to provide a greater forward propulsion.

Among birds there are a number of different types of wings adapted for different kinds of flight. The albatross, for example, has a long thin wing which is adapted for fast flying (having little drag). In contrast the broad wing of a vulture has a lot of drag and is not efficient for fast flying. However, the very large surface area of the vulture's wing gives it a lot of lift, enabling the bird to soar in very light upcurrents of air. The vulture could not glide across large distances of open sea as the albatross does, but on the other hand, it can remain aloft in light updraughts in which the albatross could not stay airborne at all. Short, rounded wings, such as those of sparrowhawks, are particularly useful for rapid manoeuvering round trees. Many aquatic birds beat their wings (in some cases partly folded) under water in order to be able to move more swiftly in pursuit of prey. Large wings cannot be used in water because it is much denser than air and thus waterbirds tend to have small wings. They therefore have to fly fast to get sufficient lift and are not very good at takeoffs, landings or quick manoeuvres.

LEFT: the thin wings of the Waved Albatross (*Diomedea irrorata*) are ideal for fast flight but give little lift. Lift is obtained by flying into the faster air stream just above the surface of the sea.

BELOW: the broad wings of this Turkey Vulture (*Cathartes aura*) catch every updraught, enabling it to soar all day long. The spread primary feathers reduce the eddies that produce drag at the wing tip.

There are 27 orders of living birds; another two orders have become extinct in the relatively recent past and the remaining ones are known only as fossils. It is conventional to consider the large number of birds (over 8,600 living species) as falling into two subdivisions, the perching birds or passerines, and the non-passerines. The perching birds are contained within a single order, the Passeriformes, which is a very large assemblage of, primarily, rather small birds. All the other orders belong to the non-passerine subdivision and are believed to be of more ancient origin.

In some classifications (as in the one used here) the penguins, order Sphenisciformes, are split off into a separate superorder—Impennes—but many taxonomists now believe that they are not so distantly related to certain orders in the Neognathae to warrant this division. The order is represented today by 15 species but some fossil forms are also known. One fossil species, *Anthropornis*, was much larger than any living penguin; it stood about 1.5 m high and probably weighed more than 100 kg. Penguins mostly occur in the colder waters of the southern hemisphere, though the Galapagos Penguin (*Spheniscus mendiculus*) is found on the equator.

The wings of penguins have evolved into flippers which enable the birds to swim powerfully. On land, however, they are not very agile though some may 'toboggan' swiftly on snow. Penguins spend most of their lives at sea and normally come ashore only to breed and moult. Their nesting sites are usually close to the sea though some birds may travel a kilometre or more inland. Depending on the species, penguins nest solitarily, in small groups or in enormous colonies of many thousands of pairs. The nest site varies from a burrow to a small pile of stones. In the two largest species, King Penguin (*Aptenodytes patagonica*) and Emperor Penguin (*A. forsteri*) there is no nest at all; the single egg is incubated on top of the parents' feet, covered by a fold of belly skin. The bird can only shuffle slowly while incubating. At a later stage, the parent carries the small chick in the same way. Some of the smaller species have a two-egg clutch and the Australian Little Blue, or Fairy Penguin (*Eudyptula minor*), may raise two broods in a year.

The Emperor Penguin breeds at high latitudes where the summer is too short for the birds to be able to lay an egg, incubate it and raise the young. To overcome this difficulty the females lay the eggs in mid-winter and leave the males to incubate them for a little over two months. The males

ABOVE: like many of the smaller penguins the Gentoo (*Pygoscelis papua*) nests solitarily in a burrow. Larger species lay only one egg but this one lays two; if food is scarce, however, just one chick may be reared. When at rest the bird lies on its belly, with its short legs and webbed feet folded behind.

BELOW: the larger penguins can tolerate lower temperatures and breed, usually in large colonies, at higher latitudes. These King Penguins, which are about to mate, lay their eggs in early spring. Like the Emperor Penguin, the young cannot complete their growth in the short summer and must attain adult size at sea.

come ashore with large fat reserves and weigh about 40 kg; during their enforced fast, however, they lose about half their weight, being around 20 kg when the chick hatches. The females return in the very early spring when the young are just hatching. The young penguins are still not fully grown when, just before winter, they leave the colony and start looking after themselves.

The superorder Neognathae includes a wide range of bird types, but from them it is possible to single out one group comprising the six orders of flightless birds. In the past some authors put most of these into a single order—the Ratites—but this assemblage is now thought to be artificial as some of its members seem not to be related closely enough; their common features of flightlessness and, typically, large size may be the result of convergent evolution of separate groups.

The first of these orders, the Struthioniformes, contains only one species, the Ostrich (*Struthio camelus*). There are several well-defined races, formerly thought to be separate species. Fossil specimens of ostriches, not necessarily identical to the present species, have been found in several places in the Palaearctic. Though now restricted to the African area, they obviously once had a more extensive distribution. More recently, they occurred in most of the semi-arid areas of Africa, though they are now extinct, or much reduced in numbers, in the northern parts of this range. A unique anatomical feature is that the birds have only two toes on each foot; one of these is much smaller than the other so that, when running, the Ostrich is effectively using only one toe. Ostriches are the largest living birds; adult males may be nearly 2.5 m in height and weigh up to 160 kg. Typically they live in parties containing a single male and about three to five females. Several females lay in a common nest and the full 'clutch' may contain as many as 50 eggs, though 20 to 30 is more usual. The male and one of the females are responsible for most of the incubation and care of the young.

The order Rheiformes contains only the rheas of which there are two living species, *Rhea americana* and *Pterocnemia pennata*. The rheas are the South American equivalent of the African Ostrich, but are considerably smaller, standing only about 1.5 m high and weighing up to 27 kg. Unlike the Ostrich, rheas have three well-developed toes. Rheas occur in the southern half of South America and inhabit the pampas grasslands especially those areas where bushes and trees also grow. They are largely vegetarians though they sometimes take insects such as grasshoppers. Certain features of their breeding behaviour are remarkably similar to those of the Ostrich. A single male defends an area and makes a scrape in the ground. In this five, six or even more, females lay their eggs. Thereafter, unlike hen Ostriches, the females take no part in the incubation or the raising of their young, leaving these duties entirely to the male.

ABOVE: the Rhea (*Rhea americana*) occurs in suitable grasslands from northern Brazil southwards. It runs swiftly and may stretch out its wings for balance. Outside the breeding season the birds gather into groups; there is safety in numbers as, at any one time, at least one bird is likely to be on the look out for danger.

BELOW: the two Ostrich sexes may easily be told apart. The male is larger than his mate and has black, not grey, plumage. If the bird gets too hot, it raises its wings, to uncover the bare thighs, allowing heat loss across their uninsulated surface. The bird's long neck and large eyes enable it to spot distant predators.

The Australasian equivalents of the Ostrich and the rheas are contained in the order Casuariiformes which has four living species. Amongst these the Emu (*Dromaius novaehollandiae*), which inhabits the open woodlands and grasslands of Australia, is the most similar to the Ostrich. The three species of cassowary are forest birds restricted to the northern parts of Queensland and New Guinea. The Emu used to be extremely common but was hunted by farmers who considered it a major pest because it tended to trample crops; despite this, however, it remains fairly common, but only in parts of its former range. With a height of about 1.8 m and a weight of as much as 50 kg, it is the second largest living bird. Some Emu populations are known to undertake migrations of considerable distance. As with the other birds so far mentioned, the male Emu takes a major share in the care of the eggs and young, though in this species only one female lays in the nest.

Cassowaries are very solidly built and have powerful feet with which they can swim strongly. They have a 'casque' (bony helmet) and their wing feathers are reduced to strong quills without any vaning. Both these characters are said to be adaptations enabling the birds to run swiftly through the undergrowth: the casque protects the head of the bird and the strong quills the flanks. Cassowaries feed primarily on seeds and berries but have also been recorded taking insects. Because of the dense nature of the habitat in which they live, close studies of their lives have not been made, but they do not seem to have communal nests; the clutch is usually only three to six eggs. Again the male is responsible for much of the care of the eggs.

There were several species of elephant bird, belonging to the order Aepyornithiformes, but all are now extinct. They were confined to Madagascar and it seems likely that they survived there until historic times; it is thought by some that the legends of the giant birds known as 'rocs', may have been based on memories or reports of these birds. Fragments

of egg shell can still be discovered. The largest of the elephant birds were rather similar to an Ostrich, but more heavily built. The largest of them all, *Aepyornis titan*, stood about 3 m tall and weighed close to 450 kg. It was the largest bird ever known. Its eggs were up to 33 × 24 cm and would have held a little over 9 litres of liquid.

: like the Ostrich and rheas, the Casuariiformes are adapted for running, having well developed legs and small wings. The forest-dwelling Australian Cassowary (*Casuarius casuarius*) shown here, is more compact than the Emu of the plains. It can pick its way through quite dense undergrowth, protected by the large, horny casque on its head.

ABOVE: the kiwi *Apteryx australis* has very reduced wings and runs swiftly through the forest on sturdy legs. It uses its feet for digging and, if molested, can inflict a serious wound with its claws. This species inhabits both islands of New Zealand but it has a more restricted distribution in the southern part of the country.

RIGHT: the Elegant Crested Tinamou (*Eudromia elegans*) lives in dry South American grasslands. It is well camouflaged and will sit tightly until very closely approached. It is about the same size as a domestic hen.

their prey. Kiwis are probably monogamous; they nest in burrows in the ground or in hollows in banks. Most females lay a single egg but some have a clutch of two.

The moas, order Dinornithiformes, are now extinct but were also restricted to New Zealand. There were probably about a dozen species, but some authors have put the total as high as 26. Moas were grazing and seed-eating birds. Some were the size of chickens but others were very large; one, *Dinornis maximus*, at a height of 3 m or more, may have been taller than the largest elephant bird, though it was not so heavily built. Moas were evidently very common birds in New Zealand and some species were hunted by the Maoris until comparatively recent times. It is even possible that one of the smaller species, about the size of a turkey, was still alive at the time of the first visits by Europeans; indeed a record of a flightless bird seen by a child around 1880 sounds strikingly like a moa.

Though they are no longer thought to constitute a homogeneous taxonomic group, these six flightless orders of the Neognathae are still commonly called the ratites—a term which refers to the keel-less sternum. In contrast, all other living birds typically have keeled sterna and are referred to as carinates. It is to the keel of the sternum that the large flight muscles are attached, and one might expect this structure to disappear during the course of evolution in those birds which abandoned flight. Undoubtedly the ratites are descendants of flying birds as many features of their skeleton can only be explained as having once been adaptations for flight.

The order Tinamiformes contains flying birds—tinamous—which are found in Central and South America.

Kiwis (*Apteryx*) belong to the order Apterygiformes. Their relationship with the other flightless birds is very uncertain. There are three species and all are confined to New Zealand. Kiwis are much smaller than the birds discussed previously, the largest weigh a little under 4.5 kg and many are less than half this weight. They are nocturnal forest dwellers and hence their behaviour is poorly known. They collect animal prey, much of it worms, from the forest floor using the long curved beak for probing into the ground. They have poor eyesight but, uniquely among birds, the nostrils are close to the tip of the bill; it is thought that they have a relatively good sense of smell and this may help to locate

Although they bear a superficial resemblance to a partridge-like gamebird, tinamous are probably most closely related to the rheas. Taxonomists consider them to be very primitive birds, perhaps close to the stock from which most modern birds evolved. Normally they spend most of their time on the ground, although they fly well if pressed. They occur in a wide range of habitats from rain forest to fairly open scrubland. All the 40 or so species are primarily vegetarian though they may also take insects. Some tinamous are polygamous. Several females may lay in the same nest and then, leaving the male to incubate the eggs and raise the young, go on to lay in the nests of other males.

The divers, order Gaviiformes, are a group of four species which are found in the cooler parts of the northern hemisphere; most of them breed in Arctic or temperate regions and move further south for the winter. Although they spend their winter on the seas and some even collect their food there in summer, most divers nest on islands in freshwater. Divers are highly specialized aquatic birds, living on a diet of fish. They have small wings and need open water for their long takeoffs and landings. Also, like many of the best swimmers, their legs are positioned too far back on the body for them to walk efficiently. In fact divers spend almost all their life in the water and only come ashore to nest; even then, they stay close to the water's edge. Divers barely make a nest but use just a small depression in the ground in which to lay usually two eggs. The chicks hatch in about 30 days and take to the water almost immediately though for the first few days they may also be carried on their parents' backs.

The grebes, order Podicipediformes, bear a superficial resemblance to small divers, but their true relationships with other birds are uncertain. In contrast to the webbed feet of the divers, those of grebes have lobes on the toes. They are expert swimmers, but move poorly on land and therefore rarely come ashore. There are about 20 species with a very wide distribution, though they are not found in Antarctica or the high Arctic. Several species are confined to small areas; for instance, the Short-winged Grebe (*Centropelma micropterum*) is found only on Lake Titicaca and adjacent lakes on the border between Peru and Bolivia. This species is the only flightless grebe.

During the breeding season grebes are found mainly on still freshwater (lakes and pools) but some move to the sea outside the breeding season. The nests are built of floating,

ABOVE: the Black-throated Diver (*Gavia arctica*) is typical of its order being well adapted to life on the water. Its head and body are perfectly streamlined and the sharp, powerful bill is ideal for catching fish. Its legs, however, are almost useless on land and it must nest on islands to avoid predators.

BELOW: like the divers, grebes often escape from antagonists by expelling air from the body and feathers, and sinking rapidly below the surface. This Western Grebe has typically striking plumage and is one of the larger species, measuring about 40 cm. It is restricted in its distribution to North America.

aquatic vegetation, anchored to other plants, and the clutch is of two to eight eggs. The young leave the nest as soon as they are hatched and are often carried on their parents' backs. Most species nest solitarily, but the Western Grebe (*Aechmophorus occidentalis*) of the United States nests in colonies which are sometimes of considerable size.

The Procellariiformes is a large order containing four different families: the albatrosses (Diomedeidae, 12 species), the shearwaters, fulmars and petrels (Procellariidae, 47 species), the storm petrels (Hydrobatidae, 18 species) and

ABOVE: the Gannet (*Sula bassana*) is a plunge diver, and is more heavily built than darters and cormorants that swim after fish. Gannets breed in only 34 colonies, all in the North Atlantic. The downy nestlings moult into a brown juvenile plumage. This is gradually lost during the four years they take to mature.

BELOW: the long legs and bills of the ciconiiforms enable them to feed in even quite deep water. Most breed colonially, often in mixed species colonies, each bird defending the area around its own nest. This Glossy Ibis (*Plegadis falcinellus*) feeds on molluscs and other invertebrates that live in the mud.

the diving petrels (Pelecanoididae, 4 species). Most of the species occur in the southern hemisphere and all are seabirds feeding on fish, squid or crustaceans. They vary in size from the Wandering Albatross (*Diomedea exulans*) with a wingspan of some 3.5 m, to the smallest of all seabirds, the storm petrels, such as the British Storm Petrel (*Hydrobates pelagicus*) which is barely heavier than a sparrow.

All procellariiforms come to land only to nest, spending the rest of the year at sea. The albatrosses, although they can walk quite well, need a long run to become airborne and most of the shearwaters walk poorly. Many nest on offshore islands where they are relatively safe in the absence of mammalian predators. Most of the smaller species gain further protection by nesting underground, in burrows or cracks in the rocks, and coming to land only during the hours of darkness so making themselves less vulnerable to attack by predatory birds. All birds in this order lay only one, white egg which is relatively large in comparison with the size of the bird; the incubation and fledging periods are unusually long. Many species nest in very large colonies.

The order Pelecaniformes contains six families of medium- to large-sized fish-eating birds: the tropicbirds (Phaethontidae, 3 species in tropical oceans), the pelicans (Pelecanidae, 6 species, fresh and saltwater, widespread but absent from cooler areas), the gannets (Sulidae, 9 species, widespread in most seas), the cormorants (Phalacrocoracidae, 26 species, widespread in both fresh and saltwater), the snakebirds or darters (Anhingidae, 4 species, freshwater in warm areas of the world) and the frigatebirds (Fregatidae, 5 species, tropical seas).

All species in the Pelecaniformes have in common a 'hind' toe which is turned forwards and connected to the second digit by a web; there are thus three webs on each foot. They collect fish by a variety of means: the gannets and tropicbirds and to a lesser extent the pelicans, dive from a height, whereas the cormorants and darters chase their prey underwater. Cormorants and darters lack the normal waterproofing of the feathers and by becoming 'waterlogged' they can stay submerged more easily, though they must stand about and dry their plumage after fishing. The frigatebirds rarely or never settle on the water (their plumage is not well waterproofed), but obtain their food by piracy, chasing other seabirds until these drop their catch. Pelecaniforms nest in a variety of sites, often in very large colonies; they lay one to several eggs, depending on the species.

The order Ciconiiformes contains seven families, the larger, better known ones being the herons, egrets and bitterns (Ardeidae, 63 species), storks (Ciconiidae, 17 species), spoonbills and ibises (Threskiornithidae, 28 species) and flamingos (Phoenicopteridae, 6 species). The order has a worldwide range but few species occur in the colder temperate or arctic regions. Most are medium- to large-sized birds with very long legs. Almost all species usually feed in water, the majority in freshwater. The food varies from fish and frogs in the case of many of the herons (crabs in some of the reef herons), to the large insects taken by the Cattle Egret (*Bubulcus ibis*). This bird follows cattle and other large animals and catches the insects as they move out of the way of the animals. Spoonbills take mainly small prey, including crustaceans, sweeping the bill through the water from side to side and snapping it shut as it comes in contact with the prey. The flamingos live in saline waters from which they filter the tiny organisms which constitute their food. The water is sucked into the mouth by the action of the tongue and the food is collected on stiff bristles on the mandibles. Although some of the cryptic species in this order, such as the bitterns, live and nest solitarily, others may nest in huge colonies. Most have clutches of two to five eggs, but flamingos usually lay only one.

The ducks, geese and swans are contained in the order Anseriformes. These well-known birds have a worldwide distribution and are mostly associated with water, either fresh or salt, though many of the geese may live largely on grasslands and only go to water to roost or to escape from enemies. There are almost 150 species. Many of them in the northern hemisphere migrate long distances to and from Arctic breeding grounds. They eat a wide variety of foods; most of the freshwater species are vegetarian but the marine species usually take animal food such as mussels and crabs. A few freshwater species, such as the Red-breasted Merganser (*Mergus serrator*) eat fish. The geese are almost exclusively grazing birds. Most species nest on the ground, though a few such as the Goldeneye (*Bucephala clangula*) use holes in trees. Many species have relatively large clutches, some of the smaller ducks laying as many as 10 eggs. The young hatch covered in down and can usually run or swim well within a few hours of hatching.

Also classified in the Anseriformes is the Anhimidae, a small family of three species, commonly called screamers. These are found in damp and marshy areas of South America. They bear a superficial resemblance to turkeys and are about

BELOW: unlike most eagles, which soar in search of food, the Little Eagle (*Hieraaetus morphonoides*) often hunts from a perch. Like other falconiforms, it feeds its young when they are small; later on it will leave the prey—small mammals such as rabbits—in the nest and the young will feed themselves from the carcasses.

RIGHT: young ducks and geese take to the water as soon as their feathers are dry. Although not fed by their parents, they are guided by them to good feeding areas, and are also protected from predators. Any of these Canada Goose (*Branta canadensis*) chicks that are left behind will have little chance of survival.

the same size. They differ from most ducks and geese in many ways, such as having barely any webs on their feet (they do not usually swim). They build nests on marshy ground and lay three to six white eggs.

The Falconiformes is a large order which contains over 270 species of mostly medium- to large-sized predatory or scavenging birds. All but two species are grouped in three families: the American vultures and condors (Cathartidae, 6 species), the hawks, buzzards, kites, eagles and Old World vultures (Accipitridae, 208 species) and the falcons (Falconidae, 58 species). This order has a worldwide distribution with the exception of Antarctica. Apart from the vultures, most species take live prey which varies in size from medium-sized animals such as small deer and monkeys, through bats and birds to insects. Many are specialized to take a particular type of prey. The Osprey (*Pandion haliaetus*, family Pandionidae) takes fish; the Secretarybird (*Sagittarius serpentarius*, family Sagittariidae), a particularly long-legged species, kills snakes and other reptiles; the Everglade Kite (*Rostrhamus sociabilis*) feeds almost exclusively on a large freshwater snail; and the Honey Buzzard (*Pernis apivorus*) feeds largely on the larvae of wasps and bees, hence its name. The vultures take carrion, though many of the Old World species, contrary to popular belief, usually feed only on freshly dead animals. The New World vultures can find carrion by smell (Old World vultures probably cannot), descending through thick forest to well-hidden corpses. Except for the large species which usually only lay one egg, most falconiforms have two or more. Nests are most frequently built of twigs and located on cliffs or in large trees but a few species nest on the ground.

The order Galliformes contains the 'gamebirds'. The 256 species are made up of the megapodes (Megapodidae, 10 species), curassows (Cracidae, 44 species), pheasants and quails (Phasianidae, 174 species), ptarmigans and grouse (Tetraonidae, 19 species), guineafowl (Numididae, 7 species) and turkeys (Meleagrididae, 2 species). The order has a worldwide distribution except for Antarctica. Some species, such as the ptarmigans, inhabit very cold areas which are snowbound during the winter; these birds have bristles on the sides of the toes which act as snowshoes and prevent them sinking into soft snow. Most galliforms are resident though the European Quail (*Coturnix coturnix*) migrates to Africa for the winter.

All gamebirds are primarily vegetarian, though the young of many species may feed on insects. The Capercaillie (*Tetrao urogallus*) of the Old World and the Blue Grouse (*Dendrogapus obscurus*) of the New World feed primarily on the buds and shoots of conifers—a seemingly rather indigestible food. Most lay their eggs in simple nests on the ground and have relatively large clutches. The megapodes or moundbuilders of Australasia do not incubate their eggs, but lay them in warm soil or rotting vegetation where the high temperature enables them to develop. On hatching, the young work their way to the surface by themselves. In at least one megapode species the Mallee Fowl (*Leipoa ocellata*) the young can fly on rudimentary wings within hours of hatching and fend entirely for themselves, there being no parental care at all.

The order Gruiformes contains about 180 species in no fewer than 12 separate families not all of which, at first sight, have much in common. Indeed taxonomists are by no means certain of the relationships of these families. Nine of them contain no more than three species each and of these many have very restricted ranges and are not well known. The three main families are cranes (Gruidae, 14 species), rails (Rallidae, some 130 species) and bustards (Otidae, 22 species). They are mostly ground living or aquatic and most nest on the ground. They eat a wide variety of foods from vegetable

ABOVE: this Helmeted Guineafowl (*Numida meleagris*) shows many of the features of its order. Habitually a ground dweller, it has strong legs and smallish wings used only for escaping predators and to enable it to roost in trees. This species is widespread in Africa and is the ancestor of domestic stocks of guineafowl.

BELOW: the cranes are a group of large terrestrial birds, whose diet consists mainly of small animals, caught with the aid of their long necks and sharp bills. They have long legs and live in open country. This species is the Blue Crane (*Anthropoides paradisea*) found in dry grassland areas in South Africa.

matter to small animals. The young of many species are able to leave the nest and move about within a few hours of hatching. Most rails lay large clutches, varying from six to twelve eggs, bustards lay from two to five and cranes normally only two. Rails have a very wide distribution and have reached many of the more isolated islands. Some, such as the Takahe (*Notornis mantelli*) of New Zealand, have reduced wings and cannot fly.

The Charadriiformes is another very diverse order containing 16 families and almost 300 species. Again, at first sight, the relationships between the families are not always obvious. The most important and best-known are the plovers and sandpipers (Charadriidae and Scolopacidae, 60 and 82 species respectively), the gulls and terns (Laridae about 80 species) and the auks (Alcidae, 19 species). Also included in the order are several minor families which include the lilytrotters, oystercatches, phalaropes, stonecurlews, sheathbills, skuas and skimmers.

There are no external characters that can be used to recognize this large and varied array of birds. However, most charadriiforms are medium- to smallish-sized birds (the extinct, flightless Great Auk (*Alca impennis*) being a notable exception), and are usually found near water, either fresh or salt. Most plovers nest away from water and many of the sandpipers which breed at high latitudes are also to be found some distance from water in the breeding season. Almost all nest on the ground but other sites include floating nests on water, in the case of lilytrotters, and cliff ledges in many of the gulls and auks; the Crab Plover (*Dromas ardeola*) nests in burrows in the sand and the Fairy Tern (*Gygis alba*) lays its egg on the branch of a tree. In most species of sandpipers and plovers the eggs are cryptically coloured and the camouflaged young leave the nest soon after hatching. In gulls and terns the young do not usually wander far from the nest. In most auks the chicks usually go to sea before they can fly properly, the interval usually varying from a day or two after hatching to perhaps three weeks, but in puffins the chicks stay longer and can fly when they leave the nest.

The order Columbiformes contains two families, the sandgrouse (Pteroclididae, 16 species) and the pigeons (Columbidae, about 285 species). Again the relationships between the families are not completely clear. Sandgrouse live in open or bushy country spending their time on the ground where they nest, laying from two to four but usually

ABOVE: the Herring Gull (*Larus argentatus*) is a typical member of the Laridae—perhaps the least specialized of the very diverse range of families that make up the Charadriiformes. It is a northern hemisphere species and occupies a wide range of inland and coastal habitats; it breeds in Europe and much of Canada.

BELOW: pigeons usually nest in trees, building a crude platform of twigs, like this Purple-crowned Pigeon (*Ptilinopus superbus*)—a fruit dove from Australia. The nests of some species are much larger. The family is distributed worldwide but is most strongly represented in the Australasian and Oriental regions.

three eggs. Many live in very arid areas and are noted for visiting waterholes in huge flocks; they may carry water to their young. Most pigeons live in wooded country and, although many do much of their feeding on the ground, they perch well and take many fruits and berries from the trees. They build a flimsy nest and lay one or two white eggs. Pigeons feed their young on 'milk', a fluid containing cells shed from the crop wall. The extinct Dodo (*Raphus cucullatus*) of Mauritius is believed to have been a huge flightless pigeon.

The well-known parrots and macaws are contained in the order Psittaciformes. The 320 or so species are widespread, particularly in the warmer areas of the world. They are primarily seed and fruit eaters, and many use their powerful bills to crack nuts; one example is the Palm Cockatoo (*Probosciger aterrimus*) from Northern Queensland and New Guinea which can open palm nuts. The Budgerigar (*Melopsittacus undulatus*) is a small parrot and is found in the more arid areas of Australia; in the wild it is always green. Most parrots nest in holes in trees, or in banks and one South American species, the Argentinian Green Parrot (*Myiopsitta monachus*) builds communal nests of twigs in trees. All parrots lay rather spherical white eggs. One species, the very rare Kakapo (*Strigops habroptilus*) from New Zealand, is virtually flightless.

The order Cuculiformes contains the African turacos (Musophagidae, 18 species) and the cuckoos (Cuculidae, about 125 species) which have a wide distribution. The turacos are frugivorous (fruit-eating) woodland birds which are sometimes known as plantain (=banana) eaters. Cuckoos are well-known from the habits of the European Cuckoo (*Cuculus canorus*) which lays its eggs in other birds' nests and leaves the foster parents to incubate and raise the cuckoo chick. However, many cuckoos raise their own young and are not parasitic. A few species such as the New World anis have communal nests. Recently it has been suggested by taxonomists that the Hoatzin (*Opisthocomus hoatzin*), once classified as a galliform, is in fact a gigantic cuckoo. This turkey-sized bird lives in lush vegetation along watersides in the northern half of South America where it feeds on leaves. The young, when disturbed, can clamber about in the bushes, using claws on their wings; if necessary they can drop from the bushes into the water and swim away from danger. The Hoatzin is put in a family by itself, the Opisthocomidae.

The 139 or so species of owl are grouped in the order Strigiformes. Owls are typically nocturnal, but some species such as the Short-eared Owl (*Asio flammeus*) and the magnificent arctic Snowy Owl (*Nyctea scandiaca*) hunt by day. Owls are renowned for good night vision, but it is perhaps their hearing which is their most exceptional asset; they can catch mice in total darkness, locating them by sound alone. There are seven species of fishing owls which have longish, unfeathered legs and catch fish in forest streams. Most owls nest in holes in trees or take over old nests of other birds but a few nest on the ground. They lay white eggs which, usually, are almost spherical. In years when there are large numbers of small mammals, some species such as the Short-eared Owl may raise five or six young, but one to three young would be more usual for most species.

The order Caprimulgiformes—nightjars, frogmouths and potoos—contains over 90 species, all of which are restricted to warmer areas, though some are summer visitors to temperate zones. Well camouflaged, they remain hidden by day and hunt insects and, in the case of the frogmouths, small animals in the twilight or the night. Most nightjars nest on the bare ground, where they lay one or two well-camouflaged eggs. The Australasian frogmouths nest on small twiggy platforms in trees and lay two to four eggs. The New World potoos lay their single egg on a stump of a tree and incubate it while sitting upright so as to look like a continuation of the stump. One very aberrant species, the Oilbird (*Steatornis caripensis*) which lives in the northern parts of South America, feeds on the large, aromatic fruits of palms and other forest trees, which it probably finds by smell. Oilbirds nest colonially in caves, finding their way in the darkness by echolocation; to do this they emit a series of clicks, audible to the human ear.

ABOVE: in addition to their well developed eye sight and hearing, owls have a silent flight and wide grasp to help them catch their prey. Most are 'sit and wait' predators, stooping either from a branch or from a slow glide if hunting in open country. This Malay Fish Owl (*Ketupa ketupu*) fishes in forest streams.

BELOW: most caprimulgiforms look very much alike. They are cryptically coloured to resemble bark and leaves and have wide mouths to snap up insects in flight. They see well at night and usually keep their sensitive eyes half closed in daylight. This Marbled Frogmouth (*Podargus ocellatus*) is from Australia.

ABOVE: swifts build their nests from airborne material, such as leaves and feathers, collected while the bird is on the wing. These are bound together with saliva to form the nest cup. This species is the House Swift (*Apus affinis*) from the New World.

BELOW: a Broad-billed Hummingbird (*Cynanthus latirostris*) feeding its young. The basic nectar diet of hummingbirds does not provide enough protein for growth and insects must be taken as well; this is particularly important for the growing nestlings.

the tiny feet. Swifts are entirely insectivorous and although they have a fairly worldwide distribution, they do not live in the colder areas in winter. They nest in a wide variety of sites, on rock faces, trees and buildings. The Palm Swift (*Cypsiurus parvus*) builds a tiny nest on a palm leaf and glues the egg to the nest with saliva to prevent it being shaken off. The cave swiftlets (*Collocalia*) nest in huge colonies in dark caves where they find their way about by echolocation. Their nests, built largely of saliva, form the basis of birds' nest soup. The hummingbirds are restricted to the New World and most are found in Central and South America. The family includes the smallest of birds—the Cuban Bee Hummingbird (*Mellisuga helenae*) which weighs only about 2.8 g. Because of their tiny size hummingbirds have high energy requirements and many become torpid at night in order to save energy. They are famed for their ability to hover and even fly backwards. Many hummingbirds live largely on a diet of nectar which they extract from flowers with their long tongues but small insects are also taken. They build tiny nests and lay one or two white eggs.

Mousebirds—sparrow-sized birds with long tails—are grouped in the order Coliiformes. The six species are found in open, wooded country in Africa where they live in small parties and roost communally. Their main diet is vegetable matter such as fruit and seeds.

The order Trogoniformes (trogons) contains about 35 species most of which are brilliantly coloured. They inhabit the tropical forests of South America, Africa and southeast Asia. Most have longish tails, but the brilliant green and red Quetzal (*Pharomachrus mocino*) from Central America has four greatly elongated upper tail coverts which extend as much as 60 cm behind the actual tail. Trogons have a habit of sitting quietly in the low branches of tall forest trees and making short sallies out after passing insects; they also eat fruits. They nest in holes in trees and lay two to four eggs.

The Coraciiformes is a diverse order of birds—about 190 species in 9 families, the most important being the kingfishers (Alcedinidae, 86 species), bee-eaters (Meropidae, 25 species), rollers (Coraciidae, 16 species) and the hornbills (Bucerotidae, 44 species). Several of the small families have very restricted ranges but the kingfishers are widely distributed and both the bee-eaters and the hornbills have a wide distribution in the warmer areas of the Old World. Most of the birds in these families have bright colouration, longish bills and tails (kingfishers have short tails), feed on animal

Two large families, the swifts (Apodidae, 65 species) and the hummingbirds (Trochilidae, about 320 species), are the main groups in the Apodiformes. Although in their own ways both groups are among the most specialized of flying birds, they do not at first sight appear to be very similar to one another. They have a number of skeletal features in common, however, especially those related to the wings and

prey (insects, small reptiles and fish) and nest in holes. Also, although there are some exceptions—notably kingfishers which live at the water's edge, most of the species are found in woodlands of the warmer areas of the world. For obvious reasons of food supply these birds do not spend the winter in very cold places. Some of the kingfishers, such as the Australian Kookaburra (*Dacelo gigas*) do not go near water and feed on terrestrial prey such as insects and lizards.

Some hornbills, such as the Silvery-cheeked Hornbill (*Bycariistes brevis*), have extraordinary nesting habits. The male seals the female into the nest hole by placing soft mud in the nest entrance and allowing it to harden. He is then wholly responsible for bringing food to his mate during the laying and incubation periods, feeding her through a tiny crack in the wall. After the young hatch, the male continues to feed the female and the small young until the young grow larger and need more food when the female breaks out to help in the foraging. At this stage young of some species may wall themselves in again with their excreta. This odd behaviour has presumably evolved to make the nest safe against raids from would-be predators such as monkeys.

The order Piciformes contains about 380 species in six families, the largest being the woodpeckers (Picidae, 209 species), the barbets (Capitonidae, 72 species) and the toucans (Rhamphastidae, 37 species). Most woodpeckers climb up the trunks of trees and extract grubs, largely from dead timber, with the help of their powerful beaks and exceptionally long tongues. The toucans are the New World equivalent of the Old World hornbills and are medium- to largish-sized tropical birds which eat a wide range of fruits and small animals. The Piciformes also includes the honeyguides (Indicatoridae, 14 species), some of which are renowned for their ability to guide certain larger animals,

This Red-bellied Woodpecker (*Centurus carolinus*) from the southern USA is demonstrating one of the characteristics of the group: the shafts of the tail feathers are specially stiffened to act as a prop for the body when the bird is climbing.

including man, to the nests of wild bees. After the nest has been broken into, the honeyguides collect their share of the spoils. They are also brood parasites, laying their eggs in the nest of other birds, such as barbets. When the young honeyguide hatches it has sharp hooks on its mandibles which it uses to kill the foster parents' own young; the hooks fall off within a few days. Most piciforms nest in holes.

The short, broad wings, long bill and hole nesting habit of the Malachite Kingfisher (*Corythornis cristata*) typify its order. This common African species dives on fish from a perch low over the water. The young birds are also reared on a fish diet.

Most species in the small suborder of broadbills are insectivorous but the Lesser Green Broadbill (*Calyptomena viridis*) from south-east Asia lives chiefly on a variety of fruits. It is the most widespread member of this rather restricted group.

Of the 8,600 or so species of living birds over 5,100 belong to the Passeriformes, an order which is usually divided into between 56 and 70 families. Not only are there a great many species, but also many of the species tend to be common in towns and gardens. Hence they are often the birds with which people are most familiar—sparrows, robins and starlings to name but a few. With such an enormous number of species it is hard to make generalizations about the order. Not surprisingly passerines have a very wide distribution; almost all possible areas, except Antarctica, have been exploited. Most passerines are small, few species being larger than the crows (Corvidae).

All passerines have an opposable hind toe (which cannot be brought forwards as in some other birds) and this is located at the same level in the foot as the other toes. For this reason, the passerines are well equipped for perching and are sometimes known as the perching birds—a somewhat misleading term when one thinks of the many other birds that also perch well. The only major habitat to which passerine birds have failed to adapt is water. Although many species

In the absence of many oscines, the Tyranni have diversified to fill most passerine niches in South America. This Vermillion Fly-catcher (*Pyrocephalus rubinus*: Tyrannidae) closely resembles an Old World flycatcher in both appearance and habits.

may collect their food on or near water, there are no truly aquatic passerines. Only the dippers (Cinclidae, 4 species) have evolved partially aquatic habits—they walk along the bottom of streams collecting insects and other small animals. No passerine has webbed feet. A few species, usually on out-lying islands, have become flightless. In spite of the enormous number of species—which might suggest very ancient origins—the passerines are considered to have evolved more recently than the other bird orders. Their fossil history is very poor, partly no doubt because of their small size.

This dauntingly large group of birds is normally divided into four suborders. The first, the Eurylaimi, contains only the broadbills (Eurylaimidae, 14 species), a group of brightly coloured tropical forest birds from Africa, and southeast Asia. The second, the Tyranni, contains about 1,060 species in 12 families, the major ones being the woodcreepers (Dendrocolaptidae, 48 species), ovenbirds (Furnariidae, 220 species), antbirds (Formicariidae, 230 species), cotingas (Cotingidae, 73 species), manakins (Pipridae, 56 species) and tyrant flycatchers (Tyrannidae, 374 species). This group of birds is firmly based in Central and South America, but a few representatives of some of the families (especially the tyrant flycatchers) occur in North America. Many Tyranni are forest dwellers, but some are found in other habitats.

The third suborder of passerines, Menurae, contains only two families and four species—the lyrebirds (Menuridae, 2 species) and the scrub-birds (Atrichornithidae, 2 species)—which are confined to Australia. The scrub-birds are small secretive birds which live in dense cover and about which relatively little is known. The lyrebirds are the largest passerines, having a body about the size of a chicken. They are noted for their magnificent voices and their superb mimicry of other birds and animals. Lyrebirds usually lay a single egg and incubate it mainly at night with the result that they have long incubation periods.

The Oscines, the largest suborder of the passerines, is the dominant group in all areas of the world outside Central and South America and is the one best known to most people. Depending on the classification used, there are between some 40 and 55 families in the group and a little over 4,000 species. The Oscines are known sometimes as 'the songbirds'. Certainly many of the most outstanding songsters belong to this suborder, but to imply that the other birds are not songbirds hardly does justice to the voices of some of them.

Some of the more important families that are included in the Oscines are larks (Alaudidae, 75 species), swallows (Hirundinidae, 79 species), bulbuls (Pyconotidae, 119

species of the 59 in the family which occurs outside the Americas, and most of the others are confined to South America. Passerines tend to make well-built nests and to have young which remain in the nest until they are ready to fly. In a few, such as some larks, the young may scatter from the nest before they are able to fly strongly. A small group of finches, the whydahs or widowbirds, are brood parasites laying their eggs in the nests of other finches.

The House Sparrow (*Passer domesticus*) is amongst the most successful of all the oscines. It adapts well to man's activities and has been introduced to many parts of the world. Its generic name forms the basis for that of the entire order of perching birds.

The Superb Lyrebird (*Menura novaehollandiae*) is the largest of the passerines, being about the size of a chicken. Its usual drab appearance dramatically alters when the bird displays; it fans its tail over its head with the two largest lyre-like feathers spread out sideways and the central ones shimmering in front.

species), thrushes, babblers, Old World warblers, tailorbirds, Old World flycatchers (all Muscicapidae, nearly 1,400 species), honeyeaters (Meliphagidae, 158 species), buntings, American sparrows, cardinals, tanagers (Emberizidae, just over 550 species), New World warblers (Parulidae, 113 species), finches (Fringillidae, 123 species), waxbills (Estrilidae, 107 species), weaver birds and sparrows (Ploceidae, 132 species), starlings (Sturnidae, 107 species) and crows (Corvidae, 102 species). This list, however, gives only a brief outline of the birds involved in this important group.

Like other passerines the Oscines have a worldwide distribution, although in South and Central America they are distinctly less important than the Tyranni. At least one sizeable family of Oscines, the Troglodytidae, must have evolved in South America: although the wren *Troglodytes troglodytes* is a well-known bird in Eurasia, it is the only

One of the characteristics of birds is that they lay eggs. These are immobile objects and, in many species, so are the young birds which hatch from them. Hence for much of the breeding cycle birds have to be relatively sedentary as they cannot wander far from, and must return to, the nest and young. Eggs and young birds are highly desirable food for many animals (including many other species of birds) and so the parents must endeavour to make the nest as undetectable or as inaccessible as possible.

A nest site has to be chosen. This may be an area in which the pair have lived in all the year round or, as in the case of many migrants, it may be a summer home alone. Often the male selects the site in which he displays and tries to attract a mate and from which he tries to exclude rivals. This territory may be a tiny space in a huge colony, as is the case with the Gannet (*Sula bassana*) or the Guillemot (*Uria aalge*) or it may be an area of many square kilometres as is the case with eagles. If the territory is large the birds will normally collect all their food within it whereas if it is small, as in the seabirds mentioned, the birds collect their food away from the territory. Once the bird has established a territory, it is likely to nest there for the rest of its life.

In most species a nest has to be built before the eggs can be laid. Some birds, however, do not make much of a nest; nightjars simply deposit their eggs on the ground and many of the waders make only a little scrape. In contrast some birds build very complex nests which take many days to complete —for example, some of the African weaver birds and the South American caciques (in the oscine family Icteridae) build nests of woven grasses which are hung from the branches of tall trees and may be as much as 180 cm long. In the case of the caciques, the nests are often placed near to a wasps' nest for greater safety; potential predators such as monkeys are likely to be frightened off by the wasps though, for some reason, the latter do not harm the birds. The Tailorbird (*Orthotomus sutorius*) from southeast Asia stitches together two large leaves and places the nest, hidden away, between them. The edges of the leaves are pierced and held together with knotted strips of grass. Still other birds construct elaborate nests of mud, allowing the mud to dry and harden. Such birds include many of the swallows, the Australian Magpie-lark (*Grallina cyanoleuca*—in the oscine family Grallinidae) and many of the South American ovenbirds which derive their names from the rock-hard structure of their nests. In some species the female alone builds the

The males of many species use displays to attract a mate. Few are more striking than that of the Greater Frigatebird (*Fregata minor*). With its wings spread and remarkable red throat pouch inflated, the male utters a loud, rattling call as a female passes over the colony. If unmated, she may land and pair with him.

nest, in others the pair cooperate; in still others, the male may build the nest alone, attract a female to it, mate with her and go straight on to build another nest in the hope of obtaining another mate. Examples of species with these polygamous males are found in several families, such as some of the wrens and some weaver birds.

Birds may nest solitarily or in colonies. Most of the small birds that depend on the concealment of their nest to escape predators nest solitarily—and take the greatest care that they are not seen approaching the nest—in order to avoid detection. Many large species which need to hunt undisturbed, such as the eagles and hawks, allow no others of their species in their territory and so they too nest solitarily. In contrast, many seabirds nest colonially; this may be partly because there is often little available space on safe offshore islands and so they must perforce nest close together. In at least some species, however, the birds gain safety in numbers by 'ganging up' to drive off an intruder. Chicken's eggs experimentally scattered on the ground inside a gull colony are less likely to be taken by predators than those similarly

The nest building ability of birds varies considerably; some make no nest at all, while others build most intricate structures. The weavers—this is an Orange Weaver (*Ploceus aurantius*)—are particularly skilful at building woven, suspended nests.

In the tropics, eggs in exposed nests, like those of the Chestnut-banded Sand Plover (*Charadrius pallidus*), may need shading from the sun to keep them at correct incubation temperature.

placed outside the colony; the gulls drive off the crows and other predators within the colony before they can collect the eggs. Some birds which feed on insects caught on the wing, such as swallows, swifts and some small falcons, also nest in large colonies, though the connection between this particular feeding habit and colonial nesting has not yet been adequately explained.

The female bird has to produce the egg or eggs. In many cases this imposes a considerable burden on her; she must eat more food than usual to meet the additional demands, and in some cases the male may help the female during the period of egg formation by bringing her extra food. Also, good supplies of calcium are needed to form the egg shells. Female birds may collect this in a variety of ways: some gather calcareous grit or snail shells, and in the Arctic, female Dunlins (*Calidris alpina*, family Scolopacidae) collect the bones of dead lemmings which they eat only during the short period in which they are forming eggs.

The number of eggs laid varies greatly from species to species but a few generalizations can be made. Within families, larger species tend to have smaller clutches than smaller species. Birds which nest in the tropics tend to have smaller clutches than those nesting in temperate areas or at high latitudes. Those seabirds that make long trips to collect

food for the young tend to have smaller clutches than those that feed near to the colony. Many of the larger birds of prey and large seabirds, such as albatrosses, have only one egg, as also do the petrels; pigeons and hummingbirds have only one or two and the plovers and sandpipers lay no more than four eggs which, being sharply pointed, fit neatly together under the incubating bird. Some of the ducks and many of the smaller gamebirds have quite large clutches—the Grey Partridge (*Perdix perdix*), for example, lays about 15 eggs. Within the passerines there is also a considerable range of clutch sizes. Birds that build 'conventional' cup-shaped nests, however, do not usually have more than about six eggs whereas many of the hole-nesting species have larger clutches. The Blue Tit (*Parus caeruleus*) which in oak woodland averages about 11 to 12 eggs, lays probably the largest clutch of any passerine.

Some birds breed colonially for a variety of reasons. The Greater Flamingos (*Phoenicopterus ruber*) rear their young in dense groups and thus they more easily defend them against predators.

ABOVE: the males of many species help their mates collect enough food to form the eggs. Here a male Common Tern (*Sterna hirundo*) brings a fish to his mate. The female will vary the size of clutch according to how much food the male brings.

BELOW: young Blackbirds (*Turdus merula*) are nidicolous and are fed by the parents until they are fully grown. Their wide mouths attract the attention of the female, and the largest chick with the longest neck is most likely to be fed first. Parental care continues after fledging.

As with nest building, the sharing of incubation varies greatly. In most species it is the female which undertakes the major part of the incubation but in some birds, the two sexes share the task more or less equally, each bird in its turn leaving the nest for several hours or even several days to collect food while the other incubates the clutch. The male's role varies with the species; he may bring food to the female, share the incubation duties, or leave his mate to look after the eggs entirely by herself. In the latter case the female may have to leave the eggs unattended while she goes to search for food. In some species, however, the brooding female lives off fat reserves laid down prior to the incubation period and sits on the eggs almost continuously until they hatch. In a few species, as in the rheas, tinamous and phalaropes, the roles are reversed and the male is left to do most or all of the incubating. In the case of some of the tinamous, the females may lay eggs for two or more males and leave them to do all the incubation and the rearing of the young.

Chicks are divisible into two kinds, nidicolous and nidifugous. Nidicolous (nest-dwelling) chicks are small, and when they hatch are often sparsely feathered or naked, usually blind and totally dependent on the parents for warmth and food; they are unable to leave the nest for many days or weeks. Most passerine chicks are nidicolous. In contrast,

LEFT: after hatching, the young of many birds grow very rapidly. This Starling chick (*Sturnus vulgaris*) hatched from its egg only 13 days previously yet is already approaching adult size and weight. In a further week it will have completed its feather growth and left the nest. Some seabirds, on the other hand, stay in the nest for several months before they reach the fledging stage.

BELOW: while all nidifugous chicks are well developed when they hatch, none are quite so precocious as some megapodes. This newly hatched Mallee Fowl chick (*Leipoa ocellata*) from Australia is completely feathered and will be able to fly within a few hours of emerging from its nest mound.

nidifugous (nest-fleeing) young are well feathered on hatching and have well developed legs and open eyes; they usually leave the nest within a few hours of leaving the egg. The domestic hen is a good example as are other gamebirds, and also ducks, plovers and sandpipers. In only a very few species of bird are the young able to fend for themselves from the moment of hatching; most still need to be guarded from predators, kept warm and either shown food or even fed. Not all young birds fall neatly into one or other of these categories—nestling gulls, for instance, are somewhat intermediate. Although young gulls can leave the nest soon after hatching, they do not wander far but wait there for the parent to bring them food.

The parents either forage close by the nest or, in the case of nidifugous species, take the young with them to good supplies of food. Some of the seabirds forage for food a long way from the nest and in consequence can feed the young only every few days. In some species, however, the feeding rate is many times an hour; a pair of Blue Tits may make over 10,000 visits to the nest during the 20 day nestling period, reaching 600 to 700 visits per day when the young are growing most rapidly. Feeding rates also vary slightly with the number of young. Parents with larger broods visit the nest more frequently than those with smaller broods though they do not seem able to increase the feeding rate sufficiently to compensate for the increase in brood size. Hence the individual young in large broods receive less food than young in smaller broods and leave the nest lighter in weight.

The young of small birds grow very rapidly; some reach adult weight at about 12 days and leave the nest soon afterwards. Most grow rather more slowly, and many of the large birds take many months to develop; the Wandering Albatross does not leave the nest for about 12 months after hatching. In some birds such as the swifts, the young birds can become torpid if the weather is unfavourable and food scarce; growth is arrested, but the birds may survive until the food supply improves. In the European Swift (*Apus apus*) the young may fly anywhere between five and seven weeks after hatching, the period depending on the weather and the insect supply during the period in question. Most birds cannot fly until they are more or less adult size and so young cannot easily escape predators once they are found. However, young gamebirds develop an intermediate set of feathers with which they can fly short distances when quite small; the young of some of the megapodes are able to fly within a few hours of hatching. Some of the young auks leave the cliff ledges where they were hatched when they are only partially grown. Although these chicks cannot fly properly the coverts of their secondary feathers develop precocially and, using these as tiny wings, they flutter from

the nesting ledge to the sea. They may manage to fly half a mile or so if the cliff is high enough.

Once the young have left the nest, they may be able to fend for themselves, as is the case for example with the swifts and certain seabirds such as gannets. Similarly shearwaters and puffins which leave their nesting colonies in the dark cannot hope to find their parents and must be able to fend for themselves. However, in many species the young are cared for after leaving the nest. During this early period young birds face a very difficult time in which they have to learn to fend for themselves. Clearly parental help at this stage greatly increases their chances of survival. The period of post-fledging care is very variable. In particular it may be short if the parents are going on to have another brood. Some of the larger birds may look after their young for a considerable period; young Tawny Owls (*Strix aluco*) leave the nest about May but their parents continue to look after them until about September. In the case of some of the very large birds of prey such as the Andean Condor (*Vultur gryphus*) the fledged chick is looked after for a year or more—so long, in fact, that the parents are not able to breed the following season.

Most small birds breed when they are one year old, but many of the larger species may not breed until they are several years old. Mute Swans (*Cygnus olor*) do not normally breed until they are about four years old, Herring Gulls (*Larus argentatus*) may take four or five years to reach maturity and other seabirds such as the Fulmar (*Fulmarus glacialis*) may not breed until the age of seven, eight or even nine.

SUBKINGDOM Metazoa
PHYLUM Chordata
SUBPHYLUM Craniata

CLASS Mammalia

Mammals

SUBCLASS

Prototheria

Metatheria (= Marsupialia) *marsupials*

Eutheria (= Placentalia) *placentals*

- NUMBER OF LIVING SPECIES: over 4,000.

- DIAGNOSTIC FEATURES: warm-blooded craniates with glandular skin which is typically hairy; young fed on milk secreted by mammary glands.

- LIFESTYLE AND HABITAT: mostly terrestrial in all habitats from forest to desert, but some fly and some inhabit freshwater or the sea.

- SIZE: 6 cm to 31 m but most are between 20 and 150 cm.

- FOSSIL RECORD: late Triassic to Recent but rare during Jurassic and Cretaceous; common in Tertiary (the 'Age of Mammals'); over 6,000 species.

LEFT: hair is one of the most obvious characteristics of the mammals. In this section of skin the long guard hairs and shorter, insulating under-fur can be seen. These may be raised by muscles, visible as pale oblique stripes. Under the skin is a layer of fat.

BELOW: in mammals the passages for air and food are separated by the palate but cross each other near the pharynx so that only one can be used at a time. Normally the airway is open but food may still be chewed in the mouth. During swallowing, food pushes up the palate and the epiglottis closes the trachea. The scroll-like turbinal bones carry a large area of sensory epithelium that heightens the sense of smell and warms incoming air.

Of all the classes of animals, the mammals are by far the most familiar. Indeed, many people mistakenly think that the word 'animal' refers only to mammals. We are of course mammals ourselves, and so are most of our domestic and farm animals. Familiarity may lead us to think of mammals as typical animals, but when we see them against a background of the animal kingdom as a whole we realise how extraordinary they are. The name of the class refers to one remarkable feature—the feeding of the young on milk produced by mammary glands that develop in the skin of the mother. The skin of mammals is provided with a variety of glands which perform many different functions. Thus most have sweat glands that assist in the control of body temperature; usually there are sebaceous glands that produce an oily secretion on the skin; and scent glands are important in many mammals for individual recognition, for marking territory, or for defence—as in skunks.

Skin glands nearly always develop in association with another mammalian characteristic—hair. A hair is made up of dead cells extruded from a tiny pocket, or follicle, in the skin. The cells are converted into keratin, a horny protein material which also forms the feathers of birds and the scales of reptiles. Sweat glands and sebaceous glands open into the sides of the follicles, and this is true of mammary glands too, for they develop in association with hairs in the embryo, and in the Prototheria milk hairs persist throughout life. Hair functions principally to maintain an insulating layer of air over the skin, thus reducing heat loss in these warm-blooded animals, but there are other subsidiary functions: specialized hairs form whiskers (vibrissae), supplied at their base with nerves so that they are sensitive to touch; other hairs form the eyelashes; the quills of porcupines are modified hairs; there are decorative hairs like the human beard; and the colours and coat patterns of mammals are mainly due to pigmentation of the hair.

Mammals maintain a relatively constant body temperature despite changes in the temperature of the environment, and so, unlike reptiles which become torpid at such times, they can be active at night or in the winter. Not all are equally efficient in this respect: some pass the winter in underground burrows or holes, often lined with bedding material, and some hibernators like hedgehogs become temporarily cold blooded. Nevertheless, when the hedgehog awakes from hibernation it quickly raises its temperature, producing heat

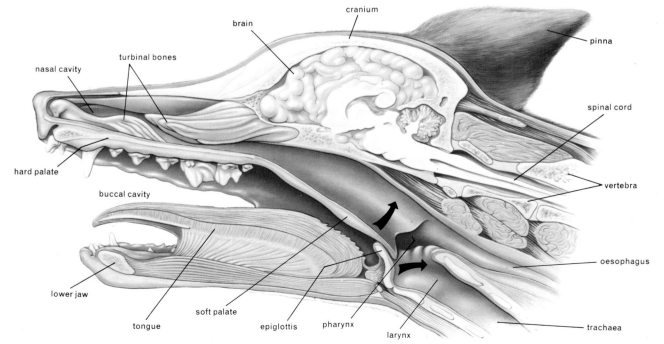

brain

cranium

pinna

turbinal bones

nasal cavity

spinal cord

hard palate

vertebra

buccal cavity

lower jaw

soft palate

tongue

epiglottis

pharynx

larynx

oesophagus

trachaea

THERAPSID REPTILE
Phthinosuchus

angular

dentary

MAMMAL
Canis

auditory bulla
(middle ear)

dentary

new joint between
dentary and cranium

quadrate

stapes

articular

ear drum

angular

ear drum

tympanic (modified angular)

malleus (modified articular)

incus (modified quadrate)

stapes

bulla (cutaway)

The articular and quadrate bones that formed the articulation of early reptile jaws also carried vibrations from the external ear drum through the stapes to the inner ear. In mammals, these functions are split: the dentary forms the whole lower jaw and the quadrate and articular have become efficient ear ossicles.

by the contraction of muscles. Whereas reptiles are ectothermic, depending upon external sources of heat, especially the sun, mammals, insulated by their fur, are endothermic and utilize internal heat: we find basking pleasant, but running is a quicker way to warm up.

Mammals are typically energetic, active creatures, and many features of their anatomy reflect this. For example, the diaphragm between the thorax and abdomen makes for more efficient breathing, and the heart is completely divided into two halves by a septum that prevents the mixing of arterial and venous blood. A high level of activity requires a greater food intake, which in its turn requires a more efficient digestive system. The mouth is different from that of reptiles in having muscular lips and cheeks (used by the young when suckling), and also in the specialization of the teeth. In reptiles all the teeth are alike, but in mammals there are incisors, canines, premolars and molars. The molars are especially complex in shape, with cusps that cut and grind the food, which is chewed before it is swallowed. To enable the animal to breathe while food is in the mouth, the air passage is separated from the food passage by a false palate and opens only at the back, opposite the windpipe; in most reptiles, air has to pass through the mouth to reach the lungs.

Another peculiar feature of mammals is the fact that the lower jaw is a single bone on each side, whereas in all other vertebrates it contains several bones. Of these only one, the dentary, bears the teeth, but it does not articulate with the skull: the jaw joint is formed between another lower jaw bone, the articular, and a skull bone, the quadrate. In mammals the dentary makes up the entire jaw and articulates with the skull directly, so the jaw joint of mammals is not homologous (that is, it does not have the same embryological origin) with that of other vertebrates. What is perhaps more remarkable is that the articular and the quadrate have been incorporated into the ear, where they form two of the three little bones (ossicles) that convey sound vibrations from the eardrum to the inner ear. The articular is the hammer (malleus) and the quadrate is the anvil (incus), while the

third bone of the chain, the stirrup (stapes) represents the sound-conducting bone of reptiles. Another bone of the lower jaw has become the tympanic, which supports the membrane of the eardrum.

Active animals need well developed sense organs. Mammals have good hearing: round the ear opening is an external ear (pinna), which in most species can be moved to catch sound. Most mammals are colourblind, but their eyes are sensitive to dim light, in accordance with their predominantly nocturnal habits. The sense of smell is usually very acute. The brain is larger and more complex than in reptiles, and mammals are more capable of learning, especially when young, so that their behaviour is less stereotyped and more adaptable. The dominant position of Man is founded upon his mammalian ancestry.

This Lion's head shows many of the features typical of mammals. The body is covered with hairs some of which are modified into eyelashes and whiskers. The nostrils lie close together, surrounded by bare skin. The external ears are large and mobile. The complex of muscles covering the face allows a variety of expressions which can help in the communication of emotions to other animals.

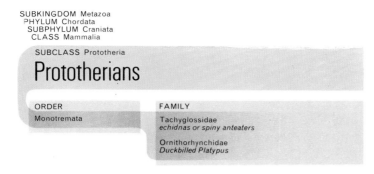

SUBKINGDOM Metazoa
PHYLUM Chordata
SUBPHYLUM Craniata
CLASS Mammalia

SUBCLASS Prototheria

Prototherians

ORDER	FAMILY
Monotremata	Tachyglossidae
echidnas or spiny anteaters

Ornithorhynchidae
Duckbilled Platypus |

- NUMBER OF LIVING SPECIES: 6

- DIAGNOSTIC FEATURES: mammals which lay eggs and have a cloaca and other reptile-like features.

- LIFESTYLE AND HABITAT: terrestrial and amphibious; restricted to Australia, Tasmania and New Guinea.

- SIZE: 65 to 80 cm.

- FOSSIL RECORD: sparse; Miocene to Recent; perhaps only 3 species; Fossil orders from late Triassic to Cretaceous may belong to the Prototheria but their position is uncertain.

The monotremes are of particular zoological interest because they are the most reptile-like of the mammals. Not only do they lay eggs, but in many anatomical details they seem to have remained at a stage through which, in the course of evolution, the other mammals have passed. Yet in most respects the monotremes are definitely mammalian: they have a glandular skin with hair, they produce milk, they have three ear ossicles, a diaphragm, a completely divided heart, and a brain which, though it possesses peculiar features, is of a mammalian type.

The two families represent adaptations to two very different modes of life. *Ornithorhynchus anatinus*, the only species of the family Ornithorhynchidae, is popularly called the Duckbill, Duckmole or Platypus (that is, flatfoot)—names which refer to some of its striking characteristics. It tunnels in the banks of rivers and obtains its food in the water. The broad, flat, hairless snout has a superficial resemblance to the bill of a duck, a resemblance which is increased by the presence on the lower jaw of numerous parallel ridges that are used to sieve the food (crustaceans, molluscs and insects) from water and mud. Unlike a duck's bill, however, the snout of *Ornithorhynchus* is not horny, but it is covered with

BELOW: the Duckbilled Platypus swims with its large, flat tail and sideways projecting, webbed feet. Under water it cannot hear or see but feels for its food using the sensitive skin of its 'bill'. It lives in freshwaters in southeast Australia.

RIGHT: the Australian Echidna is a land dweller. Its back is protected with spines, and its legs, which project at the sides of the body, end in large digging claws. The palate bears warts for grinding food, for, as in all adult monotremes, teeth are lacking.

soft skin that is highly sensitive to touch. Under water the eye and ear on each side are sealed within a common slit, which reopens on surfacing.

Young Platypuses have molar teeth, but they are replaced by horny plates in the adult. The tail is broad and flat like that of a beaver, and the feet are broad and webbed. On the forefeet the webbing extends out beyond the claws, but on land it is folded back and closed up like a fan, enabling the large claws to be used for burrowing. The body is covered with soft, close fur, as in moles, permitting backward and forward movement in the burrow with equal ease. In the breeding season the female makes a burrow containing a nesting chamber which she lines with wet plant material. She seals up the entrance, lays one to three eggs, and remains by them till they hatch in a week or ten days. A humid atmosphere is necessary, as the eggs easily lose water by evaporation. The young are hatched in a very undeveloped state, and their eyes do not open till they are 11 weeks old.

The Australian Echidna, or Spiny Anteater (*Tachyglossus aculeatus*), together with a closely related species in Tasmania and two species in New Guinea, belong to the family Tachyglossidae. In these the snout is tube-like, with a small mouth and nostrils at the end. Echidnas feed mainly on ants, which they pick up by protruding a long tongue, made sticky with saliva. Their powerful claws enable them to dig into anthills. They do not burrow, but escape from the sun's heat in holes dug out among rocks or tree roots. An echidna can also sink itself rapidly into soft ground by scraping away

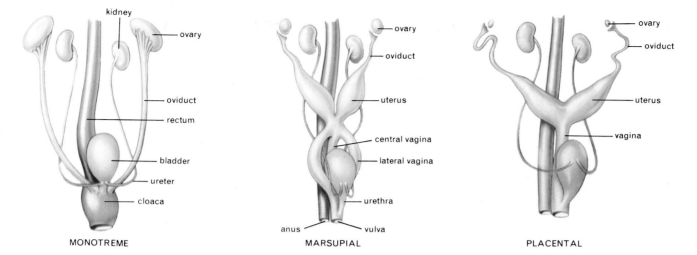

MONOTREME   MARSUPIAL   PLACENTAL

The female reproductive systems of the three mammal subclasses. In the monotremes, the oviducts, bladder and rectum all open into the cloaca, and the large ovaries yield big, yolky eggs. Marsupials have paired vaginae, and a central vagina or birth canal is formed temporarily at the time of birth. Placentals have a single vagina and, typically, a two horned uterus. They produce tiny eggs.

the earth beneath its body. Many of the hairs on the back are enlarged to form spines, moveable by muscles; normal hairs grow between the spines and on the lower surface of the body. An especially long claw on the hind foot is used to keep the spine-covered area clean. In the breeding season the female develops a shallow pocket in the surface of the abdomen, and by bending her body she lays a single egg into it. After hatching, the young remains in this 'pouch', sucking milk from tufts of hair at the base of which the mammary glands open, for there are no nipples. The mouth at this stage has not yet acquired a tubular form.

One of the archaic features of monotremes is their possession of a cloaca, a common chamber through which urine and faeces both pass, with a single opening to the exterior ('monotreme' means 'one opening'). The eggs also pass out the same way. In the male there is a penis, but this passes only sperm, not urine as in other mammals. The egg contains yolk for nourishment of the embryo, and it is surrounded by an albumen layer and a shell, laid down as it passes down the oviduct, just as in reptiles and birds. However, the egg is also supplied with nutritive material from glands in the oviduct, and this might be regarded as the beginning of the nourishment of the embryo in the uterus that is so important in more advanced mammals. The young monotreme breaks out of the shell by means of an egg tooth at the tip of its snout, like a hatching reptile or bird.

The fossil record shows that mammals were derived from a subclass of reptiles known as the Synapsida, which flourished during the Permian and Triassic periods. The oldest mammals date from the end of the Triassic, about 190 million years ago. The skeleton of one of these, *Erythrotherium*, has reptilian features found also in monotremes. During the ensuing Jurassic and Cretaceous periods, while the dinosaurs flourished, there evolved a variety of mammals, classified into several orders, but unfortunately for the most part known only from fossil teeth and jaws. They were small animals, the largest the size of a fox and most of them mouse-size or less. The monotremes are in all probability survivors from this stage of mammalian evolution, but their exact relationships are still uncertain.

The remaining modern mammals—in the subclasses Metatheria and Eutheria—are believed to go back to a common root early in the Cretaceous. They share some characters of the teeth, and differ from monotremes in details of the skull and ear; for this reason they are thought to be more closely related to each other than to the monotremes and are very often placed together as infraclasses in one subclass—the Theria.

The classification on page 254 is simplified to include only those groups with living representatives. This suffices for the discussion of mammals presented in this book but does not embrace the purely fossil groups. Inclusion of these demands a different classification. The Eutheria and Metatheria were comparative latecomers, evolving from the extinct order Pantotheria. Together with the Symmetrodonta, all these mammals form a single subclass, the Theria, in which the Eutheria and Metatheria are infraclasses. As fossil monotremes are almost unknown, their evolutionary history must be inferred from comparisons with other mammal groups. They show many reptile-like features and are placed in the separate subclass Prototheria. The extinct orders Triconodonta, Docodonta and Multituberculata resemble monotremes in some ways and also fall into this subclass, though their exact relationship is not known. Some authors consider that the multituberculates are distinct enough to be placed in a third subclass, the Allotheria.

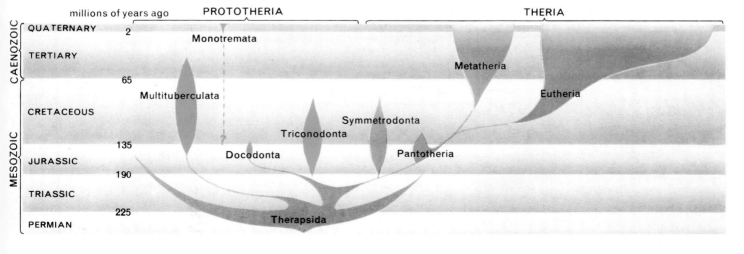

SUBKINGDOM Metazoa
PHYLUM Chordata
SUBPHYLUM Craniata
CLASS Mammalia

SUBCLASS Metatheria (= Marsupialia)

Marsupials

ORDER	FAMILY
Marsupicarnivora	Didelphidae *opossums*
	Dasyuridae *marsupial mice, Numbat etc.*
	Notoryctidae *Marsupial Mole*
Paucituberculata	Caenolestidae *opossum rats*
Peramelina	Peramelidae *bandicoots*
Diprotodontia	Phalangeridae *phalangers, possûms and Koala*
	Phascolomyidae *wombats*
	Macropodidae *kangaroos*

- NUMBER OF LIVING SPECIES: about 240.

- DIAGNOSTIC FEATURES: viviparous mammals; young born in a very undeveloped state and continue development attached to nipples, usually in a pouch; usually four molar teeth in each half jaw.

- LIFESTYLE AND HABITAT: mostly terrestrial or arboreal but a few are amphibious; restricted to Australasia and America.

- SIZE: 10 cm to 2 m, but most are 30 to 70 cm.

- FOSSIL RECORD: late Cretaceous to Recent; about 250 species.

The Metatheria are commonly known as marsupials because the female characteristically possesses a pouch (marsupium) into which the mammary glands open and in which the young spend the first part of their life. The pouch is a permanent structure, unlike that of echidnas which develops only in the breeding season. In many marsupials it opens backwards, and in some it is absent. The egg, when it leaves the ovary, contains much less yolk than the monotreme egg, and the embryo depends mainly on materials supplied by the mother while in the uterus. These materials consist partly of the secretions of uterine glands, absorbed through the surface of the embryo, but mainly they diffuse from the maternal blood into the blood of the embryo in a placenta, an area where the surface of the embryo is in close contact with the lining of the uterus.

The placenta of marsupials is simpler in construction than that of eutherians, and pregnancy lasts only a short time: from 12 days in opossums up to a maximum of six weeks in kangaroos. Consequently the young are born in a very tiny and undeveloped state. When she gives birth the mother leans back and the young climb through her fur into the pouch by their own efforts. Their hind legs are very poorly developed, and they climb by a sort of swimming action of the front legs. Their eyes are covered with skin, but they have a sense of smell. If they are lost the mother

Most marsupials spend their early life in a pouch of skin beneath the mother's body. The kangaroo bears one young at a time and keeps it in the pouch for several months. After a pregnancy of only six weeks, the new-born kangaroo is very tiny and undeveloped. It makes its own way into the pouch and attaches itself to a nipple. In the early stages the pouch is quite small but it enlarges as the young grows. In contrast with the long hind legs of the adult, those of new-born young are mere buds. When sufficiently developed the 'joey' makes temporary excursions from the pouch, but returns there to suckle until it is a year old.

makes no attempt to retrieve them. Unlike the monotremes, marsupials possess nipples. Having taken a nipple into its mouth, the pouch-young remains permanently attached there until its development is well advanced. The young, known as a 'joey', continues to be carried by the mother, either in the pouch or externally, clinging to a nipple or to her fur.

The process of birth differs from that of Eutheria, for there is no permanent birth canal (vagina). To get out of the uterus the tiny, naked young have to pass through the mother's tissue, which in preparation is loosened by the action of a hormone. That the ancestors of the marsupials laid eggs, like monotremes, is shown by the fact that a thin shell is formed round the egg as it passes down the oviduct on its way to the uterus, and in some marsupials there are the remains of an egg tooth. Male marsupials are similar to most eutherians in that the testes, during development, leave their original position near the kidneys and migrate to just below the skin of the abdomen, where they lie in a bag known as the scrotum. However, in marsupials the scrotum is situated forward from the penis, whereas in eutherians it is usually behind.

Most marsupials inhabit the Australian region, but some live in America, particularly South America. This raises an interesting problem of zoogeography. It was formerly thought that marsupials reached Australia from Asia, but no Asiatic fossil marsupials have been found to support this. According to current views on continental displacement, Australia was formerly far from Asia but it was joined to the Antarctic continent, so a possible route for the Australian marsupials was from South America through Antarctica.

All the Metatheria were formerly placed in one order, Marsupialia, but it is now thought that their diversity is better expressed by dividing them into four orders. Of these, the Marsupicarnivora is represented in both America and Australia, the Paucituberculata contains only a few shrew-like South American animals, and the Peramelina and the Diprotodontia are confined to the Australian region.

The American Marsupicarnivora are the opossums (Didelphidae). They are the oldest family of mammals living today: fossils show that they existed in North America as far back as the Cretaceous, over 70 million years ago. There are over 65 living species, mostly in South America, but the Virginian Opossum (*Didelphis marsupialis*) reaches as far north as Ontario. This species is about the size of a cat. It is well adapted for tree climbing: the five fingers of the hand are provided with claws; on the hind foot the first toe (corresponding to our big toe) is clawless, and it can be opposed to the four clawed toes when gripping a branch; hairless pads on the under surface of the hands and feet prevent friction; the structure of the limbs allows much rotation at the elbow, wrist and ankle; and, finally, the long, hairless tail can be coiled round branches.

The opossum is active only at night; by day it lies up in a hollow tree or other cavity where it makes a nest, using its prehensile tail to carry the nesting material. It is an omnivorous feeder, eating vegetable matter in addition to insects, eggs and carrion. Its brain is, for a mammal, small and very primitive. An unusual defence displayed by the opossum is 'playing possum', when the animal under stress collapses as if dead, with eyes closed and tongue hanging out. Among the South American opossums may be mentioned the mouse opossums (*Marmosa*), which are the size of mice and have no pouch. They feed on insects, birds' eggs and fruit, and sometimes they are accidentally imported to the United States in bunches of bananas. Most opossums are arboreal, but the Water Opossum (*Chironectes minimus*) is an aquatic animal resembling a small otter. When under water it closes the pouch by a sphincter muscle.

ABOVE: the Virginian Opossum is a New World marsupial which is well adapted for climbing. It grasps branches with its hands and feet, and clings with its claws and underlying friction pads. It also has a prehensile tail. Opossums are the most ancient living family of mammals and have very primitive brains.

BELOW: the Mouse Opossum (*Marmosa pusilla*) from Patagonia is one of the many species of opossum found in South America. It lives in trees, climbing with the help of its prehensile tail. Like other opossums, it is nocturnal and sleeps in a hole or deserted bird's nest. It feeds mostly on insects and fruits.

In Australia most Marsupicarnivora belong to the family Dasyuridae. They do not have prehensile feet and tails, and for the most part they live on the ground, though some can climb quite well in a squirrel-like or cat-like manner, using their claws. The smallest are the marsupial 'mice', the largest is (or was, as it is probably now extinct) the Tasmanian 'Wolf' (*Thylacinus cynocephalus*). Most dasyurids are active, rapidly moving carnivores, able to kill other small mammals and even birds by lightning rushes. The Tiger Cat (*Dasyurops maculatus*), or Dasyure, stalks prey like a cat. The Tasmanian Wolf, or Thylacine, wore its prey down by persistent pursuit. The Tasmanian Devil (*Sarcophilus harrisi*), a slower, thickset animal, is mainly a carrion feeder, with powerful jaws that can crush bones, though it will also kill snakes and poultry. The Numbat (*Myrmecobius fasciatus*) feeds on termites, digging into their nests with its front claws and picking up the insects with its very protrusible, sticky tongue.

The Marsupial Mole (*Notoryctes typhlops*), the only member of the family Notoryctidae, is a blind, mole-like animal which burrows with the aid of enlarged claws on the forefeet and a horny shield on the nose. It eats worms and insects.

The order Peramelina contains only the bandicoots (Peramelidae). They are ground-living, mainly nocturnal animals of the size of rats or rabbits, with long pointed noses, rather long hind legs and non-prehensile tails. They eat mostly insects, in search of which they dig conical pits in the ground, but some are herbivorous. The bilbies, or rabbit bandicoots (*Thylacomys*), make burrows, and also resemble a rabbit in their long ears. Other bandicoots live entirely above ground, spending the day in nests.

The order Diprotodontia is so called because of a pair of enlarged, horizontal incisor teeth at the front of the lower jaw. Nearly all are herbivorous. The possums (Phalangeridae) are a large family of arboreal animals with prehensile hands and feet, and usually also prehensile tails, like the opossums after which they are named. On the hand the first two fingers oppose the other three; on the foot the first toe

TOP: the Tiger Cat is one of the very wide range of Australian marsupials. It is about the size of a domestic cat and is carnivorous, stalking its prey then leaping onto it. The coat pattern probably assists in camouflaging it when hunting.

ABOVE: the Long-nosed Bandicoot (*Perameles nasuta*) is found in eastern Australia. It is the size of a rabbit and eats mostly insects and worms, for which it digs holes in the ground with its long snout.

LEFT: the Long-tailed Pigmy Possum (*Cercartetus caudatus*) is distributed in Queensland and New Guinea. Representatives of this group are among the smallest marsupials. There are several species in Australian forests. They are nocturnal and feed mainly on insects, but also eat blossoms and nectar.

RIGHT: the Great Grey Kangaroo (*Macropus giganteus*) inhabits open forest in eastern Australia and Tasmania. Bounding on their powerful hind legs, kangaroos can reach quite high speeds. When grazing they stand on all fours, using the long, heavy tail as an additional support.

opposes toes 4 and 5. Toes 2 and 3 are joined together so that they look like one toe with two claws and are used for grooming the fur. This feature, known as syndactyly, is present also in bandicoots, and it is possible that the Peramelina and the Diprotodontia have come from a common syndactylous ancestor.

One of the best-known phalangerids is the Brush-tailed Possum (*Trichosurus vulpecula*), which may be seen in Australian city parks. It feeds on leaves, shoots, flowers and seeds, and also insects. It has been introduced into New Zealand. The gliding possums (for example, *Petaurus*) parachute from one tree to another in the manner of flying squirrels, using a skin membrane that connects the front and hind legs on either side of the body. The Honey Possum (*Tarsipes spencerae*) feeds on nectar, pollen and insects from flowers, and it has a long, bristly tongue like that of a hummingbird. The Koala (*Phascolarctos cinereus*) has no tail, but climbs with its long arms and curved claws. It feeds almost entirely on the leaves of a few species of eucalyptus. Before the young is able to deal with such a diet it eats partly digested food extruded from the mother's anus.

The wombats, in the family Phascolomyidae, are heavy animals about 60 cm long, which dig long tunnels and come out in the evening to feed on grass, bark and roots. Their teeth continue to grow throughout life like those of rodents.

Finally we have the kangaroo family (Macropodidae), which includes, besides the large kangaroos and the wallabies, a number of comparatively small forms ranging in size from a rat to a hare. The kangaroo hardly needs description, with its enlarged hind legs, its peculiar leaping locomotion, and its large tail, used as a counterpoise or as a prop when standing. The main toe on the foot is the fourth; outside it is a smaller fifth toe and inside is a double grooming toe with two claws (second and third combined). The first toe is lacking in all except one very primitive form. The Red Kangaroo (*Megaleia rufa*) of Australian grasslands competes with sheep for food. Like the sheep it ruminates: after food has been acted upon by microorganisms in the

A Koala resting in the daytime in a eucalyptus tree. It feeds at night on the leaves, which it can store in cheek pouches, and seldom comes to the ground. A cautious, deliberate climber, it grasps the branches with its hands and feet but has no tail to assist it.

stomach it is brought back to the mouth and chewed again. The single joey first begins to leave the pouch after 190 days, and it continues to suckle till it is a year old. The Red-necked Wallaby (*Macropus rufogrisea*) was introduced into New Zealand where it is now a pest.

Thus, in Australia, the marsupials branched out in many different directions, adapted to different modes of life. There are cat-like, weasel-like and dog-like carnivores, insectivores, an anteater, and a wide range of leaf-eating and grass-eating herbivores; there are arboreal climbers, gliders, ground-living runners of various sorts, burrowers, and even a mole. In many cases there are striking resemblances to eutherians adapted to similar modes of life, but the resemblances are never exact: they indicate evolutionary convergence, but not relationship.

Before the coming of man, the only eutherians that reached Australia were bats, seals and, at a comparatively late date, rodents of the mouse family. Free from competition, the marsupials were able to fill many of the ecological niches that on other continents were occupied by eutherians. Perhaps because of the poor quality of their brains, perhaps because of the inefficiency of their reproductive system, the marsupials are hardly ever successful when competing with eutherians. The introduction of the Dingo by the Aborigines was probably responsible for the extermination of the Thylacine on the Australian continent, and dogs, cats and foxes brought by Europeans must have contributed to the decline in the numbers of numerous marsupial species—a decline which was accelerated by the direct effect of man's hunting and agricultural activities.

SUBKINGDOM Metazoa
PHYLUM Chordata
 SUBPHYLUM Craniata
 CLASS Mammalia

SUBCLASS Eutheria (= Placentalia)

Placentals

ORDER	SUBORDER
Insectivora *hedgehogs, moles, shrews etc.*	
Macroscelidea *elephant shrews*	
Scandentia (= Tupaiidae) *tree shrews*	
Primates	Prosimii *lemurs, lorises, bushbabies, tarsiers*
	Anthropoidea *monkeys, apes and man*
Dermoptera *colugos or flying lemurs*	
Chiroptera *bats*	Megachiroptera *fruit bats*
	Microchiroptera *insectivorous bats, vampire bats etc.*
Rodentia *rodents*	
Lagomorpha *rabbits, hares and pikas*	
Edentata *armadillos, anteaters and sloths*	
Pholidota *pangolins*	
Tubulidentata *Aardvark*	
Carnivora	Fissipedia *cats, dogs, bears, weasels etc.*
	Pinnipedia *seals, sealions and Walrus*
Cetacea	Odontoceti *porpoises, dolphins, sperm whales etc.*
Sirenia *Dugong and manatees (seacows)*	Mysticeti *whalebone whales*
Hyracoidea *hyraxes*	
Proboscidea *elephants*	
Perissodactyla *horses, rhinoceroses and tapirs*	
Artiodactyla	Suiformes *pigs, peccaries and hippopotamuses*
	Tylopoda *camels and llamas*
	Pecora *deer, cattle, antelopes etc.*

A Wood Mouse, *Apodemus sylvaticus*, suckles her new-born young in the protection of their nest underground. These offspring are much better developed than new-born marsupials, having received more nourishment in the uterus through the placenta.

- NUMBER OF LIVING SPECIES: about 3,800.

- DIAGNOSTIC FEATURES: viviparous mammals with well developed placenta for nourishing the young in the uterus; no pouch; usually three molar teeth in each half jaw.

- LIFESTYLE AND HABITAT: mostly terrestrial in all habitats from forest to desert, but some fly and some inhabit freshwater or the sea.

- SIZE: 6 cm to 31 m but most are between 20 and 150 cm.

- FOSSIL RECORD: middle Cretaceous to Recent; about 6,000 species.

The members of the Eutheria are often referred to as placental mammals, because in this subclass the placenta, where maternal and embryonic tissues come into intimate contact, is much more complex than in the Metatheria. The egg is very small and contains no yolk, so that from a very early stage the embryo in the uterus depends for its nourishment on materials supplied by the mother's blood. The young are born at a more advanced stage of development than in the Metatheria, but some eutherians are more precocious than others. In some, such as the horse and the Guinea-pig, the young are able to stand and follow the mother almost immediately after birth, but others, like the mouse and the rabbit, are born blind and helpless. Often the young are protected in a burrow or nest; young monkeys are carried by the mother, clinging to her fur. The rabbit, rat and sow can produce litters of ten or more, but many other eutherians give birth to only one young at a time. The sow has a long row of nipples from breast to groin, but when there are fewer young the number of nipples is reduced accordingly; thus the female elephant has one pair between the front legs and the mare a pair between the hind legs.

Eutheria have advanced beyond the Metatheria in brain structure. The cerebral hemispheres, where information from the sense organs is analysed and voluntary responses are initiated, are directly connected from one side to the other by a band of nerve fibres (corpus callosum), making better coordination possible. In Metatheria only more roundabout connections exist. The Eutheria learn faster. Of course, some are more intelligent than others: rodents and insectivores have comparatively small brains, but in carnivores, ungulates, cetaceans, and especially primates the cerebral hemispheres are enlarged and their surface is folded into convolutions, due to the greatly increased number of nerve cells. This makes for better awareness of surroundings, and greater adaptability and resourcefulness.

The oldest fossil eutherians, found in the Cretaceous of Mongolia, resemble marsupials in many details of the skeleton, showing that the Metatheria and Eutheria probably evolved from a common ancestor. Towards the end of the Cretaceous and early in the ensuing Tertiary period the Eutheria diversified into many orders, of which 18 survive today and others have become extinct. The first eutherians were small, mouse-sized animals which, unlike the opossums, lived mainly on the ground. Their teeth, with pointed cusps and sharp cutting edges, show that they fed mainly upon insects, as Insectivora still do today. That animals so diverse as horses, whales and men have come from such humble beginnings may seem incredible, yet it is an inescapable conclusion from the evidence provided by studies in palaeontology and comparative anatomy.

The diversity of eutherians may be illustrated by a survey of their feeding habits. Insectivores such as hedgehogs and shrews search for their food among herbage and ground litter. Besides insects and worms they sometimes eat vertebrates: thus hedgehogs will eat frogs and snakes, and they are partially immune to snake poison. Most bats catch flying insects such as moths, but some species are carnivorous and others feed on fruit, nectar or blood. A number of eutherians are adapted to feed on ants and termites: the pangolins and

Mammals have jaw muscles and teeth adapted to their diet. In carnivores the large temporal muscle attached to a bony crest on the skull gives a powerful bite. In herbivores the masseter is the larger muscle, moving the lower jaw from side to side or back and forth, grinding tough plant food between the flat, ridged cheek teeth. In rodents the masseter muscle is formed of several parts. Carnivores have specialized carnassial teeth for cutting through the sinews of meat and also large canines for killing prey. Rodents have no canines, but between the gnawing incisors, which continue to grow throughout life, and the grinding teeth is a gap, or diastema. Some herbivores, such as sheep and deer, have no upper incisors, and the lower ones, which are short, bite against a hard pad at the front of the upper jaw.

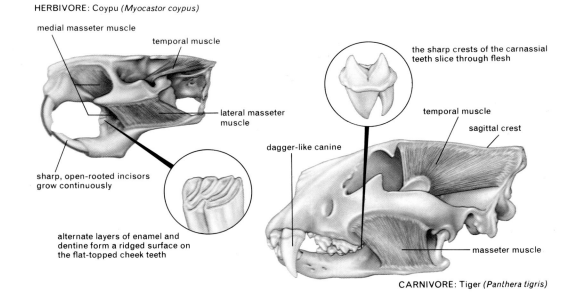

HERBIVORE: Coypu (Myocastor coypus)

medial masseter muscle

temporal muscle

lateral masseter muscle

sharp, open-rooted incisors grow continuously

alternate layers of enamel and dentine form a ridged surface on the flat-topped cheek teeth

the sharp crests of the carnassial teeth slice through flesh

temporal muscle

sagittal crest

dagger-like canine

masseter muscle

CARNIVORE: Tiger (Panthera tigris)

the American anteaters resemble echidnas in having no teeth. Typical members of the order Carnivora kill other mammals. Some, like the cats, creep close to their prey under cover till they can leap upon it; others, like wolves, follow the prey, sometimes for miles, till it is exhausted. Large canine teeth are used in killing, and at the side of the mouth there is a pair of specialized carnassial teeth which act like scissors in cutting through tough meat. Not all Carnivora are exclusively meat eaters. Thus badgers eat roots and acorns, as well as worms and mice, and most species of bear are largely vegetarian. Otters feed on fish, as do most members of the suborder Pinnipedia (seals and sealions). In the Pinnipedia the cheek teeth are modified into hook-like structures which enable them to hold their slippery prey. The toothed whales, such as porpoises and dolphins, have numerous small peg-like teeth along the jaw: they eat mainly fish and squid. The whalebone whales, which sieve planktonic animals from the sea, have lost their teeth entirely.

Among herbivores, some feed on soft food such as fruit, and have blunt-cusped molars suitable for crushing; most primates have teeth of this type. Others, which feed on harder or tougher food, such as seeds or grass, have flat-topped grinding molars which function in a horizontal jaw movement, either from side to side as in the horse and cow,

or forwards and backwards as in most rodents and also in elephants. A distinction can be drawn between browsers, such as deer, which eat leaves and shoots, and grazers which eat grass. Grass wears the teeth down, and this is compensated by having very high-crowned teeth (for example, horses) or even teeth which continue to grow throughout life (for example, rabbits), tooth material being added at the bottom while it is worn away at the top. The rodents are characterized by having a pair of enlarged incisors in each jaw, that grow permanently in this way. They are very versatile structures, used for example by squirrels for opening nuts and by beavers for felling trees. The tusks of elephants are also permanently growing incisors, whereas the tusks of the Wild Boar are permanently growing canines.

Although it is a member of the order Carnivora, the Giant Panda (Ailuropoda melanoleuca) feeds exclusively on plant matter. To adapt to the demands of chewing bamboo shoots rather than meat or fish, the cheek teeth have become enlarged and flattened.

BELOW: many fish-eating mammals, like the Leopard Seal (Hydrurga leptonyx), have numerous, sharply pointed teeth for seizing and manipulating their slippery food. As they swallow their prey whole, they have no need for any other types of teeth.

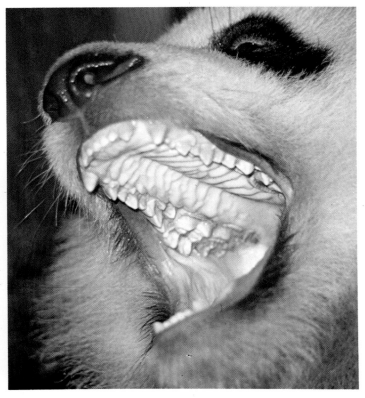

Another way to illustrate the diversity of Eutheria is to consider their locomotory adaptations. The most primitive members of the group were probably rather rat-like in body form, with an arched back, comparatively short legs and a long tail. There were five digits, all with claws, on fore and hind feet, and the under surface of the feet, where the palm and sole touched the ground, was provided with hairless pads. The claws and pads would no doubt enable the animals to climb, but they probably spent most of their time on the ground, where the claws could be used to scratch in search of insects and worms, or to dig a nesting hole. When running rapidly they would proceed by a bouncing gallop in which bending and straightening of the back played an important part, as many small mammals do today. From such generalized ancestors were derived descendants specialized in different ways—arboreal climbers, terrestrial runners and subterranean burrowers, and flying and aquatic forms.

Greater efficiency and speed of running is achieved by lengthening the limbs, especially the feet, and simplifying their movement so that they swing, pendulum-like, in a fore and aft direction. In dogs and cats the wrist and ankle are raised from the ground and the animal stands on pads underneath the toes; ungulates stand on the tips of their toes, where the claws are modified into hoofs. The number of toes is reduced, starting with the loss of the first toe (corresponding to our thumb and big toe). The dew-claw of the dog is the last remnant of its thumb. Cattle and sheep have only two toes, the third and fourth of the original five; small hooves representing toes 2 and 5 survive in deer, and pigs still have four complete toes. The horse has only one toe, the third, but fossils show that toes 2 and 4 were formerly present and have been lost. The horse has a straight back and does not bounce up and down when running, thus saving energy, and incidentally allowing man to ride. It is the most efficient long-distance runner, but the Cheetah is the fastest on short sprints. When running it is the hind legs that drive the animal forward, the front legs serving mainly for support. This difference is exaggerated in hares and rabbits, where the hind legs are larger than the front legs, and still more so in some rodents such as jerboas, which hop on their hind legs like a kangaroo.

Long legs and a digitigrade posture enable the Maned Wolf (*Chrysocyon brachyurus*) to run quickly. The ability to move at speed is particularly important in the life of a predatory carnivore.

Because of their great weight, the movements of elephants are slow and stately. Their legs are pillar-like, the five toes of each foot are very short, and the weight is supported on a cushion of elastic tissue at the base of the foot. Very different are the adaptations of burrowing mammals such as moles. The powerful front legs, with broad, spade-like hands provided with strong claws, are used to scrape away the soil while the animal pushes itself forward with its hind legs. To aid in streamlining only the hands project from the body, and they are very close to the head. Not all burrowers use the hands: the mole rats dig with their incisor teeth.

The potential top speed of an animal depends on the length of the stride. Different groups have achieved speed in different ways. Less of the foot may be in contact with the ground, so that the upper foot bones add to the leg's length, or the length of the lower bones is increased relative to that of the femur.

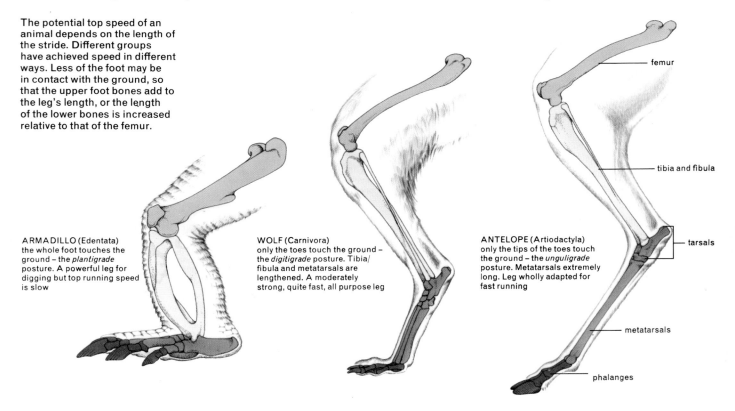

ARMADILLO (Edentata) the whole foot touches the ground – the *plantigrade* posture. A powerful leg for digging but top running speed is slow

WOLF (Carnivora) only the toes touch the ground – the *digitigrade* posture. Tibia/fibula and metatarsals are lengthened. A moderately strong, quite fast, all purpose leg

ANTELOPE (Artiodactyla) only the tips of the toes touch the ground – the *unguligrade* posture. Metatarsals extremely long. Leg wholly adapted for fast running

femur

tibia and fibula

tarsals

metatarsals

phalanges

BELOW: the Slender Loris (*Loris tardigradus*) climbs slowly and carefully in trees, clinging with its prehensile hands and feet. The tail, a balance aid for more agile climbers, has been lost.

ABOVE: a gazelle running at speed. The hind limbs drive the body forwards, and the fore feet are about to touch the ground. This unguligrade animal runs on hooves on the tips of its toes.

Squirrels climb trees by clinging to the bark with their claws. They are agile leapers from branch to branch and the flying squirrels have developed this ability further by using skin membranes which connect the front and hind legs, to glide for considerable distances from one tree to another. Some arboreal mammals have prehensile tails that can be curled round branches, and primates can grip with their hands and feet. Gibbons have very long arms, by which they can swing themselves through the trees faster than a man can run on the ground beneath. In contrast, the slowly moving sloths suspend themselves upside down from branches by their large hook-like claws.

Bats are the only vertebrates other than birds that are capable of sustained flight. The wing is a membrane of skin stretched between the very long fingers; it extends back to the hind legs and usually also between the hind legs and the tail. Flapping of the wings is performed by large muscles that move the arms. Bats were probably descended from arboreal mammals which evolved the capacity to glide.

Then there is a variety of aquatic placentals, from those which, like the otters and beavers, are equally facile in the water and on land, to the completely aquatic Cetacea and Sirenia. Intermediate are the seals and sealions, which are clumsy ashore and go there only to breed. There are several methods of swimming. Beavers paddle with their webbed hind feet, aided by undulations of their flattened tail. Sealions may be said to fly under water, with their long front flippers in which the fingers are embedded in skin; the hind feet are used for steering. Seals scull with their hind feet, shaped like a fish tail, and steer with their front feet. Whales swim with their very large tails. Their front legs are flippers, used for steering and braking, and their hind legs have disappeared. They have tail fins (flukes) which are horizontal and not vertical as in fish, and whereas fish undulate from side to side, whales move their tails up and down. The Sirenia (the Dugong and manatees), though not related to whales, resemble them in lacking hind legs and possessing large tails with horizontal fins.

The Mole (*Talpa europaea*) lives in burrows, scraping away the soil in front with its large, flat hands and pushing its pointed body forwards with the smaller hind legs.

Scent is more useful than sight to mammals that live among thick vegetation at night or in underground burrows by day, but species that live exposed in open country have very good sight, as anyone who has tried to approach deer knows. The eyes of deer, antelopes and rabbits are at the sides of the head so that, with one eye or the other, the animal can see over a very wide range, but the overlap of the two fields of view is small. To judge distances accurately it is necessary to have stereoscopic vision, a prerequisite for which is that both eyes must see the same object together; shortening the muzzle brings the eyes closer to each other so that they look forward and their fields of view overlap more extensively. We see this in the cats, which need to judge the distance of their prey when making the final spring. The most complete development of stereoscopic vision is found in the higher Primates. Although some of these, including Man, now live on the ground, they primarily evolved as arboreal animals, for which a failure to assess distances when jumping from branch to branch could have had fatal consequences. Scent is of little use for this purpose, and the reduction of the nose enabled the eyes to look directly forward. These primates are diurnal, unlike most mammals, and they have a well developed sense of colour; it is no accident that many of them are themselves more gaudily coloured than other mammals.

All mammals can hear, and most of them produce sounds. Some can hear ultrasonic sounds beyond the range of the human ear; young mice, if removed from the nest, emit ultrasonic sounds that enable the mother to find them. Most remarkable is the use of ultrasonic sound to detect objects in the animal's neighbourhood. The animal emits a series of 'clicks' which are reflected back as echoes; from the time between the click and its echo the animal knows the distance of the object. Echolocation has evolved in the bats and the whales. In some bats the sound comes out through the open mouth, in others through the nose. Elaborate skin excrescences on the face beam the sound, and the echo is received by the large ears. Even in total darkness bats can not only avoid obstacles but also catch flying insects.

Although many mammals live solitary lives, others are associated in groups in which individuals communicate by

An Otter (*Lutra lutra*) standing upright on the alert. The senses used are smell, sight and also hearing. With the moist skin on the nose—the rhinarium—it can detect the direction of scent-bearing wind. Under water the otter relies mainly on touch, using the tufts of stiff whiskers, the vibrissae, on its snout

The sense of smell is very important to most placentals: the nasal chamber in many cases occupies as much space in the head as the brain. Man, with his relatives among the primates, is exceptional in that his sense of smell is poor, and it is difficult for us to appreciate the mind of an animal that learns about its surroundings and recognizes other individuals mainly by scent. Most mammals have an area of bare skin round the nostrils, kept moist by tear fluid that runs from the eye down a tube opening into the nasal chamber. This device enables them to detect the direction of scent-bearing air currents, as when we hold up a wet finger to test wind direction. Scent is an important means of communication between individuals. In dogs and other carnivores scent glands are situated near the anus, and when two individuals meet they sniff each other in this region. A number of antelopes and deer have scent glands on the side of the face, used in scent marking the vegetation through which they pass. Urine and faeces are often used, as by dogs, to signal their presence; some lemurs and lorises smear their urine on branches with their hands. Such behaviour may seem obnoxious to us, but the secretions of the scent-glands of the civets and the Musk Deer have long been used as ingredients of perfumes.

Some bats emit ultrasonic pulses through the nostrils and fly with the mouth shut. The folds of skin on the face beam the ultra-sound and the echo is received by the large, forwardly pointing ears. By echolocation bats are able to avoid obstacles in total darkness. This is a Greater Horseshoe Bat (*Rhinolophus ferrumequinum*).

scent, by making a variety of sounds, and by bodily attitudes, gestures and facial expressions recognized through the sense of sight. Mammalian societies are so varied that it is impossible to generalize about them, and a few examples must suffice. A colony of beavers typically consists of a pair of adults and their offspring up to two years of age. They dam a stream with branches of trees that they fell, and in the pond formed above the dam they construct a lodge, roofed with branches plastered with mud. The floor of the lodge is above water level but the entrance is under water. Beavers feed on bark, and they store branches, cut into suitable lengths, as a winter food reserve. They can construct canals along which they bring branches from greater distances. Among Red Deer (*Cervus elaphus*) a number of females with their young remain together as a herd under the leadership of one of the females. In the rutting season the stags compete fiercely for possession of the female herds, but during the rest of the year they form all-male groups. In wolves, on the other hand, the mated pair remain together, the male providing much of the food for his family. More than one family may share a den, and by cooperating in hunting a pack of wolves can overpower larger prey. While the pack is away hunting one adult, often a female, is left behind to guard the pups. Within the group, a recognized order of social ranking reduces quarrelling. Social relationships are particularly highly developed in monkeys and apes, whose communities last over many generations. Monkeys spend their entire lives in a social setting. Studies of monkey societies such as troops of baboons have revealed an extremely complicated pattern of relationships between individuals, governed by rules of behaviour which the young must learn.

ABOVE: Red Deer fighting in the autumn rutting season. Only stags that are successful in these fights can gain possession of females. The bony antlers are shed after the breeding season but grow again in the following summer. Initially they are covered with skin (velvet) which peels off. Outside the breeding season the females remain in a herd, the stags forming small groups.

BELOW: higher primates can focus both eyes accurately on objects in front of the face. They can also coordinate movements of the hands with what they see. Both abilities are involved in mutual grooming, an important form of social communication in many species of monkey. One individual usually grooms another of higher social status. These are Vervet Monkeys (*Cercopithecus aethiops*).

The shrews (Soricidae) are the most numerous insectivores. They are distributed over Europe, Asia, Africa and North America, and one form in South America is the only insectivore on that continent. Shrews are mostly very small animals: one of them, the Etruscan Shrew (*Suncus etruscus*), which weighs only two grams, is the smallest mammal. They are very active creatures, pugnacious in defence of their territories, and there are records of shrews killing mice. Some species have poisonous saliva. Shrews are very voracious, eating nearly their own weight in 24 hours; they cannot survive for more than a few hours without food, and so they are active by day as well as at night. They are noisy animals and some species are believed to use high-pitched squeaks for echolocation.

ABOVE: the hedgehog *Erinaceus albiventris* from West Africa. Hedgehogs protect themselves by drawing their spine-covered dorsal skin over the head and legs. They search by scent for worms, snails and insects and sometimes eat frogs and snakes.

RIGHT: the European Mole (*Talpa europaea*) burrows with its large hands, the palms facing outwards. The body is streamlined, with no neck or projecting elbows, and the fur can lie both ways. Because they are nearly blind, moles find their food by scent and touch.

The order Insectivora contains six families of small mammals of widely different appearance, but similar in that they all have very mobile noses, small eyes and small brains. In many ways the most primitive living placentals, they are solitary, mainly nocturnal animals which feed for the most part on insects, worms and other invertebrates, for which they search on the ground using their well developed sense of smell. The moles of the northern hemisphere and the golden moles of Africa burrow underground, and there are a few aquatic insectivores. The molar teeth of insectivores have sharp, interlocking cusps, effective in breaking and cutting up the cuticles of insects, and in many insectivores the incisors are enlarged to act as forceps used in picking up prey.

Hedgehogs (such as *Erinaceus*) are easily recognized by the spines that cover their back. These are modified hairs, and the rest of the body has a normal hairy coat. Under the skin of the spine-covered area is a sheet of muscle, especially thick at the edges where it forms a band that passes over the neck and the root of the tail. By contracting this band like a drawstring the hedgehog can roll itself up, slipping the skin of the back over the head and the tail to enclose the body in a spine-covered bag. With this defence, hedgehogs do not run away when threatened, with the unfortunate consequence that large numbers are killed on the roads. Members of the hedgehog family (Erinaceidae) are found over Europe, Asia and Africa, and the European species has been introduced into New Zealand. In southeastern Asia there are some species that do not have spines.

Though shrews can dig burrows, members of the mole family (Talpidae) are much more specialized burrowers. The European Mole (*Talpa europaea*) spends most of its time tunnelling in the soil, and it is awkward on the surface. Its velvety fur allows easy passage in the burrow. Its eyes are tiny and almost functionless, and other species are quite blind. Moles feed mainly on earthworms, which they find by smell and by the acute sense of touch provided by their snouts. The soil removed in burrowing is pushed out at the surface at intervals, a habit which, when it disfigures lawns and golf-courses, incurs the enmity of man. The young are born in a nesting chamber, lined with dead leaves and dry grass, under a large mound or 'fortress'. Not all Talpidae are equally specialized for burrowing. The Russian Desman (*Desmana moschata*) is an aquatic animal with webbed hind feet and a flattened tail, it eats fish and crustaceans.

There are no true moles in Africa, but their place is taken there by the golden moles (Chrysochloridae), so called because of the metallic sheen of their fur. They differ from the true moles in that their hands act as picks, rather than shovels, being equipped with two enormous claws. They closely resemble the Marsupial Mole of Australia.

The tenrecs (Tenrecidae) are found mainly in Madagascar, but fossils show that the family probably originated on the continent of Africa, and three species still live there. In Madagascar, removed from competition, they have evolved into a variety of forms that imitate the insectivores of other parts of the world. Some have prickles like hedgehogs, others resemble shrews, and there is also a burrower and a web-footed aquatic species. The African species, known as otter shrews, are sometimes classed as family Potamogalidae.

The solenodons of Cuba and Haiti, in the West Indies, form a family by themselves (Solenodontidae). They are rat-sized animals, differing from rats in the very long nose and large claws on the feet. They use the claws to grub in the ground or break into rotten wood in search of food.

Two other families, formerly classified in the Insectivora, are now regarded as sufficiently distinct to be placed in separate orders. The elephant shrews (Macroscelididae) of Africa, forming the order Macroscelidea, are lively animals with large eyes and long hind legs. They get their name from the long, sensitive and moveable snout. The other family, the tree shrews (Tupaiidae) of southeast Asia, have attracted much scientific interest because of their possible relationship to the primates, and indeed some authorities have classified them within that order. Tree shrews are squirrel-like, able to run along branches using their claws. They have large eyes and like many primates they depend more on sight than on smell. They are active during the day and sleep at night. They also resemble primates in having colour vision. Recent research, however, tends to show that these primate-like characteristics have been evolved independently, and that tree-shrews should not be regarded as primates but should be put into an order by themselves (Scandentia). Their food consists mainly of insects, but they also eat fruit and other vegetable matter. Like insectivores, tree shrews are solitary animals, each individual defending its own territory. Maternal care for the young is minimal: the mother visits the nest only for short periods in order to feed them.

ABOVE: *Echinops telfairi* is a tenrec which has spines on its back and can roll up like a hedgehog; some other species are less spiny. Isolated on Madagascar, tenrecs have evolved to imitate insectivores from other parts of the world.

ABOVE RIGHT: an elephant shrew (*Elephantulus*). Members of the family are confined to Africa. They have long, mobile snouts, large eyes and long hind legs. They are active by day and feed chiefly on insects.

RIGHT: the Large Tree Shrew (*Tupaia tana*) from Sumatra and Borneo spends less time in trees than some other species in the order. It searches ground litter for insects and worms but will sometimes eat lizards and fruit. Tree shrews have large eyes and depend more on sight than smell.

With about 850 species, the bats (Chiroptera) form a very large order of mammals, second only to the rodents. Their power of flight has enabled them to cross ocean barriers and reach islands such as New Zealand and Samoa, inaccessible to land mammals. Four long fingers of the hand support the wing membrane like the ribs of an umbrella, but the thumb projects forward and bears a claw, used in walking or climbing. The wing involves the hind legs which are held out horizontally. They help to steady the flight by moving up and down in time with the arms; they cannot be brought forward, with the knees under the body as in other mammals, but can be turned so that the knees and toes point backwards. When at rest bats hang upside down, clinging to rocks or trees with their feet. On the ground they pull themselves forward with their thumbs and push with their backwardly directed hind legs.

Most bats feed on insects and have cheek teeth with sharp cusps like insectivores. Insects are generally caught in flight and bats take the place at night of birds such as swallows which hunt in the air by day. Experiments have shown that by using echolocation bats can find flying insects with great efficiency in total darkness. Echolocation is effective only

A Mouse-eared Bat (*Myotis myotis*: Vespertilionidae) in flight. The wing is a skin membrane, supported by four long fingers and the hind leg and tail. The thumb, with its hooked claw, projects forward and is used when walking on the ground. The ultrasonic pulses used for echolocation are emitted through the open mouth, rather than through the nose as in the horseshoe bats.

over short distances (the maximum is 90 cm in the case of horseshoe bats). If an insect comes within range the bat closes in on it, and catches it in the wing membrane, from which it is transferred to the mouth. The mouth cannot be used to snap at prey in the air, as this would interfere with the ultrasonic beam, which in most insectivorous bats emerges through the open mouth.

Not all bats catch insects in this way. For example, the Long-eared Bat (*Plecotus auritus*) of Europe has been seen hovering in trees, picking caterpillars off the leaves. Some large bats eat vertebrate prey. The False Vampire (*Megaderma lyra*) of India eats birds and smaller bats, as well as lizards and mice. In tropical America the Bulldog Bat (*Noctilio leporinus*) flies low over the surface of water and picks up fish with its large feet. By echolocation it detects the ripples made by fish swimming near the surface.

The most feared bat is the Vampire (*Desmodus rotundus*), which ranges from Mexico to Paraguay. It attacks principally cattle, horses and swine, but also dogs, poultry and humans. With its razor-sharp front teeth it removes a piece of skin from its sleeping victim and imbibes the blood, which is drawn along grooves on the underside of the Vampire's tongue. Bites are dangerous, not only because of the loss of blood, but also because this bat can carry a form of rabies.

During the day bats hide in hollow trees, rock crevices and other secluded places. A number of species gather in caves, sometimes in spectacular numbers. The floor of caves long used by bats becomes covered with a thick deposit of their excrement known as guano. Flying foxes, not having echolocation, are unable to find their way in the darkness of caves; they spend the day in trees, where they congregate in hundreds. The same 'camp' is used for many years. From it the bats travel in the evening for distances of as much as 30 km to feed.

In temperate regions only insectivorous bats occur. Most of them hibernate in the winter, but some species migrate south like insectivorous birds. Bats produce usually only one young at a birth, and in northern climates they have only one breeding season, in the spring. For a few weeks the young bat is carried by the mother, clinging to her fur, and it is provided with hook-like milk teeth that enable it to do this. Later the young are left behind in a cave while the mothers are out foraging. Female bats with young gather together in colonies separate from the males.

The order Chiroptera is divided into 16 families, in two suborders: the Megachiroptera, including the flying foxes and their allies, and the Microchiroptera, containing all other bats. Most families are tropical, but the Vespertilionidae is worldwide, reaching as far north as Lapland and Alaska, and the horseshoe bats (Rhinolophidae) are found in Europe, including Britain.

The earliest fossil bat, from the Eocene epoch, differs only in details from the living forms, and the origin of bats is unknown. It is possible that they are related to the flying lemurs, or colugos (*Cynocephalus*), gliding animals from the forests of southeast Asia. Only two species make up the order Dermoptera. Colugos have a wing membrane connecting the legs and the tail, but the fingers are not lengthened as in bats. They can glide as far as 54 m from one tree to another.

The face of a Vampire, the most notorious of all bats. After removing a piece of skin with the very sharp incisor teeth, the Vampire draws in blood along a channel in the lower lip.

TOP: Egyptian Fruit Bats (*Rousettus aegyptiacus*) roosting by day. Bats at rest hang, upside-down, from their feet. This species is unusual among Megachiroptera in having a form of echolocation.

ABOVE: the flying fox *Pteropus poliocephalus* feeding on bananas in Australia. Flying foxes roost in trees and travel far at dusk in search of fruit. They can do much damage in plantations.

Other bats eat fruit. The best known of these are the flying foxes (*Pteropus*), of which there are many species in Australia, southern Asia and Madagascar. They do not echolocate, but depend mainly on sight and smell, being active in twilight rather than in total darkness. The fruit is crushed by their flat molar teeth, and only the juice is swallowed; fibres and seeds are spat out. Flying foxes like sweet fruits favoured by man, and they can do much damage in Australian peach orchards. Some related species have specialized in feeding on the nectar of flowers, which they lap up with long, very protrusible tongues provided with bristles at the tip. A number of tropical plants depend upon bats for pollination, and have flowers which open at night. Very similar adaptations for feeding on fruit and nectar have evolved independently in tropical America among bats of a different family (Phyllostomatidae).

Several rodents, in different families, have become adapted for burrowing. The African Mole Rat (*Tachyoryctes splendens*) makes extensive tunnels with its short, powerful limbs, using its incisors to cut through roots. It has tiny eyes but many tactile whiskers.

The Rodentia, with over 1,600 species, is by far the largest order of mammals. At the front of their upper and lower jaws all rodents have a pair of chisel-like incisor teeth that never wear out because they grow continuously throughout life. Between the incisors and the grinding teeth there is a space where the mouth can be closed off by folding in the cheeks when the incisors are used to gnaw inedible material. Rodents are vegetarian, though many of them occasionally eat insects and snails. Most of them sit up and hold the food between their paws when eating. The squirrel's habit of burying nuts illustrates the propensity of a number of rodents to store excess food, which is eaten in times of scarcity in the winter or dry season. Some, such as hamsters, carry excess food in cheek pouches that extend back under the skin, often beyond the front legs.

Most rodents dig holes or burrows in the ground, using their feet or their incisor teeth. Desert species thus escape the heat and drying effect of the sun, and come out only at night. Among desert-living rodents, the jerboas (Dipodidae) of Africa and Asia and the kangaroo rats of America hop on their hind legs. Others feed underground on roots and seldom come to the surface, for example the mole rats (*Spalax*) of the Near East. Mole rats are blind, their tiny eyes being covered by skin. In contrast, squirrels have very good sight, enabling them to run through the tree tops. Squirrels are active by day, and they have colour vision. Not all squirrels are arboreal: there are ground squirrels which do not have bushy tails. Among these are the chipmunks (*Tamias* and *Eutamias*), the Woodchuck (*Marmota monax*) and the Prairie Dog (*Cynomys ludovicianus*) of North America. The flying squirrels (for example, *Glaucomys*), which glide by means of a fold of skin connecting the wrist to the ankle, are nocturnal.

On the ground, rodents usually run on the soles of the feet, but some South American forms run on their toes. These, which include the Guinea-pig (*Cavia porcellus*), agoutis (*Dasyprocta*) and the Capybara (*Hydrochoerus hydrochaeris*), have only three toes on the hind foot, ending in blunt, hoof-like claws. The Capybara is the largest rodent, the size of a small pig. It spends much of its time in the water, completely submerged except for its eyes, ears and nostrils, like a hippopotamus. It comes out to feed on marsh vegetation. Only slightly smaller, and more highly adapted for aquatic life, are the beavers. With their nostrils and ears closed, they can travel under water as far as 800 m.

Most rodents have little defence against predators, depending merely upon the speed with which they can retreat to a hole, and they are eaten in large numbers by carnivorous mammals and birds. However, the porcupines of Africa and India (*Hystrix*), with very long spines on the hind part of their bodies, are able to deter even lions and tigers. The animal runs backwards, leaving its spines embedded in the flesh of the attacker. The American porcupines belong to a different family. Their spines are shorter, but they are barbed so that they work their way more deeply into the flesh. American porcupines climb trees and some species have prehensile tails. The North American species feeds mainly on the leaves, buds and bark of coniferous trees.

In compensation for the high rate at which they are eaten, small rodents have remarkable powers of reproduction. The female House Mouse (*Mus musculus*) is ready to start breeding when she is six weeks old; she produces on average five litters a year, with an average of five young in a litter. It is therefore understandable that when conditions are favourable the population can build up to plague proportions. On the continent of Europe the Field Vole (*Microtus arvalis*) is particularly subject to population explosions. In Britain the last vole plague was in 1891-2, in Scotland, but every few years there is a 'mouse year' when the number of voles and other small rodents reaches a peak. In the case of the Norwegian Lemming (*Lemmus lemmus*), such periodical rises in population lead to spectacular emigrations in which millions of individuals die.

Most rodents are solitary, defending territories against other individuals, but many live in social communities. Rats recognize other members of their own colony by scent, and drive away strangers. The most highly organized rodent societies are the 'towns' of the American Prairie Dog, one of which may house 1,000 inhabitants. Prairie Dogs greet each other by a sort of kiss in which the incisor teeth touch. Within the town each family has its own territory, where a dominant male, two or three females and their young share a burrow, sometimes with a subordinate male. When the young mature the adults depart and start new burrows, leaving the old home to their offspring.

The Grey Squirrel (*Sciurus carolinensis*), which originated in North America, is adapted for running along tree branches, using its claws; its bushy tail helps steer the body during jumps from branch to branch. Squirrels are diurnal animals with good vision.

LEFT: kangaroo rats, such as *Dipodomys merriami*, inhabit arid country in western North America. They seldom drink, and conserve moisture by leaving their burrows only at night. They store seeds after sun-drying them. Kangaroo rats hop on their long hind legs.

ABOVE: the Capybara lives in South America and is the world's largest rodent. It spends most of its time in water with only the top of its head exposed, but comes out occasionally to feed on waterside plants. It lives in groups of up to 20 individuals.

Rodents are divided into 43 families. The largest is the Cricetidae, with over 500 species. It includes a great many mouse- and rat-like rodents from North and South America, as well as the hamsters of Europe and Asia, and the voles and lemmings, which occur in North America as well as in the Old World. Some voles, including the European Water Vole or Water Rat (*Arvicola amphibius*) and the American Muskrat (*Ondatra zibethica*), are aquatic. Rats and mice belong to the Muridae. Except for species introduced by man, this family does not occur in America, but there are more than 60 species in Australia: murids are the only terrestrial eutherians, other than man, that succeeded in getting to Australia from Asia. The only other families of rodents native to Britain are the squirrels (Sciuridae) and the dormice (Gliridae), but beavers (Castoridae) formerly lived there. Pocket gophers (Geomyidae), kangaroo rats (Heteromyidae) and tree porcupines (Erethrizontidae) are American families.

LEFT: the Varying Hare (*Lepus americanus*) from the coniferous forest belt of North America, turns white in winter. It is also called the Snowshoe Hare because of its hairy feet. Hares do not make nests and the young are well developed when they are born.

BELOW: the Pichi (*Zaedyus pichi*) of Argentina, like other armadillos, is protected by an armour of bony plates. When attacked it does not roll up but wedges itself into the ground with the edges of its carapace. It digs in sandy soil for insects, worms and carrion.

RIGHT: the Two-toed Sloth (*Choloepus didactylus*) lives in trees suspended by its hook-like claws. Sloths feed on leaves, move slowly and have a lower and less constant temperature than other mammals, ranging from 24 to 35°C.

BELOW RIGHT: the Indian Pangolin (*Manis crassicaudata*). Pangolins have a covering of large, over-lapping horny scales, and they feed mainly on ants and termites, whose nests are broken open with the large claws. Prey is picked up by using the long, sticky tongue.

Rabbits and hares were formerly classed as rodents, but they are now put into a separate order, Lagomorpha. They resemble rodents in having persistent (continuously growing) incisor teeth, but there are two pairs of these in the upper jaw, not one pair as in rodents. The Lagomorpha consists of two families: besides the rabbits and hares (Leporidae) there are the pikas (Ochotonidae). Pikas live among rocks in mountainous regions of Asia and western North America. They do not have the long ears and long hind legs that characterize the Leporidae.

The Wild Rabbit (*Oryctolagus cuniculus*) was probably originally a native of Spain and North Africa, and it was introduced into Britain as a food animal in the Middle Ages. Its reproductive powers are proverbial: a female can produce up to six litters a year and young rabbits can start to breed in the year of their birth. In view of this, the phenomenal spread of rabbits in Australia after their introduction there is understandable. The young are born in a burrow dug by the female, who makes a nest of grass, lined with fur from her own body. They are blind and helpless. On the other hand, the young of hares (*Lepus*), which do not burrow, are fully furred and have their eyes open at birth. The female hare distributes her litter among a number of 'forms', depressions among vegetation where the young lie quietly. Rabbits and hares visit their young only once in 24 hours, and then only for short periods; this is presumably so as not to betray the whereabouts of the young to predators. The American cottontails (*Sylvilagus*) make a nest like the Rabbit, but only in a shallow cavity about 15 cm deep. The Jack Rabbit (*Lepus californicus*) is a species of hare. Worldwide, there are about 60 species of Leporidae.

The order Edentata contains the armadillos, anteaters and sloths. It evolved in South America, but the Nine-banded Armadillo (*Dasypus novemcinctus*) has spread northwards as far as Kansas. The name of the order refers to the reduction in the number of their teeth: front teeth are lacking, and anteaters have no teeth at all. In armadillos and sloths the back teeth do not have any covering of enamel. Edentates have a lower body temperature than most other mammals.

In armadillos (family Dasypodidae) the back and sides of the body are covered with a protective armour made of plates of bone in the skin, covered with a layer of horny material. The armour forms bands across the back, connected by flexible skin, and some species can roll up into a ball. Armadillos have powerful claws which they use for burrowing and digging in search of food—they eat mainly ground-living insects.

Anteaters (Myrmecophagidae) use their large claws to break open the nests of termites and ants. The small mouth is at the end of a narrow, downwardly curved snout, and the insects are picked up by the protrusible tongue. The species

most often seen in zoos is the Giant Anteater (*Myrmecophaga tridactyla*), a ground-living form with very long hair on the tail. There are other smaller species which live in trees and have prehensile tails.

The sloths (Bradypodidae) are slow moving arboreal animals which eat leaves. They climb and suspend themselves by their hook-like toes. All sloths have three toes on the hind foot; three-toed sloths (*Bradypus*) have three toes on the forefoot, whereas two-toed sloths (*Choloepus*) have two. Sloths are greenish in colour, owing to algae that grow on their hair.

Other remarkable edentates lived in South America in the past. *Glyptodon*, a Pleistocene relative of armadillos, had a bony shell on its back like a tortoise, with only the head, tail and the feet protruding. *Megatherium*, a giant ground sloth, was as large as an elephant. Its skeleton shows that it could stand on its hind legs, supported by the tail, while clinging to tree trunks with its large front claws. It probably fed on leaves. Ground sloths reached the United States, and they may have survived long enough to meet man. Pieces of their dried skin, with the hair still attached, have been found in caves.

The pangolins (*Manis*) of Africa and southern Asia, formerly classified with the American edentates, are now considered to form an independent order, Pholidota. Their most striking feature is that the greater part of the body is covered with large, overlapping, horny scales, giving the animal a reptilian appearance. However, the under surface is hairy, and the dorsal scales are believed to represent rows of hairs fused together. Pangolins feed on ants and termites, and like other anteaters they have protrusible, sticky tongues and strong claws. They have no teeth, but small pebbles taken into the stomach help to grind up the food as in the gizzard of a bird. There are eight species. Some are arboreal and have prehensile tails.

The African Aardvark (*Orycteropus afer*) forms an order by itself, Tubulidentata, named from the peculiar structure of its teeth. It is a large animal, about 180 cm long including the tail. It has a long, tubular snout, long, rabbit-like ears, an arched back and strong limbs provided with claws. Its hair is sparse and bristly. The Aardvark feeds mainly on termites, picking them up with its tongue, which can be protruded 30 cm from its mouth. It is a powerful burrower, hiding underground by day and coming out to feed at night. Unlike the pangolins, the Aardvark chews its food.

The abundance of ants and termites in the warmer parts of the world has led to the independent evolution, on different continents, of mammals that specialize in feeding on these insects. The similarities shown between the various anteaters are due to evolutionary convergence, and like the parallelisms between marsupials and placentals they do not indicate relationship.

As well as terrestrial meat-eaters, like dogs and cats, the order Carnivora contains the seals and sealions, which are adapted to an aquatic life. These form the suborder Pinnipedia, the remaining Carnivora making up the suborder Fissipedia. There are seven families in the Fissipedia: the dogs (Canidae), bears (Ursidae), raccoons (Procyonidae), the weasel family (Mustelidae), the civets and their allies (Viverridae), hyaenas (Hyaenidae) and cats (Felidae). By no means all Fissipedia are purely meat-eaters: bears and raccoons, for example, are omnivorous, eating much plant material, and the Giant Panda (*Ailuropoda melanoleuca*) which feeds entirely on bamboo, seems out of place in the Carnivora. Originally, however, all carnivores were adapted to kill and eat other animals, especially mammals, and those that have given up this habit still retain traces of a predaceous ancestry.

All carnivores have large canine teeth, used for killing prey or for defence. Farther back in the jaw the flesh-eaters have a pair of large carnassial teeth with which they cut up meat, but in omnivorous species like bears these are not developed. The back teeth are flat-topped and used for crushing. They are large in bears, but in the cats they have disappeared. Carnivores cannot move their jaws sideways to grind their food as herbivores do; the jaw joint acts simply as a hinge. This joint is very strongly constructed, to withstand the strains imposed on it when the carnivore attacks its prey: a dog can be lifted from the ground while holding on to a stick with its canine teeth.

Hunting requires alertness and cunning. Carnivores are intelligent, inquisitive animals with larger brains than insectivores and rodents, and their senses are very keen. Dogs hunt mainly by scent and hearing, but cats depend more on sight. Newborn carnivores are helpless creatures, cared for by the mother, or, in the case of the dog family, by both parents in a den or other secluded spot. When the young begin to eat meat the parents bring them food; dogs regurgitate meat from the stomach for this purpose. Only gradually do the young acquire hunting skill. In their play they practise the actions that they will perform seriously as adults, and cats bring home living prey for the kittens to kill. Even adult carnivores are playful. Whereas herbivores have to spend most of the time eating, carnivores, with their nutritious meat diet, have more leisure between meals in which to laze or play.

Wild members of the dog family are found on all the continents, though the Dingo (*Canis dingo*) was brought to Australia by Man. The Common or Grey Wolf (*Canis lupus*) was formerly widely distributed in Europe, Asia and North America, reaching as far south as India and Mexico, but it has been exterminated over much of its range; the last Wolf in Britain was killed in Scotland in 1743. Hunting in packs, wolves can kill animals as large as buffalo. The Coyote (*Canis latrans*) is smaller than the Wolf and feeds on smaller prey such as rabbits. Jackals, which live in Africa and India, are more timid and feed largely on carrion, though they do

BELOW: the Kodiak Bear is the largest land carnivore, weighing half a ton and is a race of the Brown Bear. It is strong enough to enter rapids to catch salmon. Bears walk on the soles of their feet, not on their toes like cats and dogs.

RIGHT: the Small Spotted Genet ranges from South Africa up to southern Europe. It hunts birds and small mammals at night, both in trees and on the ground. Like most cats, it has claws which are retractile. Its patterned coat camouflages it when hunting.

LEFT.: the Spotted Hyaena inhabits Africa south of the Sahara region. Hunting packs can kill zebra and Wildebeest, but usually young and sick animals are taken. Much of the diet is made up of carrion. Its strong jaws can crush bones for marrow.

ABOVE: the Snow Leopard (*Panthera uncia*) lives above the tree line in the Himalayas and other Central Asian mountains, where it hunts wild sheep, goats and smaller prey. It is solitary and nocturnal. Its thick fur adapts it to the cold climate.

hunt living prey. There are several species of fox, including the Arctic Fox (*Alopex lagopus*) that turns white in winter. The African Hunting Dog (*Lycaon pictus*) lives in packs and kills antelopes and zebras.

Bears live in Europe, Asia and North America, and there is one species in South America. The most widely distributed species is the Brown Bear (*Ursus arctos*), of which there are several races, including the Grizzly. The Polar Bear (*Thalarctos maritimus*) survives the arctic winter in a hole in the snow. It is a good swimmer, and eats seals and fish. The Giant Panda, from the interior of China, is probably related to the bears, but the Red Panda (*Ailurus fulgens*) of the Himalayas is a raccoon. All other raccoons are American. They are good climbers, and one of them, the Kinkajou (*Potos flavus*), has a prehensile tail. The Common Raccoon (*Procyon lotor*) of North America eats roots and fruit as well as rodents, insects and frogs.

In the family Mustelidae, the weasels (*Mustela*) are the smallest carnivores, weighing only 60 to 110 g. The Stoat (*Mustela erminea*) is also called the Ermine, for in the colder parts of its range it turns white in winter, except for the black tip of its tail. A Stoat can kill a rabbit much larger than itself by cutting the jugular vein. Related are the European Polecat (*Mustela putorius*), of which the Ferret is a domesticated variety, and the Mink (*Mustela vison*), an American species that goes into water to catch frogs and fish. There are 17 species of otter, of which the Sea Otter (*Enhydra lutris*) is the most highly adapted for aquatic life. It is found on the coasts of the northern Pacific, as far south as California. It feeds on shellfish and sea urchins. Skunks and badgers also belong to the Mustelidae.

The civet family (Viverridae) is found mainly in Africa and southern Asia, but one species, the Feline or Small Spotted Genet (*Genetta genetta*), reaches the south of France. Some civets feed largely on fruit, but mongooses are predaceous. The Small Indian Mongoose (*Herpestes auropunctatus*) can kill snakes, and it has been introduced into Hawaii and elsewhere to destroy rats.

Hyaenas, from Africa and India, have very powerful jaws and can bite through bones. They are largely scavengers, but the Spotted Hyaena (*Crocuta crocuta*) of Africa is an efficient killer, hunting in packs. The South African Aardwolf (*Proteles cristatus*) is a hyaena, and is exceptional among carnivores in feeding almost entirely on termites.

Except for the Cheetah (*Acinonyx jubatus*), all cats have retractile claws, and most can climb trees. The Cheetah runs its prey down in the open, but other cats stalk their prey and pounce on it, killing by biting the neck. The Lion (*Panthera leo*) is the only species that lives in social groups (known as 'prides'); other cats are solitary animals. There are 37 species, distributed worldwide except in Australia and Madagascar. The extinct sabre-toothed cats (for example *Smilodon*) could kill the largest mammals by stabbing with their enormous canine teeth.

The Walrus lives at the edge of the Arctic ice. The tusks grow throughout life and are used for fighting, excavating molluscs on the sea bed and for climbing onto the ice. It locates food by touch, using the bristles that fringe the upper lips.

The Pinnipedia are so different from other Carnivora that some experts regard them as a separate order. However, details of skull structure show that the sealions and fur seals (family Otariidae) and the Walrus (Odobaenidae) are related to bears, and the true or earless seals (Phocidae) are closer to otters. The animals that perform in circuses are Californian Sealions (*Zalophus californicus*). Pinnipeds are superb swimmers; several species have been recorded as diving to depths of more than 90 m, and remaining under water for 20 minutes or more. A layer of fat (blubber) beneath the skin enables them to live in Arctic and Antarctic waters. On land they are awkward, especially the Phocidae, which cannot stand on their feet but get along in a caterpillar-like motion by arching and straightening the back. Seals come ashore to give birth and to suckle their young and most species congregate on beaches in large numbers at breeding time. The males arrive first and establish territories which they defend against other males; after the young are born the females mate with the male in whose territory they lie. The Common or Harbour Seal (*Phoca vitulina*), however, gives birth on a sandbank between one tide and the next. The largest seals are the sea elephants (*Mirounga*), which live in the Antarctic: the male can reach a length of 6 m, but the female is only half this size. Seals feed mainly on fish and marine invertebrates, but the Leopard Seal (*Hydrurga leptonyx*) has a preference for penguins. The Walrus (*Odobenus rosmarus*) feeds on the sea bottom on molluscs which it digs up with its tusks—greatly enlarged canine teeth.

The most completely aquatic mammals are the whales, porpoises and dolphins of the order Cetacea, which spend their entire lives in water and die if they are stranded on a shore. With their streamlined bodies and powerful tails, they are the fastest swimmers in the sea: a dolphin has been timed to keep up a speed of 24 to 32 kilometres an hour for long periods. Whales are also good divers, some species going down to depths of over 900 m. The Sperm Whale (*Physeter catodon*) can remain submerged for more than an hour. Being mammals, cetaceans are air-breathing animals, and when at the surface they breathe through the 'blow-hole' on the top of the head—the equivalent of the nostrils in other mammals. They have acute hearing, and communicate with each other by making a variety of sounds. They also produce ultrasonic pulses which they use for echolocation in food-finding and navigation. Cetaceans have large brains and are very intelligent. Small species like dolphins (for example, *Tursiops*) when captured become very tame and can easily be trained.

The 84 species of cetaceans are divided into two sub-orders: the toothed whales (Odontoceti) and the whalebone whales (Mysticeti). Whalebone forms a series of parallel horny plates that hang down from the upper jaw along the margins of the mouth. They form a sieve which lets water pass through but strains out the small aquatic animals (mainly shrimp-like crustaceans known as 'krill') on which

Despite its scientific name, the Southern Fur Seal (*Arctocephalus tropicalis*) is an Antarctic species. Breeding 'rookeries' form on the shores of islands. The large bulls fight each other to establish harems and mate after the pups are born.

The Minke Whale (*Balaenoptera rostrata*) is one of the smaller whalebone whales, about 9 m long. These whales feed on plankton, obtained by filling the mouth with water which is forced out at the sides through whalebone sieves. Furrows in the skin below the head allow the mouth chamber to expand.

the whale feeds. After filling its mouth the whale, by raising its tongue, forces the water out through the whalebone screen and swallows the krill. It is surprising that on such a diet some whalebone whales can grow so large: the largest, the Blue Whale (*Balaenoptera musculus*), is nearly 30 m long and weighs 120 metric tons or more—the largest animal that has ever existed. Whalebone whales perform regular migrations between feeding and breeding grounds: the Grey Whale (*Eschrichtius gibbosus*) travels every year down the Pacific coast of America to breed off Baja California.

Toothed whales feed mostly on fish and squid. Some of them are comparatively small: the Common Porpoise (*Phocaena phocaena*), for example, is only 180 cm long. The largest of the toothed whales is the Sperm Whale, which can reach a length of 20 m. The Killer Whale (*Orcinus orca*) is exceptional in that it hunts in packs, attacking seals and also other whales.

Whales have no hair, except for a few bristles near the mouth, but they have a thick layer of blubber beneath the skin. This not only acts as an insulator in cold water but it forms a reserve of energy that can be drawn upon during migrations. Whales need to be able to swim immediately from birth, and they are born in a very advanced state: the newborn Blue Whale is 10.5 m long. Young whales cannot suck, but the mother squirts milk down the throat of the young with a special muscle.

The small order Sirenia contains the Dugong and manatees. Though fully aquatic, they are not related to whales but probably have a common ancestry with the elephants. They live in shallow water near the coast or in rivers, feeding on the bottom on aquatic plants, and come to the surface every few minutes to breathe. The very muscular upper lip is used in feeding. The skin is hairless except for some bristles round the mouth; there is a thick layer of blubber. The front legs are flippers—there are no hind legs—and the tail is fish-like but with horizontal flukes as in whales. There is a pair of nipples on the breast. The Dugong (*Dugong dugon*) is found on the shores of the Indian Ocean, from Africa to Australia, and formerly lived in the Mediterranean. It is probably the origin of the mermaid legend. One species of manatee (*Trichechus manatus*) ranges from Florida to the Caribbean, another (*T. inunguis*) is found in the Amazonian region and a third (*T. senegalensis*) lives along the coast of West Africa.

Manatees are fully aquatic, with flippers but no hind legs. They spend most of the time feeding on aquatic plants on the bottom, coming to the surface every so often to breathe. To the list of their unusual features can also be added very heavy bones.

ABOVE: the Rock Hyrax (*Procavia capensis*) of Africa and south-west Asia is the cony of the Bible. It is herbivorous, living colonially in burrows among rocks. The large incisors are used for defence. The soles of the feet are moist and rubbery and enable them to climb steep, smooth surfaces.

BELOW: a cow African Elephant (*Loxodonta africana*) with two calves, one about a year old, the other about six. Females and their offspring live in family groups, the older calves helping to care for the younger ones. The African Elephant has larger ears than the Indian species and the female has tusks.

The next four orders are collectively referred to as ungulates, a name which refers to the fact that in place of pointed claws on their feet they have blunt hoofs. They are herbivorous animals with molar teeth adapted for grinding. Nearly all run on their toes, and only the elephants still have the full complement of five toes on each foot. Ungulates are born in an advanced state of development and the young are able to follow the mother very soon after birth.

The hyraxes, which make up the order Hyracoidea, are found only in Africa and at the eastern end of the Mediterranean (the cony of the Bible). Rabbit-sized, they are among the smallest of ungulates but they look like Guinea-pigs, having small, rounded ears and no tail. Pads on the under surface of the hands and feet, kept moist with skin glands, enable hyraxes to climb almost vertical surfaces of rocks and tree trunks. At the front of the upper jaw there is a pair of permanently growing incisor teeth, used as weapons.

The order Proboscidea contains the two species of elephant. The African Elephant (*Loxodonta africana*) is the largest land mammal: it can weigh more than six metric tons. It has larger ears and longer tusks than the Indian Elephant (*Elephas maximus*), in which tusks develop only in the male. The most remarkable feature of elephants is of course the trunk. This is a very long and mobile nose, with the nostrils at the end. Because of its short neck, an elephant cannot reach the ground with its mouth, but it uses the trunk to feed, wrapping it round bunches of leaves or grass, or picking up small objects with a finger-like projection at

the tip. Drinking is performed by sucking water up the trunk and squirting it into the mouth. Young elephants, however, suckle directly through the mouth.

Elephants share with Man the distinction of being the longest-lived mammals, reaching an age of 60 to 70 years. Pregnancy lasts about 21 months, the calf continues to suckle for nearly five years, and it becomes mature when about 15. Elephants' brains are larger than those of men, and the animals are very intelligent. They go about in herds, consisting of a leading female with her daughters and their offspring, to which adult males temporarily attach themselves. Elephants have huge grinding teeth. In each jaw only one of these is used at a time; as it wears down it moves forward and another tooth develops behind it to take its place. Six such teeth are used up in the course of the elephant's long life.

Fossils show that the Proboscidea originated in Africa and spread from there widely over the world, reaching even to South America. The mammoths (*Mammuthus*) inhabited northern pine forests and tundra and were hunted by Stone-age Man. Mammoth bodies are found preserved in ice in Siberia. Dwarf elephants formerly lived in Malta and other Mediterranean islands.

Horses, rhinoceroses and tapirs are the three families of the order Perissodactyla. They are called the odd-toed ungulates, because their feet have three toes (nos. 2, 3 and 4) or, in the horses, only one (no. 3); however, tapirs, though having three hind toes, have four on the fore foot as no. 5 is still present. There are only 15 living species of peris-

into a very mobile proboscis, like a short version of an elephant's trunk. Tapirs feed at night on leaves.

Rhinoceroses (Rhinocerotidae) are also much more restricted today than in the past. In Africa there are the White Rhino (*Ceratotherium simum*) and the Black Rhino (*Diceros bicornis*), both of which have two horns. In India, Java and Sumatra there are three other species, all in danger of extinction. The horns develop from the skin and are formed of agglutinated hairs; they regrow if accidentally broken off. Of the two African species, the Black Rhino has a pointed, prehensile upper lip and eats leaves, while the White Rhino has a square upper lip and eats grass. The hairlessness of rhinoceroses and elephants is an adaptation for losing heat, for in tropical climates such large mammals are liable to become over-heated. Mammoths and woolly rhinoceroses (*Coelodonta*) of ice-age Europe had a thick hair covering.

The Wild Horse (*Equus przewalskii*) formerly lived in Central Asia but now survives only in zoos. Wild asses (*E. hemionus* and *E. asinus*) still exist in North Africa and southwest Asia, and there are three species of zebra (*E. burchelli*, *E. grevyi* and *E. zebra*) in Africa. These form the family Equidae. They are inhabitants of open country, feeding on grass and adapted for running. Numerous fossil horses from North America show that the family originated on that continent from forest-living, leaf-eating animals of the size of fox terriers. They walked on pads underneath the toes, and had three toes behind and four on the front feet. As the climate became drier the prairie grasslands

BELOW: the Brazilian Tapir (*Tapirus temestris*), inhabits the tropical forests of the Amazon Basin. It hides by day, browsing at night on leaves and shoots picked with its proboscis. This flexible trunk-like structure is a modification of the nose and upper lip. Tapirs are usually found near water and their flat feet are well suited for walking on soft ground.

RIGHT: the White Rhinoceros weighs three metric tons or more. Its massive legs have three short toes ending in hooves. The thick skin is almost hairless, but the horns on the nose are formed of hairs fused together. The wide mouth has a square upper lip, adapted for feeding on grass. Only a few hundred individuals survive in the wild and these are mainly in game reserves.

sodactyls, the last survivors of an order that was much more important in the past.

Tapirs (Tapiridae) are heavy-bodied animals with short legs that live in dense tropical forest, especially near water. They have an unusual distribution: one species in Malaya and adjacent countries, and three in tropical America. However, fossil tapirs are found in North America, Europe and China, showing that the existing species are relics of a once more widespread group. In the tapirs the nose is developed

expanded and horses became adapted to live there. Their teeth became taller to withstand the heavier wear resulting from grass-eating, and their feet were modified for running on hard ground. The side toes diminished and eventually disappeared, so that the weight was supported only on the hoof of the third toe. Horses spread to Asia and Africa, but they subsequently died out in America; the wild horses found there today are descendants of domestic horses brought in by the Spaniards.

ABOVE: two widely differing artiodactyls at an African water-hole. The Impala (*Aepyceros melampus*—Bovidae) is noted for the spectacular leaps it makes when running. Only the male has horns. The Warthog, unlike most Suidae, does not dig but feeds on grass. The tusks are undeveloped in the young animals seen here.

The order Artiodactyla, or even-toed ungulates, contains the 'cloven-hoofed' animals, with two equal toes (nos. 3 and 4), supplemented in some families by side toes (2 and 5). There are nine families, grouped in three suborders.

Pigs (Suidae), peccaries (Tayassuidae) and hippopotamuses (Hippopotamidae) form the suborder Suiformes. Pigs and peccaries root in the ground with their flat snouts. They are omnivorous, eating tubers, acorns, worms, carrion, and even rats and snakes. Their molar teeth have blunt cusps adapted for crushing. The upper canine teeth of pigs form tusks which point sideways and curve upwards; they make formidable weapons. In peccaries (*Tayassu*) the canines point downwards. Pigs occur wild in Europe, Asia and Africa. The Wild Boar (*Sus scrofa*) was exterminated in Britain in Middle Ages; it is the ancestor of the domestic pig. The Warthog (*Phacochoerus aethiopicus*) is still fairly common in East and South Africa. Peccaries are found in Central and South America, the Collared Peccary (*T. angulatus*) reaching the southern border of the United States.

The Hippopotamus (*Hippopotamus amphibius*) spends the day in African rivers and lakes where it rests with only its nostrils, eyes and ears above the water, but at night it goes ashore to feed on plants, sometimes travelling a long way from the water. It marks its paths and grazing territory with deposits of dung, scattered by rapid wagging of the tail. The canine and incisor teeth are developed into tusks with which the males fight in the breeding season. In West Africa there is a second species, the Pigmy Hippopotamus (*Choeropsis liberiensis*), of the size of a pig. It is less aquatic, living in forest and entering rivers only to bathe.

The remaining families of artiodactyls are ruminants, which chew the cud. When food is first swallowed it passes into a large chamber of the stomach known as the rumen, where microorganisms (bacteria and protozoans) act upon it. It is later returned to the mouth and chewed. The microorganisms in the rumen digest the cellulose cell walls of plants on which the ruminant's own digestive juices cannot

operate, and thus the animal is able to get more nutriment from its food. On the molar teeth of ruminants the cusps are drawn out into crescent-shaped ridges with which the cud is ground up as the jaw is moved from side to side.

The suborder Tylopoda contains only the family Camelidae. Besides the camels of the deserts of Asia and Africa this includes the Llama of the mountains of South America. Camelidae originated in North America. They have long legs with only two toes. They walk, not on the hoofs, but on pads underneath the toes. There are two species of Camel: the Bactrian Camel (*Camelus bactrianus*) of Central Asia, with two humps, and the Arabian Camel (*C. dromedarius*) with one; the Dromedary is a breed of the Arabian Camel.

BELOW: Arabian Camels are well adapted for life in the desert, being able to withstand high temperatures and go without food and water for six days or more. The flat feet enable them to walk on sand; the hump is a store of fat, used as a food reserve.

The hump is a store of fat. Except for a few Bactrian Camels in the Gobi Desert there are no truly wild camels, but domestic or feral stock is widely distributed in desert regions of the earth, including Australia. Their ability to go for long periods without water and to survive on prickly desert vegetation has made them invaluable to desert-living peoples. In South America the Incas domesticated the Llama (*Lama peruana*) as a pack animal and the Alpaca (*L. pacos*) for its wool; these species do not occur wild, but there are two related wild species, the Vicuna (*Vicugna vicugna*) and the Guanaco (*L. guanacoe*).

Members of the suborder Pecora have no upper incisor teeth. When feeding, the very flexible tongue is wrapped round plants or tufts of grass, which are cut off with the sharp-edged lower incisors against a tough pad at the front of the upper jaw. There are five families of Pecora. The most primitive are the chevrotains (Tragulidae), which live in tropical forests of southern Asia and West Africa. They stand only 30 cm high. They have four toes on the foot, and long, downwardly pointing upper canine teeth.

The remaining families have horns, differing in structure from family to family. The simplest are those of giraffes (Giraffidae) and consist simply of bony knobs on the skull, covered with hair. In deer (Cervidae) they are known as antlers and, except in Reindeer, they develop only in the males. Antlers grow anew every year and at first are covered with hairy skin ('velvet'), but this peels off, leaving bare

LEFT: the Giraffe plucks leaves from high branches by using its very extensible tongue. Its height gives it an excellent view and it has very good sight. The short horns are covered with hair. On this one's neck is a tick-bird (*Buphagus*) which eats parasites.

ABOVE: the Rocky Mountain Goat (*Oreamnus americanus*) is a sure-footed resident of steep mountain slopes from Alaska to Idaho. Its thick white hair enables it to live above the snow-line where it eats lichens and alpine plants. Both sexes have beards and horns.

bone. After the rutting season the antlers are absorbed at the base and shed. The male Musk Deer (*Moschus moschiferous*) does not develop antlers, but fights with his enlarged canine teeth. In the family Bovidae the horns are permanent structures, consisting of a bony core covered by a horny sheath. Sometimes they are developed in both sexes. The North American Pronghorn (*Antilocapra americana*) is unique in that it sheds the horn-sheath annually, and for this reason it is put into a separate family, Antilocapridae.

Deer are browsing animals, feeding on the tenderer parts of plants. There are some 40 species, of which the largest is the Moose (*Alces alces*), called the Elk in Europe. The Reindeer, or Caribou (*Rangifer tarandus*), inhabits the Arctic regions of both the New and the Old World. The Giraffe (*Giraffa camelopardalis*), which stands nearly 5·5 m high, is the tallest living mammal. It feeds on the leaves of trees, especially Acacia, in East and South Africa. In the same family is the Okapi (*Okapia johnstoni*), discovered in the Congo forests as recently as 1901.

There are over 100 species of Bovidae. Wild species of sheep and goats are found from the Mediterranean area to Central Asia, and there are also two species in the Rocky Mountains of North America. The Musk Ox (*Ovibos moschatus*) lives on the tundra of northern Canada. The cattle tribe includes the American Bison, or Buffalo (*Bison bison*), the European Bison (*B. bonasus*), the African Buffalo (*Syncerus caffer*) and the Yak (*Bos grunniens*) of Tibet. The greatest variety of antelopes lives on the grasslands of Africa, from the Eland (*Taurotragus oryx*), standing 1·8 m high at the shoulders, to the little duikers (*Cephalopus*) of fox-terrier size. Wildebeests or gnus (*Connochaetes*) make spectacular annual migrations. Gazelles (*Gazella*) inhabit dry areas of Africa and Asia.

White-bearded Gnus (*Connochaetes taurinus*) grazing in association with zebras (*Equus burchelli*) in East Africa. In the Serengeti area these animals migrate in vast numbers in search of grass, going south for the rains and north in the dry season.

Finally we come to the order Primates. Although some of them, notably including Man, now live on the ground, the primates evolved as tree-living animals. They developed hands and feet adapted to grip branches, the thumb or big toe opposing the other digits. The grip was improved by the presence of pads under the tips of the fingers and toes, with ridges that in Man produce fingerprints. The claws were flattened to form nails. The limbs became long and mobile; for instance, primates can turn the hand so that the palm faces upwards. The hands were used not only for climbing but for picking up food and thus primates have lost the long neck required by animals that put their heads to the ground when feeding. As they lived in trees, sight took precedence over smell, and with reduction of the nose the eyes could come together, giving primates three-dimensional (stereoscopic) vision. Their ability to focus the eyes upon an object held in the hand, combining touch with sight, gave them a more detailed understanding of their surroundings. The hands are also used to groom the fur, including that of other individuals. Primate brains are larger than those of other mammals of comparable size.

These features have not developed to an equal extent in all primates. The suborder Prosimii contains a number of animals that have remained at a lower stage of evolution than the other members of the order. Found extensively as

ABOVE: the Aye-aye is nocturnal and lives in the dense forests of northern Madagascar. It locates wood-boring insects by tapping the bark of trees with one finger and listening for hollow sounds. It is about the size of a squirrel and is now extremely rare.

BELOW: the Bushbaby (*Galago senegalensis*) is a great leaper. With a body length of only 20 cm, it can leap up to 5 m. The hind feet have elongated ankles to give extra leverage, and both fingers and toes are equally effective in gripping small branches.

fossils in the Eocene of Europe and North America, they are now confined to the tropics of the Old World. They are divided into three infraorders: the Lemuriformes in Madagascar, the Lorisiformes of Africa and southern Asia, and the Tarsiiformes, containing only tarsiers, from the Philippines, Borneo and Sumatra. All have claws on at least some of their digits. They are mostly nocturnal animals, with a better sense of smell than other primates.

There are 20 species of Lemuriformes in Madagascar. Most often seen in zoos is the Ring-tailed Lemur (*Lemur catta*), an agile tree-climber with pointed nose and long, bushy tail. Unlike most prosimians it is active in the daytime. It feeds largely on fruit but also eats insects and eggs. The mouse lemurs (*Microcebus*) are the smallest primates; they eat mostly insects. The Indri (*Indri indri*) has no tail, but its hind legs are long. In trees it clings to vertical branches and can jump from one branch to another; on the ground it hops on its hind legs. It eats leaves. Most Lemuriformes have horizontally arranged lower front teeth that they use as a comb when grooming the fur. The Aye-aye (*Daubentonia madagascariensis*) is exceptional, for it has large, persistently growing incisor teeth like a rodent. With these it gnaws into wood to obtain wood-boring insects. The middle finger of the hand is slender and bears a claw, used to extract insects from their burrows.

The lorisiform group includes the galagos or bushbabies (*Galago*) of East Africa, which leap in trees using their long hind legs and on the ground hop like kangaroos. The lorises are slow-moving animals without tails, that live in tropical forests of southern Asia. Related to them is the Potto (*Perodicticus potto*) of African forests. Lorisiformes are nocturnal animals which feed largely on insects, but also eat a variety of fruit.

Tarsiers (*Tarsius*) have powerful hind legs and are good leapers: though its body is only 15 cm long a tarsier can jump 180 cm. Flat pads at the ends of the fingers and toes enable it to land on vertical tree trunks. It has huge eyes, and can turn the head so as to look directly backwards. It feeds on insects and small reptiles. Some of the anatomical characters show that tarsiers are more closely related to the monkeys than are the other prosimians.

Monkeys, apes and Man form the suborder Anthropoidea. There are two infraorders—the Platyrrhina of the forests of tropical America and the Catarrhina of the Old World. In the Platyrrhina the nostrils face sideways and are more

The Black Spider-monkey (*Ateles paniscus*). Spider-monkeys are the most acrobatic climbers among the Cebidae and use their arms and prehensile tails much more than other monkeys. They feed on fruit and nuts and seldom descend to the ground.

widely separated than are the forwardly or downwardly opening nostrils of the Catarrhina. The Platyrrhina comprises two families, the marmosets (Callithricidae) and the New World monkeys (Cebidae).

Marmosets are in some ways the most primitive of the Anthropoidea. They are small animals that climb like squirrels and have claws on all the digits except the big toe which bears a nail. Their brains are simpler than those of other anthropoids. The diet consists of insects and fruit. Usually two young are produced at a birth, and in some species they are carried about mainly by the male.

Most Cebidae have prehensile tails which can be curled round branches while climbing. Their thumbs, however, are small or even absent. Spider monkeys (*Ateles*) use their hands as hooks when swinging from branch to branch; they are very acrobatic, moving through the forest by leaping from tree to tree. Howler monkeys (*Alouatta*) have very large larynxes and the noise made by a troop, proclaiming the possession of its territory, can be heard miles away. Also in the Cebidae are the Squirrel Monkey (*Saimiri sciurea*), the capuchins (*Cebus*), woolly monkeys (*Lagothrix*) and the only nocturnal monkeys, the douroucoulis, or owl monkeys (*Aotes*).

LEFT: tarsiers are confined to southeast Asia. Their enormous eyes (which are, size for size, the largest in all the mammals) reflect their strictly nocturnal habits. Like the bushbabies they are all prodigious leapers.

BELOW: the Cotton-headed Tamarin (*Saguinus oedipus*—Callithricidae) has a tuft of long white hair on the head. Tamarins, like marmosets, use their claws in climbing. The young are carried by the male but are still fed by the female.

ABOVE: Yellow Baboons (*Papio cynocephalus*) at a water hole in Africa. Baboons live in tribal groups of 20–100 individuals. Several different baboon species occur throughout Africa.

BELOW: a Grey Gibbon (*Hylobates moloch*) from Borneo. Gibbons live in permanent family groups that communicate noisily with other such groups in a chanting chorus of distinctive calls.

The infraorder Catarrhina consists of the Old World monkeys (Cercopithecidae) and the Hominoidea, a group containing the anthropoid apes (Pongidae) and Man (Hominidae). The Cercopithecidae have tails which are not prehensile but are used as balancers when climbing. The tail is short in the macaques which spend most of their time on the ground. When at rest Old World monkeys sit with the body upright, supported on patches of hardened bare skin (callosities) on the buttocks. The colobuses (*Colobus* and *Procolobus*) of Africa and the langurs (for example *Presbytis*) of southern Asia live in tropical forest where they seldom come to the ground. Their thumbs are small or absent, and they use the arms for swinging, like the spider monkeys of America. They feed on leaves. The guenons (*Cercopithecus*) of African forests run on the branches of trees but do not swing with their arms. They eat shoots, fruit and insects. The Patas Monkey (*Erythrocebus patas*) lives in the African savannah and is less arboreal; the troop feeds on the ground while the leading male keeps watch for predators from a tree. Baboons (*Papio*) also live in open country, but sleep in trees or on cliff ledges. The macaques (*Macaca*) are a mostly Asiatic group, one species occurring as far north as Japan. Other species of macaque are the Rhesus Monkey (*M. mulatta*) of India, much used in medical research, and the Barbary Ape (*M. sylvana*), a colony of which lives on the Rock of Gibraltar.

Anthropoid apes resemble Man in most features of their anatomy, including the absence of a tail. They differ from him, however, in having long arms and short legs, a characteristic associated with their habit of brachiation. This is a form of locomotion in which the animal, suspended by its arms, swings hand over hand through the trees. Particularly agile brachiators are the gibbons (*Hylobates*), the smallest of the anthropoid apes. There are eight species in southeast Asia. The Orang-utan (*Pongo pygmaeus*), of Borneo and Sumatra, and the Gorilla (*Gorilla gorilla*) and Chimpanzee (*Pan troglodytes*) of Africa (together called the 'Great Apes'), being heavier, are less nimble than gibbons and make more use of their feet as auxiliary supports when climbing. The Gorilla and the Chimpanzee spend much of their time on the ground, where they walk on all fours, supporting themselves on their knuckles. They can stand upright, and the Chimpanzees have been observed to run on two legs when carrying, for example, bunches of bananas. Gibbons on the ground walk on their hind legs and use their arms for balance, but the only truly bipedal primate is Man.

The human foot is well adapted for walking. The big toe is no longer prehensile but it lies alongside the other toes; there is a heel, and the middle of the foot is arched. The spine is curved in an S-shape, so that the centre of gravity of the body is situated vertically over the feet when standing, and the head is balanced on top of the neck instead of being slung forward as in apes. The arms and hands, no longer needed for locomotion, can be used for carrying objects such as tools and weapons. Not only can the hand be used for gripping, but the thumb, longer than in apes, can make contact with the tips of all the fingers, thus enabling small objects to be handled in a very precise manner. Chimpanzees have been observed to use twigs to extract termites from holes, and bunches of leaves to obtain drinking water from tree hollows; they also throw stones and branches to repel predators. No doubt the human use of tools and weapons developed from such simple beginnings. The refinement of tools and the versatility of the hand evolved together over the last two million years.

In primate evolution there has been a progressive enlargement of the brain. It is larger in monkeys than in prosimians, and larger still in the anthropoid apes, but the brain of Man is 2.7 times as large, on average, as that of the

ABOVE: young Chimpanzees develop more quickly than human babies. They begin to walk and climb at 5 months, and a female is sexually mature at 12 years. Here a young chimp rides on its mother's back.

BELOW: the Mountain Gorilla (*G. gorilla beringei*) is confined to eastern Zaire, where about a thousand survive. Gorillas are the largest of primates, and they are strictly vegetarian.

mated pair and their offspring; each family occupies a territory that it defends against other gibbons. At the other extreme there are the large tribal groups of baboons which may contain a hundred individuals, with no permanent pairing but an elaborate social hierarchy. Members of a community show their feelings to each other by facial expressions (often made more obvious by beards and other forms of facial hair) and by a variety of vocal calls. Like the baboons, primitive man lived on the ground in open, savanna country where he was vulnerable to attack by carnivores, and banding together was necessary for defence. Not possessing the formidable canine teeth of male baboons, man defended himself with weapons held in the hand.

An important factor in human social evolution was that Man became a predator himself. Baboons and Chimpanzees sometimes kill and eat other animals, but like primates generally they are essentially vegetarian; primitive man, however, was to a large extent carnivorous. Hunting was probably a male occupation, while the females would remain at the camp site to care for the young and to collect plant food. Such a division of labour requires a capacity to share food, almost absent in other primates but characteristic of social carnivores like wolves. Whereas in other primates the mothers have to carry their young about while searching for food, the more helpless human young could stay in the comparative safety of the home, and the period of childhood could be prolonged. Nothing is known about the origin of language, by which is meant the communication of information and ideas, as distinct from the expression of emotions. No doubt the close cooperation required in primitive human society would make such a development highly advantageous, and the enlarged human brain would make it possible.

Gorilla. *Australopithecus*, a primitive form of man that lived in Africa from over three million to about one million years ago, walked upright, but his brain was about the same size as that of a Gorilla. Growth of the modern human brain has taken place in the last two million years. Our species, *Homo sapiens*, first appeared about 250,000 years ago.

Man is of course a social animal, and in this respect he resembles monkeys and apes, all of which spend their lives as members of social groups. The organization of these groups differs very greatly among the various species. Gibbons, for example, live in family groups consisting of a

As his technical knowledge advanced, Man became the most formidable of all predators. By the use of fire, Palaeolithic man was able to stampede wild horses over cliffs and to drive elephants such as mammoths into swamps where they could be killed. Man is believed to have been a contributory cause of the extinction of a number of species of large mammal in prehistoric times. Animals were hunted not only for meat but for skins and bone; Neolithic people in Britain used deer antlers as picks and shoulder-blades as primitive but effective shovels.

The domestication of sheep, goats, pigs and cattle in the Near East in the period between 9000 and 6000 BC led to a decline in the importance of hunting, and today a hunting economy survives only in a few isolated human populations, as among the bushmen and the Australian aborigines. Civilized people have continued to hunt, however, partly as a sport and partly because of the commercial value of certain animal products such as ivory and furs. Elephants have been hunted for ivory since ancient times: a population of the Indian Elephant that inhabited Syria until it was exterminated about 800 BC formed the basis of an important ivory-carving industry. The main source of ivory in recent times has been the African Elephant, which, since the development of effective firearms in the 19th century, has suffered a serious decline in numbers.

The most valuable furs are provided by species of the orders Carnivora and Rodentia, and especially those of the northern forest belt. Furs of the Ermine and the Sable (*Martes zibellina*) were traded from Russia to Western Europe in the Middle Ages, and after the discovery of North America fur trappers played an important part in exploring the interior of that continent. In recent years the farming of fur-bearing animals such as the Mink and Arctic Fox has reduced some of the pressure on the wild species.

Bushmen hunters cutting up an antelope in South Africa. Most bushmen work on cattle ranches, but a few thousand still live a roaming, hunting existence. This was the normal way of life for men everywhere before the spread of domestication.

Whales have long been hunted for their oil, which was formerly used for lighting and lubrication, and now mainly as a source of edible fats. Developed on a large scale in the North Atlantic in the 17th and 18th centuries, whaling spread to the Pacific in the 19th century. With the invention of the harpoon gun in 1869 and the use of factory ships in the present century even the largest species could be exploited. As a result, whale populations have fallen dramatically: it was estimated that by 1965 the Blue Whale was reduced in the Antarctic to 1 per cent of its former numbers.

Though Man has hunted many species, some of them to extinction, he has caused a few to multiply under domestication. The first of these was the wolf, which became the domestic dog. Perhaps the relationship began when scavenging wolves frequented encampments of hunters. Wolf puppies less than six weeks old are easily tamed and readily accept man as master. Their natural abilities would have made them useful in hunting or as guard dogs. It is thought likely that the earliest breeds of dog resembled the pariah dogs of the East and the Dingo of Australia, but a number of other breeds were soon established. Skeletons from archaeological sites in Europe reveal the existence of a small house-dog resembling a terrier, a primitive sheep dog, a primitive hound and a large, wolf-like dog which may have been ancestral to the alsatian. The ancient Egyptians had breeds resembling pomeranians, mastiffs and greyhounds.

The dog would be useful for rounding up herds of wild sheep and goats, and these species were domesticated very early in southwest Asia. Goats provided milk as well as meat and skins, and sheep were an important source of fat.

In the more primitive breeds of sheep the wool is short and covered by a sleek haircoat; increased thickness of wool has been produced by selective breeding.

Cattle are derived from the Aurochs, a large wild ox, now extinct, which was hunted by Stone-age Man and which he depicted in paintings on the walls of caves. The last Aurochs died in Poland in 1627. Besides supplying milk, meat and hides, cattle were used as draught animals to pull carts and ploughs and still serve this purpose in many areas.

Horses were probably first domesticated in Turkestan at some time before 2500 BC. They are derived from the Tarpan, which formerly lived on the Russian steppes but was exterminated in 1851. The horse gave man a greater mobility than he had ever had before, and its use in warfare, whether for riding or for pulling a chariot, has had a considerable effect on human history.

Elephants have been domesticated in India since about 2500 BC and they are still used there, particularly in the timber industry. In former times they were used in war and those with which Hannibal crossed the Alps were probably African Elephants, belonging to a small-sized race, now extinct, that inhabited North Africa.

Thus the coming of Man has had a profound influence on the other mammalian species. To the direct effects of hunting and domestication must be added those which have resulted from man-made changes to the environment, causing the widespread decline of numerous species throughout the whole animal kingdom, as well as the often undesirable increase of others. Despite the relatively short time over which all these influences have been operating, their effect has been so enormous and so wide-reaching that Man is now unquestionably the most significant of all animal species. It remains to be seen whether his intelligence is sufficiently developed to enable Man to maintain his dominant role and at the same time organize his activities well enough to live in harmony with the rest of the animal world.

BELOW: Man's domestication of animals transformed his life, giving him greater mobility and a regular supply of food. Even today, many activities—such as Australian sheep farming—still rely heavily on our relationship with other mammals.

ABOVE: logging timber with the help of an elephant. The Indian Elephant, which can pull a log weighing 2 tonnes, has long been used in southern Asia. Wild elephants are caught when 15 to 20 years old, and trained over a period of several years.

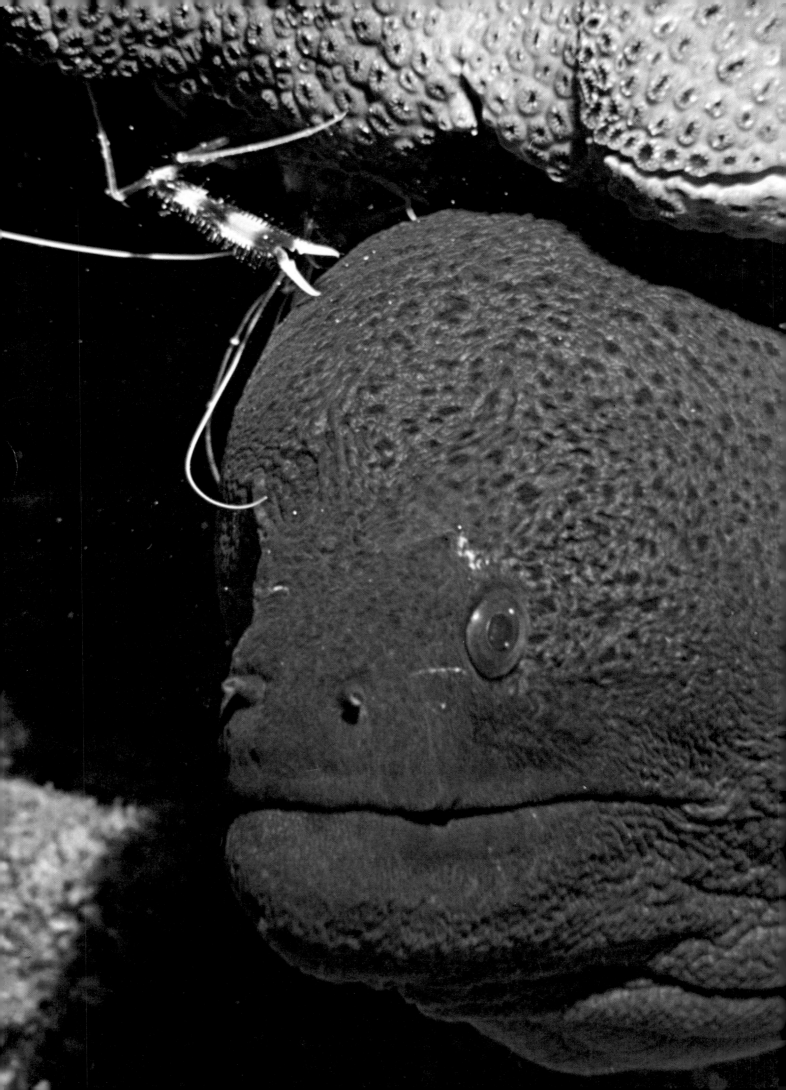

The Ways of Animals

If living organisms can be said to have a purpose it is to reproduce themselves. All other characteristics and activities of life are directed to this one end. Even Man, although he seems to have developed a greater capacity than any other animal to enjoy life for its own sake, by no means shuns reproduction! Man is unusual in producing a relatively small number of offspring per parent. At the opposite extreme there are some animals, such as tapeworms, which have become virtual egg-producing machines, possessing only the necessary minimum of other organ systems. This reflects another basic principle of animal ways—an animal rarely does anything that it does not have to do. Before it can reproduce, however, the animal must survive to maturity, and a wide and sometimes bewildering variety of adaptations has evolved to maximize the chances of achieving this. To survive and grow, an animal must be able to find a food source, and to assist the offspring in this respect many species either deposit their eggs on or near some suitable food, or else guard the eggs and sometimes even the hatched young for a period. Parental care is particularly well developed among vertebrates, especially birds and mammals.

An important prerequisite for finding food is some system for detecting its presence. Many senses are specific adaptations for this purpose, such as the echolocation facility of many bats which enables them to locate insect food in the dark. Although most animal senses are concerned with food finding, many are also of value in the detection of potential predators and mates. Thus smell is widely used by herbivorous mammals to give warning of danger, and by many species as a means of attracting mates.

Having detected a mate, an enemy or a source of food, many animals respond by physically moving in the direction of the stimulus (or away from it, if appropriate). A large number of invertebrates, however, are incapable of locomotion, being fixed to the substrate, and in such cases the appropriate action is to try to induce the stimulus to move. Both these reactions require the coordination of detectors (that is, the sense organs) and effectors (the muscles of locomotion or other movement). This is achieved through the nervous system found in all metazoans, ranging from the simple nerve net of the coelenterates to the highly centralized nervous system of Man.

Locomotion is important not only in obtaining food but also in avoiding being caught and used as food by other animals. As both predator and prey need to survive, the constant struggle to keep one step ahead has resulted in a wide range of locomotory adaptations and, more importantly, greatly increased efficiency of locomotion over that credited to archaic animals. Fleeing from one's enemies, however, is only one method of survival. Many animals adopt alternative strategies: for example, sedentary animals such as barnacles and fan worms can withdraw into a protective shell or tube. Other animals try to deceive the predator by pretending to be something else: the camouflaged mantis which resembles a leaf, or the palatable butterfly which mimics a poisonous and unpalatable species are examples of this. Some animals, however, are capable of actively defending themselves against predators: the quills of a porcupine, the stench of a skunk and the sting of a wasp are all for this purpose. The effectiveness of this defence is enhanced if the prey is able to persuade the predator that it is not even worthwhile trying to penetrate its defences. Such prey species usually have a distinctive behaviour pattern, such as the porcupine rattling its quills, or an equally well remembered visual pattern, like the black and white markings of a skunk or the yellow and black stripes which are characteristic of many insects, including wasps.

Two possible ways of increasing the chances of survival are either to live where there is a constant supply of food or to live where there is effective protection from one's enemies—or preferably both. In some cases these criteria can be met by associating with some other animal species. It is of course common for a small herbivorous animal to live on its food plant but the phenomenon of association between animal species is also widespread. Most of these associations are very one-sided, one species being totally dependent upon another for both food and protection. The malaria parasites, for example, are incapable of living anywhere other than in the liver and red blood cells of certain vertebrates and, at a different stage in the life cycle, in a mosquito.

Competition between animals for resources in any one environment reduces the individual's chances of survival and it is obviously advantageous to live where such pressure is minimal. Often, however, this is possible only in relatively inhospitable environments; for example, where it is particularly cold or arid, or somewhere that is subject to erratic food supply. Animals colonizing such places must be equipped with special strategies for survival. One widely used strategy involves restricting visits to the new environment to times when conditions are most favourable—the phenomenon of seasonal migration. Many insectivorous birds, for example, visit northern latitudes only during the summer months and leave before the onset of the relatively severe winter. Other strategies involve the development of specially tolerant physiologies, as in the desert camels, or the ability to pass into a state of torpor, as in hibernating hedgehogs, when conditions become especially hostile.

The ways of animals are many and varied, but they are all adaptations for achieving the common end—the survival of the individual to perpetuate its kind. This section looks at some of those ways.

In the universal struggle for survival, the ways of different species frequently come into conflict; here a Boomslang (*Dispholidus typus*) makes a meal of a chamaeleon.

Reproduction

Asexual reproduction may give rapid population growth but sexual reproduction produces greater variability on which natural selection can act more rapidly. The internal fertilization of higher animals enabled colonization of the land and the advancement of parental care.

It may well be true that some kinds of plants and animals are potentially immortal. Nonetheless it is certainly true that all living organisms are liable to die—if not naturally, then through disease, accident or injury, or by becoming food for others. For this reason living organisms must be capable of reproducing themselves if their kind is not to become extinct.

Asexual reproduction is the simplest reproductive method. It takes place when an organism divides into two parts, each of which grows into a complete new animal. When a protozoan such as *Amoeba* breeds asexually, each parent organism divides into two parts which are more or less equal in size. When the freshwater coelenterate polyp *Hydra* produces buds, each one initially consists of only a small part of the parent organism. In either case the growth of the new organism from part of the original is simply an extension of the ability, possessed by all animals to a greater or lesser extent, to repair their bodies by regenerating missing parts through the process of repeated cell division.

Asexual reproduction has a number of advantages. By this method one isolated animal can breed on its own. No time and energy are wasted by the processes of courtship.

As long as there is a plentiful supply of food to provide the raw materials for regrowth, the build-up in numbers as a result of asexual reproduction can be spectacular. For this reason an asexual phase is very common in the life cycles of animals in which a very small number of individuals are likely to colonize a new habitat. Such animals include both free-living and parasitic protozoans, a group in which asexual reproduction by simple cell division is very common.

When a body cell divides, there must be a mechanism whereby the information for determining the characteristics of the parent is passed on to the daughter cells. Within almost every cell there is a special region called the nucleus which contains the chromosomes. A chromosome consists of a string of genes, each of which acts as a blueprint for determining some specific attribute of the animal. Every body cell contains a complete set of genes, comprising the hereditary information for the whole organism, but only part of this is utilized by any one cell, depending on its function in the body.

The main disadvantage of asexual reproduction lies in the fact that each new organism produced can only inherit the genes of its single parent. If the parent is a well-adapted organism this is excellent in the short term, while the conditions for which the parent is adapted continue to exist. In the long term, however, these conditions must inevitably change. For example, the climate may become markedly warmer or cooler, or a host may evolve improved resistance to a parasite. Animals which breed asexually cannot readily adapt (that is evolve) in the face of rapidly changing environments. For this reason most animals breed sexually.

LEFT: production of offspring is a driving force in the lives of all animals. These hyrax (*Procavia*) young are the means by which the parents perpetuate their genes.

ABOVE RIGHT: the chromosomes of a body cell during cell division. Each chromosome has replicated itself lengthwise, giving two daughter chromosomes joined at one point. Subsequently, the two daughters separate, one moving to each daughter cell. Here the chromosomes can be arranged into eleven pairs (one such pair is arrowed), one chromosome from each pair being from the mother and one from the father.

RIGHT: when the daughter polyps of *Hydra* are almost as big as the parent, they detach themselves and become independent. They are produced asexually, by budding, and are genetically identical to their parent.

FAR RIGHT: sexual reproduction normally involves two distinct sexes. In these geckos, the male embraces the female in order to transfer sperm directly into her reproductive tracts—an obligatory procedure in land animals.

Essentially the sexual process means that each new organism is derived from two parents, and inherits from each of them some of their genes. Therefore, it is to some extent like both of its parents and is precisely like neither. The potential for greater variation of individuals, and therefore greater adaptability, is thus built into the system.

In protozoans the whole body of each partner, or at least the whole of the reproductive nucleus of the cell, must fuse with that of the partner, and therefore plays a part in producing the new generation. The tissue, at least, can be immortal, although the individual has only a limited span of existence. In multicellular organisms the varied tissues become adapted to equally varied functions, and only a small part of the body, the germ plasm in the gonads, produces the cells which play a direct part in sexual reproduction. In such an organism it is only in the germ plasm that potential immortality resides.

In normal cell division, or mitosis, which occurs during the asexual reproduction of protozoans and during normal growth in multicellular animals, each of the chromosomes duplicates itself and one full set of the new identical chromosomes passes into each new cell. In sexual reproduction, however, two cells fuse together. If normal body cells took part in the sexual process then the number of chromosomes present would double at each generation. This would clearly be an impossible state of affairs, and for this reason sex cells are formed by the gonads as a result of a different form of cell division called meiosis.

Before meiosis the normal set of chromosomes becomes arranged into pairs, each pair consisting of two chromosomes with broadly similar (but not identical) genes. During meiosis the chromosomes in each pair separate and go to different daughter cells; therefore, the resulting sex cells, or gametes, have only half the number of chromosomes of normal cells. When two gametes from different parents fuse, the resulting zygote from which the new organism will develop obtains half its chromosomes from one parent and a complementary half from the other, thus restoring the normal body cell number.

Before the gametes can fuse during the process of fertilization, they must meet. For this reason at least one of them must be capable of moving. At first sight it might seem reasonable for both gametes to be mobile, a state of affairs that does exist in certain ciliated protozoans. However, the newly formed zygote will need nutritive material as raw material for growth, and large cells packed with protein cannot move easily. Accordingly, in the vast majority of cases the female gamete, or ovum, is larger than the other, supplies most of the food, and remains still. The male gamete, or spermatozoid, is smaller and swims using a flagellum or cilia, being attracted by chemicals given off by the female gamete, a process known as positive chemotaxis. Gonads which produce ova are called ovaries, and gonads which produce sperms are called testes. Both types of gonads function not only as producers of gametes, but also as ductless glands, producing hormones which, in part at least, govern the animal's sexual development and behaviour.

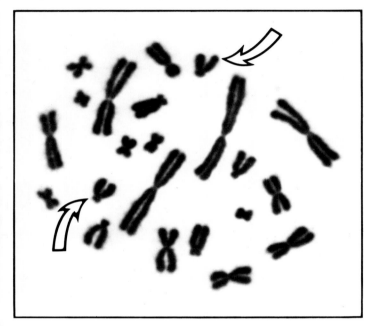

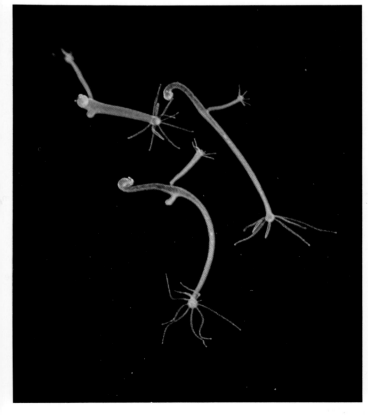

Sperms swim because animal life first evolved in water, and it was here that the evolution of the reproductive process began. In many ways reproduction is easiest in water. All that is necessary is that gametes shall be produced, and that they are liberated into the water at the same time so that fertilization can take place. Synchronization takes place in various ways. The Palolo Worm (*Leodice viridis*) is a segmented annelid of coral reefs of the Pacific Ocean. At dawn on the first day of the last quarter of the moon in October or November the posterior segments of the bodies of all of the worms break off and rise to the surface. Here they writhe, releasing ova and sperms in great concentration so that fertilization readily occurs.

Other organisms respond in more complex ways. For example, in most amphibians and fish the ductless glands of the body respond to changes in the external environmental conditions, such as the progressively increasing daylength in the spring. The gonads produce gametes, and hormones cause changes in both bodily structure and behaviour as a prelude to mating. Some of the changes in appearance have the effect of modifying the body so that either the whole, or some of its parts act as signals which are essential to co-ordinate the behaviour of the animals during the processes of courtship and mating. For example, the male Three-spined Stickleback (*Gasterosteus aculeatus*) develops a bluish back and bright red breast in spring and the female, being full of eggs, develops a rounded, silvery appearance. As careful experimental study has shown, the appearance of each sex has effects upon the behaviour of the other. The colours are emphasized and displayed during instinctive and highly ritualized behaviour patterns, at the climax of which the female deposits her eggs, and the male liberates sperms to fertilize them. In all of these cases where fertilization takes place outside the bodies of the parent animals, it is said to be external. In many such cases the zygote develops into a freely floating planktonic larva. The planktonic larvae of such animals as worms, molluscs, barnacles, crabs and many fish are vital agents in ensuring the dispersal of the species concerned.

However, even in aquatic organisms external fertilization is not the invariable rule. It is essentially a very wasteful process. The female Common Frog (*Rana temporaria*) must lay several thousand eggs every spring in order to ensure the survival of her species, and during her lifetime a female Ocean Sunfish (*Mola mola*) lays up to three hundred million. If more care can be taken of the eggs, fewer need be produced. It is for this reason that in many kinds of fish, including such popular aquarium species as the Guppy (*Lebistes reticulatus*), and in a few kinds of amphibians, for instance the Common Salamander (*Salamandra salamandra*), fertilization is internal. During mating sperms are introduced into the body of the female, so that the eggs are fertilized and can carry out part of their development before they are laid. The average female Guppy produces only forty eggs. In the dogfish and some other elasmobranchs (cartilaginous

ABOVE: European Common Frogs in their breeding pond. The male holds on to the female around her trunk and this stimulates release of eggs. When they first appear, the eggs are covered in a layer of jelly-like albumen thin enough to be easily penetrated by the sperm which the male then releases. The jelly subsequently swells to form the familiar frogspawn.

LEFT: courtship in the Three-spined Stickleback. The brightly coloured male nudges the gravid female towards his nest, where he will encourage her to spawn. After fertilizing the eggs he loses his bright colours. It is the male who guards the eggs.

ABOVE RIGHT: life on land is very hard for embryos. A Royal Python (*Python regius*) emerges from the tough-shelled egg which provided the watery environment necessary for the early development of the embryo, together with an ample food supply.

RIGHT: a greenfly giving birth to a single daughter which has been produced parthenogenetically. In good conditions, new young are born every few days.

fish) fertilization is internal, and each large yolky egg is enveloped in a tough, horny egg case, which provides it with protection after it is laid. In land animals such as insects, reptiles, birds, and mammals, which do not have a watery external environment to facilitate the locomotion of male gametes, internal fertilization becomes the rule. In most cases, except in the mammals, the zygote is enveloped in a case or shell of some kind before it is laid, not only as protection, but also to maintain the watery environment essential for the further development of the embryo.

In many kinds of animals both male and female gametes are produced by the same individual, known as a hermaphrodite. In such cases there is a possibility that gametes from the same individual will fertilize each other. This is inbreeding at its most extreme. The offspring produced as a result will not be as uniform as those produced by asexual

reproduction, for during meiosis some of the genes of the parent will have been lost, and the effects of others will have been intensified. This means that harmful genes, which are always present, are more likely to make their effects felt. Therefore, animals have evolved a variety of mechanisms that have the effect of reducing the likelihood of self-fertilization. For example, a garden snail (*Helix*) may act as either male or female during mating, but not both. Mating earthworms (*Lumbricus*), both give and receive sperms, but they are protected from the possibility of fertilizing their own gametes by the complexity of the reproductive system.

In some species the risks of self-fertilization are accepted. In tapeworms (Cestoda) each proglottid or segment of the body possesses ovaries and testes, and self-fertilization always occurs. In such a case the effects of harmful genes must have made their appearance long ago as a result, and the genes themselves must have been weeded out by natural selection. Self-fertilization in tapeworms is an adaptation to the problem, often encountered by endoparasites, of finding mates in a very restricted environment. However, another endoparasitic species, *Diplozoon*, a monogenoidean fluke which lives in the gills of minnows, has male and female flukes, which live out their lives locked together in permanent copulation. Avoiding the problems of self-fertilization by having separate male and female individuals is the method most commonly found in the animal kingdom.

In some animals sexual reproduction, having evolved, is suppressed at certain times. In greenflies (*Aphis*), for example, once winged females have reached new host plants they produce young without needing to mate—a process known as parthenogenesis. This has the advantages of asexual reproduction, and allows a rapid build-up in population as generations of wingless parthenogenetic females are produced in rapid succession. Reproduction of this kind is a degenerate form of sexual reproduction, and the need for a true sexual process remains. At least once a year a generation of both males and females is produced, and mating occurs in the normal way.

A risk in sexual reproduction is that hybridization may occur between related species. In unrelated or distantly related species the chemical attractants given out by the ova are unlikely to attract sperms of the wrong kind and, even if they do, the meeting of two very different sets of chromosomes will not lead to further development. However, in closely related species that rely on external fertilization, such as some fishes of the carp family (Cyprinidae) hybrids are sometimes produced. This represents a loss in breeding efficiency, for a hybrid is intermediate in form and is not perfectly adapted to fill either of the ecological niches of its parent species. Where internal fertilization is the rule, hybridization is sometimes prevented by relatively simple adaptations. For example, in insects the reproductive organs of the male and female are, like the rest of the body surface, covered by a rigid exoskeleton of chitin. Here the shape of the organs is characteristic of the species, and the male intromittent organs are structured so that they readily fit only those of females of the same species. In dabbling ducks of the genus *Anas*, which include such common species as the Mallard (*A. platyrhynchos*), the courtship ritual invariably includes a modified preening movement in which both the male and female point to and emphasize a coloured patch which appears on the wing. In the Mallard itself this patch, the speculum, is white and blue. Failure of the partner to give the same signal brings courtship to a sudden halt.

It is apparent that the functions of courtship behaviour are by no means simple. At its most complex it must prevent hybridization, reduce potential aggression between male and female, induce the correct mood and the correct responses in both partners, coordinate the positions of the partners, and synchronize the release of gametes. It is not surprising that courtship display is often complex, as it is in the stickleback, the ducks or in Peafowl (*Pavo cristatus*). It is rather more surprising that in some animals, such as the Housefly (*Musca domestica*) behavioural preludes to mating are virtually non-existent. Males simply seize females, and they mate. Between these extremes there is great variety in the expression of mating behaviour; but in all animals it functions to facilitate the achievement of fertilization.

Once fertilization has occurred, in most cases the food supply available to the developing embryo before hatching occurs is limited by the amount of yolk that can be packed into the egg. Since an egg is a single cell, it might be assumed that this amount must necessarily be small. However, the egg of an Ostrich (*Struthio camelus*) is about 18 cm long, and weighs about 1.7 kg. Where small eggs are normal the embryo may develop into a larva which is specially adapted for feeding and rapid growth. Those insects which exhibit complete metamorphosis are examples. The caterpillar of a butterfly or the maggot of a fly have a totally different body shape from the adults of their species. The larva is adapted

for feeding and growing, and the adult is adapted for distributing the species and breeding. Metamorphosis is the process which enables this partitioning of the life history into stages with different lifestyles.

Having a voracious, independent larval stage is not the only way of ensuring rapid growth of the young animal. An al-

ABOVE: Brook Trout (*Salvelinus fontinalis*) lay large yolky eggs that hatch into fry still bearing yolk sacs. These extensions of the gut are amply supplied with blood vessels which carry the digested yolk; this sustains the larva for the first few days of its life until it can fend for itself.

ABOVE RIGHT: a nest of a wasp (*Vespula*). The empty cells in the centre are surrounded by cells containing young larvae, which are being tended and fed by adult wasps. The sealed cells contain pupae. A newly-emerged imago is visible in a cell on the left.

LEFT: a male and female Mallard. The blue and white speculum on the wing of the female (right) is a distinctive feature of her plumage, which otherwise resembles that of females of other species. The speculum plays an important part in the courtship ritual.

RIGHT: a Topi (*Damaliscus korrigum*) suckling its young. Of all animals mammals have the best developed reproductive strategy. In addition to viviparity, this involves the extended care of the young and production of milk for their sustenance and protection from infectious diseases.

ternative is seen in those animals in which the parents provide additional food. Where fertilization is internal the eggs are sometimes retained inside the mother's body until they hatch, when she gives birth to active young. This process is called ovoviviparity. Other animals have evolved structures by means of which nutrients can be passed from the blood of the mother to that of the embryo, and waste products can pass in the opposite direction. This happens in some viviparous fishes, in some reptiles, but most notably in mammals. Among marsupials some bandicoots have structures which serve this purpose, although most members of the subclass are born small and poorly developed. However in most mammals the embryo is nourished through a placenta and the young are born relatively large and well developed, having been given an excellent start in life.

Animals which tend their eggs and young can ensure the survival of a larger proportion of their offspring than would otherwise be the case. In social insects such as bees the care of the young becomes the responsibility of many members of the colony. In mouthbrooding fishes the tiny young are taken into the mouth of a parent for protection from danger. In such cases the young are adapted to give signals which elicit the correct response from the parent. For example, a mouthbrooding fish does not regard its young as food. The best-known of these signals are those of some mammals, where the small size, domed foreheads, snub noses, and

relatively large eyes of the young have been remarked by Konrad Lorenz. The coloured gape of the young of many passerine birds has the same kind of effect, causing the adults to feed the young instinctively.

Where a link becomes established between parents and young a further possibility exists. Instead of bringing food to the young the adults have modifications to their bodies so that they can actually make food. This occurs in some fish, such as the Discus (*Symphysodon discus*), where the young feed on a special form of slime produced by the adult's skin. In pigeons a secretion of the crop of the adults, erroneously known as 'milk', is used to feed the young. In mammals modified dermal glands of the mother produce true milk, which is designed to be the ideal food for the young of the species concerned.

In relying solely on sexual reproduction with separate male and female sexes, internal fertilization, a placenta, and care of the young after they are born including feeding them on milk, placental mammals are clearly very advanced. Infant mortality is reduced to an absolute minimum. Beyond this, because the young are cared for by their parents in their earliest days they are not compelled to rely purely on instinctive behaviour for survival. They have time in which to learn, and this may be the opportunity which allows the evolution of the remarkable and highly adaptive faculty which we call intelligence.

Animal Senses

Although all animals use one or more of the five main senses, their sense organs vary widely and are adapted for different ways of life. For example, the eyes of mammals and insects have very different structures, and a bat uses its ears to perceive solid objects.

The physical structure of the world provides only a limited number of ways in which animals can discern the form and content of their environment. The five senses of man (sight, hearing, touch, taste and smell) exploit most of these—light, sound, physical contact, and the molecules of water-borne and air-borne chemicals. Amongst other animals one or two other attributes are also used, but these five still remain the most common. However, the ways in which they are exploited vary enormously, and sometimes unrecognizably. Bat echolocation is a good example of a sense which is based on the same mechanism as our own hearing, but is used in a very different way. From their mouths or noses bats emit short, high-pitched sound pulses that are inaudible to humans. By listening to the timing, intensity and pitch of the echoes, they manage to reconstruct an image of the world which is so good that they are able to swoop on a flying insect, avoid obstacles, and discriminate the sizes, shapes and textures of objects. In other words, a bat uses sound echoes in much the same way as we use light during the day, to reconstruct the form of the world around it. We should not expect, and indeed do not find, that all animals see, hear, feel or taste the world in quite the way we do.

Vision is, above all, the sense used by animals to examine objects some distance away. It provides the animal with valuable information about where to run, to fly or to land; where there is food or potential danger. The other senses can assist in this, but because light travels at high speed in straight lines and can be made to form optical images, vision is the only sense that can provide fast, accurate and unambiguous spatial information. Except for the high frequency sound used by bats, audible sound goes round corners, confusing spatial relationships. Odours, unless they are firmly stuck to the ground like an ant's trail, diffuse through the air and are dispersed. It is not surprising, then, that eyes of some kind are found in every major phylum.

In the eyes of the scallop *Pecten maximus* the image is formed by a spherical mirror, not a lens. The eyes detect movement rather than pattern, and the mollusc's response is to shut or swim away.

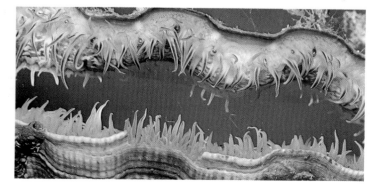

The head of the fox *Vulpes vulpes* illustrates four of the senses all mammals possess. In this active carnivore the eyes are big and equally useful by day or by night. Hearing is acute, and the ears are directionally sensitive and independently moveable. Smell is highly developed, and at night the vibrissae supplement vision.

An optical system is necessary for vision. The cornea and lens of a human eye is a camera-like refraction system which focuses an inverted image onto the retinal receptors. However, there are many other kinds of image-forming device in nature. In the eyes of most snails and worms an 'image' is formed simply by shadow; the receptors line the inside of a small cup of dark pigment with one small opening, so that any one receptor receives light from only a small part of the surroundings. These eyes are always small, and are good for sorting out regions of light and dark in the environment, but little more. However, *Nautilus*, that strange relict cephalopod mollusc from the South Pacific, has eyes that are the logical development of the eye-cup; they work like pin-hole cameras with tiny apertures without lenses, but their retinas are not much smaller than those of a human. The image formed by such eyes is dim, but it is fairly well resolved. In contrast, the eyes of more recently evolved cephalopods, such as octopuses and squids, have large lenses of excellent quality, very much like those of fish, and these are a distinct improvement on the pin-hole arrangement; they are better at resolving and much better at concentrating light. In fact the resolution of the eye of an octopus is very similar to our own, and these animals are renowned both for their acute vision and their ability to learn visual tasks. The molluscs provide another visual surprise; in the eyes of scallops the image is formed not by a lens but by a concave mirror, in much the same way as that formed by a reflecting telescope.

Insect eyes differ markedly in construction from those of vertebrates and cephalopods. In a diurnal insect, such as a bee, each rhabdom (a fused set of 9 receptors) has its own private lens, and the whole eye consists of several thousand

wavelength 300 nm

ultra-violet

the response curves of the three vertebrate cone types—blue, green and yellow—overlap and intermediate colours are interpreted from combinations of cone stimulation; hence red light elicits a strong yellow and a weak green response

Vertebrates and invertebrates are sensitive to different ranges of light wavelength. For example, most of the light reflected from an Evening Primrose flower that is visible to Man lies in the yellow part of the spectrum, and he sees a yellow flower. However, a great deal of ultraviolet is also reflected and this is visible to bees. Areas of low ultraviolet reflectance are arranged to form patterns called nectar guides. These appear darker than the rest of the petals, which reflect ultraviolet more strongly.

800 nm

a bee lacks yellow receptors and so cannot detect red light, although yellow light can be distinguished by the green receptors; the bee's ultra-violet receptors enable it to detect shorter wavelengths than those visible to a vertebrate. Photographed in ultra-violet light, the nectar guides on the yellow flower show up as bands of lower ultra-violet reflectance; these assist the bee in finding the nectaries and ensure that the nearby reproductive parts of the flower are brushed, so encouraging successful pollination

such lens-receptor combinations, giving it a characteristic faceted appearance. Compound eyes like this have a similar sensitivity to our own, because sensitivity depends principally on the f-number of each lens—just as in photography—and in a bee and man these are roughly similar, about f:2. Moths, on the other hand, with eyes that superficially resemble those of bees, are nocturnal and have evolved an

Insect eyes differ from our own in that each group of receptors has its own lens. The eyes of the robber-fly *Microstylum elegans* are particularly well developed because this dipteran catches other insects on the wing. In the flat frontal region of the eye there are much larger lenses than elsewhere, giving greater resolution. Notice also the two antennae that measure air speed.

optical system in which the images from several hundred small lenses become superimposed. This increases the amount of light each receptor receives by up to 1,000 times that falling on a bee receptor. This is not enough to make up completely for the million-fold difference between sunlight and moonlight, but it is probably as far as optics alone can help. Although insect eyes are no worse than those of vertebrates in dealing with dark or light conditions, they are poor at resolving; that is, in splitting up light according to its direction of origin. A bee can resolve stripes spaced about 1 degree apart, whereas human acuity is 60 times better—about 1 minute of arc. The difference is due to an inherent weakness in compound eye design; the tiny lenses (0.025 mm diameter) simply cannot be made to resolve better. A factor of 60 seems large, but it has to be remembered that insects are small, and their visual world is closer than ours and in angular terms correspondingly larger. Also, since they fly fast, the resolution of fine detail is relatively less important than fast reactions, just as in driving a car.

Light has other qualities besides intensity and direction. We perceive wavelength differences as colour. The three different types of cones in the retina are sensitive to blue, green and yellow light respectively. Most vertebrates, other than those that are really nocturnal for whom colour vision is an unaffordable luxury, have a similar spectral range. Insects, however, extend the wavelength range they can detect beyond the visible spectrum into the ultraviolet, so that for a bee the primary colours are green, blue and ultraviolet, but red is indistinguishable from black. Interestingly, flowering plants, which evolved in parallel with insects and exploit them for pollination, often show patterns of ultraviolet 'colour' that guide insects to the nectar.

The only animals known to be able to detect infra-red wavelengths are snakes, and, curiously, they do not use their eyes for this but rather a pair of pit-like organs just in front of the eyes. All warmblooded animals emit infra-red light, which is a component of the heat they radiate. Thus, for an animal like a snake that preys on small mammals an infra-red sense is an ideal way to detect the next meal. This can only work for cold-blooded predators; a mammal or bird produces so much infra-red radiation of its own that to use this as a way of detecting other animals would be impossible.

One final quality of light that we are usually unaware of, but which many invertebrates are able to perceive, is polarization; that is, the plane in which the waves that compose light vibrate. If you look at a blue sky through polaroid sunglasses different parts will appear lighter or darker depending on how you tilt your head. Insects in particular have receptors whose structure allows these patterns to be seen directly, and they use this information to infer the position of the sun, and navigate by it, even when the sun itself is obscured by cloud.

A number of sense organs respond to movement. Besides obvious ones like touch receptors, these include sound receptors and balance detectors, both of which must respond to very fine movements. In the hearing of land vertebrates, the small pressure changes in the air that are the physical basis of sound are intercepted by the eardrum. They are transmitted and amplified by the bones of the middle ear, from where they are transformed into vibrations in the fluid of the inner ear. There they cause motion in a long, tuned, spiral membrane—the basilar membrane of the cochlea—upon which the receptors sit. This transformation of sound into movement, rather like an old-fashioned gramophone in reverse, is remarkable in itself, but perhaps the most spectacular attribute of the ear is its sensitivity. At the

BELOW: the mammalian ear fulfils two different functions: the detection of sounds and the control of balance. Each system relies on hair cells to transform certain movement stimuli into nerve impulses. In hearing, hair cells in the cochlea translate fluctuations in air pressure into heard sounds. Balance control is based on signals from hair cells in the utricle and ampullae of the semi-circular canals. Otoliths, into which the hairs are bedded, respond to gravity and changes of inertia, so flexing the hairs and producing nerve impulses that are interpreted by the brain. The eustachian tube connects the middle ear to the throat, and it helps prevent excessive pressure differences across the eardrum.

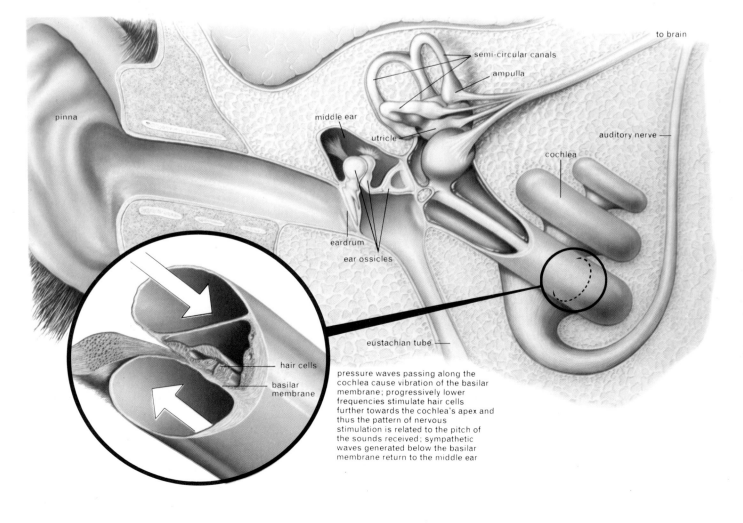

pressure waves passing along the cochlea cause vibration of the basilar membrane; progressively lower frequencies stimulate hair cells further towards the cochlea's apex and thus the pattern of nervous stimulation is related to the pitch of the sounds received; sympathetic waves generated below the basilar membrane return to the middle ear

ABOVE LEFT: this Malaysian Pit Viper (*Trimeresurus popeorum*) is able to follow the trail of warm-blooded animals by using the slit-shaped, heat-detector organs that are situated between the eyes and nostrils.

ABOVE: many insects use sound for communication. The grasshopper *Chorthippus parallelus* chirps by rubbing the hind legs against the wing bases and hears with 'ears' on the first abdominal segment.

RIGHT: in the blind cave fishes, like this *Iranocypris typhlops*, there are no eyes. The animals navigate using the lateral-line organs on the head and body that monitor variations in water flow caused by nearby objects.

threshold of hearing, a just-heard pin-drop, the actual movement of the receptors is only about 10^{-13} m. The usable range of the human ear is even more impressive. From the sound of a pin-drop to that of a low flying jet the force of the air movements changes by 140 decibels, or a factor of a hundred million million times.

Hearing has a double function in the lives of animals. Its most important role is in communication, and one finds it well developed in a few insects, such as grasshoppers, in some fish, frogs and of course birds, all of which make sounds to advertise their presence to potential mates or rivals. It is amongst the mammals, however, that the ear itself is most highly developed, as is the power of auditory communication. This is by no means confined to primates although they are peculiarly vocal. Some of the most complex and melodic 'songs' encountered in the animal kingdom are made by cetaceans, especially whales. The meaning of these songs is unknown but it is significant that their power to carry in the sea is enormous—tens and possibly hundreds of kilometres under ideal conditions. Sound can also be used to determine the locations of objects, although generally it is less precise than the use of light. Using a combination of intensity and timing differences in the sound reaching the two ears, humans can pinpoint the source of a sound to within a few degrees, and the external ear—the pinna—helps to discriminate whether a sound comes from the front or the back. It also acts as a sound collector, but less so in humans than in animals such as dogs or rabbits. In bats, dolphins and porpoises the spatial information in sound has been exploited to its limit, and the trick these animals use is to emit sound and analyse the echo. The sound has to be high frequency (up to 100 kHz in horseshoe bats—more than five times higher than the limit of human hearing) the principal reason being that only at these frequencies does the wavelength of sound become short enough (a few millimetres) to produce a detectable echo from a small object like an insect. Interestingly, natural selection has not left insects totally unprotected; some moths have a pair of simple but very sensitive ears on their thorax, and these enable them to take evasive action before a bat is near enough to pick up an echo.

The inner ear of vertebrates contains two other sense organs, concerned with balance, that make use of movement-detecting hair cells almost identical to those of the cochlea. In the utricle the hair-like projections of these cells are attached to a lump of jelly containing small heavy particles (otoliths) that pull down in the direction of gravity, enabling the organ to detect the direction of, and departures from, the vertical. The second system is the semi-circular canals; three closed fluid-filled tubes each containing a single 'pendulum' of jelly attached to a tuft of hair cells. Rotation of the head causes fluid to flow around one of the canals, depending on the axis of rotation, causing the pendulum and its receptors to bend. The degree of bending indicates how fast the head is rotating in space. In fish and amphibia there is a further sense based on the same type of cell. Along the flanks and around the head there are rows of receptors that respond to small water movements. These are the lateral lines, and they can provide a fish with a surprising amount of information. For example, a blind cave fish can avoid obstacles by using its lateral line system to measure the distortions of its own bow-wave that are produced by objects around it.

The strangest sense of all, in that it is quite unrelated to anything within our human experience, is electro-reception. Certain tropical fish (notably the mormyrids and gymnotids) inhabit waters so murky that sight is often useless, and they have perfected a sense based on small electric currents that they generate themselves. A small organ on the tail, a modified muscle or set of nerve endings, produces a pulsed electric field around the fish which is picked up by receptors on the head. Objects around the fish cause distortions on this field. They cause local differences in the electrical resistance of the surrounding water mass that in turn alter the pattern of voltage 'seen' by the fish. To give an idea of the sensitivity of the system, *Gymnarchus* can be trained to distinguish between 9 mm and 15 mm diameter glass rods surrounded by a porous pot. The difference between the two rods corresponds to a minute electrical change. From this it has been calculated that the fish should be able to detect a field equivalent to a 1.5 volt dry battery whose terminals are 100 kilometres apart!

Besides providing the basis for local navigation, electro-reception is also used for communication. A fish in one tank, connected only by wires to a second tank, will become agitated when the fish in the second tank is fed; and there are instances of aggressive behaviour and even the establishment of dominance hierarchies, all mediated by fish 'conversing' with each other using their electric organs.

Like bat echo-location and lateral line navigation, electro-reception raises the intriguing problem of how the world is perceived by an animal that navigates using a sense other than sight. Perhaps a blind person, used to interpreting the nuances in the echoes from footsteps, could give a more coherent answer than someone with sight. However, judging from the behaviour of animals that use non-visual spatial senses, their world-picture must be almost as immediate and concrete as the visual world we take so much for granted.

The chemical senses, taste and smell, are well developed in nearly all animals, largely because eating the right food is a universal requirement. Humans mainly identify airborne molecules with receptors in the nasal cavity, and water-borne molecules in food with the tongue, but this division between smell and taste is rather arbitrary as the aroma of food carries up into the nose and profoundly affects its taste. Not a great deal is known about the exact mechanism of either sense, except that there must exist a kind of lock and key arrangement on the surface of the receptors whereby appropriate molecules are trapped, and in becoming trapped excite nervous activity. What is known, however, is that the numbers of molecules involved can be very small indeed. For example, certain particularly toxic and evil-smelling substances, the mercaptans, are detected by humans when single receptors have received no more than eight molecules. Similarly, by means of its large feathery antennae, a male silkworm moth (*Bombyx*) can detect the sex attractant pheromone produced by the female at concentrations where the receptors can only be receiving single molecules.

The latter example makes the point that chemical senses can be used for communication, as well as for the testing of food. Besides attractant chemicals with a sexual function, animals from ants to fish are known to release alarm substances that warn others of danger, and chemical markers are also used to lay trails for others to follow, and to stake out territorial boundaries. A certain amount of inter-species

The electrolocating fish *Gnatho-nemus petersi* from West Africa produces brief electric pulses from an organ on its tail. These are used both to locate objects in murky water and also as a form of communication in territorial, sexual and alarm displays.

The massive antennae of the male Chinese silk moth *Antheraea pernyi* are used to detect single molecules of sex attractant given off by the females. The typical male strategy is to head up-wind when odour has been detected, 'tacking' up the plume to reach the female. Females can 'call' by inflating special scent glands.

chemical warfare even goes on. For example, the bolas-spiders lure male moths by releasing a chemical similar to the attractant pheromone of a female moth.

Compared with vision, and to a lesser extent hearing, smell is cumbersome as a sense for locating objects. If there is no wind (or water flow) a chemical substance will diffuse out uniformly, and an animal must explore its way to the source of the odour by trial and error—in the same way that we might try to trace a gas leak by smell. In a light breeze the problem is a little simpler since the odour will be distributed in a plume on the down-wind side of the source, and all the searcher has to do is to head up-wind as soon as the odour is detected. This is the main method mosquitoes use to find us; they fly up through the carbon dioxide trail we leave down-wind.

In this brief survey of animal senses each one has been considered separately, as though it governed a particular kind of behaviour independent of the others. In reality this is rarely so. For example, the control of walking is part visual, part balance adjustment and part the feel of the ground. The moth or mosquito, to make use of its sense of smell, must work out which direction 'up-wind' is, and since it is flying in the moving airstream, only the sight of the ground below can provide the information needed. Thus the rigid physical divisions between the senses are ones that we erect for convenience. In reality an animal's senses combine in ever-shifting patterns, depending on the kind of information the animal needs.

Many species of the Carnivora mark the boundaries of their territories by depositing urine on upright objects. This may be mixed with secretions from anal glands. Scent marking may serve to indicate the identity of the animal making the mark and, as the scent decays with time, when it was last visited.

Locomotion

Legs, wings, flippers and fins are among the variety of devices which animals have evolved to enable them to move actively from place to place. Suitably coordinated and powered, these structures serve to produce the necessary supportive and propulsive forces.

The methods by which animals move are almost as diverse as the animals themselves. They use oars, aerofoils, jet propulsion and all the other principles used for locomotion in human engineering, with the exception of the wheel.

The wings of birds are aerofoils, shaped much like the wings of aircraft. When a bird glides or soars, its wings are working in exactly the same way as the fixed wings of aeroplanes. A good man-made glider loses height at a rate of only 0.5–1 m per second. The most specialized gliding birds, such as gulls and vultures, lose height at similar minimum rates. Man-carrying gliders travel faster than models or birds because they are larger; it is only the rate of loss of height that is similar.

In active, flapping flight the wings assume the function of the propeller of an aircraft, in addition to the function of its wings; they drive the bird forwards through the air as well as supporting the weight of the body. A propeller pushes air backwards, thereby producing a force which drives the aeroplane forwards. The outer parts of the wings of birds make subtle changes of angle as they flap up and down, in such a way as to drive air backwards and the bird forwards. No one has actually measured the forces on the wings of birds in flight, but since the feathers are flexible the forces bend them. The directions of the forces can be deduced from the bending seen in photographs or alternatively by applying aerodynamic theory. The wings of an aircraft cannot provide enough upward force to support its weight until they are travelling forwards at a certain minimum speed. Similarly, large birds need to generate some speed before they take off. Small birds can take off from a standing start.

Several devices are used in human engineering for take-off without a runway. The most common is the helicopter, which has a rotor like an enormous propeller mounted horizontally. This pushes air downwards, producing a force which supports the weight of the helicopter. The rotor rotates perpetually in one direction, but few animal joints are capable of turning more than half a revolution. Consequently the animal analogues of helicopters use backwards-and-forwards (reciprocating) movement of aerofoils, instead of rotary movement. Small birds and insects beat their wings backwards and forwards, usually horizontally, to obtain a helicopter effect. They do this at take-off, in landing and when hovering for other purposes. For instance, humming-birds and moths hover in front of flowers while drinking nectar. When they fly fast the wing movement changes from the backwards-and-forwards beat of hovering to the up-and-down of forward flight.

Hydrofoils are the underwater equivalent of aerofoils. Forces act on hydrofoils moving through water in just the same way as on aerofoils moving through air. Indeed, evolution has converted an aerofoil to a hydrofoil in the case

By beating their wings 50 times a second, hummingbirds, like this pretty female Black-chin (*Archilochus alexandri*), are able to fly forwards, backwards or stay static, and can control their position accurately enough to drink nectar from a flower.

of the penguin's wing. Penguins cannot fly, but they swim under water by flapping their small wings. The wing action is a little different from the wing action of flying birds because the penguin's weight is supported by buoyancy, so the wings have to produce only forward thrust. The tail flukes of whales are hydrofoils which generate thrust by beating up and down. Fish beat their tails from side to side, but in other respects the tall, narrow tail fins of such fish as tunnies work in just the same way as the flukes of whales. The swimming of an eel also depends on side-to-side movements of the tail, but the forward thrust is not generated by a hydrofoil. The whole body is thrown into waves that travel backwards, pushing the water backwards and the animal forwards.

There is a variety of other methods of swimming. Water beetles row themselves along, using their large hind legs as oars. The ciliate protozoans also row, using thousands of minute oars. Most of these animals are less than a millimetre long. Their oars are the bristle-like cilia which cover the

When swimming, the Green Turtle (*Chelonia mydas*) uses its limbs in the same way as a penguin uses its wings. Here the front flippers are in the middle of their downstroke, tilted so as to drive the water back and propel the animal forwards.

whole surface of the body and beat to drive the animals along at speeds around a millimetre per second. Squids swim by jet propulsion. They have a thick, muscular wall to their gill cavity and can squirt water out at high speed. The water leaves through a nozzle which can be pointed either forwards or backwards, so that the squid can swim in either direction.

Animals use two main methods for moving overland. Some, such as mammals and insects, walk or run on legs.

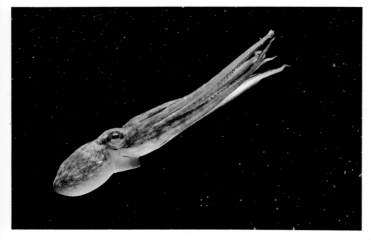

Octopuses can swim backwards by forcibly expelling a strong jet of water from the muscular mantle cavity out through a funnel which is situated under the head. The same technique is used by the more streamlined and faster squids.

Others such as snails and earthworms crawl on the under-surfaces of their bodies, using several variants of a basic mechanism that involves concertina-like lengthening and shortening. The fastest animals are those that walk and run. Earthworms (*Lumbricus*) have maximum crawling speeds of about 30 cm per minute, but centipedes run up to eighty times as fast. Mammals are much larger and run correspondingly faster. Greyhounds race at speeds around 56 km per hour, and racehorses can run faster still.

BELOW: this is purely a contest of speed. There is no reliable data on the speeds of Lions, but zebras have been filmed running at up to 14 m per second (50 kph) on the East African plains. Most horse races are won at about 16 m per second.

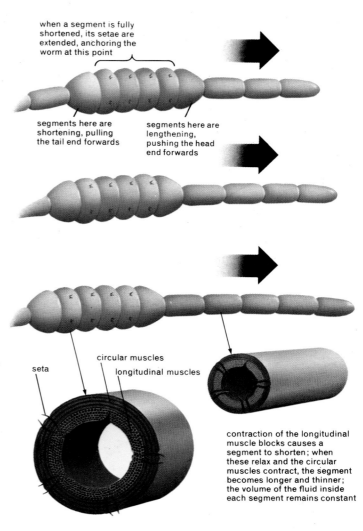

when a segment is fully shortened, its setae are extended, anchoring the worm at this point

segments here are shortening, pulling the tail end forwards

segments here are lengthening, pushing the head end forwards

seta

circular muscles

longitudinal muscles

contraction of the longitudinal muscle blocks causes a segment to shorten; when these relax and the circular muscles contract, the segment becomes longer and thinner; the volume of the fluid inside each segment remains constant

ABOVE: earthworms move by serially altering the shape of their body segments producing waves of contraction that pass backwards down the body, several such waves being present at any one time. To move backwards the pattern of contraction is merely reversed.

The Cuttlefish (*Sepia officinalis*) swims slowly forward by passing waves backwards along its fins. It also uses fast jet propulsion, like octopus and squid. Gas in the cuttlebone gives it buoyancy.

In the locomotion of most animals the various parts of the body make coordinated movements in a definite sequence. In many cases, waves of activity seem to travel along the body. This is particularly obvious in the case of the eel, which causes waves to travel backwards along its body. Water snakes and leeches swim in a similar fashion, but in the case of leeches the bending is up and down, instead of side to side. The swimming of spermatozoa and of flagellate protozoans involves similar travelling waves in their whip-like flagella. The crawling of snakes involves bends which travel backwards along the body. Earthworms keep their bodies straight when crawling on unobstructed surfaces, but the muscles of successive segments contract in turn, starting at the head, giving the appearance of waves of contraction travelling backwards along the body. The crawling of gastropod molluscs involves waves of muscular contraction in the foot. These travel backwards along the foot in limpets and winkles, but in garden snails (*Helix*) they are modified so that both the waves and the animal travel forwards.

In all these cases effective locomotion depends on the direction of the waves. If an eel made waves travel forwards along its body it would travel backwards. If a limpet reversed the waves on its foot it would also travel backwards unless it also modified the waves in the fashion of a garden snail. If a worm contracted its segments in random sequence it would make very little progress. Animals with legs generally move them in a precisely defined sequence so that waves of leg movement may seem to travel along the body, but the direction of movement does not depend on the direction of the waves.

The variations which are possible are illustrated by the centipedes and millipedes. Centipedes move the left leg of a segment forwards while the right leg is moving backwards, but millipedes move the left and right legs forwards at the same time. Many centipedes move the second left leg a little after the first, the third after the second and so on, so that waves of leg movement travel backwards along the body. Millipedes and some centipedes do the reverse, making waves of leg movement travel forwards. The centipedes that do this are ones with particularly long legs, which would cross over each other and get in each other's way if the waves travelled backwards. The cilia of ciliate protozoans also move in waves. They are closely spaced and would be apt to collide with each other if they did not move in a very orderly fashion.

Starfish walk on tiny tube feet, each of which ends in a sucker. The tube feet cover the undersurface of the body and a large starfish, such as *Asterias rubens*, has about 1,200. A starfish may walk with any of its five arms leading and the movement of the tube feet is coordinated so that at any instant all are stepping in the same direction. This much coordination is obviously essential; if the feet tried to walk at random in all directions, the starfish would remain stationary. However, there seems to be no more coordination than this essential minimum, for the tube feet seem to step in random order.

The millipede travels slowly and creates a moving pattern with its legs. Each leg moves slightly after the one behind, so that waves of movement seem to travel forward along the body.

Centipedes move the left and right leg of each segment alternately. Each leg moves just after the one before it, waves of leg movement passing backwards down the body. At higher speeds stride length is increased and fewer legs are on the ground at any one time.

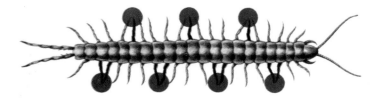

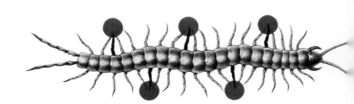

ABOVE: the minute tube feet are on the cream-coloured undersides of these starfish *Asterias rubens*. Although the feet walk in the same direction at any instant, there are no co-ordinated waves of movement.

RIGHT: the long-legged Giraffe uses a peculiar, rack-like walk, lifting the two legs of one side at nearly but not quite the same time. To go fast, however, it breaks into the galloping gait typical of other mammals.

Animals with small numbers of legs show coordination as precise as centipedes, but the mechanical principles which govern the coordination are rather different. Stability is an important consideration. A three-legged stool is stable but one with fewer legs is not, unless it has large feet. Birds and humans have large feet and can stand on two feet or even on only one. Most legged animals have smaller feet and are only stable when at least three of their feet are on the ground. Insects have six legs and generally move three at a time. They move the first and third left legs and the second right leg while the other three legs are on the ground, and then keep them on the ground while the other three legs move. Mammals with only four legs could only move one at a time if they were to remain stable, but they could not run fast if their movements were restricted in this way. A horse walking extremely slowly may sometimes move its feet one at a time, but faster gaits involve phases when only two, one or even no feet are on the ground. Stability is essential for very small animals and at low speeds; it is by no means essential for reasonably large animals moving faster, because their inertia prevents them from falling over between footfalls.

Of the various quadrupedal gaits, the walk is the one which most nearly meets the conditions for complete stability. The legs move in turn, but each tends to be lifted before the previous one is set down. In trotting, diagonally opposite legs move more or less simultaneously; the left foreleg moves with the right hind leg, and the right foreleg with the left hind leg. In galloping the two forefeet are placed on the ground (simultaneously or in turn) and then the two hind feet are placed on the ground. The canter is an intermediate gait between the trot and the gallop. Most mammals walk at low speeds, trot and then canter at intermediate speeds and gallop at high speeds. The speeds at which animals walk, trot and gallop depend on the length of their legs. Giraffes and camels do not trot but move by racking, in which the two left legs move together and the two right legs move together. Wildebeest neither trot nor rack but change directly from a walk to a canter. The fastest gait of elephants is a peculiar fast walk.

INCREASING SPEED

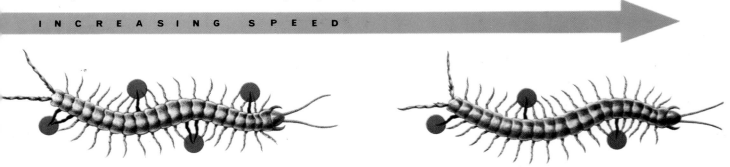

Locomotion uses energy derived from food. The less energy is used in this way, the more is available for other purposes such as growth and reproduction. Also, the power output of muscle (the rate at which it can do work) is limited. High speeds generally require more power than low ones, so means of locomotion that are economical of energy enable animals to travel faster to catch prey or to escape from predators. For these reasons animals can be expected to have evolved in such a way as to minimize the energy needed for locomotion.

The energy for locomotion is generally released by respiration, so that the rate of use of energy can be calculated from the rate at which the animal uses oxygen. This can be measured in various ways. The air breathed out by a mammal can be collected for analysis through a gas mask, or samples of the water in which a fish is swimming can be analysed to find out how fast oxygen is being removed from it. However, vertebrates moving at high speeds may develop oxygen debts, so that it is more difficult to find out how much energy they are using. In an extreme case, an athlete may hold his breath throughout a 100-metre race. In long distance races athletes depend mainly on respiration, for the body will only tolerate a limited oxygen debt. Men running in the range of speeds used in longer races use oxygen twice as fast as men cycling at the same speed. This is a measure of the energy-saving value of the wheel.

When a man (or any other mammal) runs he accelerates and decelerates in every step, so he gains and loses kinetic energy. His centre of mass rises and falls in every step, so he gains and loses potential energy. These changes of kinetic and potential energy are responsible for most of the energy cost of running, and they do not occur in cycling. Whenever the total kinetic and potential energy is increased, work is required from the muscles. Whenever the total decreases,

the energy would be lost (as heat), if it were not for the elasticity of the tendons in the leg. Tendon elasticity saves a great deal of energy in running; if it were not for the tendons, running would be about twice as expensive of energy as it actually is.

The principle is the same as that of a bouncing ball. When the ball hits the ground it is brought momentarily to rest and loses its kinetic energy, but it is distorted and so stores up elastic energy, which is converted back to kinetic energy as it leaves the ground again. Similarly, in running kinetic and potential energy are converted to elastic energy, which is converted back to kinetic and potential energy in the elastic recoil. The kinetic and potential energy have their maximum values between footfalls, when both feet are off the ground. The tendons are stretched and have elastic energy stored in them while their foot is on the ground. Even in quite slow running the force on a man's Achilles tendon may reach half a tonne, enough to stretch its overall length by nearly 2 cm.

Energy saving by elasticity is equally important in insect flight. Instead of revolving like the rotor of a helicopter, insect wings beat backwards and forwards or up and down. A flying hornet, for instance, makes about 100 cycles of wing movements (100 upstrokes and 100 downstrokes) each second. In each of these strokes the wing is accelerated to a high speed and brought to rest again: the wing tips reach speeds around 10 m per second in each stroke. The wings are light, but since the speed is so high, a good deal of kinetic energy is involved. This energy has to be given to the wings at the beginning of each stroke, and taken away at the end. Much of it is saved by elastic storage in a rubber-like substance called resilin, at the bases of the wings. If it were not, flight would need about twice as much energy as it actually does. The resilin works on the same principle as a piano string which continues vibrating for some time after a single blow. This is because of its elasticity; energy is converted back and forth between the kinetic and elastic forms in each vibration.

Locusts and grasshoppers use elastic proteins in two ways. Their elastic wing hinges make it easier to keep the wings beating, and a catapult mechanism in the legs provides the power behind the prodigious leaps for which the group is well known.

The streamlined shape of the Cod (*Gadus morhua*) is an important aid to easy movement through the water. In addition, buoyancy is provided by the swimbladder, a bag of gas in the body cavity.

There are many other adaptations in animals for saving energy in locomotion which do not depend on elastic storage. For example, the streamlined shapes of fishes allow them to pass through water with minimum resistance, and gas-filled swimbladders give many fish precisely the buoyancy they need to be able to hover effortlessly in mid water.

The energy costs of locomotion of different animals can be compared by calculating their costs of transport, that is the energy used per kilogramme of animal for each metre travelled. The cost of transport is uniformly lower for large animals than for small ones using the same type of locomotion. For animals of equal size, cost of transport is highest for running, lower for flight and lowest of all for swimming. Flight requires much more power than running, and the cost of transport is lower only because it is so fast. Compare a flying budgerigar with a running rodent, such as a kangaroo rat, of similar size. The budgerigar flies at speeds around 36 km per hour, but the kangaroo rat runs at between 1.5 and 2 km per hour. Though the budgerigar uses energy twice as fast as the kangaroo rat, the amount it uses in travelling a given distance is far less. The high cost of running may explain why mammals do not make long-distance migrations comparable to those of birds.

The Ostrich (*Struthio camelus*) is the largest of all living birds and is far too heavy to fly. Its wings are small and functionless but its legs are extremely strong and it can run very fast, reaching a speed probably greater than that of most antelopes.

The nimble Springbok (*Antidorcas marsupialis*) of South Africa, and some other antelopes, leap high as they run. Compared with the galloping gait of most mammals, this must burn up more energy but it may be used in order to confuse pursuing predators.

Camouflage
and
Mimicry

To avoid being eaten, many animals disguise themselves by using camouflage colours, shapes and resting postures. Others use warning colours to show their unpalatability, and some may mimic such warning colours or use weird markings as a bluff to deter predators.

One of the first things we notice about an animal is its colouration. To us, one animal may be attractively coloured whilst another may be dull, but in biological terms our aesthetic sense is irrelevant, and what matters is whether being red or green, for example, affects the chances of surviving and reproducing. Colouration is important to animals in three principal ways: it can affect their vulnerability to predation, their chances of obtaining food, or their success in securing a mate. The strategies which animals adopt in achieving the first two objectives include camouflage, warning colouration, mimicry and bluff.

Animals are often camouflaged so that they harmonize in colour with their background. The advantage of camouflage for most animals is that it is difficult for predators to find them. Familiar examples are the greens of caterpillars and grasshoppers which feed on leaves, and the greys and browns of moths which rest on the bark of trees. Camouflaged animals are more likely to escape detection by predators and to leave more offspring than those which are conspicuous.

The Sargassum Fish (*Histrio histrio*) lives amongst *Sargassum* weed where its irregular outline with branched processes gives it almost perfect camouflage. Away from *Sargassum* it is conspicuous and, as it swims only feebly, can be captured by predatory fishes.

Predators, however, can also be camouflaged; for example the Lion (*Panthera leo*) is ochre-coloured, thus matching the dry grass and dusty soil of dry savanna where it lives, and enabling it to stalk prey more successfully than if it were conspicuous. At the same time predators that can detect camouflaged prey have a better chance of obtaining food, and hence of reproducing, than predators that cannot detect camouflaged prey. Thus, there is a continuous race, between prey for more perfect camouflage, and predators for more effective means of detecting prey. It is these conflicting pressures that have resulted in the evolution of remarkable examples of camouflage.

Fish camouflage relies on matchin the intensity of light reflected from the body surface with that transmitted through the surrounding water. It is aided by mirrors, photophores and a narrow body.

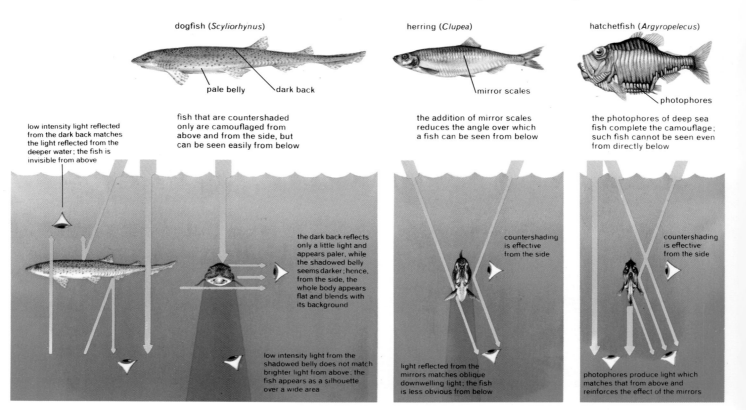

dogfish (*Scyliorhynus*)

pale belly   dark back

low intensity light reflected from the dark back matches the light reflected from the deeper water; the fish is invisible from above

fish that are countershaded only are camouflaged from above and from the side, but can be seen easily from below

the dark back reflects only a little light and appears paler, while the shadowed belly seems darker; hence, from the side, the whole body appears flat and blends with its background

low intensity light from the shadowed belly does not match brighter light from above; the fish appears as a silhouette over a wide area

herring (*Clupea*)

mirror scales

the addition of mirror scales reduces the angle over which a fish can be seen from below

countershading is effective from the side

light reflected from the mirrors matches oblique downwelling light; the fish is less obvious from below

hatchetfish (*Argyropelecus*)

photophores

the photophores of deep sea fish complete the camouflage; such fish cannot be seen even from directly below

countershading is effective from the side

photophores produce light which matches that from above and reinforces the effect of the mirrors

We normally think of camouflaged animals as being green or brown because we tend to think in terms of backgrounds of grass, trees and soil. But camouflage is equally important in the sea where it takes a very different form. For the myriad of small animals living in the surface waters—the zooplankton—the best camouflage is to be invisible, and indeed many planktonic organisms such as crustacean larvae and arrow worms are transparent.

LEFT: many animals that live in the surface waters of the sea are camouflaged through being transparent. In less oblique lighting this phyllosoma larva of a crawfish (*Palinurus*) would be virtually invisible. (×2)

ABOVE: olive Flat Periwinkles (*Littorina littoralis*) are well camouflaged on Bladder Wrack (*Fucus vesiculosus*), while the yellow variant is obviously safer on a different substrate.

BELOW: the Green Carpet Moth (*Colostygia pectinataria*) rests by day on lichen covered trees, where its bold patterning merges with the colours of the lichens.

For a fish swimming freely in the sea, danger can come from above, from below, or from the sides, and so there will be advantage if it can be camouflaged from any direction. Dogfish (*Scyliorhinus*) are dark brown or black on the upper surface and white on the belly. When viewed from above they harmonise with the dark of the ocean depths. When viewed from the side, the dark back and white belly, together with the top-lighting that produces a ventral shadow, cause the fish to merge, to some extent, with the background. The effect of this countershading is to obliterate the normal ventral shadow so that the three-dimensional fish appears flat and is less likely to be recognized as prey. But when viewed from below the same fish is still conspicuous as a dark silhouette against the bright light above.

Silvery fish, such as herrings (Clupeidae) and mackerels, (Scombridae) are rather better camouflaged than dogfish. They are also dark on the back, with the flanks and belly white. But in addition there are small mirror-like structures set vertically into the sides of the fish which reflect downwards the light from above. Such a fish is well camouflaged from all directions except immediately below where it still appears as a dark silhouette. Some of the deep sea hatchet-fish—for example *Argyropelecus aculeatus*, which lives at depths of 500 metres—are camouflaged from all directions. They are dark dorsally, but ventrally have special organs called photophores which actually produce and emit light. These are so constructed that they emit higher intensity light downwards, but progressively less intense light towards the sides. The intensity of the emitted light closely matches that of the light coming from above, and furthermore it is of the same bluish colour as the light that is found at depths of 500 metres. Similar photophores occur in many other fish and in some shrimps and squids that live at this depth.

Camouflage is effective only if the animal rests on an appropriately coloured background. Moths are commonly attracted to lights at night and often settle on walls of buildings where they are very conspicuous; in a natural situation,

however, they tend to rest on tree trunks, on shrubs, or in grass where they are difficult to find. Experiments have shown that green grasshoppers and praying mantids choose to rest on green paper or green grass in preference to other coloured backgrounds whilst brown grasshoppers and mantids prefer brown backgrounds. The grasshopper *Oedipoda miniata* has four different colour forms in Turkey, each associated with a particular type of soil which it matches in

colour. When given a choice of substrate, each colour form chooses the soil on which it is best camouflaged. Thus animals also have behavioural adaptations which serve to localize them in places where they are best concealed.

The position in which an animal holds its body when it rests is also important: it is of no use to be well camouflaged if the resting posture forms an outline easily recognizable to a predator. Moths that are grey or brown and commonly rest on the bark of trees, where they are well camouflaged, normally do so with the head facing upwards. They often have patterns of lines on their wings running vertically when they adopt this position, and these help to perfect the camouflage because lines on the bark also often run vertically. But some geometrid moths rest with the body axis held horizontally, and in these the colour markings tend to run across the insect from one wing to the other. With the moth resting sideways these markings then match with the vertically aligned marks on the bark. In some species the orientation of the settling moth is influenced by the direction of ridges and grooves on the substrate.

Settling on a particular colour of background or adopting a suitable resting position is effective provided that plenty of such places exist. But if there is a shortage of suitable places most animals eventually have to rest in a place where they are conspicuous. Some, however, are able to change their colour depending on their background. Some Arctic species such as the Ptarmigan (*Lagopus mutus*) and the Snowshoe Hare (*Lepus americanus*) are brown in summer and rest on earth and stones, but in winter they turn white to match the snow. Thus, they are well camouflaged during both these seasons. Only in spring and autumn is there a time when they may be inappropriately coloured and so be conspicuous to a hunting eagle.

Other animals can change colour more rapidly. Flatfish, for example, may be found in shallow seas on sandy, muddy or stony substrates. If a Plaice (*Pleuronectes platessa*) is moved from a sandy to a stony substrate it slowly alters its colouration over a period of a few days so that it eventually has the blotchy pattern of stones. The Cuttlefish (*Sepia officinalis*) can also change its colour as it moves from a sandy to a stony substrate, but it takes less than a second to do so. If green nymphs of the praying mantid *Miomantis paykullii* are reared on brown twigs, when they moult they change colour to emerge as brown mantids. This change can be reversed at each moult until the insect is fully grown.

Other insects have evolved so that they have very precise resemblances to particular parts of the environment. For example, stick insects (*Carausius* etc.) are slow-moving elongated insects that closely resemble the stems of vegetation. There are also some mantids that are elongated, brown stick mimics. Caterpillars which are stick mimics adopt a resting posture which makes them resemble short twigs; predatory birds can easily overlook these caterpillars unless they see them actually feeding rather than adopting this characteristic posture. Other insects are bright green and resemble leaves in shape, and some even have brownish patches resembling blemishes such as might be caused by a fungus. Yet other caterpillars are black and white and rest on the upper surfaces of leaves; though they are conspicuous they look very like bird-droppings and are therefore ignored by most insectivorous birds.

ABOVE: the caterpillar of the geometrid Swallowtail Moth (*Ourapteryx sambucaria*) adopts the guise of a twig. Both its colour and its posture assist in the deception.

BELOW: *Pycnopala bicordata*, one of the long-horned grasshoppers, is inconspicuous when resting on foliage. The resemblance to a leaf is enhanced by the mock veins and blemishes.

By contrast with camouflaged animals, whose colouration or shape harmonizes with their background, others have bright colours which make them very conspicuous. It must be of some advantage for these animals to advertise themselves. In birds such as the Mandarin duck (*Aix galericulata*), the advantage is sexual, but in other animals both sexes are brightly coloured and the colours are not used in sexual behaviour. Here the advantage is usually to advertise to possible predators that the animal is unpalatable. Such warning colours are typically red, black or yellow, and they occur in both vertebrates and invertebrates; for example, the Striped Skunk (*Mephitis mephitis*) is black and white, European wasps (Vespidae) are black and yellow, and some South and Central American poison frogs (*Dendrobates*) are red and black or yellow and black.

Warning colours can be of advantage to an animal only if predators know that they should avoid attacking those colours. Most predators do not have any inborn aversion to bright colours, such as those of a wasp, but they acquire an aversion by learning. For example, a bird which has been reared in the laboratory and has had no experience of wasps will attack and eat the first one presented to it. The wasp may sting the bird, which then refuses to attack further wasps. Similarly, if an American Monarch butterfly (*Danaus plexippus*) is offered it will be eaten, but after about ten minutes the chemicals in the butterfly cause the bird to vomit, and it will then refuse to attack Monarchs in the future. This is clearly effective, but an interesting question arises: if an animal which displays warning colours has to be eaten before the predator can appreciate the significance of those colours, how can such a feature evolve? The answer is that most examples of warning colouration occur in species which are either social (for example, wasps), or at least live in groups, so that all individuals in the group have genes in common. Thus, when one individual is killed, it is the related members of the group which benefit. If the loss of genes for bright colours is less than the loss of genes for dull colours in equally unpalatable animals, then warning colouration can evolve.

It is advantageous to distasteful animals with warning colours to be sufficiently tough to withstand the experience of being sampled by a predator. Indeed, unlike edible butterflies, Monarchs can be very roughly handled or pinched without being killed. If the butterfly can physically survive the encounter, it can afford to lose a piece of wing, for example, and this is sufficient to indicate its unpalatability or to reinforce the predator's memory of an earlier unpleasant experience. As predators seemingly can forget the warning signals after a period of time, some warningly coloured animals have displays which make them even more conspicuous and serve as reminders. Bees, for example, buzz if they are seized, and skunks erect hair and present their rear quarters towards the predator before actually ejecting a nauseous fluid.

ABOVE: the distinctive, orange and black colour of the American Monarch is easily remembered.

The eggs are laid on the toxic milkweeds which constitute the food of the larvae.

ABOVE: there are several types of coral snake in the New World. *Micruroides euryxanthus* is very poisonous with warning colours, but some other coral snakes are non-poisonous Batesian mimics.

BELOW: the striking colouration of these Red Arrow-poison Frogs (*Dendrobates speciosus*) warns would-be predators of the venom, which is lethal enough to be used on the arrow-tips of hunters.

Mimicry is the resemblance of one animal (the mimic) to another animal (the model), so that potential predators, for example, are deceived by the resemblance into confusing the two. One form of mimicry was first described by the Victorian naturalist Henry Bates. In Batesian mimicry, as it is often called, the model is distasteful and has warning colouration; the mimic has similar colours but is palatable. Batesian mimicry is of advantage to the mimic since it gives protection from predators which have learned to avoid the model. Laboratory experiments have confirmed that birds that have learned not to attack wasps will also avoid hoverfly (Syrphidae) mimics, but birds with no prior experience of wasps will eat hoverflies.

One problem for a mimic is that it cannot receive protection unless the model is present and common in the same area, otherwise instead of predators sampling and learning to avoid the model they will capture and eat the mimic and then learn to associate the mimetic pattern with edible food. Models of mimetic associations are therefore usually common animals with wide geographical ranges. Furthermore, in some mimetic butterflies only the females are mimetic whilst the males have a totally different pattern. This means that such a species can be twice as common as it could be if both sexes were mimetic, yet it can still receive the same degree of protection from the models in the area. The reason why females tend to be mimetic rather than males is that it is of advantage for females to live for a long time so that many eggs can be laid, whereas if males survive only a few days this is usually sufficient for them to fertilize several females. The Tiger Swallowtail (*Papilio glaucus*) is a North American butterfly in which the male is non-mimetic, but the female can be black, mimicking the noxious Pipe-vine Swallowtail (*Battus philenor*), or yellow and non-mimetic like the male. Where *Battus* is common, most or all of the *P. glaucus* are mimetic, but where it is absent, the female

P. glaucus are yellow as there is no advantage to be gained by being black.

If the models and mimics have populations that vary seasonally then it is usual for the mimic population to reach its peak a few weeks after that of the models. In this way local predators will have learned to avoid the models before the mimics become abundant.

Mimicry is not normally perfect, but in some cases it may take an experienced naturalist to distinguish model from mimic. Experiments suggest, however, that even a very slight resemblance to a noxious animal may give some protection from predators, and natural selection will then lead to ever more perfect resemblance as the predators in turn learn to discriminate models from poor mimics. One perfect mimicry occurs in the warningly coloured Monarch butterfly. If a caterpillar is fed on the toxic milkweed plant *Asclepias curassavica* it stores toxins from the plant so that it (and subsequently the adult butterfly) causes birds which eat it to vomit. But some caterpillars feed on non-toxic milkweeds such as *Asclepias syriaca*, and these produce palatable butterflies which are indistinguishable from the noxious ones. This form of perfect mimicry, when it occurs within a single species, is called automimicry.

Another form of mimicry originally described by the naturalist Fritz Müller is called Müllerian mimicry. Here two or more distasteful animals come to share the same colour pattern, so there is no deception involved. Because predators have to sample and perhaps kill some animals

The striking colouration of the Social Wasp warns enemies that it is unpalatable. Other unpleasant species share the same distinctive pattern so that this deterrent effect is reinforced. These are the Müllerian mimics. Batesian mimics are species which are palatable but have evolved similar colour patterns to the noxious species. Predators are fooled into thinking them distasteful too. The warning message given by black and yellow stripes is clear enough to give protection even to species that do not otherwise resemble wasps.

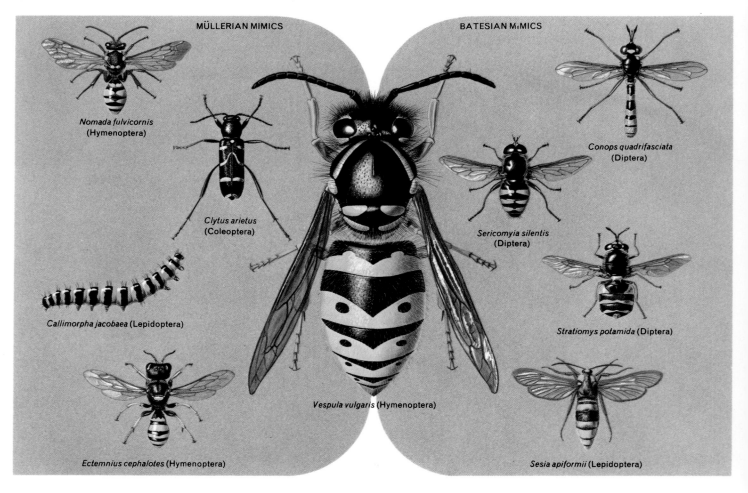

MÜLLERIAN MIMICS

BATESIAN MIMICS

Nomada fulvicornis (Hymenoptera)

Clytus arietus (Coleoptera)

Callimorpha jacobaea (Lepidoptera)

Ectemnius cephalotes (Hymenoptera)

Vespula vulgaris (Hymenoptera)

Conops quadrifasciata (Diptera)

Sericomyia silentis (Diptera)

Stratiomys potamida (Diptera)

Sesia apiformii (Lepidoptera)

before they learn to avoid the warning colours it is obviously advantageous for noxious animals to evolve similar warning signals. Social and solitary wasps and the caterpillars of the Cinnabar moth (*Tyria jacobaeae*), for example, all share the same distinctive black and yellow warning colouration.

Batesian mimicry is one form of bluff by which some animals are able to escape from predators, but there are other ways in which palatable animals can increase their chances of escape by bluff. Large Yellow Underwing moths (*Noctua pronuba*) normally rest motionless during the day amongst vegetation, but from time to time an individual may be discovered by a predator. It then flies, exposing the bright yellow patches on the hind wings. When the moth comes to rest again the yellow vanishes as the hind wings are hidden beneath the brown forewings. A predator may be startled by the sudden appearance of yellow as the moth takes off. It then pursues the moth following the yellow, but the yellow disappears, leaving the predator baffled. This flash colouration occurs in some moths and grasshoppers, and also in frogs where the colour is inside the thighs and becomes visible when they jump.

Another form of bluff is found in hairstreak butterflies (Lycaenidae). Predators which discover a resting butterfly may learn to anticipate its escape flight direction, and so increase their chances of successful capture. But some hairstreaks have short tails on the hind wings which are held up and may even wave slightly in the wind, very like antennae. Just at their base are eye-like spots, so that the rear end of the insect appears to be the head. The true antennae are concealed between the forewings. When such an insect is found by a predator it will of course fly off, but the predator may be mistaken by the false head into anticipating that it will fly in the opposite direction, and so the insect escapes. A more dramatic form of bluff occurs in some animals which are normally camouflaged but which suddenly expose large eye-like spots when they are disturbed. Examples include some moths, butterflies, praying mantids, fish and even one South American frog. The false eyes are very large and are perhaps mistaken by the potential predator as belonging to one of its own enemies such as an owl or hawk. Experiments using mealworms with small birds as predators have shown that the sudden appearance of any bright colour causes some hesitation in attack, but the more the bright colours resemble eyes, the greater is the inhibitory effect on the birds. So these large eyespots are of obvious defensive value.

ABOVE: the yellow flashes from a Large Yellow Underwing moth in flight are aimed at confusing any would be attackers. These three shots were taken at one twelfth of a second intervals.

BELOW: an Imperial Blue butterfly (*Jalmenus evagoras*) at rest. Its true head is to the left, but a false eye and false antennae on the hind wings deceive predators as to its direction of flight.

BELOW: large false eyes on the rump of the South American frog *Physalaemus nattereri* probably frighten predators which mistake it for the head of a much larger and possibly dangerous animal.

Migration

All animals move from one place to another during their lives. Some make long, seasonal round trips, while others never return to the same place. The extent and direction of such migrations depend on the availability of food and the right conditions for breeding.

Scientists cannot agree over the correct use of the word migration. Most ornithologists would only consider that a bird is a migrant if it performs a periodic round trip, usually on a seasonal basis, and usually over a relatively long distance. Many students of other vertebrates also adopt this usage. However, relatively few examples of such movements occur among insects. Hence, entomologists prefer to consider that an animal is migrating if it is moving from one breeding habitat to another. Thus, according to the entomological definition, insects migrate, but birds, which often return each year to the same breeding site, do not. On the other hand according to the ornithological definition birds migrate but insects do not. When referring to human migration we tend to use the entomological definition, as human migrants are generally regarded as those who move to a new country with the intention of remaining there.

In an attempt to resolve this conflict and confusion it has recently been suggested that, instead of arguing about the correct definition of the term 'migration', students of all animals should concentrate on the phenomenon that they are, in fact, all studying—that is, variation in the form of the lifetime track. All animals have a lifetime track, which is the path traced out by an animal between the sites of birth and death. Given this concept of the lifetime track, the most useful use of the term 'migration' is to describe those discrete units of movement from one place to another into which the lifetime track can be broken down.

Following this suggestion, the periodic movements of many birds can be described as seasonal return migrations, whereas the unidirectional movements of many insects can be described as removal migrations. Other uses of the term migration also occur.

Animals that live in the sea are confronted by a forever moving environment. It is perhaps not surprising, therefore, that most marine animals cover more of the Earth's surface during the execution of their lifetime track than do their terrestrial counterparts.

The primary production organisms of the sea, the phytoplankton, although carried long distances by oceanic currents, nevertheless probably move little in relation to the water body that carries them. The larger zooplankton, on the other hand, throughout the world's oceans, perform vertical migrations. At night they feed among the phytoplankton in the surface layers. During the day they rest in the deeper layers, where less light is available to betray their presence to predators and the lower temperature allows more efficient digestion and energy production. As the smaller zooplankton migrate up and down, their predators, the larger zooplankton, also move up and down, usually going to even greater depths than their prey. Finally, at the peak of this particular food pyramid are those larger pred-

The greatest of all travellers is the Arctic Tern (*Sterna paradisea*). It breeds during the northern summer around the Arctic Circle, south to Newfoundland and Britain. It then migrates south to enjoy the southern summer around the Antarctic ice. Such movement epitomises the ornithological view of migration.

ators, such as many squid and fish, that spend the day at great depths but at night may migrate right up into the surface waters. Zooplankton also perform seasonal vertical return migrations, spending the non-reproductive and non-feeding season at greater depths and the rest of the year in the upper layers or over a restricted depth range in the performance of daily return migration.

As different currents carry the blooms of phytoplankton and small zooplankton round the world's seas, the food supply of the larger predators, such as fish, also moves around the oceans. As a result, oceanic fish have evolved migration circuits, often carrying them in large circles, thousands of kilometres in circumference, that are geared to enable them to arrive in each area at the best time for feeding

Like most baleen whales, the Grey Whale (*Eschrichtius gibbosus*) feeds in polar waters in summer and breeds in tropical waters. Its movements are difficult to study, but it is now possible to track individual animals by the use of radiotelemetry.

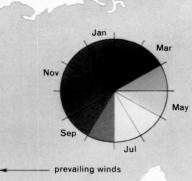

Jan
Mar
Nov
May
Sep
Jul

← prevailing winds
● recoveries of ringed birds

The seasonal return migration of the Short-tailed Shearwater (*Puffinus tenuirostris*) is typical of the annual movements of many birds, although more extensive than most. Hundreds of thousands of them gather to breed in south-east Australia, and recoveries of birds ringed there have revealed an incredible migration circuit. Not only are the birds able to exploit distant areas where fish are only seasonally abundant, but the population also spreads over a wide area, reducing competition for food. The actual migration route avoids seasonal headwinds. Shearwaters are long-lived birds and each may make this journey annually for tens of years.

or spawning. In some places, such as the North Sea, Baltic and Gulf of Mexico, smaller-scale current systems may produce smaller-scale migration circuits, but the principle seems to remain the same. Usually, these circuits involve each fish spending most of its life moving with the current, but sometimes fish may have to migrate across a current or even against a current for some distance, in order to arrive at the best place at a particularly suitable time.

As smaller fish and larger zooplankton, such as jellyfish, migrate around the world's oceans, so too do their predators. These include larger fish, squid, turtles, seals, sealions, dolphins, porpoises, and even pelagic seabirds such as petrels, shearwaters and albatrosses. Each one times its arrival in an area to coincide with the appearance not only of its own particular prey species, but also of suitable temperature and other climatic conditions. Then, as each place decreases in suitability once more, the predators move on to the next area on their particular migration circuit, often taking advantage of favourable wind or water currents when these are available. The huge baleen whales, such as the Blue Whale (*Balaenoptera musculus*) and the Humpback Whale (*Megaptera novaeangliae*), occur throughout the world's oceans and in both the northern and southern hemispheres. In summer they feed on the rich growths of zooplankton in polar waters and then, as winter approaches and ice cuts off their access to this their major food source, they migrate 3,000 to 4,000 km to mate and give birth in warm tropical waters. There the young whales need to expend less energy in keeping warm than would be necessary nearer the poles.

As oceanic fish move round their migration circuits, many move from positions over deep water to shallower water and from mid-oceanic sites to coastal waters. Surprisingly, however, only one genus of essentially oceanic fish has extended its lifetime track into freshwater. This genus, the 'freshwater' eels (*Anguilla* spp) seems to have had its origins in the tropical waters of the Western Pacific where 15 of the 17 modern species are found. Only two species occur in the Atlantic Ocean.

Some transparent elvers of the European freshwater eel *Anguilla anguilla*. Spawning is thought to occur in the Sargasso Sea. Young larvae drift across the Atlantic with surface currents and migrate up estuaries as elvers. Adults return perhaps 20 years later.

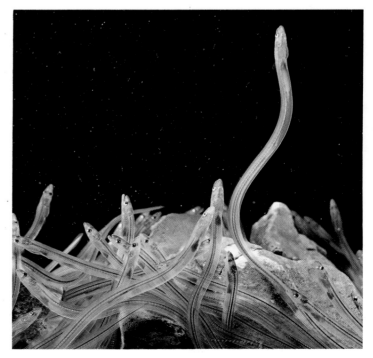

All other catadromous fish (fish that spawn in the sea, but spend a part of their lives in freshwater), such as the southern trout, or jollytails (*Galaxia* spp), are of coastal origin. One particularly noticeable feature of catadromous fish is that they are primarily distributed when at sea in warm and tropical waters. Where catadromous behaviour occurs in other groups, such as prawns, crabs and lobsters, and even in mammals such as manatees and porpoises, it is also more or less confined to warm waters.

The converse is also true. Anadromous fish and crustaceans (those that spend part of their lives in the sea, but spawn in freshwater) are confined more or less entirely to cold waters. The major groups of anadromous fish are the salmonids and lampreys, both of which inhabit temperate and sub-polar waters. No convincing explanation for this difference in geographical distribution as yet exists.

No such latitudinal variation occurs in the migration pattern of fish that spend their entire lives in freshwater. Whether they live in the sub-polar rivers of Siberia or Canada or in the tropical rivers of Africa, Asia or America, the pattern seems to be the same. Most of them perform upstream-downstream migration, swimming upstream to spawn and downstream to feed. Many fish that spend most of their lives in lakes nevertheless often migrate up into inflowing streams to spawn. However, those individuals whose lifetime tracks are very near to outflowing streams of the lake may sometimes migrate downstream to spawn.

Many stream-living fish, such as trout, feed to a large extent on members of the 'downstream drift'. This consists of an assemblage of worms, molluscs, crustaceans, mites and insect larvae that release themselves into the current whenever they need to migrate any distance to reach

LEFT: ghost crabs (*Oxypode*) use their rapid running ability during their migration. They leave their burrows in the dunes at low tide and race up and down the beach feeding in the intertidal areas.

ABOVE: a salmon (*Oncorhynchus*) leaping up a waterfall in Alaska. Salmon are prodigious leapers and when moving upstream to spawn can clear obstacles that are up to ten times their body length.

new feeding or sheltering sites. They may be carried downstream a few tens or hundreds of metres at a time, perhaps settling occasionally on their journey to test the suitability of the habitat. When adult, many of the insect members of the downstream drift fly back upstream to lay their eggs in the same regions of the stream in which their own lifetime track began. It is also thought that, when not being carried downstream by the current, worms, snails and crustaceans may also prefer to crawl upstream.

Many essentially aquatic animals, particularly those that spend part of their lives in the sea, perform some of their lifetime activities on land. Even some fish migrate between

sea and land. Grunion (*Leuresthes tenuis*) along the coast of California, migrate inshore on the first three or four high tides following new or full moon and spawn in the sand just above high water mark. A fortnight later, the next incoming tide uncovers the eggs, the young Grunion hatch and migrate offshore. The precise synchrony of these migrations, many millions of fish all performing the migration at the same time, has made them world famous. Tropical land-crabs perform the opposite migration, spending nearly all their lives on land, but migrating to the sea to spawn. Many other crabs and a wide range of other invertebrates from polychaete worms to echinoderms perform inshore-offshore migrations, spawning in water that is either deeper or shallower than that in which they feed. The lifetime track of such invertebrates may or may not extend out of the water and every gradation can be seen. At one extreme there are inshore-offshore migrants that always remain in the water; on the other hand migrants such as the land-crabs retain only the most tenuous of connections with the sea.

Though the Hippopotamus may seem a rather sedentary animal it does, in fact, make daily migrations, leaving its wallow at night to travel up to 30 km over its grazing grounds.

Spiny lobsters (*Panulirus*) may cover great distances between shallow- and deep-water ranges. Migration takes place *en masse* with thousands of adult lobsters making long single-file columns.

Animals that live on sandy shores and whose lifetime track is centred on the edge of the sea usually move up and down the shore with the tides. *Donax denticulatus*, a small tropical clam, and other molluscs use the movements of the tidal water in order to migrate, but animals such as ghost crabs and sandhoppers use their own locomotion.

On a larger scale, many vertebrates also alternate between periods at sea and periods on land. The best examples are the seals, sealions, fur seals and walruses. They feed in the sea on a wide range of planktonic and other invertebrates, fish, and in some cases even birds, particularly penguins. But during the breeding season they move onto continental or island shores, sandbanks, or ice and give birth to their young. Some species, particularly those that breed on oceanic islands, have large harems and mate on land. Other species that give birth on ice do not have harems and mate in the water.

Female sea turtles lay their eggs on the same sandy shore as two or three years previously, often after visiting one or more feeding grounds on a migration circuit of perhaps several thousand kilometres. Pelagic sea birds, such as albatrosses, petrels and shearwaters, perform similar migration circuits, returning every one or two years to breed at the same breeding colony.

Migrations between water and land are not confined to animals that spend part of their lives in the sea. Animals that spawn, feed or sleep in freshwater may also migrate onto or over land for other activities. Frogs, toads, newts and salamanders all breed in freshwater, but the vast majority spend most of the remainder of their lives on land, often several kilometres away from the pond, lake or stream to which they return each year to breed. Terrapins (fresh-water turtles) may migrate overland to different water bodies at different times of the year, as also may crocodiles and alligators. Hippos spend the daytime resting in water, but at night they come on land to graze, often travelling overland a distance of about 30 km.

ABOVE: part of a large herd of wildebeest (*Connochaetes*) crossing a river during its migration. At the beginning of the wet season wildebeest find fresh grass by travelling towards distant thunder or characteristic, dark-capped, rain-producing storm columns.

BELOW: locusts such as Migratory Locusts (*Locusta migratoria*) are grasshoppers that migrate in vast, cohesive swarms. As the swarm steamrollers across country its members settle beneath it and feed greedily, taking flight again when the tail-end passes overhead.

The best-known migrations, however, are those performed by land animals, particularly birds but also ungulates and insects. Conspicuous migrations may be seen at any point on the Earth's surface, but the nature and function of the migration differs between latitudes and between the different climatic zones.

In the tropics, least movement occurs in the rain forests. Apart from the seasonally immigrant birds that have bred in temperate and sub-polar regions, most rain-forest inhabitants have a relatively limited range, and their lifetime tracks are short. The most spectacular terrestrial migrations in the tropics occur in the arid grassland, scrub and semi-desert regions and are geared strongly to the rainfall pattern. Where the rainfall is totally unpredictable, as in Central Australia, birds such as Budgerigars and honey-eaters are highly nomadic, travelling long distances, feeding and breeding opportunistically as and where food and water temporarily become available. Such birds, and to a lesser extent some mammals, such as the Red Kangaroo (*Megaleia rufa*), can recognize and navigate to the location of distant rain. Similar behaviour is shown by a wide variety of insects, birds and mammals in other arid regions, such as various parts of Africa, Asia and South America, where rainfall is predictably seasonal within a large area, while the precise local distribution of rainfall often remains patchy and erratic. Animals adapted to such regions, such as the Desert Locust (*Schistocerca gregaria*) in and around the Sahara Desert and the herds of wildebeest, zebra and gazelle on the Serengeti-Mara plains of East Africa, have developed flexible migration circuits, often thousands of kilometres long, that take them to the general area most likely to receive rainfall at each time of year. The problem of local unpredictability is then overcome by perception of distant rain, perhaps 40 km or so away in the case of wildebeest and perhaps much further in the case of locusts.

Large terrestrial mammals with long-distance seasonal return migrations or migration circuits are confined either to the arid regions of the world or to the sub-polar regions. In both these areas food either has a low replacement rate or occurs unpredictably due to sporadic rainfall. Thus, long-distance migrations are not only found in wildebeest and zebra in the tropics, but also among a variety of bovids, horses and antelopes on the Eurasian steppes, and in the Caribou and Reindeer (*Rangifer tarandus*) in sub-polar regions. In non-arid temperate regions, the only clear-cut seasonal return migrations are between different altitudes. For example, Red Deer (*Cervus elaphus*) in Scotland and Whitetailed Deer (*Odocoileus virginianus*) in America spend the winter at lower levels than those occupied in the summer. Such altitudinal migration, however, takes place world-wide, being shown not only by these temperate mammals but also by Moose (*Alces alces*) in Canada and African Elephants (*Loxodonta africana*) on the slopes of Mount Kenya. Similar altitudinal migrations are shown by animals as diverse as birds, such as the Mountain Quail (*Orcortyx pictus*) and beetles, for example the Convergent Ladybird (*Hippodamia convergens*), both North American species.

At night, in temperate regions, largely unseen migrations are performed by bats and moths. Bats perform seasonal return migrations between hibernation and summer feeding sites. However, these migrations are usually shorter than those of birds, ranging from one kilometre to several hundred kilometres. Only a few species, such as the Free-tailed Bat (*Tadarida brasiliensis*) and the tree bats (*Lasiurus*) of North America and the Noctule (*Nyctalus noctula*) of central and eastern Europe, migrate further than 1,000 kilometres. Individual moths probably migrate in only a single direction during their lifetime, rarely travelling further than a few hundred kilometres between birth and death. However, some species, such as the Oriental Armyworm Moth (*Pseudaletia separata*) in Eastern Asia, are known regularly to travel about 1,400 km. Where a species, such as the Silver-Y Moth (*Plusia gamma*) in Europe or the Oriental Armyworm Moth, is known to migrate in different directions at different times of year, it is a different generation that flies south in autumn than flew north in spring and summer.

Easily the most conspicuous migrants among the animals of temperate regions are birds and diurnal insects. Except along the eastern and western coasts of Eurasia and America, more birds perform relatively long-distance, often trans-equatorial, seasonal return migrations than stay in a relatively confined area throughout their yearly cycle. Vast numbers of small, temperate insects may have a relatively long life-time track, taking advantage of air currents to travel across country. However, the best known insect migrants in temperate regions are butterflies. Relatively few species stay in the same local area throughout their lives. The majority seem to have a more or less linear lifetime track that is no longer than 200 to 300 km between the sites of birth and death. A few species, however, such as the Monarch (*Danaus plexippus*) in North America, and at least the autumn generations of the Red Admiral (*Vanessa atalanta*) and Painted Lady (*Cynthia cardui*) in Europe are known or suspected to have a lifetime track 2,000 to 3,000 km in length.

BELOW: the Mouse-eared Bat (*Myotis myotis*) may migrate up to 160 km between the caves where it hibernates in winter and its summer feeding area. At certain times of the year males and females gather into separate colonies.

ABOVE: these Monarch butterflies are wintering in California along the northern edge of their winter range. Here they stay in clusters on trees, making feeding sorties during fine weather. They start to copulate in January, and then migrate northwards in March.

Animal Associations

Some animals show fascinating associations with plants or other animals. In a symbiotic association both partners benefit; for example, a cleaner fish gets food and its client loses its parasites. In a parasitic association, on the other hand, only the parasite benefits.

Patterns of association between living organisms are as diverse as the organisms themselves. Modern ecology has shown us that microorganisms, plants and animals all coexist in ways involving a wide range of interactions and a considerable amount of interdependence. Even the higher green plants, which superficially seem to be essentially autonomous, have a number of tight associations with other organisms. Many of them are utterly dependent on animals for their pollination or on symbiotic fungi associating with their roots (mycorrhizae) for efficient uptake of minerals in nutrient-poor soils. They may also be attacked by an unpleasant range of host-specific pathogenic organisms including viruses, bacteria, fungi and nematodes.

Associations between organisms can occur on a variety of levels from the most casual, non-specific and transient examples to situations where a life-long partnership is obligatory. There are many intriguing instances of tight associations occurring within a species. All animal family life is a manifestation of this phenomenon, which reaches its peak of complexity in the communities of social animals such as ants, bees, or hunting dogs.

Between organisms of dissimilar species one often finds associations that involve close and relatively persistent contact between the organisms concerned as well as great species-specificity. For instance, the Pork Tapeworm (*Taenia solium*) only infects Man and the pig, and the Bee Orchid is only pollinated by a narrow range of bee species. Coupled with this close commitment, the associating organisms often show patterns of adaptation, in structure, behaviour and physiology, that can only be understood in the context of their life together.

The terminology that has been utilized in attempts to categorize animal associations is confused and complicated. Most associating animals cannot be neatly categorized, and many intermediate situations exist between apparently different forms of association. Most terminologies are based on the concepts of benefit and harm. On this basis, intimate associations in which one species reduces the natural population growth rate of the other is called a parasitic association. Where the rate is increased, the relationship is described as symbiotic. Many additional subdivisions of these two main categories have been described, but all these extra terms, such as phoresy, inquilinism, commensalism, and host-parasitoid relationships, are best considered as somehow related to parasitism and symbiosis.

Association biology is not an obscure or unworldly field of investigation. Parasites of Man, his domesticated animals and crops have immense economic, medical and social significance. Symbiotic partnerships, especially those including green plants, play a vital role in maintaining the productivity of Man's food crops.

ABOVE: a conspicuously marked clown fish (*Amphiprion*) among the stinging tentacles of a sea anemone. Its bold colouration serves as a warning relating to its dangerous symbiotic partner.

BELOW: the orange mass beneath the Shore Crab (*Carcinus maenas*) is the brood chamber of the crustacean parasite, *Sacculina*. This highly specialized parasite is a much-modified barnacle.

BELOW: the nudibranch mollusc *Hexabranchus flammulatus* moving over a coral surface on the Great Barrier Reef off Australia. Attached to its external gills is a commensal shrimp.

In symbiotic associations one or, more commonly, both partners (symbiotes) derive benefit from their shared life. Symbiotic partnerships can be obligate (absolutely necessary) or facultative (useful, but not crucial) for one or both partners. Interestingly, intimacy in the relationship between the associating organisms is not necessarily correlated with the degree of obligateness.

Microorganisms are often vital symbiotes of multicellular animals. Termites are able to digest the cellulose rich plant foods that they consume only because a part of their gut is packed with a wide range of symbiotic flagellates and bacteria. Several of the flagellates can digest cellulose and pass on useful nutrients to the insect partner. Portions of the bacterial flora can fix atmospheric nitrogen into organic compounds and make up for the deficiency of nitrogen in the termite's ingested food.

Complex patterns of advantage may occur in symbiotic associations between multicellular partners. Provision of nutrients and protection from predators are examples of the kind of advantages involved, and there are both marine and terrestrial examples of these.

Sea anemones appear to be frequent partners in symbiotic relationships, in both active and passive roles. Clown fish, for instance, live among the tentacles of large anemones on coral reefs; they even venture into the gut cavity of the anemone. Controversy continues over the method by which the fish avoid being stung by the nematocysts of their partner. By whatever means the trick is achieved, the stinging tentacles protect the clown fish from other predators. The sea anemone itself perhaps gains from the association because other fish may be lured towards the tentacles by the presence of the clown fish. In another range of associations, hermit crabs have specific partnerships with particular sea anemone species that become attached to their shells. Some crabs actively place anemones on their shells, but other species play a passive role while the anemone, in a complex ballet of movements, secures itself on its crustacean partner. The stinging cells of the anemone probably protect the crab from predators like the octopus.

On land, a group of African birds called the honeyguides utilize a specific indicating display to lead either man or the Honey Badger to a bee's nest. The bird then waits while the large mammal breaks into the nest and eats its honey. The honeyguide then flies to the untenanted nest and eats bee larvae and, more extraordinarily, bees' wax. Birds in this group have symbiotic bacteria in their gut which enable waxes to be digested.

The hermit crab *Pagurus arrosor* is a Mediterranean species found within the empty shells of some molluscs. Often these shells are protected by sea anemones which become attached to them. Here, two specimens of the anemone, *Calliactis parasitica* almost hide the shell; their stinging tentacles deter would-be predators.

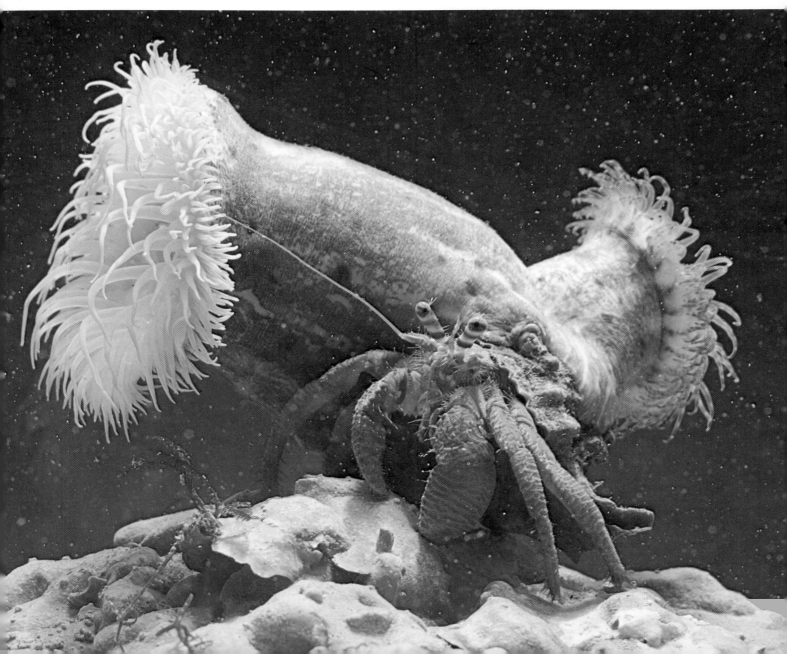

ABOVE: several bird groups are specialized morphologically and behaviourally to feed on nectar. In many cases this allows pollen to be carried between flowers and leads to successful pollination. The Orange-breasted Sunbird (*Anthobaphes violacea*) with its long, thin, down-curved bill shows this type of adaptation.

BELOW: the black, hollow bulbs at the spine bases of the whistling thorn, *Acacia drepanolobium* serve as chambers of an extended nest for the symbiotic ant species. In experiments the bulbs have developed on young plants in the complete absence of ants; in the wild, however, all bushes have their complement of ant partners.

Plant-animal symbiosis for pollination almost always involves drastic adaptive changes in flower morphology and biochemistry. These adaptations serve to attract the pollinating animal and induce it to behave in ways which achieve the transfer of pollen from the anthers of one flower to the stigma of another, thus ensuring cross-fertilization.

Insects are the most commonly utilized pollinating animals and much of angiosperm adaptive radiation can probably be linked to different patterns of pollination associations with various insect groups. A variety of rewards is used to induce insects to visit flowers. Pollen itself may be consumed as a food, but this is wasteful. More usually, carbohydrate and amino acid-rich secretions called nectars are synthesized by parts of flowers or neighbouring regions, and these act as nutrient lures for the insects.

Less commonly, other types of animals effect pollen transfer. There are flowers pollinated by birds or mammals, including bats. Where strong winds or continuous cold make an area unattractive to cold-blooded insects, it is obviously a useful strategy for the plants of that area to be pollinated by warm-blooded symbiotes; the activity patterns of the latter will be less influenced by environmental vicissitudes. There are several examples of warm-blooded pollinators. In southwestern Australia trees of the genus *Banksia* have large red or yellow inflorescences that provide an abundant food resource for a range of pollen and nectar-eating birds and at least one mammal. The Honey Possum (*Tarsipes spencerae*) has a long, narrow snout and an elongate protrusible tongue, which it uses to remove nectar from the *Banksia* flowers. Pollen is trapped on the fur of the animal and carried to female flowers on its feeding excursions.

The morphological adaptations of the Honey Possum are a useful reminder that pollination symbiosis makes adaptive demands on both plant and animal. Where pollination specificity exists it is often possible to speak of the co-evolution of the two partners. In such cases anatomical specializations by both partners can make the association between them obligate.

Other types of intimate associations have evolved between plants and animals, unrelated to pollination. Several African bushes in the leguminous genus *Acacia* have remarkable associations with ants. One has spines whose bases expand into blackened, hollow bulbs that are invariably utilized by ants to constitute multi-chambered nests. The association is cemented by the sugary secretions of extrafloral nectaries. Food and shelter are the benefits that the ants enjoy from the symbiosis; the acacia appears to gain considerable protection from browsing mammals. When a herbivore starts to nibble the leaves of the acacia, it is immediately set upon by hundreds of biting, stinging ants.

The most remarkable instances of animal-mediated pollination are those where co-evolution has moulded both partners to such a degree that an obligate relationship exists between a single insect species and a single angiosperm species. This position has been reached in the pollination of figs (*Ficus* spp). Pollen transfer is accomplished during the very complex symbiotic interrelationships that go on between the fig inflorescence and the pollen carriers—specific species of agaonid wasps which develop in the ovaries of the flowers. A central determining factor in this relationship is the structure of the inflorescence itself. A fig is an urn-shaped structure whose inner surface bears first female then male flowers. Its opening is closed by a structure called an ostiole, which is an overlapping maze of tough scales. A female wasp specific to a particular fig species will be attracted to a female stage fig by chemical clues and then force its way through the scales to get at the flowers. While struggling in, the wasp tears off its wings and the outer portion of its antennae. Many of the female flowers are

adult females fly to
new figs

adult females loaded
with pollen enter fig
through the ostiole

in forcing their way
through the narrow
ostiole, the wasps lose
their wings, which
remain projecting
from the top of the fig

the females lay
eggs in the fig
fruit, pollinating
the female flowers

the larvae develop and
pupate, preventing the
fig from ripening

wingless adult males
emerge, fertilize the
females and tunnel
through the wall of the
fruit

adult females become
covered with pollen
and leave the fruit
through the tunnel

the fig continues to
ripen normally

The relationship between some agaonid wasps, like *Blastophaga quadraticeps*, and the fruits of figs is an example of symbiosis. Here, the complex interactions ensure successful developmental cycles for both partners. The female fig wasps are winged and so can fly to female phase figs, where they lay their eggs, from the fig in which they developed. Males, which lack wings, never leave the fig of their birth.

injected with one wasp egg each by the female wasp's long ovipositors. At the same time she transfers pollen to other female flowers from the male stage fig in which she earlier developed and was fertilized.

The injection of a wasp egg near a fig ovule induces a gall to form on which the hatching wasp larva feeds. Wingless males develop first from a larval brood and fertilize the new generation of females within a fig just as male flowers begin to produce pollen within that inflorescence. In some symbioses emerging females become passively covered with pollen. In others, as in the case of *Ficus religiosa*, pollinated by *Blastophaga quadraticeps*, the female wasp has thoracic pollen pockets which she fills in the male stage fig before leaving it to pass on to one in the female stage.

Thus, in an extraordinary co-evolution, the reproductions of fig and agaonid wasp have become inextricably and harmoniously linked. Immense adaptations have taken place in both partners to bring this state of affairs into being. The genus *Ficus* dates back to the Cretaceous and the associated wasps have an even more ancient lineage, so it seems that the beautifully integrated symbiosis we see today may have been evolving for over 100 million years.

No less remarkable instances of linked adaptations occur in other plant-animal symbioses. The development of pure red-orange coloured flowers without a strong ultraviolet emission can almost certainly be linked to the evolution of bird pollination. Very many bird-pollinated flowers demonstrate these colours, which correspond to the range of colours that nectar feeding birds like hummingbirds and sunbirds can distinguish. Reciprocally, these birds have developed specialized thin, down-curved elongate beaks for obtaining the nectar.

Many orchid flowers in the genus *Ophrys* are insect mimics in shape, colour and sometimes smell. Hymenopteran insects are attracted to them for a number of reasons.

In some cases male wasps attempt to copulate with the inflorescence and in doing so pick up special packets of pollen, called pollinia, which they can transfer to the next orchid flower of the same species on which they alight. A useful symbiosis from the orchid's viewpoint, a frustrating type of parasitism from the wasp's!

Some orchid species are pollinated by hymenopteran insects. Such pollination symbioses are usually species-specific as the flower must give very precise visual and chemical signals to encourage copulatory behaviour. Here an ichneumon wasp associates with an Australian orchid and will pick up pollen in the process.

Another kind of symbiosis is called cleaner symbiosis. Although most animals take care of their own skin care and cleaning, or, in some cases, indulge in mutual grooming or preening, cleaner symbiotes show a startlingly different form of behaviour. They remove ectoparasites from the outer surfaces of much larger animals and clean injured or diseased areas of skin. In marine ecosystems the phenomenon has been most frequently observed on coral reefs, but examples are now known from temperate waters as well. The client animals (those that are cleaned) are almost always fish. The cleaners are either highly specialized fish species or shrimps.

As the cleaners are much smaller than their frequently carnivorous clients, a prime problem in their lives is avoiding being eaten by the fish which they are attempting to clean. Adaptations to avoid this possibility mostly involve distinctive patterns of colouration or equally distinctive behaviour. For example, many cleaner fish of different species share a particular marking pattern. The fish tend to be elongate, pale coloured or blue and carry an array of longitudinal dark stripes which make them extremely prominent against any background. This shared appearance (called 'guild-marking') helps to reinforce the suppression of predatory behaviour by client fish, because all fish demonstrating it perform a useful service for the client. Similarly, cleaner shrimps usually have very long legs which have alternate black and white stripes. Accentuating this conspicuousness, most cleaner fish and shrimps have displays which are performed in front of the client fish to aid recognition. Cleaner fish 'dance' and shrimps wave legs and antennae up and down.

Unfortunately for clients some non-cleaners have taken unfair advantage of this signalling system between client and cleaner. Among the reefs in the Maldive Islands a false cleaner exists. A small predatory sabre-toothed blenny has adopted the disguise of a striped, guild-marked, Cleaner Wrasse (*Labroides dimidiatus*). Using this counterfeit appearance it is able to approach large fish unharmed, but instead of cleaning them it tears pieces out of the skin or fins of the client species with its sabre teeth.

The blue and black striped Cleaner Wrasse is probably the most intensively studied cleaner symbiote in the world. It is found in coral reef ecosystems all over the Indian Ocean. Pairs of these fish inhabit quite specific locations on the reef, usually near some prominent feature like a large coral outgrowth. This location has been termed a 'cleaning station', because it appears certain that client fish like large groupers, parrot fish and surgeon fish learn the position of such stations and go to them for cleaning. Often a whole line of fish can be seen apparently waiting their turn at such places. Before a bout of cleaning activity the wrasse dances up and down in front of the client fish wriggling its body constantly. Client fish themselves exhibit quite specific posturing behaviour which often intimates to the cleaner which areas need to be cleaned. The slim cleaners enter the client's mouth cavity, probe around between the gill arches and get under the opercular plates as well as giving attention to the general body surface and fins. They remove debris, fungi and sloughing skin from wounded or infected areas and eat ectoparasites. In most instances skin and gill-dwelling monogeneans and skin-inhabiting parasitic copepods make

ABOVE LEFT: the cleaner shrimp *Hippolysmata grabhami* swims onto the surface of a Copperband Butterflyfish (*Chelmon rostratus*). The shrimp attracts fish with movements of its clearly visible white antennae, and the fish signals its willingness to be cleaned with a characteristic 'invitation posture'.

LEFT: a sweetlip fish is cleaned by two striped Cleaner Wrasse. While one approaches the head the other has already entered the buccal cavity. In this apparently dangerous position it feeds on parasites infesting the client's gills. The benefits are mutual: the Wrasse feeds with ease and the client is freed of parasites.

up much of this diet of parasites. At the end of a cleaning session, which may last many minutes, the client fish usually terminates the encounter by signalling to the cleaner. The signal may be mouth closing and opening, flicking the opercular plates or twitching and shaking the body.

Fish in the genus *Echeneis* and related forms are called remoras. They are elongate fish whose dorsal fin has become transformed into an efficient oval sucking disc on the top of its head. Toothed plates in the disc, derived from fin rays, lodge into the skin of a shark or ray and depress the floor of the disc creating a suction pressure. In this way the remoras can 'hitch-hike' on their elasmobranch partners. Some appear to use the larger fish simply as a form of transport, others scavenge from the scraps left trailing in the water after a shark kill and yet others are fully-fledged cleaner symbiotes. They remove parasites from the sharks' denticle-covered skin.

On land it is insectivorous birds that have most clearly occupied the cleaner symbiote niche. The two species of oxpeckers feed on the ectoparasitic arthropods of large herbivores on the plains of Africa, and the Egyptian Plover takes leeches from the open jaws of Nile Crocodiles. However, on the Galapagos Islands, land crabs remove ticks from sun-basking Marine Iguanas.

The Red- and Yellow-billed Oxpeckers of Africa have become very significantly dependent on cleaner symbiosis as a means of obtaining food. These intriguing birds scavenge for ticks and parasitic flies over most of the large mammalian herbivores of the plains of Africa. They have many adaptations for this associative existence. Their claws are sharp-pointed and arranged in a zygodactyl way (two toes forward—two toes back) as in the woodpeckers. This configuration enables them to run all over the surface of their client mammal, shinning up the necks of giraffes or up and down the legs of antelopes with equal ease. They remove ticks deftly using a bill which is extremely flattened from side to side so that it can shear off the tightly attached arthropods by a scissoring action. The oxpeckers are so closely involved with the large mammals that they display and mate on the backs of client animals and even build their nests out of hair which they collect from those same symbiotic partners.

ABOVE: remoras swim close to sharks while others attach themselves to the larger fish by means of ovoid, dorsal, sucking organs, as is shown here in close up by the Sharksucker (*Echeneis naucrates*). The organs have rows of spine-tipped plates that lock into the shark's skin and depress the floor of the sucking organ to generate the suction force. The precise nature of this animal association is difficult to deduce. In some instances it is probably pure phoresy—simple hitch-hiking —whether on sharks, other fishes or even on turtles or objects such as boats; in other cases it is quite certain that the remoras act as cleaner fish.

RIGHT: the oxpeckers are perhaps the best known of the avian cleaner symbionts. Here, several Red-billed Oxpeckers (*Buphagus erythrorhynchus*) are seen perched on the backs of Zululand warthogs. The birds' coral red beaks act as efficient shears for cutting off ticks and other ectoparasitic arthropods from the hides of their larger mammalian partners.

Parasites are organisms that live at the expense of other organisms, which are usually termed hosts. Parasites live in close contact with their hosts for an appreciable portion of their total life span, if not all of it. The cost to the host in such one-sided associations can take a variety of forms. Most parasites are nutritionally dependent on their hosts and remove food substances directly from the hosts' gut or by feeding on host tissues. Tissues and food losses of this nature can reduce the growth rate of hosts, diminish their reproductive rate or increase their death rate. The latter effect can be due to direct debilitating effects of the parasite or an increased susceptibility of the host to predation.

Parasites can be categorized in a number of different ways. For example, it is common to distinguish between intracellular and extracellular parasites. The former live at least part of their lives within the cells of their hosts. The latter, usually much larger than the intracellular forms, exist outside the host cells, often inhabiting tissue spaces like the lumen of the gut or the air spaces of the lungs. There are also ectoparasites which live on the outside of their hosts as distinct from endoparasites which are internal.

There can be few animals, plants or microorganisms that do not harbour parasites of some sort and remarkably few groups of organisms that do not include parasitic members. All viruses are utterly dependent on their host cells and in a sense must be regarded as the ultimate parasites. Among the animals, the Protozoa provide us with very many parasitic examples, several of which cause extremely damaging diseases in man. For example, the genus *Plasmodium* causes malaria and for a vital part of its life cycle it resides within the red blood cells of the host. In contrast to these intracellular forms, the flagellated cells of trypanosomes swim in the plasma between the red blood cells. They cause diseases such as sleeping sickness.

Among metazoans the most extensive adaptive radiation into parasitic life styles occurs among platyhelminths, some pseudocoelomate groups and the arthropods. Indeed, sometimes whole phyla may consist entirely of parasitic species. The acanthocephalans are all parasitic. They inhabit the alimentary tracts of vertebrates as adults and the haemocoeles of arthropod intermediate hosts while passing through larval developmental stages.

Most parasitic animals show a marked degree of host specificity at some parts of their life cycle. For example, the filarial worm *Wuchereria bancrofti*—a nematode which causes elephantiasis, can reach sexual maturity only in the lymphatic system of Man. Such restriction to a single host species is a rare occurrence, although, typically, a parasite can only utilize a limited range of hosts that are either phylogenetically or ecologically related.

Two other general characteristics of parasites are their great fecundity and the complexity of their life cycles. In almost all circumstances the reproductive rate of a parasite is immensely greater than that of its host. For example, *Ascaris lumbricoides*, a nematode parasite of Man, can produce about 250,000 eggs a day. Complex life histories with many different developmental stages seem to be related to the need to effect transfers from host to host.

Ectoparasites have ways of life that straddle both free-living and parasitic existences. As they live on the outside of their hosts, ectoparasites must be significantly adapted to conditions in the outside world as well as those moulded by the morphology and physiology of their host. The lampreys, ectoparasitic fish that feed on other fish, provide a good instance of this biological compromise. These primitive, elongate fish spend only part of their adult lives in contact with the host. This period consists of the feeding attacks which they make on fish such as trout. A lamprey attaches itself to a host using its complex mouth sucker. This consists of an outer sealing flange and radiating rows of sharp protuberances. Locked onto the host's skin in this way the lamprey rasps with its tongue into the side of the

ABOVE: these ticks are sucking blood from the skin around the spines of a hedgehog. Although the impenetrability of the spines forms a very efficient defence against predators, it makes it hard for the hedgehog to groom its skin. Consequently, it is usually host to a wide range of ectoparasites, particularly fleas.

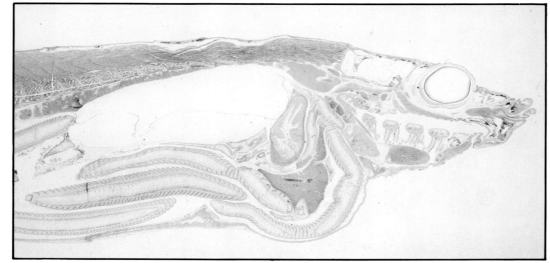

LEFT: this stained section shows the elongate, segmented larvae (also sectioned) of the tapeworm *Schistocephalus solidus* packing the body cavity of a stickleback (*Gasterosteus aculeatus*). Aquatic birds such infected fish ingest these plerocercoid larvae, which grow to maturity inside the digestive tract.

fish and sucks its blood. Such feeding sessions may last hours or even days, but between them the lamprey has no contact with its host.

Blood-feeding and keratin-eating ectoparasitic arthropods are relatively abundant. Bugs, mosquitoes and fleas spend short periods of time on vertebrate hosts sucking blood from them. Sucking lice and ticks have more persistent feeding contacts. Indeed, the whole life cycle of many sucking lice takes place on a single host individual. The human body louse (*Pediculus*) is a case in point. Like many other blood feeding parasites it possesses internal symbiotic micro-organisms that synthesize substances such as the B vitamins. Such compounds are only present at very low concentrations in blood. The chemical stability of keratin poses an enormous nutritive problem to the ectoparasites that eat it in the form of epithelial cells or feathers. It is very nearly indigestible. However, feather lice on birds overcome this problem by maintaining highly reducing conditions in their guts.

The platyhelminths show a number of distinct evolutionary trends in the establishment of parasitic modes of existence. Monogeneans show a pattern of mainly ecto-parasitic host-parasite interactions. They are usually skin- or gill-inhabiting surface parasites of fish and have direct life cycles involving only one host species. The endoparasitic digeneans have more complex, indirect life cycles, which typically demonstrate asexual reproduction in a molluscan intermediate host and sexual reproduction by adults in a vertebrate. The cestodes, or tapeworms are the most highly specialized of the endoparasitic platyhelminths. Adults inhabit the guts of vertebrates and absorb nutrients there by diffusion and active transport across a highly modified, syncytial body wall. They show no trace of a gut. Reproductive output is increased by the presence of a chain of segments (proglottids), each of which contains a full complement of reproductive organs.

ABOVE: a Rainbow Trout (*Salmo gairdneri*) being attacked by the lamprey *Lampetra fluviatilis*. All lampreys are ectoparasites of fish, feeding only on blood. They have toothed, sucker-like mouths that are adapted for holding onto the fish's skin while the rasping tongue opens a wound through which blood is sucked.

BELOW: of all the brood parasitic birds in the world, and there are over 100 species, the Cuckoo (*Cuculus canorus*) is the most intensively studied example. In this instance the young cuckoo is being fostered by a pair of Reed Warblers. Each surrogate parent is considerably smaller than its substitute nestling.

In addition to the parasitic and symbiotic associations already described there are several other types of animal association. They vary considerably in the form and degree of advantage gained by one or both partners, but generally they can be considered as subdivisions of parasitism or symbiosis or as intermediate forms of association.

The eggs and larvae of the hundreds of thousands of insect species that exist in the world are an immense potential nutrient resource for predators that can find, devour and digest them. A large number of insects themselves have utilized this food source in a remarkable way, which links reproduction and nutrition. These insects, which are called parasitoids, lay their eggs on, in or near the eggs or larvae of

A parasitoid way of life is very common among insects especially the hymenopterans. Many lay their eggs in or on other insects with the result that their larvae kill the host by feeding on its tissues. Here, a parasitoid wasp is laying its eggs in a moth caterpillar.

insect, spider, snail or isopod hosts. Thus the hatching parasitoid larva finds itself inside or close to the host and it proceeds to devour the host from within. Non-vital or renewable organs are consumed first; only later, as the parasitoid larva or larvae grow in size, are more crucial organs damaged. Ultimately, mature larvae leave the host, or pupate within it. In either case the host is almost invariably killed. If death does not occur due to the destruction of vital organs, the host dies when the larvae bore their way out

LEFT: during the day pearl fishes such as *Carapus bermudiensis* live in the well-aerated rectum of sea-cucumbers such as *Actinopygia*. They have no scales or pelvic fins and so can back into their host.

BELOW: the Stargazer Goby acts as an efficient sentry for its home-provider, the burrow-digging marine shrimp (*Alpheus*). This partnership is an example of inquilinism of a symbiotic type.

daytime and as a resting place at night. The goby operates as an 'early warning system' for the shrimp. The latter is chemically attracted to the fish and only leaves the burrow when the fish is at its mouth. Shrimp feeding occurs outside the burrow but always with one antenna in contact with the tail region of the fish. If predators approach, the goby flicks its tail before retreating into the burrow. These tail-flicking signals cause the shrimp partner to back in as well. The patterns of mutual advantage in this system of association are reasonably clear. Shrimp burrows provide the fish with an entrée into a habitat—flat sea-bottoms—which would otherwise be untenable for them. The shrimp gains the advantage of receiving a much earlier warning of approaching predators than its own very feebly developed visual receptors can provide.

The term commensalism has been used to describe several different types of association. Originally it was conceived as a relationship where one animal lived with another and ate the remains of its food—gaining nutritional advantage with little detriment to the larger host animal. Some biologists have modified this definition to include associations involving benefits other than food and to situations where the host does suffer some degree of harm.

One group of relationships are generally agreed to be commensal—those of protozoans which live attached to larger aquatic organisms. Some filter feeding ciliates and predatory suctoreans live on certain species of 'substrate animals'. However, they gain their food from the water around the host rather than from the host itself, which rarely, if ever, seems to be disadvantaged by the attached protozoans. Using such situations as a model for a definition of commensalism we could say that commensals are animals that have an intimate spatial association with specific host animals but are not nutritionally or in any additional way committed to them.

An intriguing example involving a larger commensal is shown by the pearl fish (*Carapus*) which gains protection from predators by hiding specifically in the respiratory rectum of sea cucumbers (Holothuroidea). The association is a highly specific one but involves no nutrient dependence.

The definition of commensalism above is, to some degree, allied to the phenomenon of phoresy. Many animals, such as arthropods, indulge in a range of phoretic associations. The host, often a flying insect or bird, is used simply as a transportation mechanism by the phoretic animal. In this way pseudoscorpions, for example, hitch-hike on flies by holding on to their cuticular hairs or limbs.

Caterpillars in the butterfly family Lycaenidae often enter into symbiotic associations with ants such as those of the genus *Camponotus*. The ants provide protection and sometimes food, and in return the caterpillars produce sugary excretions.

through the body wall. The hosts used by a particular parasitoid are extremely specific, and the mechanism of this specificity is double in nature. First, only in a narrow range of host species can the developing parasitoid evade the cellular defence mechanisms of the host. Second, in evolutionary response to this problem of host defences, female parasitoids are extremely particular in choosing the right hosts in which to lay their eggs. Very specific visual, tactile and chemosensory recognition must take place before egg laying occurs. Often, the females can even detect parasitoids that are already within a prospective host. A second dose of eggs would reduce the chances of survival of both broods, so the female rarely lays in hosts which have already been parasitized.

Most parasitoids are insects, and ichneumons and wasps are the best known examples. Both groups have long ovipositors for the high precision insertion of eggs into hosts. However, other taxa also have parasitoid members. *Bdellovibrio* is a parasitoid bacterium, it slowly consumes another, larger bacterium from a position inside its cell coat. *Kronborgia* is a platyhelminth that is best regarded as a parasitoid. Male and female worms live inside the haemocoele of a marine burrowing amphipod. They feed on dissolved nutrients in the host's haemolymph which they absorb across a modified body wall as they have no gut. After a long developmental period the adult worms emerge from the shrimp, inevitably killing it in the process.

Inquilines are animals that share the nest, burrow, or other home base of another, usually very specific, animal species. Both aquatic and terrestrial examples of this kind of association are relatively common. A remarkable marine instance of inquilinism is the association between *Alpheus* (a shrimp) and a goby fish. This has been seen in the Pacific, Indian Ocean, Red Sea and the West Indies. On sandy bottoms the shrimps, usually living in pairs, excavate burrows which are used by gobies as a temporary shelter in the

Pseudoscorpions often improve their powers of dispersal through temporary phoretic associations with winged insects. Here they are firmly attached to the legs of a muscid fly.

Adaptations to Extremes

A variety of animals have adapted to the world's most inhospitable environments. Deserts, ice caps, mountains, ocean depths, caves and even puddles of crude petroleum have faunas that withstand the extreme conditions of heat, cold, pressure, darkness or strong chemicals.

The world provides a wide variety of environments in which living animals are able to exist. These range from the snow-capped peaks of the Himalayas to the depths of the Pacific Ocean; from the blazing heat and drought of the Saharan summer to the bitter cold and darkness of the Antarctic winter; and, on a smaller scale, from natural puddles of crude petroleum to the anaerobic contents of the human intestinal tract.

The ways in which animals are able to cope with such environmental extremes vary both according to their sizes and to their physiology. For instance, the Polar Bear (*Thalarctos maritimus*) is able to maintain a constant blood heat, whatever the weather, but the body of a codfish swimming beneath the ice, remains at the temperature of the surrounding water (2.5°C). Large animals tend to maintain homeostasis (constant internal conditions), whereas the bodies of small animals have to be capable of withstanding fluctuations in temperature and other factors.

Mammals and birds that live in hot environments, such as deserts, where the air temperature at midday can easily exceed 50°C, exist by efficient thermoregulatory processes. Their physiological problem is really one of keeping cool without losing too much valuable water in evaporation. Nevertheless, some desert nightjars have been known to nest on the ground at temperatures of up to 60°C. Many desert birds shade their eggs from the midday sun—indeed, only the Ostrich has eggs large enough to be left unattended for any length of time without being killed by the heat.

In contrast to birds and mammals, desert lizards maintain equable body temperatures largely by behavioural means. When their bodies are cool, they bask in the sun, but as they get hot, they seek shelter in shade or in their burrows in the sand. An extreme example of behavioural thermo-regulation is shown by the Galapagos Marine Iguana

An African desert lizard (*Agama*) responding to the midday heat by facing the sun and holding itself high above the hot ground, so that much of its body is in its own shade. Some desert lizards have fringed scales which increase the amount of shade provided; it is believed that feathers originally evolved from such fringed scales.

(*Amblyrhynchus cristatus*), which is able to keep its body temperature below 40°C without resorting to shade, although the air temperature may exceed 50°C. This remarkable feat is achieved on bare lava flows exposed to the intense radiation of the equatorial sun. When sun-basking at high temperatures, Marine Iguanas orientate their bodies so that they head directly into the sun. Much of their backs and hind quarters are therefore shaded by the head, neck and shoulders. At the same time, the front legs are extended so that the fore-quarters are held well clear of the ground, allowing maximum exposure to the breeze. Incidentally, camels also tend to face the sun when at rest during the heat of the day.

Camels are well adapted to stand up to extreme heat and drought. Sweating is reduced to a minimum so that heat can be stored during the day and lost at night when the environment is cooler. The kidneys are able to concentrate urine to an extraordinary degree, and the faeces are so dry that they can be burned immediately.

Many desert arthropods are able to withstand surprisingly high body temperatures. Under experimental conditions, scorpions and tenebrionid beetles have been found to survive a temperature of 48°C for 24 hours, while the camel spider *Galeodes granti* survives up to 50°C in dry air!

Desert fishes (*Barbus* spp.) have been found living in very warm pools, but it is difficult to be certain of the actual temperatures as water is often stratified. Thus, a molly (*Mollinesia sphenops*) at Lake Amatitlan, Guatemala, lives in water at 33–36°C. Temperatures as hot as 49°C have been recorded in the surface layers, however, and the local people are convinced that the fishes are actually capable of living in boiling water!

Animals that survive in extremely cold environments include the Polar Bear (*Thalarctos maritimus*), and the Emperor Penguin (*Aptenodytes forsteri*) of the Antarctic, both of which can survive in temperatures as low as −57°C

Arctic and Antarctic invertebrates have to operate when their body temperatures are really cold. Although they are black in colour and bask in the sun, many Arctic insects are active at body temperatures far below zero, and earthworms (*Mesenchytraeus gelidus*) can bore through hard-packed snow, feeding on minute algae as they do so.

Many desert animals exercise the utmost economy in the use of water. A camel can go for long periods without drinking. The rate of urine flow is low, and little water is lost with the faeces. Not all the urea produced in metabolism is excreted, however. As in ruminants, some of it is used for

LEFT: Emperor Penguins at McMurdo Sound, Antarctica. Largest of the penguins, the Emperor incubates a single egg throughout the winter, fasting for as long as three months. The egg, and later the chick, is pressed against its parent's legs and covered by a fold of abdominal skin. Feeding is by regurgitation until the chicks reach adult size.

BELOW: Daubenton's Bat (*Myotis daubentonii*), a small Eurasian species, in deep hibernation in a cave. Moisture has covered the animal with dew drops. These bats are unharmed if their body temperature drops to just above freezing point. If the external temperature falls below 2°C they shiver to generate heat.

because they are so well insulated. Reduction of the body surface is also an important means of heat conservation, and is achieved by the development of a compact form, with a reduction of the appendages. Thus, the Musk Ox (*Ovibos moschatus*) has legs so short that it stands only 1 m high, though it is 2.5 m long. Its neck is thick, its tail very short and its ears are hidden in its furry coat.

Many animals of the taiga and tundra are able to hibernate during the winter but, in polar regions, the summer is too short for animals to lay down sufficient reserves of fat. Consequently, they must either migrate or remain active throughout the year. Hibernators include bears, ground squirrels and bats.

BELOW: the Musk Ox, one of the hardiest of animals, is well adapted to withstand the intense cold of the Arctic winter. The compact body is densely clothed in a thick underfur which, in turn, is protected by a layer of loose, straggly hair. In fierce blizzards they huddle together to conserve and share body heat.

ABOVE: *Allactaga williamsi*, like other jerboas, conserves water by sleeping with its tail held over its nose to trap exhaled moisture which it then breathes in again.

BELOW: larvae, pupae and adult mealworms in bran. These insects can live on dry food without any water, but they develop faster if provided with some moisture.

An Alpine Chough in Nepal. These birds belong to the crow family. They inhabit mountainous areas from the Iberian Peninsula and the Atlas eastwards to central Asia and China. They readily become tame near climbers' huts where food can be obtained. They have been recorded at a height of 8,200 m on Mount Everest.

the microbial synthesis of protein. The coarse hair on the camel's back acts as a barrier to solar radiation, yet it is well ventilated so that the evaporation of sweat occurs on the skin, where it provides maximum cooling to the body. At the same time, a camel avoids undue water loss by allowing its temperature to vary over a range greater than that of most other mammals. Sweating does not begin until the body temperature has risen to 40.7°C. Thus, heat is stored during the day and lost at night when the environmental termperature is lower. Eland, Addax, gazelles, Ostriches and other arid zone animals show similar adaptations. Indeed, the Addax (*Addax nasomaculatus*) can live in the most arid regions of the Sahara without drinking at all. Jerboas and kangaroo-rats live on dry desert seeds without access to water, but this is not so surprising since they avoid the daytime desert heat in deep burrows from which they emerge only at night.

Most desert insects and arachnids cannot tolerate any greater loss of water than their relatives from more humid places. However, they are often more resistant to water loss by evaporation. For example, the American Desert Scorpion (*Hadrurus arizonensis*) loses only 0.028 per cent of its body weight per hour in dry air at 30°C. The camel spider *Galeodes granti* loses 0.147 per cent at 33°C, but is unusual in that it does not die until it has lost as much as two-thirds of its total weight!

Tenebrionid beetles, including the Common Mealworm (*Tenebrio molitor*), can live on dry food without drinking. So too can the larvae of carpet beetles, clothes moths, and other pests of stored products. Many insects and other arthropods are able to withstand unfavourable conditions in a state of suspended animation, or diapause. This is characterized by temporary failure of growth and reproduction, reduced metabolism and enhanced resistance to heat, drought and other environmental conditions.

A few cases are known in which arthropods do not resist evaporative water loss, but allow their bodies to become almost completely dehydrated. This leads to 'cryptobiosis' a virtually non-living state in which metabolic activity comes reversibly to a standstill. The phenomenon is known in tardigrades, nematodes, Collembola and in the larvae of chironomid midges such as *Polypedilum vanderplancki*. These midges breed in small pools formed in shallow hollows on unshaded rocks in West Africa. During the rainy season, the hollows may fill and dry several times, but the larvae are well adapted to their unstable environment because they can absorb water and dry up many times without harm. They can persist in the state of cryptobiosis for many years and, while dehydrated, are able to tolerate over 104°C for more than an hour and can withstand immersion in liquid helium (−270°C).

Air pressure is reduced at high altitudes, but invertebrates and lower vertebrates are well able to tolerate very low atmospheric pressures. Insects and jumping spiders occur

A Yak and its new-born offspring in the Langtang Valley, Tibet. This wild ox, related to the bisons, is black, but domesticated forms are sometimes black and white.

Yaks have been recorded at higher altitudes than any other species of mammal. They stand nearly two metres high and their hair is characteristically long and shaggy.

fishes with enclosed swim bladders suffer from being brought to the surface faster than the gases in their bladders can be reabsorbed by the blood. The greatest depth from which a living fish has been recovered is 7,000 m, but brittle stars, sea lilies and sea cucumbers have been found at depths of over 10,000 m. Deep diving mammals, such as Sperm Whales (*Physeter catodon*), have been known to swim to a depth of over 1,000 m. They exhale before diving and, consequently, do not suffer from 'bends'—the formation of gas bubbles in their blood.

Abyssal animals exist in perpetual darkness, but many of them are luminescent, and their eyes are well developed. Some deep-sea squids have light-producing photophores with reflector mechanisms, lenses, mirrors and colour screens. The deep-sea squid *Lycoteuthis diadema* has as many as 22 light organs, of over ten different types.

The basis of the food chains of the abyss stems from the constant rain of dead animals showering down from the waters above, for plants are unable to exist in complete darkness. Similarly, the faunas of dark caves depend upon food brought in, one way or another, from the exterior. Unlike the animals of the deep seas, cave species almost invariably respond by losing their sight—indeed their eyes are usually vestigial or entirely absent.

The bodies of animals exposed to extreme intensities of light and ultra-violet radiation may be protected by dermal pigments: the black colour of the peritoneum of many desert lizards may be partly concerned with protection. In the eyes of day-active animals, intra-ocular oil droplets act in the same way as filters in photography. The red droplets of birds are effective at sunrise but, as the day wears on, yellow and colourless droplets become more effective. Red droplets enable marine turtles to see through the surface glare of tropical seas.

The absence of oxygen presents another problem for modern animals, and it is not certain whether life in the complete absence of oxygen is now possible—although many species can exist without oxygen for long periods of time. A small crustacean (*Thermocyclops schuurmanni*) has been found to exist for part of its life in the oxygen-free depths of African volcanic lakes, and other crustaceans can live in soda and salt lakes.

Finally, mention should be made of other extreme and unusual environments that may be inhabited by animals. A fly (*Psilopa petrolei*) inhabits puddles of crude petroleum, feeding on the dead insects that are found there, while a beetle (*Niptus hololeucus*) can live on cayenne pepper and thrive on sal ammoniac. This species has also been found inhabiting the corks of entomologists' cyanide killing bottles. It is obviously unusually tolerant of a number of unfavourable conditions!

at altitudes of over 6,000 m in the Himalayas, while some amphibians and reptiles can survive rapid exposures to simulated altitudes of 23 km! In contrast, mammals tolerate only moderately reduced pressures. The highest living species is probably the Yak (*Poephagus grunniens*) of Tibet, which has been occasionally found at an altitude of 6,500 m. The highest altitude recorded for a bird on the other hand, is 8,200 m. Alpine Choughs (*Pyrrhocorax graculus*) have followed Everest climbers as high as this.

In contrast to high altitudes, the water in the depths of the oceans is under extreme pressure. However, great pressures make little difference to marine invertebrates, because the pressure is the same on all sides and is equalized internally by the pressure of the blood and body fluids. Only

BELOW: a bottom-living brittle star (Ophiuroidea). Few echinoderms are bioluminescent, except for abyssal species which lie on the sea bed at depths of up to 6,600 m with arms raised, a characteristic feeding posture. The function of this luminescence is not known but it may be to warn predators or to attract prey.

BELOW RIGHT: the deep-sea prawn *Gnathophausia ingens* from the tropical Pacific. It has been found at a depth of nearly 2,500 km. Largest of the mysid prawns, this species grows to a length of 35 cm. It probably feeds mainly on the corpses of animals. Red coloured prawns are found only where little or no light penetrates.

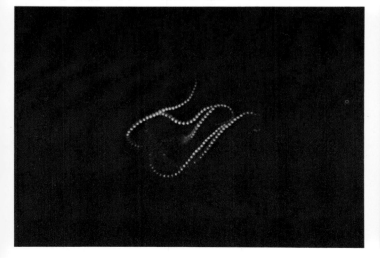

The
Conservation
of
Animals

The conservation of wildlife and the habitats upon which it depends is a subject about which there is now a greater degree of public awareness and participation than ever before. Also, partly as a result of public interest and pressure, governments have directed greater resources to the solving of conservation problems. The conservation of nature is essential to the future of man, for in the final analysis his survival will depend on a full understanding of nature, and on the ability to apply this fund of knowledge to his own wellbeing.

Despite many years of intensive research, our knowledge is still minimal in relation to the numbers of plant and animal species that exist, and much of what we do know relates to species of obvious economic importance. Out of a total of some 250,000 known species of plants, scientists understand the full biology of only a few dozen. The problems of conservation are therefore enormous. It is not simply a case of attempting to ensure the survival of as much as possible of what still remains, but also of trying to rectify some of the damage done in the past. For example, in most of the advanced industrial countries a great deal of money and expertise is now being invested in the restoration of degraded landscapes and polluted waters.

The conservation of habitats and wildlife can be justified on a number of grounds, ranging from the scientific and economic, through to moral and aesthetic reasons. But it is common sense that we should not, in our present depth of ignorance, hasten the extinction of any more species, at least until we are absolutely sure that we shall never have an economic use for them. If we permit the destruction of ecosystems and the reservoirs of plant and animal species that they contain, we may well be laying the foundations for our own ultimate extinction from starvation, or some other catastrophe. It is a sobering thought that, according to statistics of the Food and Agriculture Organisation, by 1985 at least 26 major tropical countries with a population of about 365 million will be unable to supply sufficient food to avoid gradual starvation.

Civilized man, relentlessly increasing in numbers, has impinged mercilessly on the once vast wilderness of the planet and on their plant and animal associations. But within the present century many have begun to question the morality and wisdom of his attempts to subjugate the wilderness and its inhabitants. One of the most important changes in outlook has been the realization that animals (including man himself) and plants do not live in isolation; they exist together in an environment which has physical, biological and social components. We now know that if we are to save our remaining wildlife we have to direct our attention primarily to maintaining adequate suitable habitats in which they can live.

This means, in essence, that we often have to undertake active management of both animals and habitats, because man's activities over a long period of time have upset the natural equilibrium. The African Elephant (*Loxodonta africana*), just to take one example, will not be saved merely by preventing its over exploitation by hunting and ivory poaching. The great build-up in the numbers of elephants in Kenya's Tsavo National Park in the early 1960s, with resultant over-grazing of the habitat and starvation of many animals, illustrated that point very well. When only limited areas are available to wildlife, then it may be necessary to limit their number to what the range can support and maintain in good health. Distasteful though this culling may be, it is an essential aspect of practical conservation.

Some types of habitat may require frequent management if they are to retain their wildlife interest. For example, grasslands are one of the most widespread natural vegetation types in the low rainfall areas of the world and in temperate regions, where the original forest cover has been removed by man to create pasture. Because of human activity much of the present grasslands are only semi-natural. From the conservation point of view chalk grasslands are particularly interesting because of their characteristic flora and fauna. A good example in southern England is the beautiful Adonis Blue Butterfly (*Lysandra bellargus*), whose larvae feed on Horseshoe Vetch (*Hippocrepis comosa*). In the absence of adequate natural grazing the smooth swards of chalk downland will rapidly revert to scrub, and ultimately the butterfly and its food plant decrease or vanish altogether. In order to maintain the optimum conditions it may be necessary for man to remove the scrub physically and to use sheep to graze the grass periodically, or failing that to mow the grass himself.

At the other extreme there are, of course, still extensive areas where human interference should be kept to a minimum. This category may include some alpine areas, deserts, tropical rain forests, the Antarctic continent, and the Arctic tundras and polar deserts. Unfortunately, man is steadily extending his activities to every corner of the globe. Fifteen to twenty years ago there were no obvious threats to the vast tundra region north of the Brooks Range in Arctic Alaska, but since 1968 the discovery of huge oil reserves has changed the situation. There is now industrial activity all across the North American Arctic, and it is only a matter of time before this extends offshore to the ice-covered seas of the polar basin. Sooner or later, industrial activity in these types of environment must lead to ecological degradation.

Unfortunately, economic considerations still tend to carry more weight than wildlife conservation. Even now the huge machines of the timber industry are tearing down vast areas of tropical rain forests— one of the last almost untouched great ecosystems of the world and the product of millions of years of evolution. In 1975 alone 100,000 square kilometres of rain forest were destroyed in the Amazon Basin of Brazil. Examples like these show that there is still a great need for conservation.

One of the species for which conservation measures are urgently needed—the magnificent Snow Leopard (*Panthera uncia*)

Conservation in Action

Conservation work ranges from major international effort such as Operation Oryx and the activities of the World Wildlife Fund to more local tasks such as managing woodland, clearing disused watercress beds and erecting garden nestboxes.

Conservation in its various forms can be applied at many levels, for it is not the prerogative of any particular social group, country or organization. In fact, if conservation is to succeed, then it must be the concern of a very substantial proportion of the population as well as governments. If we take conservation in its broadest sense, then there are of course hundreds of organizations of various kinds around the world that are involved in the subject in many different ways. Operating at the international level are the United Nations Environment Programme, the International Union for the Conservation of Nature and Natural Resources, and the International Council for Bird Preservation. All of these operate on a global scale and mainly at government level, providing advice and skilled manpower for the solving of conservation problems, and when necessary applying political pressure in order to create awareness and produce effective action. Certainly one of the best known of the international organizations is the World Wildlife Fund, which was established in 1961 and is devoted to the conservation of nature in all its forms. Its primary function is to raise money in order to finance essential projects. The World Wildlife Fund now operates national appeals in 26 different countries, and up to the end of 1977 had financed an impressive total of 1,874 conservation projects.

In individual countries there is a multiplicity of organizations varying from statutory bodies to charities and local citizens groups. For example, in Australia the Australian Conservation Foundation was set up in 1965 as a private non-profit making organization; its function is to promote throughout Australia an understanding of conservation. In the United States the Department of the Interior has the Federal Fish and Wildlife Service, whilst non-statutory bodies include the Nature Conservancy, National Wildlife Federation, the Sierra Club, and the National Audubon Society. The United Kingdom has the official Nature Conservancy Council (formerly the Nature Conservancy) which was established in 1949, and now has a network of more than 150 National Nature Reserves covering 120,465 hectares. Non-statutory bodies include the Society for the Promotion of Nature Conservation, which is the umbrella organization for the county naturalists' trusts. These trusts now have a combined membership of about 112,000 (distributed between 40 individual trusts), and a total of 1,080 reserves covering 36,818 hectares. The Royal Society for the Protection of Birds has a membership of more than 250,000 and a total of 72 reserves covering 31,694 hectares. These are only a few examples of the types of organizations involved in conservation activities, and there are many others throughout Europe, and in Africa and other countries.

Many of the organizations mentioned only came into existence after the second world war, but conservation had

ABOVE: Yellowstone National Park, Wyoming, USA, was the first national park to be established (1872). Its hot springs are world famous, but the park also contains a wide variety of animal habitats ranging from rugged mountains to peaceful valleys.

BELOW: the American Bison (*Bison bison*) once existed in vast herds, totalling perhaps up to 60 million animals. By 1890, hunters had reduced this figure to a mere 500. Under protection since the early 1900s, the population has now risen to about 25,000.

its beginnings long before then. The concept of national parks was born in the United States, and the first such park to be established in the world was Yellowstone National Park in Wyoming in 1872. The naturalists' trusts movement in the United Kingdom began with the formation of the Norfolk Naturalists' Trust in 1929. One of the earliest examples of conservation in action took place in 1906 in Montana, USA. The last sizeable herd of American Buffalo (*Bison bison*) was threatened when the government announced plans to open the Flathead Indian Reservation to settlement. The Canadian government offered to buy the herd and it was eventually transferred to the Wood Buffalo National Park in the Canadian north, although the operation took several years to accomplish as the Buffalo were difficult animals to transplant.

to 2,000 kg the task was by no means easy, but it was done. Apart from the difficulties of catching the animals, transferring them to the park required the bridging of two rivers and the construction of several kilometres of trail. Altogether twelve animals were transferred to the national park, and by 1972 their numbers had increased to 18.

Another urgent situation arose in 1961, and on this occasion the cause of concern was another endangered species, the Arabian Oryx (*Oryx leucoryx*). At that time the last known surviving herd was just inside the boundaries of the former Aden Protectorate. The fact that a mechanized hunting party had travelled 600 miles from Quatar on the Arabian Gulf in order to raid this herd, decided the Fauna Preservation Society in London to make plans to rescue some of these animals and move them to safety elsewhere.

When a species has dwindled in numbers to a dangerously low level, and a serious threat to its continued existence arises, it may be necessary as in the case of the Buffalo, to take immediate and drastic action. Probably the most classic case of this type involved not an individual species, but the entire animal population of a very large area. This rescue exercise, known as 'Operation Noah', was certainly one of the biggest in terms of logistics. The cause of this operation was the commencement of construction of the massive Kariba Dam across the Zambesi River in Rhodesia in 1955. When completed, this created the largest man-made lake in the world. The idea of 'Operation Noah' was to rescue as many animals as possible from the area that was to be flooded. The operation was started in December 1958, and between then and June 1963 some 6,500 animals had been taken out of the doomed area.

The population of the White Rhinoceros (*Ceratotherium simum*) in East Africa is so low that it is classed as an endangered species. In 1960 a severe outbreak of poaching threatened the survival of this species in Uganda, and so plans were made to transfer some animals from areas where they still survived, into the Murchison Falls National Park. Despite doubts by some ecologists about introducing rhinos into the park, it was decided that the situation warranted decisive action. Since adults of this species can weigh up

ABOVE: a rhinoceros, drugged with an anaesthetic dart, about to be towed on a raft to safety during Operation Noah in 1955. As the waters of Lake Kariba rose behind the dam, many animals were rescued from the water or from islands that later disappeared.

BELOW: camels make useful mounts for anti-poaching patrols. Much of Africa's wildlife is protected, but poaching is still a problem. Furs, hides and elephant tusks can fetch high prices from unscrupulous dealers, and modern weapons have made poaching relatively easy.

In April 1962 'Operation Oryx', financed by grants from the World Wildlife Fund and the Shikar Safari Club, succeeded in catching three oryx in the protectorate. These animals, together with others donated by the late King Saud of Saudi Arabia, the ruler of Kuwait, and the Zoological Society of London, formed the nucleus of the World Herd of Arabian Oryx at Phoenix Zoo, Arizona, USA. About 80 captive animals of this species are held in six collections as a precaution against the final extinction of the animal in the wild, assuming that any still survive there at all.

The saga of the Arabian Oryx leads to the question of the role of zoos and other collections of live animals in the field of conservation. Theirs is clearly an important role, for it is from stock bred in captivity that rare species can be reintroduced into their former range, assuming that suitable areas of habitat still exist. If it comes to the worst, then at least zoos and private collections provide the last opportunity of preventing a species from becoming totally extinct. In some instances, however, even zoos may not be able to save a species from extinction, for the very simple reason that it may prove impossible to breed it in captivity. An example is the Monkey-eating Eagle (*Pithecophaga jefferyi*), which is an endangered species, but has never been bred in captivity.

Despite the problems there have been a number of successes in the field of captive breeding for the purposes of reintroduction. A famous example of this is Père David's Deer (*Elaphurus davidianus*), which was brought to Britain in 1900 from the last remaining captive population in China. A herd was established at Woburn Park in Bedfordshire, and subsequent to its arrival in Britain the species became extinct in China. This deer was reintroduced into China from British stock in 1960, but appears to exist only in the Peking Zoo since its original habitat in the wild is now unsuitable. The two largest herds of this species are still those at Woburn Park and Whipsnade Zoo. Another example is the European Bison (*Bison bonasus*) which became extinct in the wild but was bred in good numbers from zoo stock.

ABOVE: by 1961 the Arabian Oryx (*Oryx leucoryx*) had been reduced by hunting to a small herd in the southern Sahara desert. But at Phoenix Zoo, the World Herd now consists of approximately 80 animals, and it is hoped that this species has been saved.

BELOW: Père David's Deer stags (*Elaphurus davidianus*) in Woburn Park, England. The last remaining animals in China, in the Emperor's Palace in Peking, were destroyed in the Boxer rebellion. However, animals from the Woburn herd have since been reintroduced.

ABOVE: these young Peregrine falcons (*Falco peregrinus*) are being reared by hand at Cornell University. The serious depletion of the numbers of this bird has been checked to some extent by the work of the ornithological laboratory there.

BELOW: the Hawaiian Goose (*Branta sandvicensis*), also called the Né-né because of its low, moaning call, is confined in the wild to the islands of Hawaii and Maui. In 1950 it was extinct on Maui but was reintroduced by the release of a small captive-bred population.

Another outstanding success, this time involving a bird, was with the Hawaiian Goose or Né-né (*Branta sandvicensis*). By 1950 the population of this species in the Hawaiian Islands, to which it is endemic, was believed to be less than 50 individuals. In 1950 three were brought to the Wildfowl Trust at Slimbridge in England, and from these a substantial stock was gradually bred. In the summer of 1962 a total of 30 birds was sent by air from Slimbridge to New York, and then on to Honolulu. Late in July they were released in the volcanic Haleakala Crater on the island of Maui, where the species had been extinct for many years; the reintroduction was successful and the Hawaiian Goose is once again established in the wild.

That superb bird of prey the Peregrine (*Falco peregrinus*) has had a bad time almost throughout its range due to a combination of persecution, egg collecting, the taking of young birds by falconers, and the insidious effects of pesticides that seriously disrupted its breeding success. In North America the situation has been grim indeed, and in the United States there are now no indigenous Peregrines left east of the Rocky Mountains. As a result of this serious situation the Peregrine Fund was established at the Cornell University Laboratory of Ornithology in 1970. The fund produced its first 20 Peregrines in 1973 and up to the autumn of 1977 had raised a total of 229 young from breeding pairs in their facilities; of these, 133 have been released into the wild. In addition, the Peregrine Fund has carried out or given support to field studies in Alaska, Colorado, Mexico, Greenland, Scotland, Spain and Ungava, and hopes to extend this work to Australia and South America.

Some animals, either because of their habits or the nature of the terrain in which they live, present very special problems from the point of view of conservation. If it is to be fully effective, conservation has to be based on accurate scientific knowledge of the species concerned, and acquiring this essential knowledge for some species can be very difficult. A case in point is the Polar Bear (*Thalarctos maritimus*). One of the problems with this species, apart from the fact that it lives in some very remote areas, is that in its extensive wanderings over the northern pack ice it recognizes no international boundaries. Effective conservation of this species required an international approach, and this was achieved in 1973 with the Agreement on Conservation of Polar Bears, to which the governments of Canada, Denmark, Norway, United States, and the Soviet Union are signatories. This agreement represents the first and so far the only international agreement for environmental protection in the Arctic. It affords almost complete protection to the Polar Bear, allowing merely a small amount of subsistence hunting by indigenous peoples.

Technology is regarded by many as a serious threat to the global environment and its wildlife. On the other hand it has also allowed some sophisticated methods to be used to further the cause of conservation in a very efficient manner. A very obvious example is the use of computers to process huge quantities of data, and to produce maps showing the distribution of plants and animals over wide areas.

At a more sophisticated level modern technology has been used to gather vital information on the movements of Polar Bears. It is very difficult to follow the wanderings of a white animal that ranges over hundreds of square miles of remote polar seas and pack ice, and over a decade ago scientists considered the possibility of placing radio collars on Polar Bears, so that the signals emitted by the collars could be picked up by a satellite in orbit approximately 970 km above the earth. This idea finally came to fruition in June 1977 when two bears were immobilized and fitted with radio collars at a point which lay some 80–90 km north of Point Barrow in north-west Alaska. Every fourth day the collars sent a signal to the Nimbus 6 satellite and the process of plotting the movements of the animals began. One of these bears first headed east, then reversed direction and travelled west towards Soviet territory. By mid-November 1977 it was nearly 1,000 km to the west of Point Barrow. A similar approach was used earlier in studying the movements of Grizzly Bears (*Ursus arctos*) in Yellowstone National Park, where bears were fitted with small radio transmitters. The signals emitted could be picked up by ground based observers armed with the requisite equipment, and in this way the extent of the areas ranged over by individual bears was gradually established.

All these conservation activities are large scale, international and national operations. But there are also many opportunities for individuals to participate according to their ability and level of interest. Some people are unable to devote much time to conservation, but willingly make financial contributions without which the voluntary conservation movement could not operate. Even those with limited time can make positive contributions without ever leaving their gardens. The erection of nestboxes of various types to encourage breeding birds, the planting of trees and shrubs attractive to insects, and the construction of ponds to provide a suitable habitat for frogs and newts are just a few examples of what can be done. Keen gardeners should

ABOVE: the Adonis Blue butterfly (*Lysandra bellargus*) feeds on Horseshoe Vetch which is found only on chalk grasslands. For conservation, such grasslands need to be grazed regularly, or else they revert to scrub, and both plant and butterfly vanish.

resist the temptation to eradicate completely Stinging Nettles (*Urtica dioica*) from their land, since this plant provides food for the caterpillars of beautiful butterflies such as the Comma (*Polygonia c-album*), Peacock (*Inachis io*) and Red Admiral (*Vanessa atalanta*). With a little careful thought and planning the conservation value of even the smallest garden can be considerably improved.

For those who wish to participate in a more active way there is plenty of scope. Nature reserves very often have to be wardened regularly, and in many instances this task is performed by voluntary wardens. In the UK the Nature Conservancy Council, the Royal Society for the Protection of Birds, and the county trusts all make extensive use of voluntary wardens. The RSPB uses a large number of volunteers to guard the breeding Ospreys (*Pandion haliaetus*) in the Scottish Highlands, and the National Nature Reserve at Tring Reservoirs in Hertfordshire is wardened entirely by volunteers on behalf of the Nature Conservancy Council.

BELOW: modern technology can help as well as cause problems for the conservationist. A helicopter is used to reach Polar Bears on the vast Arctic region between Alaska and Siberia. When these two recover from the anaesthetic, administered in a dart fired from the air, they can be tracked as they move across the icy wastes.

Perhaps even more interesting, particularly to younger people, are the extensive opportunities available for participating in the active management of nature reserves and other areas. Most of us live in areas that have been constantly modified by man over a long period of time and where changes are still taking place. In addition, nature conservation is itself dealing with dynamic systems and living organisms that are continually changing and evolving. This means that most of the interesting places in our own area must be physically managed if we wish them to retain their primary conservation interest. This interest may be the existence of a certain type of habitat, or the presence of a particular species of plant or animal.

A good example of the type of habitat that may require management is deciduous woodland, much of which is being clear felled and replaced by conifers. In many parts of Britain deciduous woodland used to be managed for economic reasons under a system known as coppice with standards, but this form of management became uneconomic many years ago. The wood may have been oak standards with hornbeam coppice, and areas of coppice were cut in rotation on a 10–15 year cycle. A wood managed in this way therefore had every stage of growth from freshly cut to mature coppice represented within it, and the associated flora and fauna were usually diverse and rich. Some woodland reserves are now being managed in this way once again, but this time for conservation purposes. Needless to say this labour intensive operation is usually done by volunteers.

Other practical management tasks may involve the clearance and restoration of ponds and lakes that have silted up and become overgrown with vegetation. Hundreds of ponds have disappeared in this way in Britain over the past 30 years, and they represent a type of habitat that we can ill afford to lose. A healthy pond attracts not only amphibians such as the frog, but is essential to many invertebrates such as dragonflies and water beetles. A rather more unusual form of aquatic habitat now found in a few nature reserves under private control are watercress beds that were once operated commercially, but subsequently abandoned. They usually have clear unpolluted waters fed by springs, and are consequently rich in invertebrates that provide food for ducks, waders and other birds. To maintain them in an

ABOVE: an Avocet (*Recurvirostra avosetta*) in the RSPB reserve at Minsmere, Norfolk, England. The conditions at this reserve are maintained to provide an ideal breeding ground for the birds— shallow water with muddy shores and special nesting areas.

optimum condition for wildlife requires the regular clearance and control of the watercress.

Another form of conservation in action involves the conversion of sites from one form of use to a state where they can contribute to conservation, or even the creation of new habitats from scratch. In the former category can be included the improvement of disused flooded gravel pits by planting trees and shrubs round the margins, and by installing floating platforms to provide nesting sites for birds. The Royal Society for the Protection of Birds used the services of the Royal Engineers to create new areas for breeding Avocets (*Recurvirostra avosetta*) at their Minsmere reserve in Suffolk, but many less ambitious schemes have been carried out by local voluntary conservation groups.

BELOW: clearing away watercress in a small nature reserve. Such work helps to create a habitat for frogs, newts, small fish and insect larvae. This kind of small-scale conservation work is of great value in maintaining the wildlife of the countryside.

Endangered Animals

Over 450 species and subspecies of tetrapods are particularly in danger of extinction. Commercial exploitation and destruction of habitats are the main causes but competition with introduced species has also contributed to the decline of some animals.

In 1844 an Icelandic fisherman, together with a few companions, went out to the island of Eldey off the coast of Iceland and killed the last two Great Auks (*Alca impennis*) in the world. In 1914 the last Passenger Pigeon (*Ectopistes migratorius*) died in captivity, yet in the first half of the nineteenth century their migratory flocks had darkened the skies in North America. On Laysan Island, Hawaii, in 1923, an American biological survey team were present when the last three Laysan Honeycreepers (*Himatione sanguinea freethi*) were killed in a sandstorm, because introduced rabbits had previously destroyed the vegetation that provided cover. The last survivors of the Wake Island Rail (*Rallus wakensis*) were almost certainly eaten by hungry Japanese soldiers in World War II. These are just a few of the milestones in a long and depressing catalogue of extinctions involving birds. It is customary to take the year 1600 as the starting point for calculating the rate of extinction, and this coincides with the great expansion of European exploration and settlement. In the case of birds, some 165 species and subspecies became extinct between 1600 and 1975. They

ABOVE: snaring lemurs in traps baited with mangos is practiced by the natives of Madagascar, as they regard these animals as a delicacy. This and other forms of hunting have contributed to the fact that several lemurs have become endangered, particularly the White-footed Sportive Lemur, (*Lepilemur mustelinus leucopus*).

BELOW: whaling is now big business, largely due to the demands of the pet food and cosmetic industries, and sophisticated equipment has been developed to process whale carcasses rapidly. As a result several species have seriously declined in numbers, particularly the Blue Whale (*Balaenoptera m. musculus*).

disappeared at the rate of 16 from 1600–1700, 22 from 1701–1800, 79 from 1801–1900, and 48 from 1901 to 1975.

Turning now to mammals, we find that since the disappearance of the Aurochs or Wild Ox in about 1627, no native wild species of European mammals have become extinct in recent centuries. This contrasts greatly with the situation elsewhere. For example, the West Indies have lost at least ten species since 1800; Australia has lost nine, South America one, and Africa, Asia and North America two each.

These examples serve to illustrate that the process of extinction as a result of man's activities has been going on for some time, but reached a peak in the 19th century, and the situation has not improved much since then. Most recent extinctions are due, directly or indirectly, to the activities of man in the form of commercial exploitation, sport hunting, the destruction or degradation of natural habitats, and the introduction of plants and animals into areas where they do not naturally occur. This last activity has had a particularly severe effect in insular or island areas.

For example, the Hawaiian Islands now contain 25 per cent of all endangered fauna listed by the United States Department of the Interior. Habitat destruction, commercial exploitation, and the depredations of introduced species, have collectively had a disastrous effect on the unique land-snail fauna of Hawaii. At one time there were 41 species in the genus *Achatinella*, which is confined to the island of Oahu. However, 14 of these are now extinct and a further 25 species are endangered. The situation is almost as bad on some of the large continental land masses. In North America alone 62 species of fish have died out since the white man arrived there, and some 109 animal species are currently in danger of extinction in the United States.

At this point it is necessary to define what is meant by an endangered species, since what may well be endangered in a local or national context, may not be so at all at the international level. The main sources of information on threatened species are the Red Data Books published and periodically updated by the International Union for the Conservation of Nature and Natural Resources (IUCN) which is based in Switzerland. These books view the situation at the international level, which is what we are concerned with here, and in them six categories are recognized, including Endangered and Vulnerable. Endangered species are those in danger of extinction: their survival is unlikely if the causal factors continue operating. Vulnerable species are those likely to become endangered if the factors responsible for their decline continue to operate.

At the present time, worldwide, about 473 species and subspecies of birds, mammals, amphibians and reptiles alone are recognized as existing in such small numbers as to be especially vulnerable to extinction—that is to say they are in either the Endangered or Vulnerable categories. For the present purpose discussion is primarily confined to those in the Endangered category, of which there are, as shown in the table on page 350, about 131 species and subspecies of birds, 162 of mammals, and 58 of amphibians and reptiles. The Red Data Books are constantly under revision, so the above figures should be regarded as only approximate. It must also be remembered that in other categories there are many species for which only inadequate information concerning their status is available, and it is quite probable that some of these will prove to be endangered.

The proportion of endangered island forms is high because of their particular vulnerability. No less than 87 (66 per cent) of the endangered species and subspecies of birds listed in the table come into this category. In the case of the amphibians and reptiles, 21 (36 per cent) of the species and subspecies in the table are island forms, of which 11 are races of tortoise endemic to the Galapagos Islands. So far as mammals are concerned, 34 (21 per cent) of the listed forms are found on islands, and they include six lemurs, the

Commercial exploitation of new areas of land leads to the destruction of animal habitats. In the Amazon basin, for example, large tracts of forest are being removed, with the consequent loss of habitat for many animal species. Removal of rain forest may also lead to climatic changes, the effects of which have yet to be researched.

Indri (*Indri indri*), Verreaux's Sifaka (*Propithecus verreauxi*) and the Aye-Aye (*Daubentonia madagascariensis*), all of which are confined to Madagascar. In arriving at the above figures, forms occurring in Australia and on the two main islands of New Zealand are not included.

A considerable number of species in the Endangered category were brought to this state as a result of excessive exploitation (i.e. hunting and collecting), and many of them were rendered vulnerable in the first instance through destruction of their natural habitats. An examination of the data relating to the primary causes of the reduction in num-

bers of the endangered species is a revealing exercise. In the case of mammals we find that habitat destruction or modification has been a responsible factor in 70 per cent of the cases, and excessive exploitation at 66 per cent is also important. Birds have suffered from habitat destruction in 73 per cent of the cases, and in this context forest clearance has been a serious problem in the case of 12 species of Psittaciformes. Excessive exploitation has also been involved in 42 per cent of the bird species. Turning to the amphibians and reptiles, habitat destruction at 40 per cent has been particularly serious in respect of salamanders,

CHECKLIST OF ENDANGERED ANIMALS

The zoogeographical regions in which the animals occur are indicated by letters:
A – Australasian, E – Ethiopian, N – Nearctic, Nt – Neotropical, O – Oriental, P – Palaearctic.

AMPHIBIANS
8 species

URODELA

Ambystoma macrodactylum
Santa Cruz Long-toed Salamander (N)

Batrochoseps aridus
Desert Slender Salamander (N)

Typhlomolge rathbuni
Texas Blind Salamander (N)

ANURA

Discoglossus nigriventer
Israel Painted Frog (P)

Pelobates fuscus insubricus
Italian Spade-foot Toad (P)

Bufo houstonensis
Houston Toad (N)

Bufo periglenes
Orange Toad (Nt)

Rana pipiens fisheri
Vegas Valley Leopard Toad (N)

REPTILES
50 species and subspecies

CHELONIA

Batagur baska
River Terrapin (O)

Pseudenys ornata callirostris
South American Red-lined Turtle (Nt)

Testudo e. elephantopus
South Albemarle Tortoise (Nt)

Testudo elephantopus abingdonii
Abingdon Saddle-backed Tortoise (Nt)

Testudo elephantopus becki
North Albemarle Saddle-backed Tortoise (Nt)

Testudo elephantopus chathamensis
Chatham Island Tortoise (Nt)

Testudo elephantopus darwini
James Island Tortoise (Nt)

Testudo elephantopus ephippium
Duncan Saddle-backed Tortoise (Nt)

Testudo elephantopus guentheri
South-west Albemarle Tortoise (Nt)

Testudo elephantopus hoodensis
Hood Saddle-backed Tortoise (Nt)

Testudo elephantopus microphyes
Tagus Coved Tortoise (Nt)

Testudo elephantopus nigrita
Indefatigable Island Tortoise (Nt)

Testudo elephantopus vandenburghi
Cowley Mountain Tortoise (Nt)

Chelonia mydas
Green Turtle

Eretmochelys imbricata
Hawksbill Turtle

Lepidochelys kempii
Atlantic Ridley Turtle

Lepidochelys olivacea
Pacific Ridley Turtle

Dermochelys coriacea
Leathery Turtle

Podocnemis expansa
South American Turtle (Nt)

Pseudemydura umbrina
Short-necked Turtle (A)

CROCODYLIA

Alligator sinensis
China Alligator (P)

Caiman c. crocodilus
Spectacled Caiman (Nt)

Caiman crocodilus apaporiensis
Rio Apaporis Caiman (Nt)

Caiman crocodilus fuscus
Magdalena Caiman (Nt)

Caiman crocodilus yacare
Paraguay Caiman (Nt)

Caiman latirostris
Broad-nosed Caiman (Nt)

Melanosuchus niger
Black Caiman (Nt)

Crocodylus acutus
American Crocodile (N/Nt)

Crocodylus cataphractus
African Slender-snouted Crocodile (E)

Crocodylus intermedius
Orinoco Crocodile (Nt)

Crocodylus moreletti
Morelet's Crocodile (Nt)

Crocodylus p. palustris
Marsh Crocodile (O/P)

Crocodylus rhombifer
Cuban Crocodile (Nt)

Crocodylus siamensis
Siamese Crocodile (O)

Osteolaemus tetraspis
Dwarf Crocodile (E)

Tomistoma schlegelii
False Gavial (O)

Gavialis gangeticus
Indian Gavial (O)

SQUAMATA

Phelsuma edwardnewtoni
Rodriguez Day Gecko (E)

Brachylophus fasciatus
Fiji Banded Iguana (A)

Crotaphytus wislizenii silus
Blunt-nosed Leopard Lizard (N)

Cyclura pinguis
Anegada Ground Iguana (Nt)

Cyclura r. rileyi
Watlings Island Ground Iguana (Nt)

Gallotia simonyi
Simony's Lizard (P)

Ameiva polops
St. Croix Ground Lizard (Nt)

Aniella pulchra nigra
Black Legless Lizard (N)

Bolyeria multocarinata
Round Island Boa (E)

Casarea dussumieri
Keel-scaled Boa (E)

Epicrates inornatus
Puerto Rican Boa (Nt)

Thamnophis sirtalis tetrataenia
San Francisco Garter Snake (N)

Naja oxiana
Central Asian Cobra (P)

BIRDS
131 species and subspecies

STRUTHIONIFORMES

Struthio camelus syriacus
Arabian Ostrich (P)

RHEIFORMES

Pterocnemia pennata taropacensis
Puna Rhea (Nt)

PODICIPEDIFORMES

Podiceps taczanowskii
Junin Grebe (Nt)

Podilymbus gigas
Giant Pied-billed Grebe (Nt)

PROCELLARIIFORMES

Diomedea albatrus
Steller's Albatross (P)

Procellaria parkinsoni
Black Petrel (A)

Pterodroma cookii cookii
New Zealand Cook's Petrel (A)

Pterodroma hypoleuca axillaris
Chatham Island Petrel (A)

Pterodroma phaeopygia sandwichensis
Hawaiian Dark-rumped Petrel (A)

Pterodroma aterrima
Reunion Petrel (E)

PELECANIFORMES

Sula abbotti
Abbott's Booby (O)

CICONIIFORMES

Nipponia nippon
Japanese Crested Ibis (P)

ANSERIFORMES

Anas oustaleti
Marianas Mallard (O)

FALCONIFORMES

Gymnogyps californianus
Californian Condor (N)

Accipiter francescii pusillus
Anjouan Sparrowhawk (E)

Aquila heliaca adalberti
Spanish Imperial Eagle (P)

Buteo galapagoensis
Galapagos Hawk (Nt)

Chondrohierax uncinatus minus
Grenadan Hook-billed Kite (Nt)

Chondrohierax uncinatus wilsoni
Cuban Hook-billed Kite (Nt)

Eutriorchis astur
Madagascan Serpent Eagle (E)

Pithecophaga jefferyi
Monkey-eating Eagle (O)

Rostrhamus sociabilis plumbeus
Everglade Kite (N)

Falco punctatus
Mauritius Kestrel (E)

GALLIFORMES

Crax blumenbochii
Red-billed Curassow (Nt)

Oreophasis derbianus
Horned Guan (N)

Pipile p. pipile
Trinidad Piping Guan (Nt)

Tympanuchus cupido attwateri
Attwater's Prairie Chicken (N)

Catreus wallichii
Cheer Pheasant (O)

Crossoptilon mantchuricum
Brown-eared Pheasant (P)

Lophophorus lhuysii
Chinese Monal (P)

Syrmaticus ellioti
Elliot's Pheasant (P)

Tragopan caboti
Cabot's Tragopan (P)

Tragopan melanocephalus
Western Tragopan (O)

GRUIFORMES

Grus americana
Whooping Crane (N)

Grus leucogeranus
Siberian White Crane (P)

Gallinula chloropus sandvicensis
Hawaiian Gallinule (A)

Notornis mantelli
Takahe (A)

Rallus poecilopterus
Barred-wing Rail (A)

Trickolimnas sylvestris
Lord Howe Wood Rail (A)

Rhynochetos jubatus
Kagu (A)

Chloriotis nigriceps
Great Indian Bustard (O)

CHARADRIIFORMES

Haematopus chathamensis
Chatham Island Oystercatcher (A)

Chaxadrius novaeseelandiae
New Zealand Shore Plover (A)

Numenius borealis
Eskimo Curlew (N)

Himantopus novaezelandiae
Black Stilt (A)

COLUMBIFORMES

Caloenas nicobarica pelewensis
Palau Nicobar Pigeon (O)

Columba inornata wetmorei
Puerto Rican Plain Pigeon (Nt)

Columba palumbus azorica
Azores Wood Pigeon (P)

Columba t. trocaz
Madeira Long-Toed Pigeon (P)

Columba trocaz bollii
Canary Islands Long-toed Pigeon (P)

Ducula galeata
Marquesan Pigeon (A)

Ducula oceanica teroakai
Truk Micronesian Pigeon (A)

Hemiphaga novaeseelandiae chathamensis
Chatham Island Pigeon (A)

Nesoenas mayeri
Pink Pigeon (E)

Streptopelia picturata rostrata
Seychelles Turtle Dove (E)

PSITTACIFORMES

Amazona arausiaca
Red-necked Parrot (Nt)

Amazona guildingii
St. Vincent Parrot (Nt)

Amazona imperialis
Imperial Parrot (Nt)

Amazona versicolor
St. Lucia Parrot (Nt)

Amazona vittata
Puerto Rican Parrot (Nt)

Coracopsis barklyi
Seychelles Lesser Vasa Parrot (E)

Cyanoramphus auriceps forbesi
Chatham Island Yellow-crowned Parakeet (A)

Cyanoramphus malherli
Orange-fronted Parakeet (A)

Cyanoramphus novaezelandiae cookii
Norfolk Island Parakeet (A)

Eunymphicus cornutus uvaeensis
Uvea Horned Parakeet (A)

Pezoporus wallicus flaviventris
Western Ground Parakeet (A)

Psephotus pulcherrimus
Paradise Parakeet (A)

Psittacula echo
Mauritius Parakeet (E)

Rhynchopsitta pachyrhyncha terrisi
Mexico Thick-billed Parakeet (Nt)

Strigops habroptilus
Kakapo (A)

STRIGIFORMES

Tyto soumagnei
Soumagne's Owl (E)

Otus ireneae
Mrs Morden's Owlet (E)

Otus rutilus capnodes
Anjouan Scops Owl (E)

CORACIIFORMES

Halcyon cinnamomina cinnamomina
Guam Micronesian Kingfisher (O)

PICIFORMES

Campephilus imperialis
Imperial Woodpecker (N)

Campephilus principalis bairdii
Cuban Ivory-billed Woodpecker (Nt)

Campephilus p. principalis
Ivory-billed Woodpecker (N)

Drycopus javensis richardsi
Tristram's Woodpecker (P)

Sapheopipo noguchii
Okinawa Woodpecker (P)

PASSERIFORMES

Xenicus longipes
New Zealand Bush Wren (A)

Atrichornis clamosus
Noisy Scrub-bird (A)

Xenopirostris damii
Van Dam's Vanga (E)

Xenopirostris polleni
Pollen's Vanga (E)

Troglodytes aedon guadeloupensis
Guadeloupe Wren (Nt)

Troglodytes aedon mesoleucus
St. Lucia Wren (Nt)

Cinclocerthia ruficauda glutturalis
Martinique Trembler (Nt)

Ramphocinclus b. brachyurus
Martinique White-breasted Thrasher (Nt)

Ramphocinclus brachyurus sanctaeluciae
St. Lucia White-breasted Thrasher (Nt)

Cichlhermina cherminieri sanctaeluciae
St. Lucia Forest Thrush (Nt)

Copsychus sechellarum
Seychelles Magpie Robin (E)

Erithacus komadori subrufa
Southern Ryukyu Robin (P)

Phaeornis obscurus myadestina
Kauai Thrush (A)

Phaeornis obscurus rutha
Molokai Thrush (A)

Phaeornis palmeri
Small Kauai Thrush (A)

Turdus poliocephalus poliocephalus
Grey-headed Blackbird (A)

Picathartes gymnocephalus
White-necked Rock Fowl (E)

Picathartes oreas
Grey-necked Rock Fowl (E)

Acrocephalus caffer aquilonis
Eiao Polynesian Warbler (A)

Acrocephalus caffer longirostris
Moorea Polynesian Warbler (A)

frogs, toads and lizards, especially urban development in California and the Bahamas where there are a large number of very localized species with small populations. Excessive exploitation at 62 per cent has seriously affected turtles, alligators, caimans and tortoises.

As we are now well aware, wild animals require suitable habitat in order to survive. It is the habitat that provides living space, cover, food and water, and it is an enormously complex web of interacting environmental factors. Unfortunately, throughout history man, through activities such as urban and industrial developments, farming, deforestation and so on, has consistently disrupted the delicate ecosystems upon which animals depend for their survival.

The table of endangered species shows the zoogeographical region or regions to which each belongs. The regions are defined by the following letters—A for Australasian, E for Ethiopian, N for Nearctic, Nt for Neotropical, O for Oriental and P for Palaearctic. This information is not given for the oceanic whales or turtles. Because of the vast number of endangered species it is not possible to describe them all; on the pages that follow, therefore, only a selection of the most interesting ones are dealt with in more detail.

Bebrornis rodericana
Rodrigues Brush-warbler (E)

Trichocichla rufa ssp.
Long-legged Warbler (A)

Bowdleria punctata wilsoni
Codfish Island Fernbird (A)

Dasyornis broadbenti littoralis
Western Rufous Bristlebird (A)

Petroica traversi
Chatham Island Black Robin (A)

Pomarea m. mendozae
Hivon Flycatcher (A)

Pomarea mendozae nakuhivae
Nakuhiva Flycatcher (A)

Pomarea n. nigra
Tahiti Flycatcher (A)

Terpsiphone corvina
Seychelles Black Paradise
Flycatcher (E)

Rukia ruki
Truk Greater White-eye (A)

Zosterops albogularis
White-breasted Silver-eye (A)

Apalopteron f. familiare
Mukojima Bonin Honeyeater (P)

Meliphaga melanops cassidix
Helmeted Honeyeater (A)

Moho braccatus
Kauai O'o (A)

Dendroica p. petechia
Barbados Yellow Warbler (Nt)

Leucopeza semperi
Semper's Warbler (Nt)

Hemignathus lucidus affinis
Maui Nukupu'u (A)

Hemignathus lucidus hanapepe
Kauai Nukupu'u (A)

Hemignathus obscurus procerus
Kauai Akiaoloa (A)

Hemignathus wilsoni
Akiapola'au (A)

Loxioides bailleui
Palila (A)

Loxops coccinea ochracea
Maui Akepa (A)

Paroreomyza maculata flammea
Molokai Creeper (A)

Paroreomyza maculata maculata
Oahu Creeper (A)

Psittirostra psittacea
O'u (A)

Pyrrhula pyrrhula murina
Sao Miguel Bullfinch (P)

Foudia flavicans
Rodrigues Fody (E)

Foudia rubra
Mauritius Fody (E)

Calleas cinerea cinerea
South Island Kokako (A)

Strepera graculina crissalis
Lord Howe Currawong (A)

Corvus kubaryi
Marianas Crow (O)

Corvus tropicus
Hawaiian Crow (A)

MAMMALS
162 species and subspecies
METATHERIA

Onychogalea fraenata
Bridle Nail-tailed Wallaby (A)

Onychogalea lunata
Crescent Nail-tailed Wallaby (A)

Gymnobelideus leadbeateri
Leadbeater's Possum (A)

Antechinomys laniger
Eastern Jerboa Marsupial (A)

Thylacinus cynocephalus
Thylacine (A)

INSECTIVORA

Solenodon paradoxus
Haitian Solenodon (Nt)

CHIROPTERA

Macroderma gigas
Australian False Vampire Bat (A)

Hipposideros ridleyi
Ridley's Leaf-nosed Bat (O)

Myotis grisescens
Gray Myotis (N)

Pteropus rodricensis
Rodrigues Flying Fox (E)

Pteropus tokudae
Guam Flying Fox (O)

PRIMATES

Lemur m. macao
Black Lemur (E)

Lemur macao rufus
Red-fronted Lemur (E)

Lemur macao flavifrons
Sclater's Lemur (E)

Lemur macao sanfordi
Sanford's Lemur (E)

Lepilemur mustelinus ruficaudatus
Red-tailed Sportive Lemur (E)

Lepilemur mustelinus leucopus
White-fronted Sportive Lemur (E)

Indri indri
Indri (E)

Propithecus verreauxi
Verraux's Sifaka (E)

Daubentonia madagascariensis
Aye-Aye (E)

Callithrix flaviceps
Buff-headed Marmoset (Nt)

Saguinus o. oedipus
Cotton-top Tamarin (Nt)

Leontopithecus r. rosalia
Golden Lion Tamarin (Nt)

Leontopithecus rosalia chrysomelas
Golden-headed Tamarin (Nt)

Leontopithecus rosalia chrysopygus
Golden-rumped Tamarin (Nt)

Saimiri oerstedii
Red-backed Squirrel Monkey (Nt)

Cacajao c. calvus
White Uakari (Nt)

Cacajao rubicundus
Red Uakari (Nt)

Lagothrix flavicauda
Yellow-tailed Woolly Monkey (Nt)

Brachytales arachnoides
Woolly Spider Monkey (Nt)

Mandrillus leucophaens
Drill (E)

Cercocebus g. galeritus
Tana River Mangabey (E)

Macaca silenus
Lion-tailed Macaque (O)

Colobus badius rufomitratus
Tana River Red Colobus (E)

Simias concolor
Pig-tailed Langur (O)

Pygathrix nemaeus
Douc Langur (O/P)

Hylobates klossi
Kloss's Gibbon (O)

Hylobates moloch
Grey Gibbon (O)

Hylobates pileatus
Pileated Gibbon (O)

Pongo pygmaeus
Orang-Utan (O)

Gorilla gorilla beringei
Mountain Gorilla (E)

LAGOMORPHA

Pentalagus furnessi
Ryukyu Rabbit (P)

Romerolagus diazi
Volcano Rabbit (N)

Caprolagus hispidus
Assam Rabbit (O)

RODENTIA

Sciurus niger cinereus
Bryant's Fox Squirrel (N)

Dipodomys heermanni morroensis
Morro Bay Kangaroo Rat (N)

Reithrodontomys raviventris
Salt-marsh Harvest Mouse (N)

Plagiodontia aedium
Cuvier's Hutia (Nt)

CETACEA

Platanista indi
Indus dolphin (O)

Balaenoptera musculus 3 ssp.
Blue Whale

Megaptera novaengliae
Humpback Whale

Balaena mysticetus
Bowhead

Eubalaena glacialis 3 ssp.
Black Right Whale

CARNIVORA

Canis lupus irremotus
Northern Rocky Mountain Wolf (N)

Canis rufus
Red Wolf (N)

Canis s. simensis
Northern Simien Fox (E)

Vulpes velox hebes
Northern Kit Fox (N)

Selenarctos thibetanus gedrosianus
Baluchistan Bear (O/P)

Ursus arctos nelsoni
Mexican Grizzly Bear (N)

Mustela nigripes
Black-footed Ferret (N)

Lutra felina
Marine Otter (Nt)

Lutra platensis
La Plata Otter (Nt)

Lutra provocax
Southern River Otter (Nt)

Pteronura brasiliensis
Giant Otter (Nt)

Aonyx microdon
Cameroon Clawless Otter (E)

Viverra megaspila civettina
Malabar Large Spotted Civet (O)

Macrogalidia musschenbroecki
Brown Palm Civet (A)

Hyaena hyaena barbara
Barbary Hyaena (P)

Felis lynx pardina
Spanish Lynx (P)

Felis pardalis albescens
Texas Ocelot (N)

Felix concolor cougar
Eastern Cougar (N)

Felis concolor coryi
Florida Cougar (N)

Prionailurus irimotensis
Iriomote Cat (P)

Neofelis nebulosa
Formosan Clouded Leopard (O)

Panthera leo persicus
Asiatic Lion (O)

Panthera t. tigris
Bengal Tiger (O)

Panthera tigris virgata
Caspian Tiger (P)

Panthera tigris altaica
Siberian Tiger (P)

Panthera tigris sondaica
Javan Tiger (O)

Panthera tigris amoyensis
Chinese Tiger (P)

Panthera tigris balica
Bali Tiger (O)

Panthera tigris sumatrae
Sumatran Tiger (O)

Panthera pardus panthera
Barbary Leopard (P)

Panthera pardus nimr
South Arabian Leopard (E)

Panthera pardus tulliana
Anatolian Leopard (P)

Panthera pardus orientalis
Amur Leopard (P)

Panthera pardus jarvisi
Sinai Leopard (P)

Panthera uncia
Snow Leopard (O/P)

Acinonyx jubatus venaticus
Asiatic Cheetah (P)

Zalophus californianus japonicus
Japanese Sea Lion (P)

Monachus monachus
Mediterranean Monk Seal (P)

Monachus tropicalis
Caribbean Monk Seal (Nt)

Monachus schauinslandi
Hawaiian Monk Seal (A)

SIRENIA

Trichechus inunguis
Amazonian Manatee (Nt)

PERISSODACTYLA

Equus przewalskii
Przewalski's Horse (P)

Equus hemionus khur
Indian Wild Ass (O)

Equus hemionus hemippus
Syrian Wild Ass (P)

Equus asinus
African Wild Ass (E)

Equus zebra 2 ssp.
Mountain Zebra (E)

Equus grevyi
Grevy's Zebra (E)

Tapirus pinchaque
Mountain Tapir (Nt)

Tapirus bairdii
Central American Tapir (Nt)

Tapirus indicus
Malayan Tapir (O)

Rhinoceros sondaicus
Javan Rhinoceros (O)

Rhinoceros unicornis
Great Indian Rhinoceros (O)

Didermoceros sumatrensis
Sumatran Rhinoceros (O)

Ceratotherium simum cottoni
White Rhinoceros (E)

ARTIODACTYLA

Babyrousa babyrussa
Babirusa (O)

Muntiacus feae
Fea's Muntjac (O)

Dama mesopotamica
Persian Fallow Deer (P)

Cervus duvauceli 2 ssp.
Swamp Deer (O)

Cervus e. eldi
Manipur Brown-antlered Deer (O)

Cervus eldi siamensis
Thailand Brown-antlered Deer (O)

Cervus nippon taiouanus
Formosan Sika (P)

Cervus nippon keramae
Ryukyu Sika (P)

Cervus nippon mandarinensis
North China Sika (P)

Cervus nippon grassianus
Shansi Sika (P)

Cervus nippon kopschi
South China Sika (P)

Cervus elaphus corsicanus
Corsican Red Deer (P)

Cervus elaphus wallichi
Shou (O)

Cervus elaphus barbarus
Barbary Deer (P)

Cervus elaphus hanglu
Kashmir Stag (O)

Cervus elaphus yarkandensis
Yarkand Deer (P)

Cervus elaphus bactrianus
Bactrian Deer (P)

Odocoileus virginianus leucurus
Columbia White-tailed Deer (N)

Odocoileus hemionus cerrosensis
Cedros Island Deer (N)

Ozotoceros bezoarticus celer
Argentine Pampas Deer (Nt)

Antilocapra americana peninsularis
Lower California Pronghorn (N)

Antilocapra americana sonariensis
Sonoran Pronghorn (N)

Taurotragus d. derbianus
Western Giant Eland (E)

Bubalus mindorensis
Tamaraw (O)

Bubalus depressicornis
Lowland Anoa (A)

Bubalus quarlesi
Mountain Anoa (A)

Bos sauveli
Kouprey (O)

Bos grunniens
Wild Yak (O/P)

Cephalophus jentinki
Jentink's Duiker (E)

Hippotragus niger variani
Giant Sable Antelope (E)

Oryx leucoryx
Arabian Oryx (P)

Alcelaphus buselaphus tora
Tora Hartebeest (E)

Alcelaphus buselaphus swaynei
Swayne's Hartebeest (E)

Neotragus m. moschatus
Zanzibar Zuni (E)

Gazella subgutturosa marica
Sand Gazelle (E/P)

Gazella dorcas massaesyla
Moroccan Dorcas Gazelle (P)

Gazella dorcas saudiya
Saudi Arabian Dorcas Gazelle (E/P)

Gazella dorcas pelzelni
Pelzeln's Gazelle (E)

Gazella gazella arabica
Arabian Gazelle (E/P)

Gazella cuvieri
Cuvier's Gazelle (P)

Gazella leptoceros
Slender Horned Gazelle (E/P)

Gazella dama mhorr
Mhorr Gazelle (P)

Gazella dama lozanoi
Rio de Oro Dama Gazelle (E)

Capricornis s. sumatraensis
Sumatran Serow (O)

Hemitragus javakari
Arabian Tahr (E)

Capra ibex walie
Walia Ibex (E)

Capra p. pyrenaica
Pyrenean Ibex (P)

Capra falconeri megaceros
Straight-horned Markhor (O)

Ovis ammon musimon
Mediterranean Mouflon (P)

. . . . a small selection of the animals that are in imminent danger of extinction

URODELA

SANTA CRUZ LONG-TOED SALAMANDER (*Ambystoma macrodactylum*). This species is on the verge of extinction, but conservation measures, such as the excavation of breeding ponds, that have been taken give good prospects for its survival. It is restricted to a handful of mid-coastal localities in the vicinity of Santa Cruz County, California, and has been very seriously threatened by land developments.

There is at present no evidence of a serious decline in either of the two endangered species of plethodontids, but their highly specialized habits render them particularly susceptible. The DESERT SLENDER SALAMANDER (*Batrochoseps aridus*) is found in an area of less than one kilometre square in Hidden Palm Canyon in the Santa Rosa Mountains of California, and the total population is probably no more than 500. It spends the summer under limestone sheeting in the walls of the canyon. The TEXAS BLIND SALAMANDER (*Typhlomolge rathbuni*) is even more remarkable in that it occurs only in deep wells and underground streams in Hayes County, Texas, where it feeds on small aquatic invertebrates. It is threatened by well-capping, drainage and pollution. Conservation measures have been instituted for both these species.

ANURA

ISRAEL PAINTED FROG (*Discoglossus nigriventer*). This frog was only discovered in 1940 and may now be extinct. It inhabits freshwater swamps on the eastern shore of Lake Huleh in northern Israel, and possibly adjacent parts of Syria. Nothing is known of its breeding biology, and its extinction may have been caused by swamp drainage, and the lack of cattle grazing which apparently created favourable ecological conditions for it.

HOUSTON TOAD (*Bufo houstonensis*). Most of the isolated populations of this toad which existed in south-central Texas seem to have gone, and the numbers remaining are not accurately known. It frequents Loblolly Pine areas on sandy soils, and may miss several consecutive breeding seasons if the spring rainfall is poor. It is being genetically swamped by hybridization with two other species of toad, and has also suffered from habitat destruction.

CHELONIA

RIVER TERRAPIN (*Batagur baska*). This freshwater turtle ranges from Bengal to Sumatra where it is found in tidal areas and the associated estuaries of major rivers. It nests on sandbanks above the tide mark. Its eggs are considered to be far superior in taste to those of any sea turtle, and it is the annual harvesting of eggs, together with slaughter of the turtles themselves and ecological changes in the environment, that have led to serious population declines.

The eleven races of the LAND TORTOISE (*Testudo elephantopus*) on the various islands in the Galapagos mostly inhabit dry transitional or moist vegetation zones. They all suffered previously from overhunting but are now rigidly protected. On two islands habitat destruction by introduced goats has had a serious effect,

Green Turtle

and the ABINGDON SADDLEBACK TORTOISE (*T. e. abingdoni*) may well be extinct. On James Island, in addition to overgrazing by goats, the nests and young of *T. e. darwini* are being destroyed by feral pigs.

GREEN TURTLE (*Chelonia mydas*). This turtle is widely distributed in seas where the temperature never falls below 20°C. It spends most of its life in areas of submarine vegetation, and performs long and little understood migrations to the sandy beaches where it nests. Its numbers have been drastically reduced by over-exploitation for meat, hides, eggs and other products. In addition, there has been massive killing of turtles in the trawl nets of the wide ranging fishing fleets. In

some areas, such as the Bahamas and Florida, it is now extinct, but in others, such as Australia and the Hawaiian Islands, there are still good populations.

SHORT-NECKED TURTLE or SWAMP TURTLE (*Pseudemydura umbrina*). This species is found in temporary swamps on coastal plains in the Upper Swan and Bullbrook area in southwest Australia, and it is estimated that there are now 200–300 on two official reserves. Its specialized habitat has never been common and was being reduced by clearance and drainage. The survival of the juveniles immediately after hatching depends mainly on rainfall, and they take from 6 to 9 years to reach maturity.

Land Tortoise

PROCELLARIIFORMES

STELLER'S ALBATROSS (*Diomedea albatrus*). This is one of the most attractive of the albatrosses, and it has been described as the world's rarest living seabird. The total population at present is probably not much more than 200 individuals, and the only known definite breeding locality is the volcanic Torishima Island in the Japanese southern Islands of Izu. This albatross formerly bred in immense numbers on this and several other islands, but from late in the last century and into the 1930s the activities of feather hunters brought the species to the verge of extinction. Although now rigidly protected, it is unlikely that this albatross will ever again approach its former abundance. One of the limiting factors may well be food supply, since the northwest Pacific has been heavily fished for a very long time. Pollution, too, has no doubt had its effect on this bird.

PELECANIFORMES

ABBOTT'S BOOBY (*Sula abbotti*). Although the total population of this seabird may currently be about 8,000 birds, it is nevertheless very much a threatened species, due to man's destruction of its habitat. Although at one time it had a somewhat wider distribution, its only known breeding station now is Christmas Island in the eastern Indian Ocean. It has somewhat specialized habits and nests almost exclusively in the forest above the 150-metre contour. Another feature of importance in terms of its ability to survive is that it has the slowest breeding rate of any species in this order; it lays only one egg every two years. The problem facing Abbott's Booby is that by an unfortunate coincidence its main forest nesting areas stand on the site of substantial deposits of high grade phosphate. This is being mined, thus requiring clearance of the forest. Conservation measures have been drawn up which it is hoped will enable this interesting booby to survive.

Monkey-eating Eagle

Steller's Albatross

CICONIIFORMES

JAPANESE CRESTED IBIS (*Nipponia nippon*). This is one of the larger species of ibis and is now a very rare bird indeed. It is now confined to Sado Island, where it frequents pine woodlands and paddies, and the population numbers only some eight birds. It used to be much more widespread in Japan and Korea, but was almost exterminated between 1870 and 1890. Breeding sites were, however, discovered again in the 1930s, but deforestation during and following World War II almost led to its final demise.

FALCONIFORMES

MONKEY-EATING EAGLE (*Pithecophaga jefferyi*). This, one of the world's rarest eagles, has always been confined to the Philippine archipelago, and has suffered from overhunting, collecting for zoos, and from the extensive destruction of its forest habitat. The present population is at the most 500–600 birds mostly on the island of Mindanao, but with a handful on Leyte and possibly a few on Luzon. The nest of this eagle is a huge platform of sticks high in a forest tree, and is used year after year if the birds are left free from disturbance.

GRUIFORMES

TAKAHE or NOTORNIS (*Notornis mantelli*). A certain amount of mystery surrounds this flightless species, in that the reasons for its decline are not

Takahe

at all clear. It is, however, believed that climatic changes may have been an important factor. It was believed to have been extinct since the late nineteenth century until, in 1948 it was rediscovered in the South Island of New Zealand where it lives in high altitude tussock grasslands. The present population is probably about 250 birds and it is the subject of intensive research to ensure its survival. The main current threats are predation by introduced Stoats and grazing competition from introduced Red Deer.

Seychelles Magpie Robin

PASSERIFORMES

SEYCHELLES MAGPIE ROBIN (*Copsychus sechellarum*). This tame species is now confined to Frigate Island in the Seychelles, where the population at present is about 40 birds, and is believed to be slowly increasing under protection. It used to be found also on the islands of Mahe, Praslin, La Digue, Marianne Felicite, and also on one of the islands in the Amirantes, but it is now extinct on all of these. The Seychelles Magpie Robin has suffered particularly from the depredations of feral cats, and also from competition by introduced mynah birds.

METATHERIA

BRIDLE NAIL-TAILED WALLABY (*Onychogalea fraenata*). This wallaby had not been seen for several decades until 1974, when a viable population was located in central Queensland in an area barely 100 kilometres square. Within this area, where it inhabits thick scrub, it appears to be fairly common. Unfortunately, the area as a whole may be subject to settlement for ranching purposes, and the conversion of the habitat to pasture for the grazing of beef cattle.

EASTERN JERBOA MARSUPIAL (*Antechinomys laniger*). Now found only in northern Victoria, west of the divide in New South Wales, and in southern Queensland, this is a species about which far too little is known for conservation purposes. The size of the present population is quite unknown and virtually no information on its general biology and breeding habits is available.

THYLACINE or TASMANIAN WOLF (*Thylacinus cynocephalus*). This is the largest of the living carnivorous marsupials. It was once widespread on the mainland of Australia and New Guinea, but has been extinct there for many thousands of years. It may still survive in Tasmania from where there are still occasional reports, though none of them has been verified. The most recent report was in early 1978 when two policemen stated that they had seen one near the town of Derby. The main cause of its extinction on the mainland was probably competition with the introduced Dingo.

INSECTIVORA

HAITIAN SOLENODON (*Solenodon paradoxus*). This insectivore is confined to the Dominican Republic, where its habitat consists of rocky, scrub and forested areas. The size of the population is not known, but the evidence suggests that it is declining due to a combination of land development, deforestation, collecting for zoos, and disturbance and predation by dogs.

Haitian Solenodon

CHIROPTERA

RODRIGUES FLYING FOX (*Pteropus rodricensis*). This fruit bat is endemic to Rodriguez Island in the Indian Ocean, where it requires extensive areas of mixed natural vegetation to ensure an adequate supply of the ripening fruits on which it feeds. Now only some 2 per cent of its natural habitat is left and the species has suffered from starvation and direct killing by man and cyclones. From over 1,000 in 1955 the population has now shrunk to 100–125 individuals.

Black Lemur

PRIMATES

On the island of Madagascar no less than six lemurs are endangered. The BLACK LEMUR (*Lemur macao macao*) inhabits very humid forest and is now confined to the Sambirano area and the islands of Nossi-Be and Nossi-Komba. Despite the loss of much of its natural habitat this species has adapted to some of the plantations that have replaced the forest. The most abundant lemur in Madagascar 30 years or so ago was the RED-FRONTED LEMUR (*Lemur macao rufus*), but its population is now on the decline due to destruction of habitat and intensive hunting. A similar comment applies to SANFORD'S LEMUR (*Lemur macao sanfordi*), which is now confined to a few remnant rain forest areas. SCLATER'S LEMUR (*Lemur macao flavifrons*) may well be extinct as none has been seen for many years. It occurred in a narrow belt of coastal forest in the northwest of the island. The RED-TAILED SPORTIVE LEMUR (*Lepilemur mustelinus ruficaudatus*), which used to be numerous in all the western forests has now been depleted to the point of extinction, mainly due to loss of habitat. The most endangered subspecies of this genus is the WHITE-FOOTED SPORTIVE LEMUR (*Lepilemur mustelinus leucopus*), which is adapted to very dry conditions and the endemic vegetation of south Madagascar, where its habitat is being rapidly destroyed by bad land use, fire, cattle and goats. Unfortunately, in all these cases there is almost no information at all on the present status of the population levels.

In Kenya the TANA RIVER MANGABEY (*Cercocebus g. galeritus*) which may now number about 1,000–1,500 individuals, and the TANA RIVER RED COLOBUS (*Colobus badius rufomitratus*) of which there may be up to 2,000 left, both inhabit restricted riverain gallery forest. Both these Old World monkeys (Cercopithecidae) have suffered

from the adverse ecological effects of hydroelectric dams and bad land-use practices on their habitat. In India the LION-TAILED MACAQUE (*Macaca silenus*) has always been confined to the Western Ghats, but it is now extinct in the northern parts of its range. The remaining populations in the south of its range are now isolated on some half dozen separate hill ranges, where it lives only in the indigenous evergreen and semi-evergreen rain forests up to about 150 m. It has suffered from forest clearance for agricultural purposes and from the effects of hydroelectric projects. A casualty of modern warfare is the DOUC LANGUR (*Pygathrix nemaeus*) in Laos and Vietnam, where it lives in the rain forest from sea level up to 2,000 m. It has been hunted for food by hungry soldiers, and its habitat has been affected by bombing and defoliants. Just how many still survive is not known.

Douc Langur

ORANG-UTAN (*Pongo pygmaeus*). The total population of this species in East Malaysia, Borneo and Sumatra may still number somewhere between 10,000 and 30,000, but it is nevertheless endangered due to continuing destruction of its habitat by the rapacious demands of the timber industry. To a lesser extent it also suffers from the killing of adult females to obtain live young for the animal trade, despite the fact that it is totally protected. It inhabits primary and secondary rain forest in lowland and hill areas up to about 1,800 m.

MOUNTAIN GORILLA (*Gorilla gorilla beringei*). This animal is now confined to interconnected mountain ranges in western Rwanda, southwest Uganda and eastern Zaire, where it lives mainly in high altitude rain forests and bamboo woodland. The total population is thought to be only about 500. Much of its remaining range is in official reserves, and in the Kahuzi-Biega National Park it occurs commonly in secondary forest. The population decline of this Gorilla is primarily due to various types of human disturbance, which have also resulted in the driving of other animals such as buffalo and elephant to abnormally high altitudes. There they have overgrazed the vegetation that would otherwise have been available for the Gorillas.

RODENTIA

SALT-MARSH HARVEST MOUSE (*Reithrodontomys raviventris*). This interesting little mouse formerly had a wider distribution, but is now confined to relatively undisturbed salt and brackish

Salt-marsh Harvest Mouse

marshes that border San Francisco, San Pablo and Suisun Bays in California, USA. The size of the present population is unknown, but it is certainly low and is now broken up into small isolated groups. Much of its habitat has disappeared as a result of urban and industrial development, and extensive areas of marsh have been diked to form salt ponds, thus preventing the mice from taking refuge at high tide. It lives in the old nests of the Song Sparrow (*Melospiza melodia*) above high water level.

CETACEA

The plight of the whales has received an enormous amount of publicity, and strenuous efforts have been made to protect those species whose populations have shrunk to a dangerously low level, and to establish a level of annual harvesting of the other species that will not result in over-exploitation. Ever since the invention of the harpoon gun and steam whale catcher in the

Orang-utan

1860s, and more recently the advent of pelagic factory ships and other sophisticated technical aids, the slaughter of whales has been phenomenal. All the species listed as endangered are now protected by member countries of the International Whaling Commission.

The BLUE WHALE, the largest animal known, has three subspecies—*Balaenoptera m. musculus*, which occurs in the northern hemisphere, *B. m. intermedia*, which is found in Antarctic waters, and *B. m. brevicauda* (Pygmy Blue Whale), which is found mainly in the Indian Ocean but has also been seen off the west coast of South America. These whales migrate between their summer feeding grounds in polar waters and the winter breeding grounds in temperate regions. The original population of Blue Whales in the Antarctic was considered to be about 200,000 and this had been reduced by whaling activities to some 4,000 by 1963.

The present world population of the HUMPBACK WHALE (*Megaptera novaengliae*) is thought to be a little over 7,000, which cannot be considered as very satisfactory, and there has been little evidence of recovery over the past decade, except perhaps in the North Atlantic. Although of oceanic distribution in the northern and southern hemispheres, it comes close to the coasts for breeding purposes.

Another threat now hanging over both the Blue Whale and the Humpback Whale is the recent growth of commercial fishing for krill (*Euphausia superba*) which is a staple item in their diet. There is an urgent need for detailed research into the possible long term effects of this developing activity.

The BOWHEAD (*Balaena mysticetus*) is confined to Arctic waters and has four geographically isolated populations in the North Atlantic and North Pacific. It has enjoyed total protection since 1935 except for a small amount of subsistence hunting by eskimos, and the populations appear to be slowly recovering from the low levels that they had reached by the end of the nineteenth century. The BLACK RIGHT WHALE has also been fully protected since 1935 except for subsistence hunting. It has three populations—*Eubalaena g. glacialis* in the North Atlantic, *E.g. japonica* in the North Pacific and *E.g. australis* in the southern hemisphere. There is now evidence of a recovery in numbers except in the North Pacific and the Antarctic.

Humpback Whale

Giant Otter

CARNIVORA

The MARINE OTTER (*Lutra felina*) formerly occurred along the Pacific coast of South America
from central Peru to Cape Horn, but is now
confined to a few bays and inlets near estuaries
on the coasts of Peru and Chile. It has long been
persecuted by fishermen because of alleged
damage to prawn stocks. There are about 200–
300 in Peruvian waters according to the latest
population estimates, but there are no figures
available for other areas.

Unlike the Marine Otter the remaining three
endangered species in the Neotropical region
have all suffered from overhunting for their fur.
There are no reliable population estimates for
any of them, but the evidence shows quite
clearly that all three have been drastically reduced in numbers. The LA PLATA OTTER (*Lutra
platensis*) inhabits Brazil, Paraguay, Argentina
and Uruguay, but is apparently most numerous
in Paraguay where the nature of the terrain has
made hunting difficult. In some parts of its
range river pollution has affected the fish stocks
on which it feeds. The SOUTHERN RIVER OTTER
(*Lutra provocax*) is found in estuaries, rivers and
lakes in central and southern Chile, and may
possibly still survive in western Argentina. Unfortunately, its numbers have declined to the
point where it is now rarely seen. Perhaps the
most interesting species of all is the GIANT OTTER
(*Pteronura brasiliensis*), which has the dubious
distinction of being the most endangered species
of Amazonian mammal, and despite a certain
degree of legal protection it is still hunted. Its
former distribution is not known precisely, but
it is now found in greatly diminished numbers in
Brazil, Argentina, Uruguay, French Guiana,
Surinam, Guyana, Venezuela, Peru and Colombia. Its habitat is rivers and the networks of
lagoons and backwaters associated with them.
It favours particularly those rivers with peat-
stained slow flowing waters, and during periods
of flooding it follows spawning fish into the
smallest creeks and drains. This otter is diurnal
in habit, and where it is not persecuted it allows
people to approach closely. It is also rather social
and in undisturbed areas it may be found in
groups of up to twenty.

Members of the cat family have suffered almost
throughout their entire range in Europe, Asia,
Indonesia and North America, from persecution

and loss of habitat. One of the common excuses
for their destruction has been alleged predation
on domestic livestock. The EUROPEAN LYNX
(*Felis lynx*) ranges from central and northern
Europe eastwards to Siberia and Mongolia, and
its various subspecies are apparently managing
to maintain themselves fairly well. The Spanish
race (*F. lynx pardina*), however, is in a less
fortunate position and survives only in inaccessible parts of southern Spain and Portugal.
Widespread destruction of its preferred forest
habitat has caused it to move to less favourable
areas, where the present population is probably
under 1,000 animals. This race of the lynx is one
of the smallest and most attractive.

Other 'big cats' of the Palaearctic region are
not so fortunate. The CASPIAN TIGER (*Panthera
tigris virgata*) frequents reed beds, valley thickets,
and cork and tamarisk forests in the Soviet
Union, and may possibly still survive in Iran.
It has been subjected to overhunting and has
also been affected by the loss of both prey species
and habitat, and the size of the surviving population is not known. The same factors have
brought the SIBERIAN TIGER (*P. tigris altaica*) of

the Soviet Union, China and Korea into the
endangered category, and again it is not known
how many still survive. In North Africa the
BARBARY LEOPARD (*P. pardus panthera*) of Algeria,
Morocco and Tunisia, has been hunted for its
skin, and there is no reliable information on the
numbers still surviving in the mountains. Further to the east the SINAI LEOPARD (*P. pardus
jarvisi*), formerly found in the desert mountain
ranges of the Sinai Peninsula, is possibly now
extinct, another victim of loss of prey and
predator control.

The situation is no better in North America.
The TEXAS OCELOT (*Felis pardalis albescens*), now
confined to Texas and north-east Mexico, is
declining steadily and now numbers perhaps
1,000 or less. It frequents mountainous areas
and riparian scrub, and clearance of the latter
vegetation in many areas has contributed to its
decrease. The FLORIDA COUGAR (*Felis concolor
coryi*) was formerly shot by stockmen whenever

Florida Cougar

the opportunity arose because of its predation on
livestock. It is now protected and appears to be
re-occupying some of its former range. Further
north the EASTERN COUGAR (*F. concolor cougar*)
still survives in the forests of northeast United
States and southeast Canada, but there is little
data on the present population levels.

Siberian Tiger

There are only three species of monk seal in the world and all are now endangered. It is unfortunate that really very little is known about their ecology and habits, although the Hawaiian species has been fairly well studied. Since the CARIBBEAN MONK SEAL (*Monachus tropicalis*) has not been seen for about 25 years, it seems highly probable that it is now extinct. The MEDITERRANEAN MONK SEAL (*Monachus monachus*) is still found along the coasts of North Africa, the Lebanon, Turkey, Cyprus, Bulgaria, Greece, Yugoslavia, Italy, Sicily, the Balearics and Madeira, but both its range and numbers have decreased. The present population is considered to be somewhere between 400 and 800 and the decrease has been mainly due to persecution by fishermen and human disturbance, particularly at the breeding caves. The future prospects for this species are not at all good, as increasing human pressures (particularly the tourist industry) and the high level of pollution in the Mediterranean can only be detrimental.

Hawaiian Monk Seal

The HAWAIIAN MONK SEAL (*Monachus schauinslandi*) is known only from the Hawaiian Islands and adjacent waters, and is probably a relict population closely related to the Caribbean species. It breeds only on islands in the northwest of the chain, such as Pearl and Hermes Reef, Laysan and Lisianski Islands, and it cannot tolerate very much human disturbance. It was probably abundant when the Europeans discovered Hawaii in 1778, but was rapidly decimated by sealers and whalers and was nearly extinct at the beginning of the present century. Under protection the population had increased to around 1,000–1,500 animals in the early 1970s, but by 1976 it had decreased again to 700–800.

ARTIODACTYLA
Out of the total of 19 endangered forms of deer in the family Cervidae no less than 10 belong to the Palaearctic region, and it is more than probable that four of those are already extinct in the wild. These are the FORMOSAN SIKA (*Cervus nippon taiouanus*) of lowland grasslands on Taiwan; the SHANSI SIKA (*C. nippon grassianus*) of forested and mountainous areas in China; the SOUTH CHINA SIKA (*C. nippon kopschi*), which also inhabits mountainous country; and the YARKAND DEER (*C. elaphus yarkandensis*) of riparian woodlands in China. In all four cases overhunting and loss of habitat have been the main factors in their decline. The NORTH CHINA SIKA (*C. n. mandarinensis*) is probably also near extinction for

the same reasons. The PERSIAN FALLOW DEER (*Dama mesopotamica*), which formerly ranged from Syria and Palestine, through Iraq and southern Iran, is somewhat specialized in that its habitat is riparian woodland thickets. It has suffered from a number of factors including loss of habitat, competition from domestic animals and various types of development, and is now probably confined in the wild to a few localized areas in Iran.

The CORSICAN RED DEER (*Cervus elaphus corsicanus*) may well be extinct in Corsica, but survives in Sardinia with a population of no more than 200. Its habitat is woodland meadows, and the causes of its decline are overhunting and loss of habitat to agriculture. In North Africa the total population of the BARBARY DEER (*C. elaphus barbarus*) in Algeria and Tunisia is about 400. It is, however, confined to fairly dense cork-oak, pine and oak forests on the borders of those countries, and still seems to be declining in numbers in Algeria. The last of the Palaearctic

forms is the BACTRIAN DEER (*C. elaphus bactrianus*), which inhabits tugai forests in the Soviet Union and Afghanistan, and currently numbers about 800 individuals, many of which are in protected reserves.

Argentine Pampas Deer

Formosan Sika

In the Oriental Region the SHOU (*C. elaphus wallichi*) of Tibet and Bhutan is very probably extinct as a result of overhunting in its temperate forest habitat. Hanging on by a thread, if it survives at all, is the MANIPUR BROWN-ANTLERED DEER (*Cervus e. elidi*) of wetland areas in India, which was down to only 14 animals in 1975. FEA'S MUNTJAC (*Muntiacus feae*) is a victim of warfare. It lives in the montane evergreen forests of south Burma and west Thailand, and has been under severe hunting pressure, particularly from Burmese insurgents living off the land. In India the KASHMIR STAG (*Cervus elaphus hanglu*), which lives in riparian and montane coniferous forests, has been reduced by hunting and loss of habitat to agriculture to less than 1,000.

Of the two Nearctic deer on the endangered species list, the COLUMBIA WHITE-TAILED DEER (*Odocoileus virginianus leucurus*) now numbers some 400–500 in flood-plain forests and meadows in Washington and Oregon, while the CEDROS ISLAND DEER (*O. hemionus cerrosensis*), confined to the island of that name in Baja California, has a much lower population. The present population of the only endangered Neotropical deer, the ARGENTINE PAMPAS DEER (*Ozotoceros bezoarticus celer*) of coastal plains in Argentina and Brazil, is about 200.

Where to see Animals

Wildlife can be found almost everywhere. There are virtually no totally barren areas on Earth and those that do exist, such as the Antarctic ice cap and the peaks of the Himalayas, require a great deal of effort to visit. Not surprisingly, wild creatures choose to live in much the same areas (and for similar reasons) as man. It is therefore virtually impossible not to have wildlife nearby. Our homes and gardens are occupied by insects, small mammals and birds. But within a particular area the number of different species that can be seen is limited. To see less familiar animals it is necessary to travel to other parts of the world, particularly the national parks and reserves, or to visit zoos and museums. In recent years this has become considerably easier as a result of the increasing speed and cheapness of air travel. A number of tour companies now offer holidays specifically designed for those who wish to view the wildlife of faraway places. For example, it is now possible to fly to Guayaquil in Ecuador and, following in the wake of the *Beagle*, take a boat to the Galapagos Islands to see the remarkable and unique fauna that initiated Charles Darwin's ideas on evolution. Wildlife tours also visit Antarctica and Ulan Bator, as well as the better-known safari parks of Kenya and Tanzania.

Unfortunately, much of the world's wildlife has suffered immense depredations in its contact with man. At first the decline was due to persecution, direct hunting with primitive weapons for the pot. The arrival of more sophisticated and more effective instruments of death accelerated the decline of many animals. Alongside this direct persecution came an even more powerful form of destruction. The growth of the world's population has accelerated during the present century to such an extent that there is virtually no inhabitable part of the planet that is not being exploited in some way or another. The result has been to increase demands on the natural resources of every part of the earth. Forests must be cleared, plains ploughed, marshes drained—all to produce food to sustain the human population. Many animals have simply lost a place to live; their habitats have been destroyed and they are now extinct. Only those species that can exist alongside man survive in large numbers. The others become more and more isolated into smaller and smaller fragments of habitat. Eventually, their populations reach a level below which inbreeding and consequent genetic deficiencies set in. The Tiger is a case in point. Its population in India is no more than 2,000 animals. Most are found in reserves which are under threat from one major project or another. Each reserve is small and, more importantly, isolated from the next, often by hundreds of miles of unsuitable Tiger habitat. In Africa it is now difficult to find any large mammal outside the boundaries of the national parks, though they once ranged freely over much larger areas.

While wildlife is in decline, with animals progressively more confined to areas that we have set aside (often they are unsuitable for any other use), the interest in wildlife has boomed. More and more people are travelling to see animals and this, in itself, puts additional burdens on the park and nature reserve resources. The status and location of many of the larger and more spectacular species is well known—thus they are easy to visit and to watch. It is paradoxically easier to see a Tiger today than it was 20 years ago when they were far more numerous. Those who join tours to see Gorilla, Asiatic Lions, Indian Rhinos and even Oil-birds, are generally rewarded with success. Thus, for the wildlife watcher the decline in the number of these animals has made them easier, not more difficult, to see. However, it is sad that it is their very rarity that has made such animals the most popular.

National parks and nature reserves are being established throughout the world. They are effective if they are large enough and well enough protected for wildlife to flourish. They are ideal places to see animals. But there are other places too. Zoos are less exotic and offer a less satisfactory experience, but they contain a large number of animals that can be seen in a relatively short time. Here too, however, there are good and bad organizations. The well established zoos of the world are becoming more and more aware of their responsibilities to the wild populations of animals on which they depend for their supplies of exhibits. They are resisting the temptation to take rare animals from the wild and concentrating instead on breeding their own. In some cases, such as Lions, breeding in captivity is embarrassingly easy. On the other hand, the captive breeding of Cheetahs is exceedingly difficult. Yet most zoos have Cheetahs on show.

The better zoos have animals on view that few can ever expect to encounter in the wild. 'Twilight Worlds', where day is turned into night, provide an opportunity to see nocturnal animals of the tropical jungles that hardly anyone ever sees wild, even if they try. Berlin Zoo has over two thousand species on show and Mexico City Zoo welcomes over five million visitors a year.

Yet another place to see wildlife is a museum. It sounds contradictory, for all the animals are dead, but the great museums have galleries full of animals that simply cannot be seen anywhere else. They have the species that cannot be kept in zoos—insects that are too small, whales that are too large. They have the great rarities and, of course, the great extinct animals of the past. Several museums now boast large dioramas —panoramic displays of animals in their usual habitats. The days of rows of animals in glass boxes are rapidly disappearing. Not only modern animals but also fossils are displayed, together with reconstructions of the animals from which they originated and imaginative presentations of the habitats in which the animals lived millions of years ago. Some museums are paying more attention to the biology as well as the natural history and ecology of animals. The British Museum (Natural History) in London, for example, opened a Human Biology section in 1977, which proved to be immensely entertaining and informative, particularly for children.

Watching Laughing Gulls (*Larus atricilla*) at the
Brigantine National Wildlife Refuge, New Jersey, USA.

National Parks
and
Nature Reserves

Northern Africa

1. AMBOSELI MASAI GAME RESERVE, Kenya. 1,259 square miles. A starkly beautiful dry landscape at the heart of Masai country beneath Mount Kilimanjaro. The Masai still herd their cattle here alongside one of the most important populations of Black Rhinoceros left anywhere. All the usual large mammals of the African plains can be seen, including Elephant, Buffalo, Lion, Leopard, Cheetah, Giraffe, Wildebeest and zebras. Bird life is exceptionally rich and varied with the Taveta Golden Weaver a local speciality. The reserve is an easy drive from Nairobi with rough track onwards to Tsavo West.

4. GARAMBA NATIONAL PARK, Zaire. A large area of poor savanna country on the Sudan border, formerly the world stronghold of the northern White Rhinoceros. This seriously endangered animal still exists here, but its chances of survival are slim. Elephant, Hippopotamus, Lion, Leopard, Wild Dog and Chimpanzee are among the other attractions.

5. MASAI MARA GAME RESERVE, Kenya. The northward extension of Tanzania's Serengeti across the Kenyan frontier. Most of the spectacular Serengeti species including Wild Dog and Cheetah can be seen, and it is a splendid place in its own right.

6. MOLE GAME RESERVE, Ghana. A West African savanna reserve with a good game lodge. This is a good area for Lion, Buffalo, Roan, Reedbuck, Bushbuck, Oribi, Crocodile and baboons.

7. NAIROBI NATIONAL PARK, Kenya. 44 square miles. Probably the most visited park in Africa, it is right next to the airport and only a few miles from Kenya's capital. Many grassland animals, such as Cheetah, Lion, Wildebeest, Kongoni, Giraffe, Eland and zebras are present and can be easily seen. Bird life is abundant and varied including Crested Hawk Eagle, Secretarybird, Ostrich, the usual East African vultures, Black Crake and several hundred other species.

ABOVE: Ol Tukai Lodge is in the main game viewing area of Amboseli, one of the most popular Kenyan reserves.
RIGHT: Hippopotamuses are found in most of the rivers, lakes and swamps which lie within the African reserves.

2. AWASH NATIONAL PARK, Ethiopia. Splendid area of savanna country with formidable scenery and a good network of tracks. Accommodation is available in the caravan camp. Soemmering's Gazelle is a speciality; also waterbuck, both kudus, Grevy's Zebra and Cheetah.

3. DINDER NATIONAL PARK, Sudan. Open savanna country in the Nile country with a rest-camp and a good road. Lion, Elephant, Buffalo, Roan, Giraffe and kudus can be seen, and the park also has Black Rhinoceros, Leopard and Cheetah.

8. NAKURU NATIONAL PARK, Kenya. Created chiefly as a bird reserve at the world famous Lake Nakuru to protect the millions of flamingos that come to feed—arguably the most impressive bird spectacle on earth. The introduction of fish has led to the appearance of a huge flock of White and Pink-backed Pelicans and there are always a hundred plus other species to see. The surrounding land has good populations of large mammals including waterbuck, Olive Baboon and Leopard. It is just outside the town of Nakuru and only a few hours by road from Nairobi.

9. NIMULE NATIONAL PARK, Sudan. An excellent underdeveloped wildlife area offering animals in unspoilt surroundings. White Rhinoceros is the star attraction but Elephant, Hippopotamus and Buffalo are also present. The animals are said to be remarkably approachable. In order to visit this park it is necessary to make private safari arrangements at present.

10. SAMBURU-ISIOLO GAME RESERVE, Kenya. 115 square miles. A spectacularly beautiful area on the borders of the Northern Frontier District in the arid lands of northern Kenya Samburu, as it is generally known. There is an excellent lodge beside the Uaso Nyiro River where crocodiles and Leopard can be seen. Beware, the monkeys steal cameras! Specialities include Grevy's Zebra, Reticulated Giraffe and Beisa Oryx, found along with the more widespread Lion, Black Rhinoceros, Buffalo and plentiful Gerenuk. Other inhabitants include interesting birds such as the ornamental Vulturine Guinea-fowl and the Somali Ostrich.

11. TSAVO NATIONAL PARK, Kenya. 8,034 square miles. One of the largest national parks in the world and a classic savannah wildlife zone. Almost every organized safari visits the famous Kilaguni Lodge where a huge variety of mammals and birds can be seen; they come to feed at meal times. The park is famous for Elephant, but Lion and Leopard, Buffalo and Black Rhinoceros, Fringe-eared Oryx and Gerenuk, as well as many other antelope, are present. Tsavo National Park is easily reached by road from Nairobi and Mombasa.

TOP: the shallows of Lake Nakuru (60 sq km) abound with flamingos.
ABOVE: a group of Impala seeking shade among the acacias of Samburu.
BELOW: zebras and Maribou Storks congregating at a waterhole in Tsavo Park.

12. VIRUNGA VOLCANOES NATIONAL PARK, Zaire. 3,124 square miles. A huge park extending for 150 miles along the borders with Uganda and Rwanda, both of which have adjacent parks. There is a huge variety of landscape though most visitors stay at Rwindi in the grassland zone. Here Lion, Topi, Kob, Buffalo and hyaenas are all to be seen. The rivers are alive with Hippopotamus (26,000 at the last count) and have the most viewable of Giant Forest Hogs. In the rain forest that dominates most of the park the Mountain Gorilla is the speciality, but there are Chimpanzees as well.

13. 'W' NATIONAL PARK, Upper Volta—Dahomey—Niger. A national park overlapping the boundaries of these three sub-Saharan countries. Savanna gives way to sahel in the north with forests along the rivers. Elephant, Buffalo, Roan, Kob, Lion, Hippopotamus, hartebeests and crocodile are the main animals to be found here. Very few visitors reach this park, but camps are available.

Southern Africa

1. ETOSHA NATIONAL PARK, South West Africa. 8,000 square miles. A muddy depression called Etosha Pan lies in the centre of the most remote regions of Africa. It is about 80 miles long and centres on a huge marsh that periodically, after rain, floods the surrounding countryside creating a hub of life in an otherwise dry landscape. At the waterholes game gathers, sometimes in huge numbers, along with its attendant predators. Lion, Leopard, Cheetah, Black Rhinoceros, Springbok (once an incredible pest in the southern part of Africa), Elephant, Gemsbok, Giraffe, Hartmann's Mountain Zebra, kudus and hyaenas can all be found. Despite its remoteness it has an excellent choice of lodges and camps. There are regular flights from Windhoek.

BELOW: zebras, Gemsbok and Springbok at a waterhole by Etosha Pan.
RIGHT: Lions are a regular sight in the parks of southern Africa.

2. KAFUE NATIONAL PARK, Zambia. 8,650 square miles. A well-watered landscape with woodland in the south and grassland to the north which the visitor to this vast national park can only dip into. There is an excellent hotel and several camps and a good network of tracks. The park is best known for the large variety of antelopes that includes Sable, Roan, Oribi, Eland, Lechwe, Sitatunga, kudus and some duikers. However there are also Elephant, Buffalo, Hippopotamus, Black Rhinoceros, as well as Lion, Leopard and the increasingly scarce Wild Dog.

3. KRUGER NATIONAL PARK, South Africa. 8,000 square miles. A dry savanna landscape with rolling hills and much open bush country. This is South Africa's largest national park and one of the most visited parks in the world. Among its many attractions are Elephant, Hippopotamus, Giraffe, Lion, Leopard and seventeen species of antelope including Roan, Sable and Kudu. Pride of place goes to the White, or Square-lipped, Rhinoceros which was reintroduced here after being wiped out last century. There is an excellent lodge and road system.

4. MOREMI WILDLIFE RESERVE, Botswana. A small tribal reserve that is part of the huge and rich Okavango Swamp area. Large mammals are seasonally abundant and Lechwe a speciality. The bird life is immensely rich.

ABOVE: the vast expanse of Ngorongoro Crater, looking down from the rim.

5. NGORONGORO CRATER CONSERVATION AREA, Tanzania. 2,500 square miles. Although the park is huge, the crater itself, in the centre, is comparatively small. No area of similar size is so rich in large animals. Most of the East African specialities can be seen in its 102 square miles, including Lion, Elephant, Buffalo, Wildebeest, Hippopotamus, Black Rhinoceros, Leopard, Cheetah, Giraffe and zebras. The crater can only be explored by 4-wheel drive vehicles and accommodation is at the Crater Lodge which is perched spectacularly on the lip of the crater.

6. SERENGETI NATIONAL PARK, Tanzania. 5,600 square miles. Much has been written about the Serengeti and many films have extolled the wildlife found there. It is thus everyone's idea of what Africa is like. Unfortunately, this tends to give the impression that, because so many animals can be found here, they are not at risk. However, it is now difficult to find large mammals outside the parks and reserves of East Africa. Visitors will see most of the large and spectacular mammals that grace our television screens, such as Elephant, Lion,

ABOVE: the Buffalo herds are an impressive spectacle on the Serengeti plains.

BELOW: the White or Square-lipped Rhinoceros is a main feature of the Umfolozi Reserve.

BELOW: a Giraffe and Buffalo at Nyamandhlovu Pan in Wankie Park.

Black Rhinoceros, Leopard, Buffalo, Giraffe, huge herds of Wildebeest (seen best at the time of their annual migration), Topi, Grant's and Thomson's Gazelles, Impala, Eland, zebras, waterbucks and many more.

7. UMFOLOZI GAME RESERVE, South Africa. One of the few wildlife areas in Africa where walking is allowed. This reserve is the home of the White Rhinoceros and holds the Black Rhinoceros as well. There is also good population of antelopes including Nyala. A hutted camp is available for visitors.

8. WANKIE NATIONAL PARK, Rhodesia. 5,540 square miles. This huge national park is second only to Victoria Falls as a tourist attraction in Rhodesia. There are savannas, forests, marshes and waterholes all with their own differing fauna. Large mammals are as numerous and as varied as in many of the better-known parks to both north and south. Elephant, Lion, Leopard, Cheetah, Buffalo, both White and Black Rhinoceros, Roan, Gemsbok, Steenbuck and others can all be seen.

Asia

1. ASTRAKHAN RESERVE, USSR. The great delta of the River Volga enters the Caspian Sea here and the reserve might well rank with the great European deltas of the Guadalquivir, the Rhone and the Danube. It boasts herons, egrets, Glossy Ibis, Greylag Geese and vast hordes of migrant waterfowl. Special permission is required to visit.

2. BARSA-KELMES RESERVE, USSR. A sandy island situated below sea level on the northwestern shore of the Aral Sea. Over 200 species of birds have been recorded, including large waterbirds, such as White Pelican, as well as many migrants. Wolf and Jerboa still exist, and Saiga, Persian Gazelle and the rare Onager have been re-introduced. Difficult of access.

viewing system has to be seen to be believed—suffice it to say that tourists are ushered within a dozen yards of these predators on foot.

5. KANHA NATIONAL PARK, India. The home of the most easily viewed wild Tigers in the world and the only ones that can be regularly seen in daylight. Kanha is a forest area with large open meadows that are the nearest thing to an African plain to be found in Asia. Here, herds of Chital, Blackbuck and Barasingha roam openly allowing a close approach and photography. Tiger viewing is from the backs of elephants.

6. KASIRANGA WILDLIFE SANCTUARY, India. An area of forest and grassland regularly inundated by the Brahmaputra where nearly half of the world's Indian Rhinoceros population lives. There are also Elephant, Tiger, a good-sized herd of Barasingha, Hog Deer, some truly wild Asiatic Buffalo and some exotic birds. A choice of lodges is available and exploration on elephants is easily arranged.

ABOVE LEFT: Gir Wildlife Sanctuary is the last home of the Asiatic Lion, smaller and much rarer than its African counterpart. LEFT: the distinctive buildings of Tiger Tops Jungle Lodge, Chitawan. BELOW: Chital, native to India and Ceylon, seen here at Kanha. An alternative name for this species is Axis Deer.

3. CHITAWAN NATIONAL PARK, Nepal. Situated in the grassy jungles of the Nepalese *terai*, Chitawan was once the hunting reserve of the kings of Nepal, a fact that saved the area from clearance and settlement. It is a principal home of the Great Indian One-horned Rhinoceros and of a healthy population of Tigers. There are Chital, Hog Deer, Leopard, Langur and most of the world population of the fish-eating crocodilian, the Gavial. Tiger Tops Jungle Lodge offers a truly magnificent wildlife experience. The park can be explored on the back of an elephant and there is one of the most attractive of safari lodges.

4. GIR WILD LIFE SANCTUARY, India. This open dry forest area is known primarily as the last home of the Asiatic Lion, but has many other attractions as well. There are Sambar, Sloth Bear, Leopard, Nilgai, Chital, Chinkara, Four-horned Antelope and a good variety of birds. The area is easily reached and the Forest Guest House is a magnificent old place. The Lion

9. SARISKA WILD LIFE SANCTUARY, India. A fine forested hill area where Tiger can be seen by spotlight after dark with a fair amount of regularity. Sambar, Chital, Porcupine and Jungle Cat can also be viewed by the same method. There is a reasonable lodge and game viewing vehicles are available.

10. TAMAN NEGARA NATIONAL PARK, Malaya. Established in 1938 as King George V National Park, Taman Negara is a large area of mountain forest in peninsular

Malaysia where wildlife, that was once more widespread, can still be found. It is mostly visited by bird enthusiasts and the up-river approach to the jungle lodges is extremely attractive. However, there are large mammals and the enthusiastic and hard working visitor may be fortunate enough to encounter Elephant, Tiger, Gaur and Sambar. It would need a very long stay to stand any chance of seeing the Sumatran Rhinoceros for which this park represents one of the last strongholds.

11. UDJONG KULON NATURE RESERVE, Java. A tiny island at the western end of this densely-populated island. Although there are rumours from elsewhere, as far as it is known this is the last home of the Javan Rhinoceros with a population of some 40 animals. The terrain is so difficult that even full-time students spend months here without a single sighting of this rare animal. Also in the forests there are Javan Tiger, Banteng, Leopard and chevrotains.

12. WILPATTU NATIONAL PARK, Sri Lanka. A sandy scrubland area with thickets and forests broken by lakes and marshes on the west coast of Ceylon. The attractions are Elephant, Chital, Sambar, Wild Asiatic Buffalo, Sloth Bear, Leopard and many birds. There is accommodation for a small number of visitors and an adequate network of tracks that facilitate exploration. Guides are provided.

LEFT: sunrise at the Keoladeo Sanctuary reveals a representative tableau of cranes and other waterfowl.

ABOVE: the dense trees of the Kinabalu rain forests offer a habitat that is well suited to the Orang-utan.

7. KEOLADEO GHANA BIRD SANCTUARY, India. Also known simply as Bharatpur, this former shooting preserve of the local maharajas is without doubt one of the most magnificent waterbird sanctuaries in the world. The initiated can see over 150 species in a day and birds are everywhere in great variety. Eagles perch, full to the brim with food. Vast flocks of ducks and geese feed on the lagoons. Sarus Cranes dance and honk and the only regularly viewable Siberian White Cranes in the world visit in winter.

8. KINABALU NATIONAL PARK, Sabah. This northern state of the Island of Borneo still has huge areas of rain forest where wildlife can exist. It is, however, being progressively lumbered and this is a major danger to the principal forest dwelling species. The Kinabalu Park is centred on the mountain of the same name that rises to over 4,000 m. Here is one of the major strongholds of the fast disappearing Orang-utan. There are gibbons and other monkeys and probably a few Sumatran Rhinoceros.

North America

1. BEAR RIVER MIGRATORY BIRD REFUGE, Utah, USA. Centred on the Bear River where it empties into the Great Salt Lake, this is a magnificent area for migrating swans, geese and duck, and also a notable breeding haunt of the same species. Immense variety of species occur and can be seen from a series of embanked tracks.

2. EVERGLADES NATIONAL PARK, Florida, USA. One of the richest marshland habitats in North America and the home of innumerable waterbirds many of which get no further into the United States. Also a resort of Puma, Black Bear, Manatee (though difficult to see here), White-tailed Deer and Alligator. Roseate Spoonbill, Snail Kite and Limpkin are among the major avian attractions and keen visitors may wish to visit the adjacent Florida Keys for terns and boobies.

3. GLACIER NATIONAL PARK, Montana, USA. Together with Waterton National Park across the border in Canada, this park forms one of the most spectacular landscapes in North America. Black and Grizzly Bears are present and there are Bighorn Sheep, Mountain Goat, Porcupine, Beaver, Marmot, Moose and Elk (Wapiti). It is a true wilderness area in which to search for wild animals. Cabins are available.

BELOW: bridges such as this part of the Anhinga Trail bring otherwise hidden parts of the Everglades Park into view.

LEFT: flat prairie, coniferous forest and rocky, snow-covered peaks are characteristic habitats of the Grand Teton Park.

4. GRAND TETON NATIONAL PARK, Wyoming, USA. A Rocky Mountain park with impressive scenery and major wildlife attractions that include Grizzly Bear, Elk, Moose, Mule Deer, Marmot and many birds. There is plentiful and varied accommodation. The adjacent National Elk Refuge, as its name implies, is is a resort of Elk (Wapiti) particularly in winter when these animals come to Jackson Hole in large numbers.

5. JASPER NATIONAL PARK, Alberta, Canada. A much frequented Park with Grizzly Bear, Mule Deer, Bighorn Sheep, Mountain Goat, Caribou and Moose, among others.

6. MOUNT MCKINLEY NATIONAL PARK, Alaska, USA. An area rich in wildlife. Unfortunately, the pressure of visitors has led to the banning of private cars, and travel round the park is by the efficient internal bus service. Herds of Caribou are notable, though the Dall or White Sheep are also important. Moose, Grizzly Bear, Wolverine, Wolf, Marmot, Pika and Porcupine all occur and there are large numbers of breeding shorebirds in season.

7. WOOD BUFFALO NATIONAL PARK, Alberta, Canada. One of the world's largest national parks and the home of both Wood and Plains Buffalo as well as a number of hybrids. This vast area is typical of the Canadian plains and difficult to explore. It is also the home of the last Whooping Cranes, although no visitor is likely to see these rare birds. Other attractions include Moose, Elk, Caribou, Black and Grizzly Bears, Wolf, Porcupine, Canadian Lynx and Wolverine. Back-packing, as it is called here, is the only satisfactory method of exploration.

8. YELLOWSTONE NATIONAL PARK, Wyoming, USA. The first of the world's national parks and one of the great playgrounds of the American west, Yellowstone still has its attractions for those who seek wildlife. Its forests and mountains are the home of Black and Grizzly Bear (someone is hurt every year feeding these apparently 'tame' animals) as well as Moose, Bison, Elk, Pronghorn, and the attractive Bighorn Sheep. Bobcat, Puma, Lynx and Coyote also live here, but are less easy to see. The geysers are also worth seeing.

9. YOSEMITE NATIONAL PARK, California, USA. A huge spectacular canyon covered with conifer forests. Seventy-eight mammals have been recorded.

South America

10. ANTARCTICA. A continent in its own right, this is an enormous area that is difficult to explore. Wildlife is more or less confined to the coastal regions, where birds and sea mammals are the attractions. Most visitors approach via South America and explore the Antarctic Peninsula which includes the South Orkneys and Shetlands, Paradise Bay and Deception Island. Here there are whales, a variety of seals, including the voracious Leopard Seal, several species of penguins, albatrosses, Giant Petrels and the abundant storm petrels. Almost the only method of visiting is by joining an organized cruise.

11. EMAS NATIONAL PARK, Brazil. Rolling prairie landscape with savanna-type woodland and riverine forest that is exceptionally rich in wildlife. There is no accommodation and the tracks that penetrate it are few and rough. However, it is undoubtedly the best place in the country, perhaps in South America, to see the elusive mammals of the continent. Maned Wolf, Giant and other armadillos, Giant Otter (in grave danger), Giant Anteater, Marsh Deer and Pampas Deer are all present and mostly viewable. Birds include rheas, seriemas and King Vulture. Termites are superabundant and their numerous mounds dominate parts of the landscape.

12. GALAPAGOS NATIONAL PARK, Ecuador. These world-famous islands are the home of a unique colonizing fauna and flora that were part of the inspiration of Charles Darwin's theories of evolution. Unfortunately, much has already been destroyed, particularly by poachers and by the introduction (accidental and otherwise) of alien animals. Goats and pigs are particularly important in this respect. Giant Tortoise, Marine and Land Iguanas and Darwin's finches are perhaps the most interesting of the species found here, but seabirds (including the only equatorial penguins and albatrosses) abound and can be approached easily. Visitors are recommended to join a cruise out of Guayaquil, although there is an airport.

13. HENRI PITTIER NATIONAL PARK, Venezuela. Situated on the coast and including an area of the densely-forested coastal ranges, this is one of the richest bird reserves in the world. Howler Monkey and Three-toed Sloth may be worth watching out for, but most visitors go there to see the birds—over 500 species of them. The accommodation is simple and there is a navigable road.

14. IGUAÇU NATIONAL PARK, Brazil. Based on the famous and spectacular Iguaçu Falls, this huge park consists of subtropical forest in which monkeys, including the Black Howler Monkey, are the main interest. Capybara, Jaguar and Puma are also present along with a huge variety of birds. It can be easily reached by aeroplane and there is accommodation at the hotel near the Falls. An adjacent park in Argentina shares the same name.

The dense forests around the Iguaçu Falls abound with a variety of wildlife.

A visitor befriends a Land Iguana on the island of Fernandina in the Galapagos.

15. PAMPAS GALERAS NATURE RESERVE, Peru. Reached from Lima this reserve is dedicated to the protection of one of the last remnants of the wild Vicuna.

16. TAYRONA NATIONAL PARK, Colombia. A beautiful beach-bound park on the Caribbean coast with an offshore coral reef that is alive with tropical fish. There is also a considerable mountain. This varied area is rich in birds, in the world's richest of all bird countries. Three species of turtle breed along the beaches and monkeys inhabit the forests. There are Andean Condors nearby. There is a road and several camping sites.

5. COTO DOÑANA NATURE RESERVE, Spain. World Wildlife Fund money purchased this vast wilderness at the mouth of the River Guadalquivir just in time. Though mainly known as one of the finest bird areas in Europe and the last stronghold of the Spanish Imperial Eagle, there are other attractions as well. There are Fallow and Red Deer, Wild Pig, and many reptiles. The Pardel Lynx also lives here, but it is difficult to see.

6. DANUBE DELTA, Romania. Though neither a reserve nor a national park, the Danube Delta on the Black Sea is one of the major wildlife regions of Europe. White Pelican, Glossy Ibis, innumerable herons and egrets, duck, geese, shorebirds and many smaller birds find a refuge here. Birds are in view at all times and while there are few other attractions (for example, Wild Pig) even the non-bird-minded will be enthralled. There is a good transportation system by boat and accommodation in hotels.

Europe

1. ABRUZZO NATIONAL PARK, Italy. A large mountainous area near Rome that boasts spectacular scenery and is a natural attraction to the people of the capital. Its forests are the home of Brown Bear, Chamois (both are of distinct races), Wolf, Wild Cat and a good collection of birds including Golden Eagle. There are hotels and pensions and a series of mountain roads and tracks to facilitate exploration.

2. BIALOWIEZA NATIONAL PARK, Poland. This famous area is the last remnant of original European lowland mixed forest. It has been protected since 1919 and became a national park in 1947 mainly to protect the very rare European Bison or Wisent. The two World Wars had their effect on these animals, but they still exist here and in the adjacent forest of the Soviet Union. Moose, Tarpan and Lynx are also present along with an interesting avifauna. The park can be visited with ease, even by westerners, and there is good accommodation available.

3. CAIRNGORMS NATIONAL NATURE RESERVE, UK. Situated on a gently rolling plateau, this Scottish reserve holds both Red Deer and Reindeer. Birds are abundant, and there are several birds of prey, notably Golden Eagle, Peregrine Falcon and Osprey. There are footpaths but no roads.

4. CARMARGUE ZOOLOGICAL AND BOTANICAL RESERVE, France. The Carmargue is the delta of the River Rhone on the Mediterranean coast near Marseilles and a large part of it is included in the Reserve. It is here that the bird-watcher venturing abroad for the first time gets a taste of the exotic. Flamingos breed here—the only viewable place in Europe—and there are great mixed heronries, harriers, Black Kite, eagles and a few hundred others. Accommodation is plentiful and most of the birds can be seen from public roads. Access is by permit.

ABOVE: White Pelicans are among the many bird species that congregate in large numbers on the Danube Delta.
LEFT: the Ibex has suffered severely from over hunting and is now protected in parks such as the Gran Paradiso.

7. GRAN PARADISO AND VANOISE NATIONAL PARKS, Italy—France. These alpine parks are part of the same general region though on different sides of the political boundary. The park boundaries have been carefully drawn to exclude the heavily-settled valleys, though ski resort development is a continual threat even at high altitude. The mountains are the main attraction, however, with their populations of Alpine Ibex, Chamois and marmots. Otherwise this is another of those bird places with the Golden Eagle finding a stronghold. The scenery here is very beautiful.

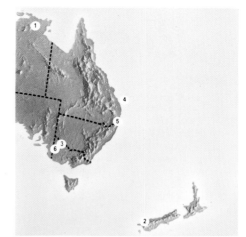

Gannet colonies are a spectacular sight on the cliffs around Muckle Flugga lighthouse in the Shetlands.

8. MINSMERE BIRD RESERVE, UK. An area of pools, marshland, woods and heath that is visited by over 250 different species of birds each year. There is a breeding colony of the European Avocet—one of two re-established in Britain after an absence of 125 years.

9. NAARDERMEER, Holland. A splendid little bird reserve near Amsterdam where Spoon-bills and a variety of other waterbirds nest.

10. NEUSIEDLERSEE AND SEEWINKEL RESERVES, Austria. A huge inland lake surrounded by large reed beds and one of the most famous of European bird haunts. Three hundred species have been recorded.

11. ORDESA NATIONAL PARK, Spain. A moun-tain reserve high in the Pyrenees against the French border where many bird-watchers seek out their first Lammergeier. It is also a noted haunt of Chamois and Ibex. There are good tourist facilities and wildlife for those prepared to walk a little.

12. SHETLAND, UK. Though not a park, the spectacular cliffs of Shetland, several of which are included in nature reserves of one sort or another, are one of the great wildlife spectacles of Europe. Thousands of sea-birds cram onto the ledges and in summer all is movement and action before one's eyes. The Hermaness Reserve in the far north is as spectacular as any, though the cliffs of Noss are more easy to visit.

13. STELVIO NATIONAL PARK, Italy. An alpine park with a good population of Brown Bear, Red Deer, Roe Deer, Chamois, Marmot and Golden Eagle.

14. HIGH TATRA NATIONAL PARK, Czecho-slovakia. A splendid wilderness area high in the Tatra Mountains where Chamois, Wolf, Brown Bear, Lynx and Marmot can still be found. Plentiful tourist facilities just outside the park proper.

Australasia

1. ARNHEMLAND, Australia. The wild nor-thern region of Australia where the fauna and flora has more affinities with that of New Guinea. Rich in bird life, but also one of the last haunts of both Freshwater and Estuarine Crocodiles. It holds a specialized mammal population of species difficult to find to the south.

2. FIORDLAND NATIONAL PARK, New Zealand. The largest national park in the country situated in the remote fastnesses of South Island. Its mountains and forests are the home of native birds, including the Takahe, that have been eliminated else-where. Also present are Kakapo, Blue Duck and Pukeko along with the introduced Red Deer. There are good hotels and mountain huts and an excellent walking trail.

The visitor to Fiordland Park finds himself among impressive scenery.

3. HATTAH LAKES NATIONAL PARK, Australia. An area of lakes and eucalypt forests in northern Victoria, where both Red and Grey Kangaroos can be seen along with a wealth of birds.

4. HERON ISLAND MARINE NATIONAL PARK, Australia. A tiny part of Australia's great Barrier Reef, but one of the richest. Here, fish of immense variety amaze visitors—especially those who are able to see them with the aid of artificial light. There are beautiful coral formations, turtles and abundant seabirds as well. Simple accom-modation is available on the Island.

5. LAMINGTON NATIONAL PARK, Australia. A sub-tropical forest region that holds many of the more interesting marsupials of Australia. Also one of the very few places where secretive and often nocturnal crea-tures have become accustomed to man and are prepared to show themselves in daylight. The park is thus almost the only spot to see such animals as the Tiger Cat, Ringtail Possum, Pretty-faced Wallaby and so on. There is also a rich avifauna that includes Brush Turkey, Satin and Regent Bower-birds and Paradise Riflebird.

6. WYPERFELD NATIONAL PARK, Australia. Excellent open Mallee-scrub area for kangeroos, megapodes, parrots and their allies, and Emu.

The wide expanse of Wyperfeld is an important area for Australian wildlife.

Zoos

Europe

1. AMSTERDAM ZOO, Holland. Covering 25 acres, it was founded in 1838 and includes a formidable 1,500 species. The zoo has a good breeding record and is notable for the number of rare species which it exhibits. In general, the zoo fraternity frowns on keeping rare species in captivity, but at Amsterdam the attempts at breeding are a sufficient justification.

2. ANTWERP ZOO, Belgium. A great zoo established in 1843 by the Société Royale de Zoologie d'Anvers and covering a mere 24 acres. This was one of the leaders in the new approach to keeping and exhibiting animals. Its architecture is often stunning and the means by which animals are kept separate from people are often highly ingenious. A first class zoo with a well-deserved reputation.

ABOVE: flamingos in an attractive ornamental setting at Antwerp, one of Europe's most appealing zoos.

3. BARCELONA ZOO, Spain. Established in 1889 on 30 acres, this is the world's second-most-visited zoo and, with just under five million visitors a year, the most popular in Europe. It has 350 species from a variety of lands exhibited in modern open conditions.

4. BASEL ZOO, Switzerland. Though it covers only 17 acres, this outstanding zoo boasts a collection of 450 species and has a remarkable record of breeding 'difficult' species. Its accomplishments include young of Gorilla and Pygmy Hippopotamus, and the collection is housed with typical Swiss efficiency and cleanliness. An outstanding zoo in every way.

5. BERGEN AQUARIUM, Norway. Principally, a show place for the extraordinary diversity of sea life produced by the Norwegian fishing industry, and it is a great pity that other important fishing ports have not followed its lead. Established in 1961 it houses an excellent and representative collection of North Sea fish species, together with seals from the Arctic that are seldom kept in captivity.

6. BERLIN ZOO, Germany. Established in 1844 and situated in the West German part of Berlin, this is the world's greatest collection of wild animals. It has been completely rebuilt following war-time devastation and its 74 acres now house no less than 2,100 species. Everything has been constructed on the mammoth scale with all of the advantages of modern methods of displaying animals.

BELOW: the Tigers at Berlin Zoo have an enclosure built of large rocks which provide both a 'stage' and shelter.

7. CHESTER ZOO, UK. Opened by the North of England Zoological Society in 1930, this zoo has established a first-rate reputation for itself. There are over five hundred species represented in the collection and, in general, its exhibits are housed in modern open environments including a particularly fine Elephant House. Outstanding are the Gorillas which include both Mountain and Lowland subspecies displayed in natural groupings, as well as Orang-utans. There is a nocturnal house. Only a few miles from the centre of Chester and visited by over a million visitors a year.

Visitors to Copenhagen Zoo gather round the Polar Bear enclosure.

8. COPENHAGEN ZOO, Denmark. Established in 1859 and covering only 25 acres, Copenhagen Zoo has a collection of over 800 species on display. Among these not so many years ago was a Sumatran Rhinoceros, but this has now died and is unlikely (in view of the present status of this highly endangered animal) ever to be replaced. Musk Ox and other arctic species are usually on show.

9. EDINBURGH ZOO, UK. Owned by the Royal Zoological Society of Scotland and opened in 1913. The zoo has a good collection of animals on display in a wooded hillside setting, where its urban surroundings have made it suitable for quarantine purposes. Its attractions include a tropical house, the Carnegie Aquarium (with turtles) and a feral population of Black-crowned Night Herons that flight out to feed in the local countryside. Tigers breed freely and the zoo has made a considerable amount of money by selling young animals surplus to its requirements. Its educational work is first class and a model to many others concerned with the serious work that zoos have the obligation to fulfil.

10. FRANKFURT ZOO, Germany. The zoo at Frankfurt-am-Main is generally regarded as one of the most progressive and successful of present-day zoos. The buildings are modern and well designed to make the inmates feel at home in natural groups, as a result of which the breeding record here is second to none. Its latest feature is an exoticarium of revolutionary design.

11. HAMBURG ZOO, Germany. Established in 1907 Hamburg was the pioneer bar-less zoo and is justifiably named Carl Hagenbeck Tierpark. Hagenbeck's influence can be seen in every major zoo in the world, though it all started here. Post-war rebuilding has maintained its pre-eminence.

12. HÖÖR ZOO, Sweden. Near Lund, this zoo covers 250 acres and devotes itself to breeding the native Swedish fauna. Although it is not Sweden's major zoo (which is at Stockholm) it has an excellent reputation with Wolf, Moose and deer.

13. LONDON ZOO, UK. Together with its Park at Whipsnade the Zoological Society of London's zoo in Regent's Park is one of the world's greatest collections of living animals. While most of the zoos of other countries are subsidized (and cheap to enter) London Zoo has maintained its independence. It has been modernized over the years and now has splendid modern ape and cat houses, the excellent nocturnal house and a good aquarium. Giant Panda and Gorilla can be seen among the 1,400 species in the collection. Whipsnade is a more pleasant place to see animals in a park-like setting. It has fewer species, but many of them breed.

14. MUNICH ZOO, Germany. Established in 1910 the zoo is set in beautiful parkland and includes two 'extinct' forms that have been 'reconstructed' by regressive breeding, the Forest Tarpan and the Auroch. Over 650 species inhabit the 173 acres of this zoo.

15. PARIS ZOO, France. Another of the great zoos of the world. Established in 1793 in a mere nine acres, a second establishment was created in 1934 which has a larger area and expanding collection. Here, animals are displayed without bars—the method that has become so popular throughout the world following the designs of Hagenbeck

at Hamburg. Breeding is significantly successful including the rare Okapi.

16. ROME ZOO, Italy. A large zoo of 42 acres where 800 species are kept, including an excellent display of antelopes. There is a large aviary of typically revolutionary Italian design, as well as a representative collection of the world's animals.

17. SLIMBRIDGE, UK. Owned and run by The Wildfowl Trust, the collection at Slimbridge established first by Peter Scott is the greatest of its kind in the world. Two and a half thousand birds of 170 species (mostly wildfowl) are on display and, additionally, wild geese (and others) can be seen on the surrounding grassland. Very few of the world's ducks, geese and swans cannot be seen here. The contribution of the Trust to conservation is epitomized by the remarkable success of gathering together a small flock of the highly endangered Hawaiian Goose and establishing a free-breeding population some of which have been returned to the wild. Today, these delightful little geese wander freely around Slimbridge belying their world status. However, such is the scientific back-up here

The Lion terraces at London Zoo.

that the staff have now discovered a genetic fault that is the result of in-breeding from such a tiny initial nucleus. There are lessons for other zoos here.

The Big Pen at Slimbridge.

18. VIENNA ZOO, Austria. Established in 1752 this is the oldest of the great zoos of the world and its architecture reflects the glories of the Austro-Hungarian Empire. Despite this its collection of over 700 species is well housed along modern lines. There is an aquarium as well as the more obvious of the larger zoo stars.

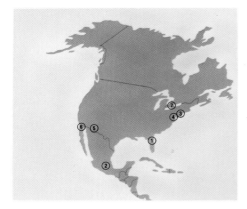

North America

1. MARINELAND, USA. Marineland was founded in 1938 near St. Augustine in Florida and has since served as the model on which all of the dolphinariums and other marinelands have been based. It houses over 150 species including the spectacular Killer Whale, dolphins and sharks. While others have imitated its commercial performing animal exhibits, few have the scientific overtones and purposes of the original institution.

Performing Dolphins at Marineland.

2. MEXICO CITY ZOO, Mexico. Though it covers only 33 acres and has only 250 species on show, Mexico City Zoo is the world's most attended zoo. Over five million visitors a year flock to see its animals. Admittance is free, as it is in the other important municipal zoos in Bombay, Chicago, Havana, St. Louis, Kansas City, New York and Washington. Mexico Zoo has a good record of breeding.

3. NEW YORK, USA. New York currently boasts three zoos and an aquarium, though here we are concerned with the two institutions owned by the New York Zoological Society: the New York Aquarium established in Brooklyn in 1896 and the Zoological Park in the Bronx founded in 1899. The Aquarium, which houses 150 species, has been fully modernized and is a model that many others have followed. The zoo is one of the world's greatest and houses over 1,000 species in its 252 acres.

4. PHILADELPHIA ZOO, USA. An old-established zoo which owns 42 acres and houses a collection of 500 species. The zoo has an excellent reputation for breeding and has had considerable success with the larger apes. This is an important major zoo with a good collection of large animals on view.

5. PHOENIX ZOO, USA. Established in 1962, covering 112 acres and the home of 250 species of desert animals, Phoenix in Arizona is principally known as the home of the World Herd of Arabian Oryx. When the last few were threatened with extinction in the wild at the hands of ruthless mechanized poachers, the Fauna Preservation Society and the World Wildlife Fund set out to capture some of the remaining animals and establish them at Phoenix. The result has been successful and the World Herd is steadily increasing—a species has been saved. Meanwhile wild Arabian Oryx have apparently disappeared for ever.

Arabian Oryx at Phoenix Zoo.

A panoramic view of part of the scenic San Diego Zoo.

6. SAN DIEGO ZOO, USA. Though founded only in 1922 the San Diego Zoological Society quickly established its zoo in Balboa Park as one of the world's great zoos. The collection includes well over 1,000 species many of which breed under the most ideal of conditions. The North American fauna figures largely, but there are rare animals from many different parts of the world.

7. TORONTO ZOO, Canada. One of the world's largest modern zoos, the Metro Toronto Zoo's outdoor exhibits, native fauna preserves and picnic areas cover over 700 acres. Most of the 3,500 specimens from 400 species are housed in enormous buildings, each with a different zoo-geographical theme. There are train tours of the Canadian Animal Domain.

Asia and Australia

1. COLOMBO ZOO, Sri Lanka. Covering 42 acres and housing about 500 species, Colombo Zoo has a good reputation for breeding animals and is generally regarded as one of the finest zoos to be found in the whole of Asia.

2. DELHI ZOO, India. For a few pennies the delightful gardens of Delhi Zoo can be explored. There are large enclosures which house a representative collection of the major elements of the Indian sub-continent's fauna—a fauna which is as endangered as any in the world. Outstanding perhaps is the collection of 'White' Tigers, but these are poorly housed in comparison with 'Jim' who has a delightfully natural enclosure and who is the model for many of the Tiger photographs that appear in books. Every evening in winter hordes of wild duck (Shoveler and Pintail predominate) pour in to roost on the ponds where there is a substantial population of free-flying storks and herons.

3. MELBOURNE ZOO, Australia. The Royal Melbourne Zoological Gardens, opened in 1857, was the first zoo in Australia. Its innovative displays cover 55 acres and present a wide range of fauna in natural surroundings. Some 3,300 specimens from nearly 400 species include an extensive range of Australian fauna and a fine collection of cats, primates and reptiles.

4. MOSCOW ZOO, USSR. Established in 1864, Moscow Zoo was recently recreated on a completely new site. The buildings and compounds are totally modern in outlook and the animals are well displayed without bars. The authorities are enlightened in their attitudes towards trading and loans, and even Giant Pandas have been borrowed by zoos in the west.

One of the famous 'White' Tigers which are a star attraction at Delhi Zoo.

LEFT: Peking Zoo houses the world's most important collection of the Giant Panda.

5. PEKING ZOO, China. Founded in 1906, Si Shih Men in Peking covers 130 acres and has up to 500 species of animals on show. Though entry is not free (except to those under one metre in height) the zoo is immensely popular with several million visitors each year. The animals are housed in the modern style dispensing with bars and include (of course) the largest collection of Giant Pandas in captivity anywhere. There are good educational aids and interesting information boards. Tourists fortunate enough to enter China will have no difficulty in visiting the zoo.

6. TARONGA ZOO, Australia. This attractively sited zoo occupies 70 acres overlooking Sydney Harbour, and exhibits some 5,000 specimens from 750 species, including many species of unique Australian animals. Founded in 1916, it is most famous for its bird collection. Its sister zoo, the Western Plains Zoo at Dubbo, was opened in 1977 and follows the modern trend towards open range exhibits without bars.

7. TOKYO ZOO, Japan. Established in 1882 Ueno Zoological Gardens cover 32 acres with 450 species on show. It is a large major collection limited (as is London) by space. As a result it established the Tama Zoological Park covering 84 acres in 1958 outside the city. The older main zoo is ranked third in the world according to the number of people who pass through its turnstiles, with nearly four million visitors a year.

Animals and visitors alike enjoy the wide open spaces of the Western Plains Zoo.

Museums

Europe

1. BRITISH MUSEUM (NATURAL HISTORY), UK. Though housed in a monolithic piece of Victoriana in Kensington, London, the 'Natural History Museum' is gradually overcoming the inherent difficulties and displaying progressively more of its exhibits in modern form. Many displays are now centred around a 'natural' landscape showing the relationships between different animal groups; use is being made of photographs and other display material, and there are films and special shows. Children's projects are encouraged, giving youngsters specific items to find and understand rather than drifting from one gallery to the next in wonderment and, later, boredom. The Whale Hall is particularly impressive. Behind the scenes the Museum is a maze of Victorian research rooms housing one of the world's greatest collections. In the 1970s the Bird Room was moved to Tring in Hertfordshire, where a larger floor area has facilitated research. Based on the collections from the Empire, it is one of the most comprehensive in the world.

2. MUSÉUM NATIONAL D'HISTOIRE NATURELLE, France. Established in Paris at the end of the eighteenth century, this was the first of the great museum collections and the pattern for the many other European museums that followed. It came at the end of the era of encyclopedias and was a local and physical symbol of the movement that sought to classify all knowledge. Excellent galleries and research material.

The impressive entrance hall of the British Museum (Natural History).

3. NATIONAL MUSEUM, Denmark. Copenhagen Museum houses a fine collection of natural history material and is particularly strong on its public displays.

4. RIJKS MUSEUM, Holland. A huge museum with a strong natural history collection drawn from various parts of the world. There are specialist collections from the former Dutch overseas Empire.

North America

5. AMERICAN MUSEUM OF NATURAL HISTORY, USA. Situated in New York, the American Natural History Museum is the largest natural science museum in the world. It has great attractions to the general public and the specialist alike. The latter would find it difficult to perform any formal systematic work on the animals of the New World without reference to its collections. In particular its collection of Neotropic (South American) birds is the most comprehensive available. Its staff are engaged in long-term fauna studies and supports expeditions to distant parts of the world. Its publications include several notable monographs of individual families and its renovated galleries seek to educate the visitor rather than to present unexplained wonders. Outstanding among the various displays are the Hall of North American Mammals, the Akeley Memorial Hall of African Mammals and the Oceanic Mammals.

6. BUFFALO MUSEUM OF SCIENCE, USA. A truly modern museum established in 1929 with an excellent collection of natural history material and an outstanding educational programme. It offers loans, movies, evening and day classes and even has displays which feature living animals.

7. MILWAUKEE PUBLIC MUSEUM, USA. This is the largest encyclopedic natural history collection in the world, though it was founded only as recently as 1882. Behind the scenes are these huge collections, while

Africa

17. NATIONAL MUSEUM, Kenya. The National Museum, situated in a quiet suburb of Nairobi, houses an excellent collection of East African material with perhaps birds forming the bulk.

18. NATIONAL MUSEUM, South Africa. Situated in Bloemfontein, this excellent natural history museum has a fine collection of material from southern Africa and an outstanding display of fossil dinosaurs including some magnificent reconstructions.

ABOVE: one of the spectacular African dioramas at the Milwaukee Museum.

BELOW: setting up a Baringo fossil elephant in the National Museum at Nairobi.

to the public it presents a wealth of exhibits arranged on ecological or time principles. Its African mammalian dioramas are outstandingly well presented.

8. MUSEO DE HISTORIA NATURAL, Mexico. Established in Mexico City this museum has been modernized in recent years to present the natural history of the largest country in Central America from an ecological viewpoint. Excellent dioramas and other public displays.

9. NATIONAL MUSEUM OF CANADA, Ottawa. While its origins date from 1842 the museum itself was opened in 1911 with a strong natural history content. There are excellent collections from the Canadian Arctic.

10. SMITHSONIAN INSTITUTION, USA. Established in 1846 in Washington DC the Smithsonian has the motto 'For the increase and diffusion of knowledge among men'. As a result it has built up a reputation as one of the world's great scientific institutions, and there seems to be hardly any aspect of natural history that it does not cover. It established the National Zoological Park in 1899 and the Museum of Natural History in 1911. It supports a considerable number of field research projects, including one on radar studies of birds in Egypt and another on radio telemetry studies of Tigers at Chitawan in Nepal. It organizes natural history tours for the public. It publishes significant works of biological interest and has a massive internal research programme based on one of the world's most comprehensive collections of animals. This go-ahead museum is showing the way to many other more conservative organizations.

South America

11. MUSEU PAULISTA/MUSEU YPIRANGA, Brazil. Created in 1885 in Sao Paulo, this museum is famous for its botanical and zoological collections and is one of the four most important sources of such material in South America.

Asia

12. INDIAN MUSEUM, India. Established in Calcutta in 1814, this is India's oldest

museum and the largest and most comprehensive in Asia. It has a strong zoology section and is the base for much original research work.

13. LENINGRAD, USSR. The Russian collections from Siberia and Alaska were the foundations on which this early great collection of natural history was based. Today it still remains the greatest collection in the Soviet Union, though its public displays could be improved.

14. NATIONAL MUSEUM, Singapore. This museum, formerly named after the great explorer and administrator Raffles and established in 1823, houses the best collection of South-east Asia. It acts as a clearing house for information and almost without exception a rare specimen collected in this part of the world will have ended up in the Raffles Museum. Its libraries are full of interest and its study collection material is second to none.

15. NATIONAL SCIENCE MUSEUM, Japan. Established in Tokyo in 1872 this is now one of the largest collections of natural history material in the world. It covers the whole field and has good public displays.

16. OSAKA MUSEUM OF NATURAL HISTORY, Japan. One of the world's largest collections and second only to Tokyo and the huge American collections in its research material.

Australia

19. AUSTRALIAN MUSEUM, Australia. Situated in Sydney and dating from 1849, this is the most important natural history museum in the country and, among other attractions, houses the Waterhouse Collection of Australian butterflies.

The imposing facade of the Australian Museum in Sydney.

Glossary

This glossary gives concise definitions of most of the technical words used in this book. Many words, such as the names of larvae, are restricted in use to one particular group of animals and are defined in the main text when they are first used. In these cases the word does not appear in the glossary. However, where such words are defined early in long pieces of text and subsequently used again, as in the arthropod section, they are defined in the glossary for convenience.

Each entry starts with a word (the entry-word) or phrase in bold type followed by the definition. Sometimes related words or phrases are also defined, and alternative definitions of the same word may be given. Where the alternative definitions are quite different, they are numbered; they are not numbered if the alternative definitions reflect only minor variations in meaning.

Cross-references within the glossary are indicated by italics. In general, all such italicized words are defined elsewhere in the glossary. Sometimes, however, italicized words differ from entry-words in the parts of speech used. In such cases it is the root of the italicized word that should guide the reader. For example the adverb *morphologically* can be interpreted from the noun entry-word **morphology**.

Frequently the meaning of a word is clarified by comparison with other entries; such cross-comparisons are given at the end of the entry. Where the meaning of a word is further elaborated in the text, or further examples are given, the relevant page reference follows the entry.

A

abdomen the part of the body posterior to the *thorax* in insects and *tetrapods*; similar parts in other animals

aberrant not conforming to the usual type

abyss the deepest parts of the ocean, 3 km or more deep

acellular 1. (of an organism) not made up of separate *cells*, mainly used of single-celled organisms; 2. (of *bone*) solid; not having any small cavities within

acoelomate without a body cavity or *coelom*

adaptation *evolution* to occupy an *ecological niche* efficiently; the result of such evolution; **adaptive radiation** *evolution* from a common ancestor which results in the formation of several *species*, each occupying a distinct *ecological niche*

adductor muscle one that brings two parts of a body together

adipose fatty, fat-filled

aestivation period of dormancy or quiescence during a hot or dry season usually summer; see also *hibernation*

albumen a *protein*, typically found as the white of eggs; related proteins (albumins) occur in the blood

alga (*pl* **algae**) a primitive plant without any supporting tissue; the group includes single-celled green algae and brown seaweeds

alimentary of feeding and digestion; **alimentary canal** the tube that runs from the mouth to the anus and is concerned with the digestion and absorption of food; the gut; **alimentary system** the alimentary canal and its associated glands and other organs

alternation of generations a feature of many *life-cycles* where there are two or more reproductive stages each of different bodily form; usually only one of these stages reproduces *sexually*, the others reproduce *asexually*, e.g. the *polyp* (asexual) and *medusa* (sexual) stages in jellyfish (Scyphozoa)

alveolus (*pl* **alveoli**) a very small hollow or pit

amber the fossilized resin of coniferous and related trees

amoeboid (of a cell) capable of changing its shape by means of *pseudopodia*

amphibious capable of living on land or in water

amplexus position adopted by many mating amphibians in which the male grasps the female around her abdomen

ampulla (*pl* **ampullae**) a small bladder or *sac*

anadromous (of fishes and others) migrating from the sea into fresh water to spawn e.g. salmon (see p. 320); see also *catadromous*

anaerobic (of *respiration*) in the absence of oxygen

analogous (of *organs*) having similar function, but not necessarily having similar origins; see also *homologous*

anatomy the study of the internal structure of organisms, originally by dissection, now almost synonymous with *morphology*

angiosperm a plant belonging to the division Angiospermae or flowering plants

annulus (*pl* **annuli**) a ring; **annular** ring-shaped

antenna (*pl* **antennae**) the paired *appendages* on the heads of crustacea, myriapods and insects; similar structures on annelids (strictly *tentacles*)

anterior nearer or towards the head end of an animal

antibody a *protein* produced by an organism, in respose to a foreign substance (antigen), with the effect of neutralizing the foreign substance

anus the opening to the exterior at the *posterior* extremity of the gut; **anal** of or towards the anus

appendage (in arthropods) one of the paired jointed structures arising from each segment; similar structures in related animals

aragonite a crystalline form of calcium carbonate

arboreal living in trees

asexual reproduction reproduction not involving *meiosis* and *fertilization* (see p. 294)

asymmetrical without any plane of *symmetry*

atoll *annular* island or group of islands surrounding a lagoon, composed of coral (see p. 79)

atrium (*pl* **atria**) 1. one of the chambers of the heart in vertebrates; 2. chamber surrounding the *pharynx* in lower chordates; **atriopore** the opening of the atrium to the exterior (see pp. 184, 188)

auditory of the ear or sense of hearing

auricle one of the chambers of the heart receiving blood from the body

autotrophic able to synthesize food from simple compounds absorbed directly from the environment, using, for example, light as a source of energy; see also *heterotrophic, photosynthesis*

axial arranged along the long axis of an animal or organ

axopodium (*pl* **axopodia**) a long *pseudopodium* with a central stiff rod

B

barbel fleshy, thin protruberance from around the mouths of many fishes

basal of the lowest layer or segment; at the bottom or base

basement membrane a sheet of non-living material, secreted by a layer of *epithelial* cells which thus lie on the membrane

bathyal of the deep sea, but above the abyss; **bathypelagic** bathyal but swimming above the bottom

benthic of the sea bottom; **benthos** benthic *fauna*

biconvex having both sides or faces convex

bifid symmetrically forked into two branches

bilateral on or of both or two sides; **bilaterally symmetrical** having a single plane of *symmetry*

binary fission mode of *asexual reproduction* in which an organism's body (often only a single cell) divides into two more-or-less equal halves

binocular vision the ability to focus both eyes onto an object at the same time; see also *stereoscopic vision*

bioluminescence the generation of light by an organism using *enzymes* at normal body temperatures

biome a major community of plants and animals, characterized by a

particular vegetation type and occupying a large geographical area (see p. 34)

biota the combined *flora* and *fauna* of a region; **biotic potential** the potential natural rate of increase of a population

bipedal walking on two legs

biramous (of an *appendage*) having two branches

bisexual (of a *species*) having two distinct sexes

bivalve having two shell *valves*, usually with a hinge between; a member of the class Bivalvia

blubber a thick layer of *adipose* tissue in the *dermis* of marine mammals

body cavity see *coelom*

bone the hard *connective tissue* of which the skeleton of most vertebrates is composed; one of the individual bony elements of a vertebrate skeleton

brachiate to move about by swinging by the arms

brackish (of natural waters) between fresh and saline

branchial of the *gills*

brood 1. the eggs or offspring of a single female from one mating; **brood chamber, brood pouch** specialized region of an animal (not always the female) for carrying the brood; 2. to look after a brood

browse to feed on *sessile* organisms (usually, but not always, plants)

buccal of the mouth

bud an offshoot from an organism which eventually will become detached as an independent organism; **budding** reproduction by producing buds

bursa (*pl* **bursae**) a cavity, usually with thickened walls, frequently *eversible*

byssus series of *proteinaceous* threads used by many *bivalves* for attachment to the *substrate* (see p. 115)

C

caecum (*pl* **caeca**) a blind pouch off the *alimentary canal*

calcareous limy, chalky, composed of calcium carbonate

calcify to deposit calcium carbonate in a *tissue*

calcite a crystalline form of calcium carbonate; **calcitic** formed of such crystals

calyx (*pl* **calyces**) cup-shaped supporting structure

canine of dogs; **canine tooth** the first tooth on the *maxilla* (second upper jawbone) of mammals, usually enlarged for stabbing; also the corresponding tooth on the lower jaw

carapace a shield of *bone* or *chitin* on the dorsal surface, often covering the sides as well

carbohydrate sugar, starch, cellulose or any other similar material composed of carbon, hydrogen and oxygen

carbonaceous organic matter converted to elemental carbon, usually as part of the process of *fossilization*

carnassial tooth a cheek tooth specialized for shearing flesh

carnivore a flesh-eating animal, one which eats other animals (not necessarily a member of the mammal order Carnivora); **carnivorous** flesh-eating

carrion dead animal matter as a source of food

cartilage a stiff, translucent *connective tissue*; gristle

caste one of the specialized forms of worker in a social insect colony (especially in termites)

catadromous (of fish and others) migrating from fresh water to the sea to spawn, e.g. freshwater eels (see p. 320); see also *anadromous*

caudal of or towards, the tail

cell the basic unit of all animals and plants comprising a volume of *cytoplasm*, bounded by a *plasma membrane* and with one or more *nuclei*

cement 1. modified bone holding a tooth in the jaw; 2. an adhesive secreted by an animal, usually from a special cement gland

cephalic of, or towards, the head

cephalothorax combined head and *thorax*

cerata finger-like projections from the back of many nudibranch molluscs, often armed with *cnidoblasts* captured from coelenterates (see p. 112)

cerebral hemispheres the paired expansions of the forepart of the brain, most highly developed in mammals

chaeta (*pl* **chaetae**) a *chitinous* bristle of a worm; *seta*

chaparral characteristic *biome* of regions with a warm-temperate climate, but with little, if any, summer rain; the vegetation usually consists of scattered low shrubs and other drought-resistant plants (see p. 38)

chela (*pl* **chelae**) the pincer-like terminal structure on the limbs of many arthropods; **chelate** bearing chelae

chelicera (*pl* **chelicerae**) the *chela*-like first *appendage* of chelicerates (arachnids and king-crabs) (see p. 137)

chemoreceptor a sensory *cell* or *organ* which responds to certain chemicals in its environment (e.g. a taste-bud)

chitin a hard, nitrogen-containing, complex *carbohydrate* forming the external skeleton of arthropods and others, and the *cuticle* in many other animal groups (see also *sclerotin*)

chlorophyll the green pigment found in most plants and some protozoans, responsible for trapping light energy; see also *photosynthesis*, *primary production*

chloroplast an *organelle* containing chlorophyll

choanocyte a cell bearing a collar, surrounding a *flagellum*, found in sponges and some protozoans (see p. 69)

chromatin the complex of *protein* and *DNA*

chromosome one of the thread-like objects found in the cell *nucleus*, made of *chromatin* (see p. 294)

cilium (*pl* **cilia**) a small thread-like *organelle* on the surface of many cells, usually in large numbers, used by the cell for locomotion or to create water currents; **ciliary feeder** an animal that uses cilia to create water currents, which bring food particles to it; see also *deposit feeder*, *filter feeder*, *suspension feeder*

cirrus (*pl* **cirri**) a stiff, slim finger-like projection

class a *taxonomic rank*, below *phylum* and above *order* (see p. 48)

clavicle one of the bones of the *pectoral* girdle, between the shoulder joints on each side; the collar-bone

cleidoic closed; used of an egg to indicate it is isolated from the external *environment*, usually by a shell (see pp. 191, 216)

cloaca (*pl* **cloacae**) a chamber at the *posterior* end of the *alimentary canal* into which the *excretory* and *genital* systems also open

cnidoblast a specialized stinging cell found only in coelenterates (see p. 75)

cochlea the organ of sound detection in the *inner ear* of vertebrates (see p. 302)

cocoon a protective case of silk surrounding a *pupa* or group of eggs or *larvae*

coelom body cavity; fluid-filled cavity between the main organ systems within the body; see also *enterocoel, haemocoele, pseudocoelom, schizocoel* (see p. 59)

cold-blooded see *ectotherm*

collagen a *protein* which forms fibres particularly found in bones, tendons and other supporting tissues

colon the large *intestine* of vertebrates; the posterior part of the intestine of insects

colony 1. a group of individuals (*zooids*) *physiologically* or *anatomically* linked to each other; 2. a *population* of animals living and breeding together to their mutual advantage

commensal (of a *species* or individual) living in association with another and benefiting from it, without benefiting the other (see p. 333)

compound eye a group of simple light *receptors* (ommatidia) collectively forming a single *organ* of sight (see pp. 135, 300)

cone a colour-sensitive light *receptor* of a vertebrate eye (see p. 301)

conjugate (of Protozoa and *gametes*) to come together; **conjugant** one of two conjugating protozoans

connective tissue *tissue* having a supporting, space-filling or transporting function, comprising a *matrix, cells*, and usually fibres

continental drift the movement of the continents relative to each other, over the surface of the earth (see p. 27)

continental shelf the area of

relatively shallow water, down to 200 m, adjacent to the continental land mass

contractile capable of contracting (becoming smaller or shorter)

convergence see *evolutionary convergence*

convex with the central region above the level of the peripheral regions; slightly rounded

copulation sexual union or intercourse; **copulatory organ** one facilitating copulation e.g. a *penis*; **copulatory bursa** a *bursa* to receive a copulatory organ

coracoid one of the bones of the *pectoral* girdle of *tetrapods*, originally bracing the shoulder joint against the *sternum*

corpuscle 1. a single cell within the body fluid, e.g. blood; 2. a small body, e.g. that comprising an organ of touch or taste

cryptobiosis the suspension of virtually all biological activity in the face of extreme conditions, usually freezing or *desiccation* (see p. 336)

cryptozoic living in *habitats* normally hidden to man, e.g. in caves or in the soil

ctenoid comb-like

cusp one of the projections from the surface of a tooth

cuticle an external layer, frequently made of *chitin*, secreted by the *epidermis*

cyst 1. a closed (usually), fluid-filled *sac*, frequently under pressure; 2. a resistant, dormant, encapsulated form assumed by some invertebrates at critical times of the *life cycle*

cytoplasm all the living matter of a *cell* outside the *nucleus*

D

decompose to decay, break down, break up; **decomposer** an organism which facilitates decay

degenerate to become less complex or specialized or smaller, either within a *life cycle* or during *evolution*

dentine a *connective tissue* constituting the bulk of a vertebrate tooth, mainly composed of apatite (calcium fluorophosphate) and *protein*

dentition the set of teeth of a particular animal or *species*

deposit feeder an aquatic animal feeding on *detritus* which has settled on the bottom; see also *ciliary feeder*, *filter feeder*, *suspension feeder*

dermis (*pl* **dermes**) the *connective tissue* immediately underlying the *epidermis* of vertebrates

desiccation drying out, loss of water

detritus dead organic matter which accumulates on the bottom of aquatic *environments* or in soil

deuterostome an animal in which the anus is equivalent to the single opening of coelenterates; the mouth forms a second opening (see p. 57)

diapause a period of dormancy, notably found in insect *pupae* (see p. 336)

diaphragm 1. a specialized sheet or *septum*, especially that in Bryozoa; 2. the muscular sheet separating the *thoracic* and *abdominal* cavities in mammals

diductor muscle a muscle that causes two parts of the body to separate, such as the shell *valves* of brachiopods

differentiation 1. the process by which various groups of cells in an early embryo become specialized to form distinct tissues and organs; 2. divergent evolution

digest to break down *ingested* food into soluble matter which can be absorbed through the wall of the intestine

digit a finger or toe; **digitigrade** walking on the toes (see p. 264)

dilate to swell or enlarge in volume

dioecious having the two sexes separate: bisexual

diploblastic having only two basic layers of tissue—*endoderm* and *ectoderm* (see pp. 59, 74)

diplosegment a pair of segments which externally appear as one; characteristic of the diplopods

dispersal the spread or distribution of the offspring of an organism, or of some chemical produced by it

distal away from the body (as of a limb or appendage) or from a defined point within or on the body; see also *proximal*

diurnal active during the day

divergence see *evolutionary divergence*

diverse (of a community or *taxon*) including numerous, widely differing *species*

diverticulum a blind, usually elongate, pouch branching off a canal (usually the *alimentary canal*)

DNA, deoxyribonucleic acid a complex molecule whose structure contains all the heredity information necessary for an organism

dominance hierarchy order of precedence established within social animals to determine which has priority for some resource

dorsal of the back

duodenum the part of the vertebrate *intestine* immediately following the *stomach*

E

eardrum (*tympanum*) the thin membrane in the *middle ear* which vibrates in response to airborne sound (see p. 302)

ecdysis (*pl* **ecdyses**) the shedding of the old *exoskeleton* of an arthropod

echolocation the use of sound for the detection of prey and for orientation; animal sonar (see p. 300)

ecological niche the specific rôle occupied by a particular *species* in its *habitat*

ecosystem an *environment* and all the organisms living and interacting within it (see p. 34)

ectoderm the outer *tissue* layer of a coelenterate; the comparable layer in the early *embryos* of other metazoans

ectoparasite a *parasite* which remains on the outer surface of its *host* (see p. 330)

ectotherm an organism whose internal body temperature is determined wholly by the external *environment*; **ectothermic** cold-blooded

egg tooth a small hard nodule on the tip of the snout of newly hatched reptile, bird or monotreme, which is used to break out of the egg

elytron (*pl* **elytra**) 1. the heavily *sclerotized* forewing of a beetle or bug, used as a cover for the hindwing; 2. one of the plate-like scales on the *dorsal* surface of some polychaetes

embryo a young organism within an egg, or its parent, not recognizably similar to the adult; **embryology** the study of embryos and their development

enamel the hard outer layer of a vertebrate tooth; similar to *dentine* but almost entirely composed of apatite

endemic (of a *species*) occurring naturally in a region; native; usually implies exclusive occurrence

endoderm the inner tissue-layer of a coelenterate; the comparable layer in the early *embryos* of the other metazoans

endoparasite a *parasite* which lives within the body of its *host* (see p. 330)

endoskeleton a skeleton formed from hard *connective tissue*, for example, *bone*, within the body of an organism

endostyle the groove along the ventral mid-line of the *pharynx* of urochordates and cephalochordates, secreting *mucus*; the comparable organ in hemichordates (see pp. 185, 189)

endotherm an organism whose internal body temperature is, within broad limits, independent of the external temperature by virtue of its own *metabolism*; **endothermic** warm-blooded

enterocoel a *coelom* which initially arises as a pouch from the *endoderm* of the embryo; see also *schizocoel*

environment the total of all external influences (biological, chemical and physical) acting on an individual

enzyme a *protein*, produced by a living cell, which facilitates a specific chemical reaction

epidermis (*pl* **epidermes**) the outer *epithelial* tissue layer(s) of the skin; **epidermal** of the epidermis

epipelagic living in the upper layers of the open sea

epithelium (*pl* **epithelia**) a type of *tissue* comprising one or more layers of cells on a *basement membrane*

eversible capable of being extruded and retracted, usually by being turned inside out; **evert** to extrude

evolution the process whereby the

overall characteristics of a population or *species* change over many generations; **convergent evolution, evolutionary convergence** evolution of initially dissimilar, unrelated *taxa* to become similar; **divergent evolution, evolutionary divergence** evolution of related, initially similar *taxa* to become different; **evolutionary radiation** see *adaptive radiation*

excrete to eliminate waste from the body; **excretory system** the excretory organs and associated ducts

exhalant (of a *siphon* or water-current) carrying water and matter out of or away from an animal

exoskeleton a skeleton formed by secretion from the *epidermis*, external to the animal itself, e.g. a shell or *cuticle*

extant (of a *taxon*) living today; not *extinct*

extensile capable of increasing in length

extinct (of a *taxon*) no longer existing

F

faeces the undigested and indigestible remains of food eliminated from the intestine

family a *taxonomic rank*, below *order* and above *genus* (see p. 48)

fang 1. a tooth or *appendage* modified for carrying poison; 2. a canine tooth of a *carnivorous* mammal

fauna all the animals within a particular region

fenestra (*pl* **fenestrae**) an opening in an otherwise continuous *bone* surface (see p. 218)

fertilization the union of two *gametes* or *nuclei*; **self-fertilization** union of gametes produced by the same individual; **internal fertilization** union of gametes within the body of the female; **external fertilization** union of gametes outside the body (see p. 295)

filament a long thread-like structure

filiform thread- or whip-like

filter-feeder an animal which obtains its food by drawing in water and straining off food particles from

it; see also *ciliary feeder*, *deposit feeder*, *suspension feeder*

fin a fold of skin, usually with some skeletal support; **fin ray** a soft *cartilaginous* rod, one of those forming the *distal* skeleton of a fish fin; **pectoral fin** one of a pair immediately behind the gills; **pelvic fin** one of a pair just anterior to the *anus* (in some fish, these may lie anterior to the pectoral fins)

fission cleavage; division into two or more approximately equal parts; usually used of single *cells*

flagellum (*pl* **flagella**) a whip-like *organelle* used by a cell for locomotion or to cause water-currents

flame-cell an excretory flask-shaped cell, bearing a *flagellum* or a group of *cilia* inside the cavity, found in various worms; the beating cilia give the effect of a flickering flame

fledge (of birds) to acquire the adult feathers (as opposed to juvenile down); **fledgling** a young bird which has recently fledged or is in the process of fledging

flora all the plants in a particular region

fluke 1. a type of platyhelminth worm especially that parasitizing the liver; 2. the lateral, flat lobes of a whale's tail

fluviatile of rivers

foetus an advanced *embryo* that is recognizably a miniature of the adult

follicle a small cavity or sheath, especially one associated with a mammalian hair

food chain a series of organisms, each of which feeds upon the member or members immediately below it in the chain (see p. 36)

food pyramid a *food web* in which there is less *diversity* in both *species* and number at the top than at the bottom

food web interconnected *food chains*

foregut (in arthropods) the anterior part of the *alimentary canal*, derived from *ectoderm* and thus lined with *cuticle*

fossil evidence of a dead organism as revealed in a rock; **fossilization** the conversion of a dead organism to a fossil; **fossil record** the history of life on earth as revealed by fossils

fouling organism one which lives

in or attached to submerged man-made structures, causing damage or interference

free living not parasitic

frugivorous feeding on fruit

fusiform torpedo-shaped, streamlined

G

gamete a sex-cell; see also *ovum*, *sperm*

ganglion (*pl* **ganglia**) a mass of nerve *tissue* containing the cell-bodies of the *nerves*

ganoid (of a fish *scale*) bearing a layer of enamel; **ganoid fish** one bearing ganoid scales (see p. 199)

gelatinous having a jelly-like consistency

gene the unit of hereditary information; **genetic** hereditary (see p. 294)

geneology the combined study of the embryonic and evolutionary development of an organism

genital of the reproductive organs; **genital pore, genital aperture** see *gonopore*; **genitalia** the reproductive organs—the *gonads* and associated ducts

genus (*pl* **genera**) a *taxonomic rank* below *family* and above *species* (see p. 48)

germ plasm those cells in a *gonad* which produce the *gametes*; the cells of an *embryo* which ultimately develop into the *gonads*; sometimes confined to the *nuclear* material in such cells

germinal (of cells or tissues) capable of division or reproduction

gill plate-, feather-, or comb-like *respiratory organ* of an aquatic animal; **gill slit** cleft in the *pharynx* of a chordate; **gill bar** bar between the gill slits in *protochordates*; **gill chamber** space between the body and the *carapace* in many crustacea, housing the gills; **gill pouch** pouch-like gill slit of a hag or lamprey

gizzard the muscular, grinding part of the *alimentary canal*; particularly that of birds, which usually contains small stones to assist in the crushing of food

gnathobase specialized grinding

protuberance on the basal segment of many arthropod limbs, assisting in the preparation of food

gnathochilarium special grinding apparatus of some myriapods

gnathostome a jawed vertebrate (see p. 191)

gonad the organ which produces the *gametes*; **gonoduct** duct carrying the gametes to the outside; **gonopore** external opening of a gonoduct; **gonozooid** (in colonial animals) a *zooid* devoted to reproduction

gravid carrying ripe eggs; pregnant

grub a legless insect *larva*

gut the *alimentary canal*

H

habitat the preferred immediate *environment* of an organism

haemocoele a body cavity formed by expansion of the blood system; found particularly in molluscs and arthropods

haemoglobin the red, oxygen-carrying *protein* found in the blood of vertebrates, annelids and a few other animals

haltere the hindwing of true flies (Diptera) modified into a balancing organ (see pp. 157, 162)

haptor an organ of attachment, usually hooked

hedonic gland specialized skin gland found in amphibians, with a musky secretion, used during courtship and mating

herbivore an animal that feeds on plants; **herbivorous** plant-eating

hermaphrodite having both male and female gonads in the same individual

heterocercal (of a tail) with the vertebral column extended into an upper lobe, and with a smaller, lower lobe (see p. 195)

heterotrophic obtaining food and energy in the usual animal fashion, by ingesting and breaking down complex organic matter; see also *autotrophic*

hexapod an animal with six legs, notably an insect; also used for the larvae of some arachnids

hexaradiate *radially symmetrical* with three planes of *symmetry* (i.e.

six equivalent radii); used of Anthozoa

hibernation period of dormancy, *torpidity* or quiescence during a cold season, usually winter (see p. 335) see also *aestivate*

hindgut (in arthropods) the posterior part of the *alimentary canal*, derived from *ectoderm* and thus lined with *cuticle*

hinge tooth a ridge on the hinge region of one valve of a *bivalve* shell which fits into a groove on the other valve

histology the study of *tissues*

holdfast that part of an animal or plant by which it attaches itself permanently to the *substratum*

homeostasis the maintenance, by an animal, of its internal environment in as constant a state as possible, despite any changes in the external *environment*

homing the return by an animal to its original location when it is displaced, either naturally, artificially or as part of its own regular migrations

homologous (of *organs*) having similar structure, by virtue of having similar *geneological* origins; see also *analogous*

hoof a modified claw, which surrounds the terminal toe bone and bears the weight of the animal; found in ungulate mammals

hormone a chemical, produced in one part of the body, and having some specific effect in another part

host the organism within or upon which a *parasite* lives; **final host** the host in which the parasite reproduces sexually; **intermediate host** any host (there may be one or several) in which a parasite lives before reaching the final host; **paratenic host** a host in which the reproductive stage of the parasite can live, but not reproduce (see p. 330)

humerus (*pl* **humeri**) the bone of the upper arm between the shoulder and the elbow

hybrid the offspring of a cross between parents of different species, strain or variety

hydrostatic pressure pressure due to depth in water, or pressure that can be expressed as such

hydrostatic skeleton an effective skeleton utilizing body fluid under muscle-generated pressure, used by soft-bodied animals

hypophysis (*pl* **hypophyses**) a glandular protruberance below the brain that becomes the pituitary gland in higher vertebrates

I

igneous rock rock which has formed by the solidification of molten rock; see also *metamorphic rock*, *sedimentary rock*

ilium (*pl* **ilia**) the bone of the *pelvic girdle* that is attached to the *sacrum*

immune capable of withstanding the invasion of the body by foreign substances (antigens) by means of *antibodies*; **immune system** the *organs* and *tissues* which produce the antibodies; **immunology** the study of immunity

incisor tooth one of the teeth at the front of a mammal jaw

infra-red radiation heat radiation; a form of electromagnetic radiation that has longer wavelengths than those of visible light and shorter wavelengths than those of radio waves (see p. 302)

ingest (of food) to take beyond the mouth

inhalant (of an opening) taking in a current of water

inner ear that part of the vertebrate ear which detects sound vibrations (the *cochlea*) and senses the animal's orientation and balance (the *semi-circular canals* and *utricle*) (see p. 302)

innervate to supply an organ with a connection from the main *nervous system*

insectivore an animal that feeds principally on insects and other small invertebrates; not necessarily a member of the mammal order Insectivora

instar a stage in the life history of an arthropod, between successive moults (*ecdyses*)

intermediate host see *host*

interspecific competition competition for food or some other *resource* between different *species*

interstitial living in the spaces (interstices) between the particles of the *substratum*

intertidal living between the tide marks

intestine the part of the *alimentary canal* that follows the *stomach*

intromittent organ a male *copulatory organ* for carrying *sperm* into the body of the female

introvert a *proboscis* which is normally retracted but can be everted at will

ischium (*pl* **ischia**) the most posterior of the bones of the tetrapod *pelvic girdle*

kelp large seaweeds found just below the low water mark

keratin a *protein*, produced by dying skin cells, which forms an outer, impervious layer, and also hair, scales, feathers, horns and the baleens of whales; **keratinization** the process by which the outer skin cells are impregnated with keratin

kHz, kilohertz a measure of frequency equal to 1,000 cycles per second

kingdom the highest *taxonomic rank*

labium (*pl* **labia**) 1. a fused pair of appendages of insects, and some related forms forming a shelf below the mouth; 2. lip-like organ around the mouth of molluscs and other invertebrates; **labial** of the lips or labia

lacuna (*pl* **lacunae**) a small space

land bridge an isthmus or corridor of land between two land masses, usually hypothetical and invoked to explain some animal distribution patterns (see p. 26)

larva (*pl* **larvae**) the young stage of an animal, with a distinctly different body form to the adult

lateral of the sides of an animal or organ; away from the midline

lens clear, solid, structure in the eye, which focusses light onto the light *receptors*

life cycle the sequence of stages from *fertilization* to fertile adult, which itself produces *gametes* that repeat the cycle; a fertile adult may develop directly from the *zygote* formed by fertilization, but in some cases one or more *asexually* reproducing stages may precede the fertile adult; in a few cases, only

asexual reproduction is involved in the life cycle

ligament a fibrous and elastic connection between bones or other hard parts

littoral of the *intertidal* zone or the zone immediately above it (the splash zone); of the seashore

living fossil a species which has *morphologically* changed very little over many tens of millions of years

lophophore an array of *cirri* or *tentacles* each covered in *cilia*, borne either in a circle or horseshoe about the mouth (as in bryozoans and phoronids), or on a pair of more-or-less spirally arranged arms (as in brachiopods)

lung-book a respiratory organ, consisting of a series of plates (through which blood can circulate) in a chamber

macrophage 1. an animal feeding on large particles; see also *microphage*; 2. a body-fluid cell, specialized for *phagocytosing* foreign particles

malpighian tubule one of the excretory tubules of insects

mammary gland the milk-producing gland of mammals

mandible 1. the lower jaw of a vertebrate; 2. the hard appendages on either side of the mouth in crustacea, myriapods and insects; 3. any similar structure in other animals

mantle the shell-secreting sheet of tissue in molluscs and brachiopods; **mantle cavity** the cavity between the mantle and body

maquis a local (Mediterranean) version of *chaparral*

mastax the specialized grinding part of the *alimentary canal* in rotifers and some other animals

matrix 1. the non-living material between the cells of a *connective tissue* (which may be fluid in blood); 2. the material from which a *fossil* is extracted

maturation ripening; attainment of reproductive age

maxilla (*pl* **maxillae**) 1. one of the pairs of *appendages* around the mouth of crustaceans, myriapods and

insects; 2. the second tooth-bearing bone of the upper jaw of jawed vertebrates; see also *premaxilla*

medusa (*pl* **medusae**) the stage in the life cycle of many coelenterates that resembles a jellyfish (see p. 74)

meiosis a form of *cell* division, similar to *mitosis*, but in which the number of *chromosomes* per cell is reduced by half (see p. 295)

membrane a thin sheet of *tissue*, usually comprising two apposed *epithelia*; see also *basement membrane*

mesenchyme a loose agglomeration of space filling cells; the loose *mesodermal* tissue in an *embryo* before it consolidates into *organs*

mesoderm the layer of cells between the *ectoderm* and *endoderm* in *triploblastic* animals (see p. 59)

mesoglea the gelatinous *connective tissue* between the *endoderm* and *ectoderm* of coelenterates

mesothorax the second of the three segments in an insect's *thorax*

metabolism the biochemical and physiological processes that take place within an organism's *cells*

metameric segmentation see *segmented*

metamorphic rock rock which has undergone change since it was first formed; the change has usually been accomplished by heat or pressure or both; see also *igneous rock*, *sedimentary rock*

metamorphosis a relatively rapid change in the *morphology* of an animal, usually that between *larva* and adult

metathorax the third of the three segments in an insect's *thorax*

microphage an animal which feeds on small particles suspended in water; see also *macrophage*

microplankton the smaller constituents of the *plankton*, comprising the main food of *microphages*

microvillus (*pl* **microvilli**) minute finger-like projection from the surface of a *cell*

middle ear that part of the hearing apparatus between the *eardrum* (tympanum) and the *inner ear*; it is air-filled and contains the auditory *ossicle* or ossicles which carry sound vibrations from the eardrum to the *cochlea* (see p. 302)

midgut 1. (in arthropods) the middle part of the *alimentary canal*, between the foregut and hindgut; it is derived from *endoderm* and comprises the part of the gut where most of the digestion and absorption take place; 2. (in protochordates) that part of the alimentary canal following the *pharynx*

mitosis *cell* division in which each daughter cell receives an exact copy of the *chromosome* set of the parent cell (see p. 295)

molar tooth one of the cheek or chewing teeth of a mammal; molars differ from *premolars* in that they are not preceded by milk teeth

monogamous having only a single mate; see also *polygamous*

monophyletic (of a group—usually a specific *taxon*) all being evolved from a common ancestor; see also *polyphyletic*

morphology the study of the form and shape of organisms; now almost synonymous with *anatomy*; **morphological** pertaining to an organism's form or shape

motile capable of locomotion

moult 1. (in arthropods) *ecdysis*; 2. (in reptiles) the shedding of the outer, dead *keratinized* layer of skin; 3. (in birds and mammals) the replacement of the *plumage* or *pelage* by a new one, usually gradually

mucus a slimy and viscid fluid produced by many animals; **mucous** composed of, or producing mucus; **mucous membrane** a mucus-secreting *epithelium*, especially that lining the *alimentary canal* of vertebrates; **mucoid** pertaining to mucus

multicellular composed of many distinct cells, as opposed to *unicellular* or *acellular*

multiple fission mode of asexual reproduction; similar to *binary fission*, but in multiple fission many similar offspring are formed

muscle a *tissue* specialized for contraction and subsequent relaxation

mutation a random, spontaneous change in the hereditary make-up of an organism

myotome one of the blocks of muscle, serially arranged along the body of a cephalochordate, urochordate larva or vertebrate

N

nasal of the nose or sense of smell; **nasal cavity** the cavity above the mouth-cavity in mammals, lined by *olfactory epithelium*; **nasal organ** the olfactory organ of fishes

natural selection the cause of evolutionary change; the process by which those individuals of a population who are better adapted to their *environment* survive to leave more offspring than those less well adapted

nauplius (*pl* **nauplii**) the early *larva* of many crustaceans, bearing only three pairs of *appendages*

nekton actively swimming aquatic *fauna*; see also *plankton*

nematocysts the stinging *organelles* in the *cnidoblasts* of coelenterates (see p. 75)

neoteny an evolutionary process by which sexual maturity is attained in an animal showing otherwise *larval* characteristics

nephridium (*pl* **nephridia**) one of the excretory organs of many invertebrates

neritic living in coastal as opposed to offshore waters

nerve a cell specialized for transmitting information in the form of electrical pulses, consisting of a *cell* body and one or more extended fibres; a group of these fibres; **nerve cord** the brain and spinal cord of vertebrates and their *geneological* predecessors; **nerve net** a mesh of interconnecting nerve fibres comprising the nervous system of many simple metazoans; **nervous system** the nerves and associated cells and *organs* of an organism

neural of *nerves* or the *nervous system*

New World the Americas or the Western Hemisphere

niche see *ecological niche*

nocturnal active principally at night

nomenclature the process and practice of naming organisms (see p. 49)

notochord the stiff skeletal rod extending along the body (or tail) of *protochordates*, hags and vertebrate *embryos* (see p. 180)

nuclear membrane the thin membrane surrounding the *nucleus* of an animal or plant *cell*

nucleic acids *DNA* and related compounds

nucleus (*pl* **nuclei**) the region of an animal or plant *cell* containing the *chromosomes* and bounded by a nuclear membrane

nymph the early stages in the *life cycle* of those insects that do not have distinct *larvae* and *metamorphosis*

O

ocellus (*pl* **ocelli**) a simple light-sensitive *organ* such as the simple eyes of many arthropods, but also those of urochordate *larvae*

oesophagus the gullet; that part of the gut between the mouth and the *stomach*

Old World Eurasia, Africa and Australasia; the Eastern Hemisphere

olfactory of the sense of smell; **olfactory organ** special organ or area bearing *receptors* reacting to air- or water-borne chemicals

omnivore an animal which feeds on both plant and animal material

oocyte the cell which becomes the female *gamete* (egg); **oocyst** a *cyst* which forms around a *fertilized* egg, or the equivalent in protozoans

operculum (*pl* **opercula**) a stiff lid or flap which either closes an opening or protects soft tissues behind it

opisthosoma the most *posterior* portion of a body that is divided into regions

opposable (of a *digit*) capable of being brought to bear against the other digits

oral of the mouth; **oral surface** that surface bearing the mouth

order a *taxonomic rank*, below *class* and above *family* (see p. 48)

organ a collection of *tissues* with a distinct form and some specific function

organelle a component of a *cell* with some specific function

ornamentation the superficial detail on the surface of a *shell*, or other external skeleton

osmosis the passage of water through a membrane into a region of high concentration; **osmoregulation** the maintenance of the correct concentration and balance of salts inside an animal

osphradium a *chemoreceptor* organ in the *mantle cavity* of many molluscs

ossicle a small bone, especially one of those in the *middle ear* of *tetrapods*; one of the *calcitic* plates of an echinoderm

ossify to be converted into *bone*

osteology the study of bones

otolith a small concretion, usually made of calcium salts, in the *inner ear* of vertebrates—part of the *organ of balance*; an object having a similar function in an invertebrate; see also *statolith*

ovum (*pl* **ova**) the animal egg or female *gamete*; **ovary** the egg-producing organ or female *gonad*; **ovicell** a *broad pouch*; **oviduct** the female *gonoduct*

oviparous laying eggs; **oviparity** being oviparous; see also *ovoviviparous, viviparous*

ovoviviparous producing eggs, but retaining them within the oviduct until they hatch, thus giving a superficial appearance of *viviparity* (see p. 298)

ozone a form of oxygen, much more reactive than the normal form, found especially at high altitudes; it is formed by the action of ultraviolet light on ordinary oxygen

P

palaeontology the study of *fossils*

palate the roof of the mouth; **false palate, secondary palate** the roof of the mouth in crocodiles and mammals, separating the mouth cavity from the *nasal cavity*

palp a sensory *appendage* (or branch of an appendage) usually close to the mouth

pampas the temperate grasslands *biome* of southern South America

papilla (*pl* **papillae**) a small protuberance

parapodium (*pl* **parapodia**) the lateral *appendages* of polychaete worms (see p. 127)

parasite an organism which obtains its food directly from the *tissues* of another living organism, usually without killing it (see p. 330)

paratenic host see *host*

parenchyma loose, space-filling *tissue*

parthenogenesis reproduction, from an *oocyte*, but not involving *meiosis* and *fertilization* (see p. 297)

pathogenic harmful; causing disease

pectoral 1. (in *tetrapods*) of the shoulder region; **pectoral girdle** the shoulder girdle; 2. (in fish) of the region immediately behind the head; **pectoral fin** see *fin*

pedal of the foot, or base, or point of attachment

pelage the hairy covering of a mammal

pelagic of the open sea; see also *neritic*

pelvic 1. (in *tetrapods*) of the hip region; **pelvic girdle, pelvis** the hip girdle; 2. (in fish) of the region immediately in front of the anus; **pelvic fin** see *fin*

penis (*pl* **penes**) a male *intromittent organ*, usually that of mammals, which also carries the urinary duct

peristalsis the propulsion of food or other material along the gut or blood vessel by means of muscle contractions

peritoneum the *epithelium* lining a true *coelom*

phagocytosis the process by which food particles are engulfed by single *cells*; a particle is surrounded by *pseudopodia* which merge beyond it

pharynx (in *protochordates* and fish) the first part of the *alimentary canal* perforated by the *gill slits*; the *homologous* part in vertebrates between the mouth cavity and the *oesophagus*; the throat; a specialized, frequently muscular region of the *alimentary canal* immediately behind the mouth, found in many invertebrates

pheromone a substance produced by one individual with a specific effect on the behaviour of another individual of the same *species*

phoresy the use of one organism by another for transport (see p. 333)

photosynthesis the making of food from simple molecules (carbon dioxide and water) using light as an energy source; see also *chlorophyll, primary production*

phyllopod a leaf-like *appendage*

phylogeny evolutionary history, ancestry; **phylogenetic** ancestral

phylum (*pl* **phyla**) a *taxonomic rank*, below *kingdom* and above *class* (see p. 48)

physiology the study of the internal functioning of organisms

pinna (*pl* **pinnae**) the outer ear of mammals

pinocytosis the process by which liquids are imbibed by cells; channels sink in from the surface, and small vesicles are pinched off from these

placenta (*pl* **placentae**) an organ in which maternal and *embryonic* or *foetal* tissues are closely applied to facilitate the exchange of food, oxygen and waste; **placental mammal** a eutherian; **placentation** the means and process of attachment of a placenta to the *uterus* (see p. 298)

plankton small aquatic organisms which drift at the mercy of water currents; see also *nekton*

planospiral (of a shell) coiled but with the coils all in one flat plane

plasma membrane the thin membrane of fat and *protein* surrounding an animal *cell*

plasmodium (*pl* **plasmodia**) the result of multiple nuclear division and growth of a single cell, without the cell itself actually dividing

plexus a mesh of interconnected nerves or blood-vessels

plumage the feather covering of a bird

pollination the transfer of pollen from the male part of one flower (the stamen) to the female part of another (the pistil)

polyembryony the production of two or more individual young from a single *fertilized* egg

polygamous the mating of one male with several females (polygyny) or one female with several males (polyandry)

polymorphism the occurrence within a single colony or population of two or more distinct forms

polyp the stage in the *life cycle* of many coelenterates that resembles a coral or sea anemone, usually *sedentary* or attached to a float

polyphyletic (of a group—usually a specific taxon) having evolved from two or more distinct original ancestoral groups; see also *monophyletic*

population a group of interbreeding animals, of the same *species*, in some biologically well defined area

posterior away from the head end of an animal

precocial, precocious in some respect being more advanced than would be normal at a given age

predator an animal which actively seeks and kills other animals (prey) for food

prehensile capable of grasping or seizing

premaxilla (*pl* **premaxillae**) the first tooth-bearing bone on the upper jaw of jawed vertebrates; see also *maxilla*

premolar tooth one of the cheek (post-canine) teeth of mammals; premolars are usually preceded by a set of milk teeth. See also *molar tooth*

preoral in front of the mouth

primary production the fixation in green plants of light energy from the sun in a form which may be eaten by animals (consumers) (see pp. 10, 36); see also *photosynthesis*

proboscis (*pl* **proboscides**) a protrusion from the *anterior* end of an animal

progeny offspring, whether produced *sexually* or *asexually*

prosoma the most anterior part of the body in those animals with distinct body regions; used particularly of arachnids

protein complex chemical comprising a series of many simple nitrogen-containing molecules (amino acids) linked in chains; the basic substance of which living material is made

prothorax the first of the three segments in an insect's *thorax*

protochordate a cephalochordate or urochordate

protonephridium simple, primitive excretory tube, usually with flame cells; see also *nephridium*

protostome an animal in which the mouth corresponds to the single opening of coelenterates; the *anus* forms as a second opening (see p. 57)

protrusible eversible; capable of protrusion

proximal towards or near to the body or to a defined point within or on the body; see also *distal*

pseudocoelom a body cavity formed by the swelling and sometimes coalescing of *mesenchyme* cells, or one that is not otherwise a true (*peritoneum*-lined) *coelom*

pseudopodium (*pl* **pseudopodia**) an extended or protruded portion of protoplasm, frequently temporary

pubic bone, pubis (*pl* **pubes**) one of the bones of the *pelvic girdle*, between the hip joints on either side, sometimes but not always meeting in the midline

pulmonary of the lungs

pupa (*pl* **pupae**) the quiescent stage between *larva* and adult in many insects

quadrupedal walking on four legs; see also *bipedal*

radially symmetrical having many planes of *symmetry*, all passing through the same axis

radiation see *adaptive radiation*,

radius one of the bones in the *tetrapod* forelimb, between the elbow and the wrist at the side of the first *digit*

rank see *taxonomic rank*

raptor *predator*, usually used of birds; **raptorial** adapted for catching prey

receptor an *organ* or *organelle* specially adapted for the sensing of some specific stimulus

rectum the most posterior part of the vertebrate *alimentary canal*, modified for producing *faeces*; comparable parts in other animals

regeneration the regrowth of an *organ*, limb or even much of the body after accidental loss

relict form, relict species a *living fossil*

resource anything an animal needs for its continued existence or reproduction, including both consumables (e.g. food, oxygen) and non-consumables (e.g. breeding sites)

respiration the exchange of oxygen and carbon dioxide between an organism and its environment; **respiratory organ** an organ facilitating respiration (such as a gill or lung); **respiratory system** the respiratory organs and associated structures

retina light-sensitive *nervous tissue* capable of receiving an image

retractile capable of being retracted or pulled back, usually into the body; see also *eversible, introvert*; **retractor muscle** a muscle which retracts a proboscis or other organ

rumen the large 'first stomach' or fermentation chamber in ruminants

sac a sack- or pouch-like structure

sacrum that part of the vertebral column which is attached to the *ilium* of the *pelvic girdle*; usually consisting of several fused *vertebrae*

sage-brush a local (American) form of *chaparral*, named after the dominant plant (see p. 38)

salivary gland a gland opening into the mouth region whose secretion (saliva) usually has some digestive function

sarcophagous feeding on flesh, used especially of an external *parasite*

savanna the *biome* of warm, tropical and sub-tropical grassland, frequently with scattered trees (forest-savanna) (see p. 38)

scale (originally) a sheet of bone, overlaid by modified skin; modern vertebrates may retain only the bone covered by hardly-modified skin, as in many fish, or retain the modified skin as horny scales, having lost the underlying bone, as in many *tetrapods*

scavenger an animal feeding on *carrion* or *detritus*

schizocoel a *coelom* which forms by the *mesoderm* splitting into two layers; see also *enterocoel*

sclerite an *exoskeleton* plate, especially of an arthropod; **sclerotin** the constituent of arthropod sclerites, principally *chitin* and *protein* and frequently waxes or calcium salts; **sclerotization** the process of formation and hardening of a new arthropod *exoskeleton*

scrotum the bag of skin containing the mammalian *testes*

scute an *epidermal scale*, or the epidermal part of a scale, usually consisting of *keratin*

sebaceous gland a mammalian *epidermal* gland, opening into a hair *follicle* and secreting sebum, a waxy fluid which waterproofs the hair

sedentary remaining in one place

sediment a deposit; **sedimentary rock** rock derived from hardened and compressed sediments (see p. 18); see also *igneous rock*, *metamorphic rock*

segmented (of a body) divided into a series of more-or-less equivalent sections along its length; **segmentation** being segmented; **metameric segmentation** segmentation with serial repetition of various *organs* and organ systems

self-fertilization see *fertilization*

semi-circular canals part of the inner ear of vertebrates, responsible for the detection of changes in orientation (see p. 303)

seminal vesicles receptacles in the male for *sperm* storage

sense organ a set of *receptors* for some specific stimulus, with associated structures

septum (*pl* **septa**) a sheet of tissue separating two cavities, especially those between adjacent *segments*

sessile permanently attached to the *substrate*

seta (*pl* **setae**) a hair-like structure, usually made of *chitin*; especially those of annelids, also called *chaeta*

sexual dimorphism condition in which the two sexes are *morphologically* different (other than in the *genitalia*)

sexual reproduction reproduction involving *meiosis* and *fertilization*

shell a hard permanent *exoskeleton*, usually *calcitic*, but may be of *chitin* or silica

sinusoidal in the form of a sine curve; wave-like

siphon a tube of *tissue* which directs a water current

soft-bodied without a hard skeleton

solitary not *colonial*

somatic of the body, as opposed to the *germ plasm*

spawn eggs; **spawning** the act of laying eggs

species the fundamental *taxonomic rank*; a group of populations of organisms which can interbreed and are reproductively isolated from other such groups; a group thought by *taxonomists* to be so (see p. 48)

sperm, spermatozoon (*pl* **spermatozoa**) the male *gamete*, usually small and *flagellated*; **spermatophore** a small packet of spermatozoa

spicule a small spike; a small spiky skeletal element, especially of a sponge

spine 1. an extension of the skeleton; 2. a large *seta*; 3. the vertebral column

spiracle 1. the first *gill-slit* of fish, usually distinct in form from the others; 2. the external opening of an insect *trachaea*

spongin a hard *protein* forming part of the skeleton of many sponges

statolith a small *calcitic* concretion which forms part of an invertebrate organ of balance; see also *otolith*

stereoscopic vision perception of an object in three dimensions given by the ability to coordinate, in the brain, the information received by the two eyes; see also *binocular vision*

sternum (*pl* **sterna**) the bone along the ventral *thorax* of *tetrapods*, to which many of the ribs are attached

stolon a cord or tube of *tissue* connecting the members of some animal colonies

stomach the swollen or sac-like part of the vertebrate alimentary canal that follows the *oesophagus*; similar structures in invertebrates

stylet small, hard, sharply-pointed object, used for piercing

sublittoral of the shallow sea below low tide level

substrate, substratum the surface upon which or in which an animal lives

suctorial capable of exerting suction; bearing suckers

suspension feeder an animal feeding on *detritus* and other organic matter suspended in the water; see also *ciliary feeder*, *filter feeder*, *deposit feeder*

suture the junction between two bones or parts of a shell (but not at a hinge)

swarm a mass of animals flying or swimming together, frequently for reproductive purposes

swim bladder a gas-filled bladder in most bony fish, used for maintaining neutral buoyancy

symbiosis (*pl* **symbioses**) an association of organisms from which both participant *species* derive some benefit; **symbiote** a participant in such an association (see p. 325)

symmetrical having one or more planes of symmetry, that is planes (or surfaces) which divide the animal into two halves which are approximately mirror images

syncytium (*pl* **syncytia**) a *tissue* in which there are no cell walls; a *plasmodium*

systematics the study of the kinds and diversity of organisms, their interrelationships and classification

synthesis the building up from simpler parts; used especially of complex molecules

T

tagma (*pl* **tagmata**) a group of *segments*, modified for some function; **tagmosis, tagmatization** the evolutionary process of developing tagmata (see p. 134)

taiga a transition *biome* between coniferous forest and *tundra*

tarsus (*pl* **tarsi**) the ankle or ankle-bones

taxon (*pl* **taxa**) a group of organisms at any level, with some formal identification; **taxonomy** the study of classification; **taxonomic rank** the position in the classification hierarchy (see p. 48)

temperate (of climate) not extreme

temporal of the temple, the region of the skull behind the eyes and above the point where the lower jaw articulates with the upper jaw

tentacle an extended, flexible organ used either in the acquisition of food or for bearing sense organs

terrestrial living on land

territory an area or space defended by an individual or pair against others of the same species

test an internal skeleton that lies just below the *epidermis*, thus resembling an external skeleton

testis (*pl* **testes**) the male *gonad*, producing *sperm* and usually male *hormones* as well

tetrapod any member of the essentially land-dwelling vertebrate groups; i.e. the amphibians, reptiles, birds and mammals

thorax the middle region of the body, lying behind the head, and in front of the *abdomen*; (in insects) that part bearing the legs; (in *tetrapods*, especially mammals) that part containing the heart and lungs

tibia (*pl* **tibiae**) one of the bones in the lower hind leg of a *tetrapod*; the 'shin' bone of a mammal

tidemark the upper (or sometimes the lower) limit of the tide; frequently variable in position

tissue a collection of usually similar *cells*, all with some similar basic function

torpid (of an *endotherm*) the lethargic condition due to low body temperature, usually when hibernating

toxin a poisonous substance

trachaea (*pl* **trachaeae**) an air-carrying tube, usually with *annular* skeletal supports

triploblastic having three basic layers of *tissue*, with *mesoderm* in addition to the *ectoderm* and *endoderm* of a *diploblastic* animal (see pp. 59, 74)

trochophore larva a planktonic *larva*, bearing one or more locomotor rings of *cilia*

trophic of food or feeding

trunk the body, excluding head, limbs and tail

tubercle a small rounded protuberance or swelling

tubicolous living in a tube

tubule a small tube

tundra a *biome* where vegetation growth is severely restricted by low temperature; characteristic of alpine and polar regions (see p. 39)

tympanum, tympanic membrane the *eardrum* of *tetrapods*; **tympanic organ** (in insects) the organ of hearing

typology *taxonomic* philosophy which holds that all organisms can be regarded as variants of a series of major *morphological* patterns

U

ulna one of the bones of the forearm of *tetrapods*, between the elbow (forming the 'funny-bone') and the wrist, on the side of the fifth *digit*

ultrasonic of high-frequency sound

ultraviolet radiation a form of radiation beyond the violet end of the visible spectrum (see p. 301)

umbo (*pl* **umbones**) the small bulge on a *shell*, representing the original shell of the very young stage or *larva*

unicellular having a body composed of only one *cell*

urea a fairly soluble nitrogen-containing waste product, principally found in mammals and aquatic animals

uric acid a more-or-less insoluble nitrogen-containing waste product of many land animals

uterus the part of the female reproductive system which retains the eggs, especially a part in which the fertilized eggs are allowed to develop

utricle a *sac* containing an *otolith*, in the *inner ear* of vertebrates, for balance (see p. 303)

V

vacuole a small fluid-filled cavity inside a *cell*

vagina a passage by which a *uterus* (or a corresponding part of the female reproductive system) communicates with the exterior

valve 1. one of the two parts of a *bivalve* shell; 2. a set of partial *septa* in a tube, which allow fluid to pass one way only

vascular system a system of tubes for carrying fluid (for example, blood) about the body

vector an organism which carries a *parasite* from one host to another

vegetative reproduction (of colonial animals) *asexual reproduction* similar to budding, but the resultant *zooid* remains attached to the parent

velum (*pl* **vela**) a veil-like membrane

venation the arrangement of veins in an insect wing

ventral of the lower surface

venom a *toxic* secretion, used for paralysing or killing prey

vermiform worm-like in shape

vertebra (*pl* **vertebrae**) one of the bones of the vertebrate backbone; **vertebral column** spine: backbone

vesicle small bladder or cavity

vestigial (of an organ) no longer having any function and consequently diminished in size

viscera the organs within the body cavity; **visceral mass** the contents of the shell of gastropods and some cephalopods

viviparous giving birth to young (or *larvae*) as opposed to eggs; see also *oviparous*; **viviparity** being viviparous

W

warm-blooded see *endotherm*

warning colouration distinctive colouration of a distasteful or poisonous animal (see p. 314)

Y

yolk sac a *sac* containing the egg yolk, especially in a newly-hatched *larva*

Z

zoogeography the study of the geographical distribution of animals (see p. 26)

zooid one of the members of an invertebrate *colony*

zygote the cell formed as a result of *fertilization*

Index

Entries in this index are of the following types:
● common names of animals and groups
● scientific names of animals and groups
● names of larval forms
● group names no longer taxonomically current but retained as descriptively useful; for example 'agnathans'

Numbers in bold type refer to pages on which the animal or group is illustrated or is mentioned in the labelling or legends within an illustration. All other numbers refer to pages on which the animal or group is mentioned either in the main text or, though it is not illustrated, in a caption.

Some scientific names are followed by the corresponding common name, or by a synonym, in parentheses. This indicates that additional references to the species or group will be found under the index entry for the common name or synonym. Less frequently the converse occurs: that is, the common name is followed by the scientific name or common synonym in parentheses.

For many of the entries of scientific names above the rank of genus, the actual reference on the page indicated may be to an anglicized version of the name: for example, for 'Insecta' read also 'insects'.

Acknowledgements

The publishers would like to thank the following individuals and organizations for their kind permission to reproduce the photographs in this book.

G. V. Adkin
13 above, 19, 233, 241 above, 243 above, 244 above, 251 centre, 267 below

Malcolm Aird
344 below

Heather Angel
12, 22 centre, 24 above, 24 below left and right, 36 below, 36-7, 42, 44-5, 68 left, 73 right, 76 below, 77 above, 88 above, 99, 110, 111 left, 114, 142 below, 143 below, 146 below right, 147, 151 below, 154, 157 below, 160 below, 165 left, 169, 170, 171, 174 above right, 177 below left, 179 above left, 188, 192, 193, 207 left, 208, 209 above, 211 above, 212 below, 213 below, 220, 222 below left, 299 above, 300 below, 301 above left and right, 313 above right, 320 below, 326 below, 319 below

Animals Animals
(Ranjit Sinh) 366 centre right

Animals Animals/Oxford Scientific Films Ltd
321 below, (M. Conte) 248 below, (D. Langur) 354 below right, (Z. Leszczynski) 203 above, 206, 226 right, (C. Palek) 247 above (Stouffer Prpductions) 298 above, 355 centre, (Kojo Tanaka) 16

Aquila Photographics
181 above

Ardea
18 above, 41, 266 below, 375, below left, (W. J. Bailey) 37 inset left, (F. Bala) 370 above, (I. Beames) 262, 267 above, 283 below, 305 below right and left, 315 above, 364 above, (H. and J. Beste) 238 above and below, 242 below, 244 below, 249 above, 256 above, 260 below, (J. R. Boames) 362 above, (A and E. Bomford) 284 above, (G. K. Brown) endpapers, 282 right, (W. Curth) 270, (J. P. Ferrero) 260 above, 275 below, (K. Fink) 30 above, 225 above, 226 left, 280 above, 340, 342 below, 356 below, 357 above, (R. Fleming) 336 above right, (P. Green) 204, 229, (C. Haagner) 57 below, 237 above, 276 above, 286 above, 311 below left, 334 below, (D. Hadden) 79 above, (J. L. Mason) 126 below, 140 below, 210, 219 below left, 308 above, 335 above, 368 above, (P. Morris) 235 right, 265 below, 269 below, 344 above, (B. L. Sage) 357 centre, (P. Steyn) 284 above, (B. Stonehouse) 335 above left, (R. and V. Taylor) 43, 53 above and below right, 68 right, 72 above, 112 above, 117 above, 180 below, 187 right, 195 below, 196 above and below, 306 below, (A. D. Trounson and M. C. Clampett) 241 below, 245 above, (R. Waller) 39 below left, (A. Warren) 369 below, (W. Weisser) 51 below

Australian Marine photographic Index/ Neville Coleman
195 above, 324 below

D. J. Bayliss/RIDA Photo Library
20 above, 22 above, 104 below right, 121 above

Stephen Benson
371 below left

Bio-Arts
106, 107, 232, 242 above, 253 above, 298 below, 330 below, 333 below

S. C. Bisserot
53 left

The Trustees of the British Museum (Natural History)
182

Dr. A. C. Campbell
177 above right

Dr. J. C. Chubb
94 above and below, 95

S. A. Clark
93 below right

Bruce Coleman Ltd
185 above, 253 below, 285 below left, 286 below, 296, 319 above, 332 above, 374 below left, (K. Balcomb) 318 below, (J. and D. Bartlett) 240 below, 250 below, 259 above, 269 above right, 274 below, 365 below, (S. C. Bisserot) 112 above, 212 above, 323 below, 358-9, (M. N. Boulton) 292, 337 above, 366 centre left, (J. M. Burnley) 209 below

left, 283 above right, 342 above, (J. Burton) 1, 20 below, 36 centre, 131 above, 152, 177 above left, 184, 185 below, 211 below, 215 above left, 219 above, 251 below, 300 above, 304, 324 centre, 328 above, 336 below, (R. and C. Calhoun) 246 below, 273 above, 306 above, (B. Campbell) 216 below, (R. I. M. Campbell) 377 centre, (B. J. Coates) 353 below left, (N. Coleman) 120 above, (G. Cox) 254, (S. Dalton) 156 above left and above right, 157 above right, 310, (J. Dermid) 29, 259 above, (O. E. Duscher) 249 below right, (F. Erize) 32 left, 51 centre, 239 below, 273 above, 278 above and below, 281 left, 357 above, 364 below left, (M. Freeman) 40 left, 275 above left, 356 above, (J. Foott) 108, 323 above, (C. B. Frith) 31 above, 137, 216 above, 221 below, 289 above, 295 above right, 302, 352 above, (S. Halvorsen) 250 above, (Udo Hirsch) 215 above right, (T. L. Hout) 335 below, (D. Hughes) 149 above, (P. Jackson) 370 below, (C. Laubscher) 326 above, (L. Lyon) 280 below, 287 below, (I. MacPhail) 353 below centre, (J. Markham) 331 below, (N. Myers) 275 above right, 363 above, (M. T. O'Keefe) 279 below, 356 centre, (C. Ott) 40 right, (Oxford Scientific Films Ltd.) 331 above, (J. Pearson) 14-15, 322 above, (G. Pizzey) 256 below, (G. D. Plage) 245 above, (H. Reinhard) 199 above and below left, 201 below, 205, 252 below, 315 below right, (V. Serventy) 223 above left, 258 below, 354 above, (J. Simon) 369 above, (M. F. Soper) 239 above, (Stouffer Productions) 266 above, 274 above, 320 above, (Sullivan and Rogers) 219 below right, (N. Tomalin) 194 below, 360, (R. J. Tulloch) 371 above, (D. and K. Urry) 318 above

Colorific Photo Library Ltd
(Nina Leen) 345 above, (Life Magazine/J. Dominic) 355 below, (T. Spencer) 343 above, (Time Life/G. Silk) 311 below right

Dr. D. Corke
151 above

G. Cubitt
6-7, 10, 26, 243 below, 282 left, 343 below, 348 above

A. P. Davis
104 centre

P. E. J. Dyrynda
96, 97

Earth Scenes/Oxford Scientific Films Ltd
(B. P. Kent) 136, (Mark Newman) 368 below

Dr. E. P. Evans
295 above

R. Fletcher
134, 158 right, 162 above, 163, 164, 224 below right, 313 below

Lloyd Frampton
46

Dr. D. George
167 left

Dr. P. E. Gibbs
122

Goodman Photographic
176

Dr. D. W. Halton
87 above left and below right

Robert Harding Associates
51 above, 328 below, 338-9, 373 below right

D. Hosking
224 below left, 252 above

E. Hosking
38 left, 235 left

JACANA Agence de Presse
(Arthus Bertrand) 265 centre, 283 above left, 287 above, (C. Carre) 80, (J. P. Champroux) 251 above, (H. Chaumeton) 70, 101 left, 103, 117 below right, (J. P. Hervy) 246 above, (P. Laboute) 118, (C. Nardin) 222 above, (J. Robert) 181 below, 307 below, (K. Ross) 113 above, (P. Summ) 21 left, (G. Trouillet) 263 right, (Varin-Visage) 248 above, 265 above, 285 above and below right, 322 below, 329 centre, 365 centre, (Rene Volot) 36 above, (F. Winner) 109, (Ziesler) 271 below

Dr. J. B. Jennings
83 centre

Dr. W. J. Kennedy
22 above, 104 below left, 121 below

F. W. Lane
372 above

Dr. G. F. Leedale/Biofoto Associates
63 below left, 85 below

E. Lapan/Lawrence Berkeley Laboratory
66

K. E. Lucas
71 right, 102 below, 175 above

W. MacQuitty
365 above

R. Manuel
98 right

Dr. E. B. Matthews
93 below left

Dr. G. Mazza
76 above, 78 right, 119 left and right, 120 below, 168, 197 above, 200 above, 299 below, 307 above, 325

Harry Millen/Taronga Zoo
375 below right

Milwaukee Public Museum
377 below

Jean Morris
281 right, 329 below

P. Morris
21 right, 23, 128 below, 194 above, 207 right, 221 above, 272 above, 311 above, 329 above, 349, 354 below left

Natural History Photographic Agency
214, (H. R. Allen) 264, (D. Baglin) 261 above, (A. Bannister) 77 above, 78 below, 161, 215 below, 225 above, 271 above, 301 below, (J. E. Blossom) 268 left, (S. Dalton) 157 above left, 165 right, 178, 213 above, 240 above, 303 above, 317 above, (F. Greenaway) 296 above, (Jeffrey) 37 inset right, (P. Johnson) 32 right, 247 below, 353 below right, 362 below, 364 below right, (J. Kroener) 34, (L. H. Newman) 271 centre, (E. H. Rao) 227, 366 above, (K. H. Switak) 221 centre, (M. W. F. Tweedie) 149 below, (P. Wayne) 367

Natural Science Photos
(C. Banks) 217 below, (I. Bennett) 113 below, 175 below, 179 below left, 236 above and below, (S. Brown) 348 below, (M. Chinery) 38 above right, (N. Fain/Marineland of Florida) 199 below right, (J. A. Grant) 163 above, 330 above, (R. Kemp) 234, 294, (G. Kinnis) 141 centre, (D. B. Lewis Hull) 131 below, (G. Montalverne) 81 left, 148 below, (L. E. Perkins) 324 above, (M. Price) 30 below, (P. H. Ward) 156 below right, 160 above, (C. E. Williams) 135 left

Oxford Scientific Films Ltd
39 above left, 65 above left, 82 below, 83 below, 88 below, 111 right, 130 below, 133 138, 139, 141 below, 145 below, 158 left, 159 left and right, 177 above centre, 179 above, 186, 201 below left and right, 297 below, 303 below, 305 above, 312, 314 below, 317 centre, 327, 332 below left, 333 above, 334 above, 336 above left, 337 below left and right, 371 below right, (J. A. L. Cooke) 18 right, 38 below right, 78 left, 93 above right, 135 right, 142 above right, 143 above, 145 above right, 148 above, 150 above and below, 153, 162 below, 315 centre

New South Wales Government, London
377 below

J. Norris Wood
297 above, 329 above

N. Picozzi
13 below

C. Roessler
57 centre, 73 left, 202 above and below

G. R. Roberts
260 centre, 261 above

B. L. Sage
39 below, 345 below, 346 below, 347 above and below, 353 above

Dr. I. Sazima
317 below

Dr. R. E. Schroeder
321 above

Seaphot
31 below, 100, 173, 174 above left, 185 centre, 309 left, (R. Chesher) 128 below, 179 centre, 332 below right, (D. Clarke) 177 below right, 187 left, 197 below, (P. David) 203 below, 313 above left, (W. Deas) 79 below, 124, 174 below, (D. George) 166, (R. Johnson) 290-1, (L. P. Madin) 81 right, (R. Salm) 352 below, (H. Schumacher) 123

Dr. C. R. C. Sheppeard
4-5

Spectrum Colour Library
372 below, 373 left, 375 above, 376

T. Stack
200 below

Dr. R. C. Tinsley
83 above, 85 below, 87 above right, above centre, below left and centre

M. I. Walker
57 above, 62 below, 65 below right, above and below left, 90, 91, 92 left and right, 130 above

D. Wallin
180 above

C. James Webb
63 below right, 84 left and right, 93 above left

K. Westerskov
72 centre and below, 74, 98 left, 104 above, 263 left

Dr. G. R. Williamson
279 above

Dr. D. P. Wilson
101 right, 126 above

ZEFA
11, 132, 258 above, 346 below, 374 below right

The Zoological Society of London
373 above right

Picture research by Mary Corcoran

ILLUSTRATORS

Richard Bonson
42-43, 60-61, 155, 307, 308-309

Hilary Burn
121, 178 right

John Clarke
60-61, 62 left, 66 top, 68 top, 74, 80 top, 82, 88 top, 90, 94, 96 left, 98, 100, 102, 106, 108 top, 109 left, 110 top, 114 top, 118 top, left, 122 top, 124 left, 126, 132 top, 134, 136 left, 137, 138 left, 143, 144 top, 150, 151, 154, 165 below, 166 top, 168 top, 170 top, 172 top, 180, 184 top, 188 top, 190 top, 192 left, 193 left, 194 left, 198 top, 206, 216, 229, 230, 254, 256, 258, 262

Fairey Surveys
27, 28-29, 33

Roger Gorringe
49 below, 88 below, 89, 99, 105, 115, 132 centre, 133, 147 top, 153, 178 left, 232-233, 234-235, 264

Frank Kennard
58, 62 right, 69, 75, 80 left, 91, 95, 97, 101, 108 below, 118 right, 122 right, 133, 124 right, 125, 127, 166 below, 167, 168 below, 170 centre, 171, 172-173, 182, 183, 184 below, 185, 186, 188-189, 218, 254, 255, 257 top, 263, 301, 302

Mick Loates
192, 192 below, 193 right, 194 right, 195, 199, 204-205, 312

Alan Male (Linden Artists)
135, 136 right, 138 below, 139, 144 below, 147

Robert Morton (Linden Artists)
207, 220, 226, 227, 228

Colin Newman
63, 71, 83, 86, 96 right, 103, 109 right, 111, 114 below, 145, 160, 165 top

Alasdaire Robertson
231

Milne/Stebbing Illustration
24-25, 64, 66-67, 107, 110 below, 119, 137 right, 316, 327

Studio Briggs
19 top, 48, 49 top, 52, 56, 190-191, 257 below

Brian Watson (Linden Artists)
35, 319, 362, 364, 367, 368, 370, 371, 372, 374, 375, 376

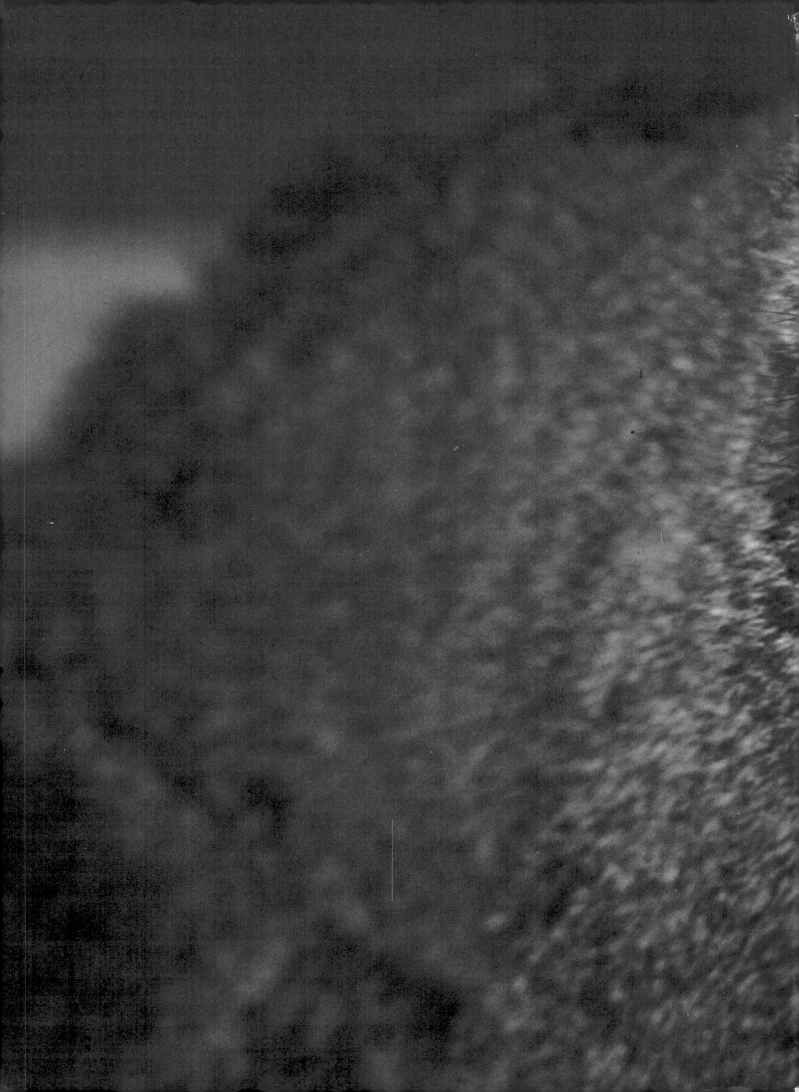